Lecture Notes in Mathematics 2050

Editors:
J.-M. Morel, Cachan
B. Teissier, Paris

For further volumes:
http://www.springer.com/series/304

Bo'az Klartag • Shahar Mendelson
Vitali D. Milman

Editors

Geometric Aspects
of Functional Analysis

Israel Seminar 2006–2010

 Springer

2006–2010

Editors
Bo'az Klartag
Tel Aviv University
Tel Aviv, Israel

Vitali D. Milman
University of Tel Aviv
Tel Aviv, Ramat Aviv, Israel

Shahar Mendelson
Technion, Israel Institute of Technology
Technion City, Haifa, Israel

ISBN 978-3-642-29848-6   ISBN 978-3-642-29849-3 (eBook)
DOI 10.1007/978-3-642-29849-3
Springer Heidelberg New York Dordrecht London

Lecture Notes in Mathematics ISSN print edition: 0075-8434
                                ISSN electronic edition: 1617-9692

Library of Congress Control Number: 2012942081

Mathematics Subject Classification (2010): 80M35, 26A51, 32-XX, 46-XX, 60-XX

Printed on acid-free paper

Springer is part of Springer Science+Business Media (www.springer.com)

Preface

Since the mid-1980s the following volumes containing collections of papers reflecting the activity of the Israel Seminar in Geometric Aspects of Functional Analysis appeared:

1983–1984 Published privately by Tel Aviv University
1985–1986 Springer Lecture Notes in Mathematics, vol. 1267
1986–1987 Springer Lecture Notes in Mathematics, vol. 1317
1987–1988 Springer Lecture Notes in Mathematics, vol. 1376
1989–1990 Springer Lecture Notes in Mathematics, vol. 1469
1992–1994 Operator Theory: Advances and Applications, vol. 77, Birkhäuser
1994–1996 MSRI Publications, vol. 34, Cambridge University Press
1996–2000 Springer Lecture Notes in Mathematics, vol. 1745
2001–2002 Springer Lecture Notes in Mathematics, vol. 1807
2002–2003 Springer Lecture Notes in Mathematics, vol. 1850
2004–2005 Springer Lecture Notes in Mathematics, vol. 1910

The first six were edited by Lindenstrauss and Milman, the seventh by Ball and Milman and the last four by Milman and Schechtman.

As in the previous volumes, the current one reflects the general trends of the Theory. Most of the papers deal with different aspects of Asymptotic Geometric Analysis understood in a broad sense. It includes classical topics in the geometry of convex bodies, inequalities involving volumes of such bodies or, more generally, log-concave measures, valuation theory, probabilistic and isoperimetric problems in combinatorial setting. A special attention is given to the study of volume distribution on high dimensional spaces. Additional direction is the characterization of some classical constructions in Geometry and Analysis (like the Legendre and Fourier transforms, derivation and others) is represented by a few papers. This leads also to an unexpected use of fractional linear maps and one paper intensively study these maps and present their use in the Convexity Theory. In many of the papers Probability Theory plays an important role and probabilistic tools are used intensively. There are also papers on related subjects. All the papers here are original research papers and were subject to the usual standards of refereeing.

As in previous proceedings of the GAFA Seminar, we also list all the talks given in the seminar as well as talks in some related workshops and conferences. We believe this gives a sense of the main directions of research in our area.

We are grateful to Miriam Hercberg for taking excellent care of the typesetting aspects of this volume.

Tel Aviv, Israel   Bo'az Klartag
Haifa, Israel   Shahar Mendelson
Tel Aviv, Israel   Vitali Milman

Contents

The α-Cosine Transform and Intertwining Integrals on Real Grassmannians

Semyon Alesker

Abstract In this paper we describe the range of the α-cosine transform between real Grassmannians in terms of the decomposition under the action of the special orthogonal group. As one of the steps in the proof we show that the image of certain intertwining operators between maximally degenerate principal series representations is irreducible.

1 Introduction

In this paper we describe the range of the α-cosine transform between real Grassmannians in terms of the decomposition under the action of the special orthogonal group. As one of the steps in the proof we show that the image of certain intertwining operators between maximally degenerate principal series representations of the group $GL_n(\mathbb{R})$ is irreducible.

Let V be a Euclidean space of dimension n. Let us denote by $Gr_i(V)$ (or just $Gr_{i,n}$) the Grassmannian of real i-dimensional subspaces. The α-cosine transform T_{ji}^{α}, where α is a complex number, is a linear operator $T_{ji}^{\alpha} : C^{\infty}(Gr_{i,n}) \rightarrow C^{\infty}(Gr_{j,n})$ which is given explicitly by a kernel for $Re\alpha > -1$ (defined below), and is obtain by the meromorphic continuation for other values of α. The problem of description of its range for $\alpha = 1$ was solved in [2] by Bernstein and the author. Various particular cases for $\alpha = 1$ and various i, j were solved previously by Goodey and Howard [11, 12], Goodey et al. [13], Matheron [24, 25] in connection to convex and stochastic geometry. However the method of the paper [2] does not generalize to $\alpha \neq 1$. The injectivity of the α-cosine transform for $i = j = 1$ and α being an odd positive integer was proved

---

S. Alesker (✉)
Sackler Faculty of Exact Sciences, Department of Mathematics, Tel Aviv University,
Ramat Aviv, 69978 Tel Aviv, Israel
e-mail: semyon@post.tau.ac.il

B. Klartag et al. (eds.), *Geometric Aspects of Functional Analysis*, Lecture Notes
in Mathematics 2050, DOI 10.1007/978-3-642-29849-3_1,
© Springer-Verlag Berlin Heidelberg 2012

by Schneider [30]. The case of general α was studied by Rubin [28] for $i = j = 1$ (equivalently for $i = j = n - 1$) where he constructed inversion formulas in those cases when the α-cosine transform is injective. The case of general Grassmannians and α being a positive integer was considered by Goodey and Howard [12] where they studied the injectivity of the α-cosine transform. More general case when either i or j is equal to one was studied by Rubin [28] where he gave some inversion formulas for the α-cosine transform (when it is injective). Some related results for the Grassmannians were considered by Spodarev [31]. For some history and references about the α-cosine transform see also Koldobsky [19] where the applications of the α-cosine transform to convexity are discussed. We refer to Rubin [27] (especially Sects. 3.3 and 3.4 there), for other results about α-cosine transform and their relations to PDE and harmonic analysis.

It was observed in [2] for $\alpha = 1$ and generalized (easily) in this paper for other values of α that for $i = j$ the operator T_{ii}^{α} can be rewritten to commute with the action of the full linear group $GL_n(\mathbb{R})$ (where the spaces of functions $C^{\infty}(Gr_{i,n})$ are interpreted as spaces of sections of certain $GL_n(\mathbb{R})$-equivariant line bundles over the Grassmannians). Then it essentially coincides with the standard construction of intertwining integrals (see e.g. [35,36]). The study of the case of various i, j reduces to the case $i = j$ using the identity (which was communicated to us by B. Rubin) $T_{ji}^{\alpha} = cT_{jj}^{\alpha} \circ R_{ji}$ where R_{ji} is the Radon transform between Grassmannians, and c is a constant. One has also to use the description of the range of the Radon transform R_{ji} due to Gelfand et al. [10] (see also [14]).

Let us define the operator T_{ji}^{α}. Let $E \in Gr_{i,n}$, $F \in Gr_{j,n}$. Assume that $i \le j$. Let us call by *cosine of the angle* between E and F the following number:

$$|\cos(E, F)| := \frac{\text{vol}_i(Pr_F(A))}{\text{vol}_i(A)},$$

where A is any subset of E of non-zero volume, Pr_F denotes the orthogonal projection onto F, and vol_i is the i-dimensional measure induced by the Euclidean structure. (Note that this definition does not depend on the choice of a subset $A \subset E$). In the case $i \ge j$ we define the cosine of the angle between them as cosine of the angle between their orthogonal complements:

$$|\cos(E, F)| := |\cos(E^{\perp}, F^{\perp})|.$$

(It is easy to see that if $i = j$ both definitions are equivalent.)

For any $1 \le i, j \le n - 1$ one defines the α-cosine transform

$$T_{j,i}^{\alpha} : C(Gr_{i,n}) \longrightarrow C(Gr_{j,n})$$

as follows:

$$(T_{j,i}^{\alpha} f)(E) := \int_{Gr_{i,n}} |\cos(E, F)|^{\alpha} f(F) dF,$$

where the integration is with respect to the Haar measure on the Grassmannian. In Theorem 4.15 we describe the range of T_{ji}^{α} for $\alpha \notin \mathbb{Z}$ and for $\alpha \in \mathbb{Z}_+$ in terms of the decomposition under the action of the special orthogonal group $SO(n)$. A slight refinement of the method allows to obtain such a description for $\alpha \neq -n, -(n-1), \ldots, -1$ though it is not presented here.

Thus the main new case is $i = j$ which reduces to some purely representation theoretical statements. Let us describe those which have independent interest in representation theory. Remind that a *degenerate principal series representation* of a reductive group (say $GL_n(\mathbb{R})$) is a representation induced from a character of a parabolic subgroup. Such representations were studied extensively in representation theory since there are one of the main sources of construction of representations of the reductive groups (see e.g. [3, 9, 15–17, 21–23, 26, 29, 38]). Let us introduce more notation. Let $1 \leq k \leq [n/2]$. Let

$$P_k := \left\{ \begin{pmatrix} c & b \\ 0 & a \end{pmatrix} \mid c \in GL_{n-k}, a \in GL_k \right\}.$$

For $\alpha \in \mathbb{C}$ let

$$\chi_{\alpha}^{\pm} : P_k \longrightarrow \mathbb{C}$$

be given by

$$\chi_{\alpha}^{+}\left[\begin{pmatrix} c & b \\ 0 & a \end{pmatrix} \right] = |\det a|^{\alpha},$$

$$\chi_{\alpha}^{-}\left[\begin{pmatrix} c & b \\ 0 & a \end{pmatrix} \right] = \text{sgn}(a) |\det a|^{\alpha}.$$

Let

$$\mathcal{L}^{\alpha, \pm} := \text{Ind}_{P_k}^{GL_n} \chi_{\alpha}^{\pm}$$

(the induction is not unitary!). In this paper we prove the following result which is crucial for the study of the α-cosine transform.

Theorem 4.2. *Assume that* $\alpha \neq 1, 2, \ldots, n-1$. *Then the representation* $\mathcal{L}^{\alpha, \pm}$ *has a unique composition series. If* $\alpha < n/2$ *then the Gelfand-Kirillov dimension of consecutive irreducible subquotients is strictly increasing. If* $\alpha > n/2$ *then it is strictly decreasing.*

We also get the following corollary.

Corollary 4.3. *Let* $\alpha, \beta \neq 1, 2, \ldots, n-1$. *Assume that either* $\alpha > n/2$ *and* $\beta < n/2$, *or* $\alpha < n/2$ *and* $\beta > n/2$. *Let* $\varepsilon, \delta = \pm$. *Then up to a multiplication by a constant there is at most one intertwining operator from* $\mathcal{L}^{\alpha, \varepsilon}$ *to* $\mathcal{L}^{\beta, \delta}$. *Such an operator has an irreducible image.*

Theorem 4.2 and Corollary 4.3 combined with the results by Howe and Lee [15] imply the description of the range of the α-cosine transform for suitable α.

The main tools of this paper are the Beilinson–Bernstein localization theorem [4] and the description of the category of perverse sheaves on complex matrices with the rank stratification due to Braden and Grinberg [7].

Acknowledgements We express our gratitude to T. Braden for the explanations of the results of [7], and to B. Rubin for communication us Proposition 2.2 and important remarks on the first version of the paper. We are grateful to A. Beilinson, J. Bernstein, A. Braverman, and D. Vogan for very useful discussions. We thank A. Koldobsky for useful discussions and some references.

2 Some Preparations

In this section we describe some preliminary results about the α-cosine and Radon transforms. Let V be an n-dimensional Euclidean space.

Lemma 2.1. *Let $Re(\alpha) > -1$. Fix $E_0 \in Gr_i(V)$ Then the integral*

$$\int_{Gr_i(V)} |\cos(E, E_0)|^\alpha dE$$

converges absolutely.

Proof. We may assume that α is real and $\alpha > -1$ and $i \leq n/2$. Let us denote $G = SO(n)$. Let K denote the stabilizer of E_0 in G. Let us fix an orthonormal basis $e_1, \ldots, e_i$ in E_0. Let us also fix an orthonormal system $\xi_1, \ldots, \xi_i$ in $E_0^\perp$. Let us denote by A the subset of $Gr_i(V)$ consisting of subspaces of the form $E = \text{span} < \cos(\theta_1)e_1 + sin(\theta_1)\xi_1, \ldots, \cos(\theta_i)e_i + sin(\theta_i)\xi_i >$. Clearly A is isomorphic to the i-dimensional torus T^i. For a subspace $E \in A$ of the above form $|\cos(E, E_0)| = |\prod_{k=1}^i \cos(\theta_i)|$.

Next one has a decomposition $Gr_i(V) = K \cdot A$. Since the function $E \mapsto |\cos(E, E_0)|$ is K-invariant, lemma follows from the fact that the integral $\int_{T^i} |\prod_{k=1}^i \cos(\theta_i)|^\alpha d\theta_1 \ldots d\theta_i$ is absolutely convergent for $\alpha > -1$.   □

The following proposition is due to Rubin (private communication).

Proposition 2.2. *Let $Re(\alpha) > -1$. Then*

$$T_{ji}^\alpha = c(\alpha) \cdot T_{jj}^\alpha \circ R_{ji}$$

where $c(\alpha)$ is a constant depending on α, i, j only.

Remark 2.3. It follows from the proof of this proposition that in fact

$$c(\alpha) = \left(\int_{L \in Gr_i(\mathbb{R}^j)} |\cos(L, L_0)|^\alpha dL \right)^{-1}$$

where $L_0 \in Gr_i(\mathbb{R}^j)$ is any fixed subspace. Note that the integral converges absolutely by Lemma 2.1.

Proof of Proposition 2.2. For any $F \in Gr_j(V)$ we have

$$T_{ji}^\alpha(f)(F) = \int_{E \in Gr_i(V)} f(E)|\cos(E, F)|^\alpha dE.$$

Lemma 2.4. *Let* $E \in Gr_i(V)$, $F \in Gr_j(V)$, $j < i$. *Then*

$$|\cos(E, F)|^\alpha = c(\alpha) \int_{L \in Gr_j(E)} |\cos(F, L)|^\alpha dL.$$

Let us postpone the proof of this lemma and let us finish the proof of Proposition 2.2. We have

$$T_{ji}^\alpha(f)(F) = c(\alpha) \int_{E \in Gr_i(V)} dE f(E) \int_{L \in Gr_j(E)} |\cos(F, L)|^\alpha dL$$

$$= c(\alpha) \int_{L \in Gr_j(V)} dL |\cos(F, L)|^\alpha \int_{E \supset L} f(E) dE$$

$$= c(\alpha) \int_{L \in Gr_j(V)} dL |\cos(F, L)|^\alpha (R_{ij} f)(L) = c(\alpha)(T_{jj}^\alpha \circ R_{ji})(f)(F).$$

Let us now prove Lemma 2.4.

We may assume that F does not intersect $E^\perp$. Let us fix an orthonormal basis $\xi_1, \ldots, \xi_j$ in F. Let η_p be the orthogonal projection of ξ_p to E. Let $F' \in Gr_j(E)$ be the image of F under the orthogonal projection. Let us fix an orthonormal basis $\xi'_1, \ldots, \xi'_j$ in F'. Then

$$|\cos(E, F)| = |\cos(F', F)| = |\det(\xi'_p, \xi_q)| = |\det(\xi'_p, \eta_q)|.$$

Let $St_j(E)$ denote the Stiefel manifold of j-tuples of orthonormal vectors in E. It has the unique normalized Haar measure. We have:

$$\int_{L \in Gr_i(E)} |\cos(L, F)|^\alpha dL = \int_{<u_p> \in St_j(E)} |\det[(u_p, \xi_q)]|^\alpha du$$

$$= \int_{<u_p> \in St_j(E)} |\det[(u_p, \eta_q)]|^\alpha du. \qquad (1)$$

We have

$$|\det[(u_p, \eta_q)]| = |\det[(u_p, \sum_l (\eta_q, \xi'_l)\xi'_l)]|$$

$$= |\det[\sum_l (\eta_q, \xi'_l)(\xi'_l, u_q)]| = |\det([(\eta_q, \xi'_l)][(\xi'_l, u_q)])|$$

$$= |\det[(\eta_q, \xi'_l)] \det[(\xi'_l, u_q)]|$$
$$= |\cos(F, F')| \cdot |\cos(F', \text{span} < u_p >)|$$
$$= |\cos(E, F)| \cdot |\cos(\text{span} < u_p >, F')|.$$

Substituting this into (1) we get:

$$\int_{L \in Gr_j(E)} |\cos(L, F)|^\alpha dL = |\cos(E, F)|^\alpha \int_{L \in Gr_j(E)} |\cos(L, F')|^\alpha dL.$$

Thus Lemma 2.4 follows with the constant

$$c(\alpha) = \left(\int_{L \in Gr_j(E)} |\cos(L, F')|^\alpha dL \right)^{-1}$$

(note that the last expression is independent of $F' \in Gr_j(E)$). □

We will also need later on the following result about the Radon transform due to Gelfand et al. [10] (see also [14]).

Proposition 2.5. *For $j < i$ the Radon transform $R_{j,i} : C(Gr_{i,n}) \to C(Gr_{j,n})$ is injective iff $i + j \geq n$ and has a dense image iff $i + j \leq n$.*

We will need also a simple lemma which can be easily checked (where the representations $\mathcal{L}^{\alpha,\pm}$ were introduced in the introduction).

Lemma 2.6.
$$(\mathcal{L}^{\alpha,\pm})^* = \mathcal{L}^{-\alpha+n,\pm} \otimes |\det(\cdot)|^k$$

Now let us consider the case $i = j$ in more detail. We will rewrite the α-cosine transform so that it will commute with the action of the group $GL_n(\mathbb{R})$. Let us remind some basic definitions. Let E be an i-dimensional real vector space.

Definition 2.7. An α-density μ on V is a $\mathbb{C}$-values function on the set of bases in E, $\mu : \{\text{bases in } E\} \to \mathbb{C}$ such that for any $g \in GL(E)$ one has $\mu(g(e_1, \ldots, e_i)) = |\det g|^\alpha \mu((e_1, \ldots, e_i))$ for any basis $e_1, \ldots, e_i$ of E.

Note that the space of α-densities is one dimensional since the set of bases is a principal homogeneous $GL(E)$-space. Also the space of Lebesgue measures on E is naturally isomorphic to the space of 1-densities. Indeed let ν be a Lebesgue measure. Then ν defines a 1-density such that its value on a basis $e_1, \ldots, e_i$ is equal to the measure ν of the parallelepiped spanned by these vectors. This map from the Lebesgue measures to 1-densities defines an isomorphism of (one-dimensional) vector spaces.

Let $L_\alpha \to Gr_i(V)$ be the line bundle over the Grassmannian $Gr_i(V)$ whose fiber over a subspace E is equal to the space of α-densities on E. Naturally L_α is a $GL(V)$-equivariant line bundle over $Gr_i(V)$. The representation of $GL(V)$ is the space of sections of L_α is isomorphic to $\mathcal{L}^{\alpha,+} \otimes |\det(\cdot)|^{-\alpha}$ (corresponding to the subgroup P_{n-i}).

Let $M'_\alpha \to Gr_{n-i}(V)$ denote the line bundle over the Grassmannian $Gr_{n-i}(V)$ whose fiber over $F \in Gr_{n-i}(V)$ is equal to the space of α-densities over V/F. Finally let us define $M_\alpha := M'_\alpha \otimes |\omega|$ where $|\omega|$ is the line bundle of densities over $Gr_{n-i}(V)$. The representation of $GL(V)$ is the space of sections of M_α is isomorphic to $\mathcal{L}^{-\alpha+n} \otimes |\det(\cdot)|^{-n+i}$ (corresponding to the subgroup P_i).

Let us define an intertwining operator

$$T_i^\alpha : C^\infty(Gr_{n-i}(V), M_\alpha) \longrightarrow C^\infty(Gr_i(V), L_\alpha)$$

as follows. For $E \in Gr_i(V)$ and $f \in C^\infty(Gr_{n-i}(V), M_\alpha)$ set

$$(T_i^\alpha f)(E) = \int_{F \in Gr_{n-i}(V)} pr^*_{E,F}(f(F)),$$

where $pr_{E,F}$ denotes the natural map $E \to V/F$ and $pr^*_{E,F}$ is the induced map $|\wedge^i (V/F)^*| \to |\wedge^i E^*|$. Clearly T_i^α is a non-trivial operator commuting with the action of $GL(V)$.

Let us fix a Euclidean metric on V. Then we can identify $Gr_{n-i}(V)$ with $Gr_i(V)$ by passing to the orthogonal complement. Also for each subspace $E \in Gr_i(V)$ the space of α-densities can be identified with $\mathbb{C}$ if $1 \in \mathbb{C}$ corresponds to the α-density which is equal to one on each orthonormal basis in E. This defines trivializations of the line bundles L_α and M_α. These trivializations commute with the action of the orthogonal group $O(n)$. We have the following easily proved observation (for $\alpha = 1$ it was noticed in [2]):

Claim 2.8. *With these identifications the map T_i^α coincides with the α-cosine transform T_{ii}^α.*

Corollary 2.9. *The operators T_{ji}^α admit a meromorphic continuation with respect to α to the whole complex plane.*

Proof. By Proposition 2.2 we can write for $Re(\alpha) > -1$

$$T_{ji}^\alpha = c(\alpha) T_{jj}^\alpha \circ R_{ji}$$

where the constant $c(\alpha)$ is given by the formula in Remark 2.3. By Claim 2.8 the operators T_{jj}^α coincide with the standard intertwining integrals. But they admit a meromorphic continuation with respect to α to the whole complex plane (see e.g. [35, 36]). Thus it remains to check that $c(\alpha)$ admits a meromorphic continuation to the whole complex plane. But this follows from the previous fact and the expression for $c(\alpha)$ from Remark 2.3.

3 The Beilinson–Bernstein Theorem

We remind the Beilinson–Bernstein theorem on localization of g-modules following
[5]. We denote by capital letters the Lie groups, and by the corresponding small
letters their Lie algebras.

Let G be a complex reductive algebraic group. Let P be a parabolic subgroup of
G. Let $P_1 := [P, P]$ denote the commutator subgroup, and let $T_P := P/P_1$. Thus
t_p denotes the Lie algebra of T_P.

Let us define the set of roots of t_p in g. Choose a Levi subgroup L of P, and let
C be the connected component of its center. Then C acts on g by the adjoint action,
and we obtain the roots of c in g (which do not always form a root system). The map
$C \hookrightarrow P \to T_P$ is a finite covering of T_P. Hence to every root of c in g corresponds
a root of t_p in g. These roots are independent of the choice of L. Let $R(t_p) \subset t_p^*$
be the set of roots of t_p in g. The set $R(t_p)$ is naturally divided into the set of roots
whose root spaces are contained in n and its complement. Let $R^+(t_p)$ be the set of
roots of t_p in g/p. If α is a root of t_p in g then the dimension of the corresponding
root subspace g_α is called the multiplicity of α. Let ρ_p be the half sum of the roots
contained in $R^+(t_p)$ counted with their multiplicities.

Let B be a Borel subgroup of G contained in P. The map $B/B_1 \to P/P_1$ gives
the canonical surjection $T_B \to T_P$. It dualizes to the inclusion $t_p^* \hookrightarrow t_b^*$.

We say that $\lambda \in t_p^*$ is *dominant* if for any root $\alpha \in R^+(t_b)$ we have $< \lambda, \alpha^V > \neq$
$-1, -2, \ldots$. We shall say that $\lambda \in t_b^*$ is *B-regular* if for any root $\alpha \in R^+(t_b)$ we
have $< \lambda, \alpha^V > \neq 0$.

Via the inclusion $t_p^* \hookrightarrow t_b^*$ we view the elements of t_p^* as elements of t_b^*, and
under this identification let us define an element $\rho_l := \rho_b - \rho_p \in t_b^*$.

For the definitions and basic properties of the sheaves of twisted differential
operators we refer to [5, 18]. Here we will present only the explicit description of
the sheaf $\mathcal{D}_\lambda$ in order to agree about the normalization.

Let X be the flag variety of G of type P (then $X = G/P$). Let $\mathcal{O}_X$ denote the
sheaf of regular functions on X. Let $U(g)$ denote the universal enveloping algebra of
g. Let U^o be the sheaf $U(g) \otimes_{\mathbb{C}} \mathcal{O}_X$, and $g^o := g \otimes_{\mathbb{C}} \mathcal{O}_X$. Let $\mathcal{T}_X$ be the tangent sheaf
of X. We have a canonical morphism $\alpha : g^o \to \mathcal{T}_X$. Let also $p^o := \operatorname{Ker} \alpha = \{\xi \in$
$g^o | \xi_x \in p_x \forall x \in X\}$. Let $\lambda : p \to \mathbb{C}$ be a linear functional which is trivial on p_1
(thus $\lambda \in t_p^*$). Then λ defines a morphism $\lambda^o : p^o \to \mathcal{O}_X$. We will denote by $\mathcal{D}_\lambda$ the
sheaf of twisted differential operators corresponding to $\lambda - \rho_p$, i.e. $\mathcal{D}_\lambda$ is isomorphic
to $U^o/\mathcal{I}_\lambda$, where $\mathcal{I}_\lambda$ is the two sided ideal generated by the elements of the form
$\xi - (\lambda - \rho_p)^o(\xi)$ where ξ is a local section of p^o. Let $D_\lambda := \Gamma(X, \mathcal{D}_\lambda)$ denote
the ring of global sections of $\mathcal{D}_\lambda$. We have a canonical morphism $U(g) \to D_\lambda$. Let
us also denote by $\mathcal{D}_\lambda - mod$ (resp. $D_\lambda - mod$) the category of $\mathcal{D}_\lambda$- (resp. $D_\lambda-$)
modules.

In this notation one has the following result.

Theorem 3.1 (Beilinson–Bernstein). *1. If* $\lambda + \rho_l \in t_b^*$ *is dominant then the functor*
$\Gamma : \mathcal{D}_\lambda - mod \rightarrow D_\lambda - mod$ *is exact.*
2. If $\lambda + \rho_l \in t_b^*$ *is dominant and is regular then the functor* Γ *is also faithful.*

This theorem for the full flag variety was proved in [4]. The general case is discussed in [5]. In order to apply this theorem in our situation we need two more facts.

Lemma 3.2. *Let* $g = gl_n(\mathbb{C})$. *The canonical morphism* $U(g) \rightarrow D_\lambda$ *is an epimorphism and preserves the natural filtrations by the order on both algebras.*

This lemma implies that the structure of any D_λ-module considered as $U(g)$-module remains the same (length, composition series, etc.).

Proof of Lemma 3.2. By Proposition III.6.2 of [5], Lemma 3.2 will be proved if the moment map $\pi : T^*X \rightarrow g^*$ is birational on its image and the image is normal (see [5] for the definition of the moment map). The image of the moment map is a closed subvariety of the nilpotent cone in g and is called the Richardson class of p (here g^* is identified with g via the invariant bilinear form on g). But by the result of [20] all the Richardson classes for the Lie algebra $gl_n(\mathbb{C})$ are normal. Also it is easy to see that the moment map for any parabolic subalgebra of $gl_n(\mathbb{C})$ is birational.   □

Note also that always the functor Γ has a left adjoint functor (called the localization functor) $\Delta : D_\lambda - mod \rightarrow \mathcal{D}_\lambda - mod$. It is defined as $\Delta(M) = \mathcal{D}_\lambda \otimes_{D_\lambda} M$.
The next lemma is proved in ([5], Proposition I.6.6).

Lemma 3.3. *Suppose* $\Gamma : \mathcal{D}_\lambda - mod \rightarrow D_\lambda - mod$ *is exact. Then the localization functor* $\Delta : D_\lambda - mod \rightarrow \mathcal{D}_\lambda - mod$ *is the right inverse of* Γ:

$$\Gamma \circ \Delta = Id.$$

Furthermore

(1) Γ *sends simple objects to simple ones or to zero.*
(2) Γ *sends distinct simple objects to distinct ones or to zero.*

The following lemma is obvious.

Lemma 3.4. *1. The positive roots* $R^+(t_b)$ *of* t_b *in* g *are* $(0, \ldots, 0, -1, 0, \ldots, 0, 1, 0, \ldots)$.
2. $\rho_b = \frac{1}{2}(-(n-1), -(n-3), \ldots, n-1)$.

Now let us consider the case of the group GL_n. Remind that

$$P_k := \left\{ \begin{pmatrix} c & b \\ 0 & a \end{pmatrix} \mid c \in GL_{n-k}, a \in GL_k \right\}.$$

Let $\lambda : t_b \to \mathbb{C}$ be a character of the form

$$\lambda = (\underbrace{\alpha, \ldots, \alpha}_{n-k \text{ times}}; \underbrace{0, \ldots, 0}_{k \text{ times}}) + \rho_{p_k}.$$

Lemma 3.5. *1. $\lambda + \rho_l$ is regular iff $\alpha \neq 1, \ldots, n-1$.*
2. $\lambda + \rho_l$ is dominant iff $\alpha \neq 2, 3, \ldots$.

Proof. We have

$$\lambda + \rho_l = (\alpha - \frac{n-1}{2}, \alpha - \frac{n-3}{2}, \ldots, \alpha + \frac{n-1}{2} - k; \frac{n-1}{2} - k, \frac{n-1}{2} - k - 1, \ldots, \frac{n-1}{2}).$$

Now the result follows from the definitions by a direct computation.   □

4 Main Results

The following proposition was proved in [15] though in this paper we will present a different proof.

Proposition 4.1. *If $\alpha \notin \mathbb{Z}$ then the representations $\mathcal{L}^{\alpha,\pm}$ are irreducible.*

Also in [15] it was computed the length and the K-type structure of all irreducible subquotients of $\mathcal{L}^{\alpha,\pm}$ for all $\alpha \in \mathbb{C}$. We will prove the following result.

Theorem 4.2. *Assume that $\alpha \neq 1, 2, \ldots, n-1$. Then the representation $\mathcal{L}^{\alpha,\pm}$ has a unique composition series. If $\alpha < n/2$ then the Gelfand-Kirillov dimension of consecutive irreducible subquotients is strictly increasing. If $\alpha > n/2$ then it is strictly decreasing.*

From this theorem we immediately get the following corollary.

Corollary 4.3. *Let $\alpha, \beta \neq 1, 2, \ldots, n-1$. Assume that either $\alpha > n/2$ and $\beta < n/2$ or $\alpha < n/2$ and $\beta > n/2$. Let $\varepsilon, \delta = \pm$. Then up to a multiplication by a constant there is at most one intertwining operator from $\mathcal{L}^{\alpha,\varepsilon}$ to $\mathcal{L}^{\beta,\delta}$. Such an operator has an irreducible image.*

In particular we get the following corollary for the operators T_i considered in Sect. 1.

Corollary 4.4. *Let $\alpha \neq 1, 2, \ldots, n-1$. Then the image of the operator*

$$T_i : C^\infty(Gr_{n-i}(V), M_\alpha) \longrightarrow C^\infty(Gr_i(V), L_\alpha)$$

is irreducible.

Before we prove Theorem 4.2 and Proposition 4.1 notice that the case $\alpha > n/2$ reduces by duality to the case $\alpha < n/2$ using Lemma 2.6 and the fact that the Gelfand-Kirillov dimensions of a representation and its dual are equal (this follows from Theorem 1.2 in Vogan's paper [34] which allows to compute the Gelfand-Kirillov dimension of a representation using the K-type structure).

Thus we will assume that $\alpha \neq 1, 2, 3, \ldots$. Let $\varepsilon = \pm$. Consider the character $\lambda = (\underbrace{-\alpha, \ldots, -\alpha}_{n-k \text{ times}}, \underbrace{0, \ldots, 0}_{k \text{ times}}) + \rho_{p_k}$. Let ${}^{\mathbb{C}}Gr_{n-k}$ denote the complex Grassmannian of $(n - k)$-dimensional subspaces in the complexified space ${}^{\mathbb{C}}V := V \otimes_{\mathbb{R}} \mathbb{C}$. Set $\kappa := 0$ if $\varepsilon = +$, and $\kappa := 1$ if $\varepsilon = -$. Let us describe the $O(n, \mathbb{C})$-equivariant sheaf $\mathcal{M}$ of $\mathcal{D}_\lambda$-modules on ${}^{\mathbb{C}}Gr_{n-k,n}$ corresponding to $\mathcal{T}^{\alpha,\varepsilon} := \mathcal{L}^{\alpha,\varepsilon} \otimes |\det(\cdot)|^{-\alpha} \otimes \operatorname{sgn}(\det(\cdot))^\kappa$. (It is slightly more convenient for the notational reasons to work with the last representation rather than with $\mathcal{L}^{\alpha,\varepsilon}$ itself.)

First notice that the Euclidean structure on V defines a non-degenerate symmetric quadratic form on the complexification ${}^{\mathbb{C}}V$. Let us denote this form by B. The $O(n, \mathbb{C})$-orbits on ${}^{\mathbb{C}}Gr_{n-k,n}$ are classified by the rank of the restriction of the form B to a subspace (this fact is well known and can be easily checked). In other words two subspaces $E, F \in {}^{\mathbb{C}}Gr_{n-k,n}$ belong to the same orbit if and only if $\operatorname{rk} B|_E = \operatorname{rk} B|_F$. Let U denote the open orbit consisting of the subspaces such that the restriction of the form B on them is non-degenerate. Let us fix a subspace $E_0 \in U$. Note that $O(n, \mathbb{C})$-equivariant coherent $\mathcal{D}_\lambda$-modules on U correspond to representations of the group of connected components of the stabilizer of E_0 which is equal to $O(n - k, \mathbb{C}) \times O(k, \mathbb{C})$. Consider the $O(n, \mathbb{C})$-equivariant $\mathcal{D}_\lambda$-module $\mathcal{M}_0$ corresponding to the representation $(A, B) \mapsto \operatorname{sgn}(\det A)^\kappa$ where $(A, B) \in O(n - k, \mathbb{C}) \times O(k, \mathbb{C})$. Let $j : U \hookrightarrow {}^{\mathbb{C}}Gr_{n-k,n}$ be the identity (open) imbedding. One easily checks the following claim.

Claim 4.5.

$$\mathcal{M} = j_* \mathcal{M}_0.$$

In order to study the composition series of the $\mathcal{D}_\lambda$-module $\mathcal{M}$ it is enough to study its pull back to a transversal to the minimal $O(n, \mathbb{C})$-orbit on ${}^{\mathbb{C}}Gr_{n-k,n}$. This minimal orbit consists of subspaces such that the dimension of the kernel of the restriction of the form B to them is equal to $\min\{k, n-k\}$. Let us describe explicitly a transversal to it. First note that we may assume that $k = \min\{k, n - k\}$. Indeed suppose that $k > l := \min\{k, n - k\}$. Fix a $(k - l)$-dimensional subspace F such that the restriction of the form B to F is non-degenerate. Replacing V by V/F and k by l we may assume that $k = l$. We may also assume that $n = 2k$ replacing V by certain $2k$-dimensional subspace.

Let us denote by H_k the space of complex symmetric $(k \times k)$-matrices. We can present $V = M \oplus N$ where M and N are certain isotropic k-dimensional subspaces. Furthermore we can choose bases in M and N such that in these bases the matrix of B is equal to $\begin{bmatrix} 0 & I \\ I & 0 \end{bmatrix}$. Any matrix $A \in H_k$ can be considered as a matrix of an operator $A : M \to N$ with respect to these bases (which we will denote by the same letter A). Then the graph of this operator $graph(A)$ is a k-dimensional subspace in V. When A runs over H_k it defines a transversal to the minimal orbit. Moreover the stratification of ${}^C Gr_k(V)$ by $O(n, \mathbb{C})$-orbits induces on H_k the usual stratification by the rank of a matrix.

By the Riemann-Hilbert correspondence the $\mathcal{D}'$-module $\mathcal{M}$ corresponds to certain twisted perverse sheaf. For the details on the Riemann-Hilbert correspondence we refer to [18]. Slightly oversimplifying, note only that at a neighborhood $\mathcal{O}$ of any point any twisted perverse sheaf can be presented by a usual (untwisted) perverse sheaf $P_{\mathcal{O}}$, and at the intersection $\mathcal{O}_1 \cap \mathcal{O}_2$ of two such neighborhoods $\mathcal{O}_1 \cap \mathcal{O}_2$ the isomorphism $P_{\mathcal{O}_1}|_{\mathcal{O}_1 \cap \mathcal{O}_2} \to P_{\mathcal{O}_2}|_{\mathcal{O}_1 \cap \mathcal{O}_2}$ is given by a twist by certain rank one local system $L_{\mathcal{O}_1, \mathcal{O}_2}$ on $\mathcal{O}_1 \cap \mathcal{O}_2$. These local systems $L_{\mathcal{O}_1, \mathcal{O}_2}$ must satisfy certain compatibility conditions on the triple intersections of neighborhoods (see [18]).

Let us denote by S_j the locally closed stratum of H_k of matrices of rank $k - j$, $j = 0, \ldots, k$. Thus S_0 is the open stratum of non-degenerate matrices. For the details about the topology of S_j we refer to [7], Sect. 3. We need only the fact that $\pi_1(S_0)$ is isomorphic to $\mathbb{Z}$, and a generator is a loop around S_1.

We will identify $\mathcal{D}'$-modules with twisted perverse sheaves via the Riemann-Hilbert correspondence, and we will denote them by the same letters.

Let us choose a small ball $B \subset H_k$ with the center at the origin 0 such that the twisted perverse sheaf $\mathcal{M}$ restricted to B is isomorphic to a usual (untwisted) perverse sheaf. Let $h : S_0 \cap B \hookrightarrow B$ denote the open imbedding. Then $\mathcal{M}|_B$ can be described as follows. First note that $\mathcal{M}|_{S_0 \cap B} = \mathcal{M}_0|_{S_0 \cap B}$ is a local system on $S_0 \cap B$. Then $\mathcal{M} = h_*(\mathcal{M}_0|_{S_0 \cap B})$. It is easy to see that $\mathcal{M}_0|_{S_0 \cap B}$ is rank one local system with the monodromy $\mu = e^{i\pi(\alpha + \kappa)}$ around the generator of $\pi_1(S_0 \cap B) \tilde{\to} \pi_1(S_0)$. It is sufficient to study the composition series of $\mathcal{M}|_B$. For convenience we will replace B by H_k (since they are diffeomorphic as stratified spaces).

Let us denote by $\mathcal{L}_0$ the rank one local system on S_0 with the monodromy $\mu = e^{i\pi(\alpha + \kappa)}$ around the generator of $\pi_1(S_0)$. Let $\mathcal{L} := j_* \mathcal{L}_0$. We have to study the length and the composition series of $\mathcal{L}$.

In order to do that let us remind the description in terms of quivers of the category of perverse sheaves on H_k stratified by $\{S_j\}$ due to Braden and Grinberg [7]. Let us define categories $\mathcal{A}_k$ and $\mathcal{B}_k$ of quivers following [7] for an integer $k \geq 1$. An object (A_*, p_*, q_*) of $\mathcal{A}_k$ is a collection of $k + 1$ finite dimensional complex vector spaces and linear maps between them as follows:

$$A_0 \underset{q_1}{\overset{p_1}{\rightleftarrows}} A_1 \underset{q_2}{\overset{p_2}{\rightleftarrows}} \ldots \underset{q_k}{\overset{p_k}{\rightleftarrows}} A_k$$

satisfying the following conditions:

(A1) $1 + q_j p_j$ and $1 + p_j q_j$ are invertible for $j = 1, \ldots, k$;
(A2) $p_j q_j = q_{j+1} p_{j+1}$ for $j = 1, \ldots, k - 1$.

A morphism $f : A \to A'$ of quivers from $\mathcal{A}_k$ is a collection of linear maps $f_j : A_j \to A'_j$ such that $f_j p_j = p'_j f_{j-1}$ and $q'_j f_j = f_{j-1} q_j$. Then $\mathcal{A}_k$ is an abelian category. A sequence of quivers $A' \to A \to A''$ is exact if and only if the sequences $A'_j \to A_j \to A''_j$ are exact for all j. For an object $A \in A_n$ define maps $\zeta_j : A_j \to A_j$ by $\zeta_j = 1 + q_{j+1}p_{j+1}$ if $j < k$ and by $\zeta_j = 1 + p_j q_j$ if $j > 0$. These definitions agree due to condition (A2). The maps $\{\zeta_j\}$ define an automorphism $\zeta_A : A \to A$ for any object $A \in A_k$. This ζ defines an automorphism of the identity functor on A_k. More precisely if $f : A \to A'$ is a morphism in A_k then $f\zeta_A = \zeta_{A'} f$. The following statement was proved in [7], Propositions 4.1, 4.2.

Proposition 4.6. *(1) A_k is a direct sum of abelian categories*

$$A_k = \bigoplus_{\lambda \neq 0} A_k^{(\lambda)},$$

where $A_k^{(\lambda)}$ is the full subcategory of objects A for which $\zeta_A - \lambda$ is nilpotent.

(2) If $\lambda \neq 0, 1$, then $A_k^{(\lambda)}$ is equivalent to $A_0^{(\lambda)}$. The equivalence is given by sending A to (A_0, ζ_0).

An object of B_k is a collection (B_*, p_*, q_*) consisting of finite dimensional complex vector spaces $B_0, \ldots, B_k$ and linear maps between them

$$B_0 \underset{q_1}{\overset{p_1}{\rightleftarrows}} B_1 \underset{q_2}{\overset{p_2}{\rightleftarrows}} \ldots \underset{q_k}{\overset{p_k}{\rightleftarrows}} B_k$$

which satisfy the following conditions. Let $\zeta_j = 1 + q_{j+1}p_{j+1}$ for $0 \leq j < k$, and $\eta_j = 1 + p_j q_j$ for $0 < j \leq k$. We require that

(1) All the maps ζ_j and η_j are invertible;
(2) $\zeta_j^2 = \eta_j^2$, for $j = 1, \ldots, k-1$;
(3)

$$p_j \zeta_{j-1} = -\zeta_j p_j,$$

$$\zeta_{j-1} q_j = -q_j \zeta_j,$$

$$p_{j+1} \eta_j = -\eta_{j+1} p_{j+1},$$

$$\eta_j q_{j+1} = -q_{j+1} \eta_{j+1}, \text{ for } j = 1, \ldots, k-1.$$

The morphisms of quivers in B_k are defined in the obvious way. Then B_k is also an abelian category. Define $\bar{\zeta}_j = (-1)^j \zeta_j$ for $j < k$, and $\bar{\zeta}_k = (-1)^{k+1} \eta_k$. It is easy to see that $\bar{\zeta}$ is the automorphism of the identity functor of B_k.

For any object $B = (B_*, p_*, q_*) \in B_k$ each vector space B_j decomposes as

$$B_j = B_j^+ \oplus B_j^-$$

where $\zeta_j|_{B_j^+} = \eta_j|_{B_j^+}$ and $\zeta_j|_{B_j^-} = -\eta_j|_{B_j^-}$. The operators ζ_j and η_j preserve this decomposition. Let B_k^+ be the full subcategory of B_k of objects B with $B_0 = B_k = 0$

and $B_j^- = 0$ for $j = 1, \ldots, k - 1$. Easily all the maps p and q must vanish for $B \in \mathcal{B}_k^+$. Let us define $\mathcal{B}_k^-$ to be the full subcategory of $\mathcal{B}_k$ of objects with $B_j^+ = 0$ for $j = 1, \ldots, k - 1$.

The following statement was proved in [7], Proposition 4.4.

Proposition 4.7. *The category $\mathcal{B}_k$ splits as a direct sum*

$$\mathcal{B}_k = \mathcal{B}_k^+ \oplus \mathcal{B}_k^-.$$

Let us define $\mathcal{B}_k^{(\lambda)}$ to be the full subcategory of $\mathcal{B}_k^-$ of objects B for which $\bar\zeta_B - \lambda$ is nilpotent. Let us also introduce category $\mathcal{A}_0 = \mathcal{B}_0$. The objects of this category are pairs (A_0, ζ_0) where A is a finite dimensional complex vector space, and $\zeta_0 : A_0 \to A_0$ is an invertible transformation. The next proposition was proved in [7], Proposition 4.4.

Proposition 4.8. *The category $\mathcal{B}_k^-$ decomposes as*

$$\mathcal{B}_k^- = \bigoplus_{\lambda \neq 0} \mathcal{B}_k^{(\lambda)}.$$

If $\lambda \notin \{-1, 0, 1\}$ then $\mathcal{B}_k^{(\lambda)} \simeq \mathcal{A}_0^{(\lambda)}$. Furthermore there are equivalences of categories

$$\chi_k^{(1)} : \mathcal{B}_k^{(1)} \xrightarrow{\sim} \mathcal{A}_{\lceil k/2 \rceil}^{(1)} \text{ and } \chi_k^{(-1)} : \mathcal{B}_k^{(-1)} \xrightarrow{\sim} \mathcal{A}_{\lfloor k/2 \rfloor}^{(1)}.$$

Here we denote by $\lfloor x \rfloor$ the maximal integer not greater than x, and by $\lceil x \rceil$ the minimal integer not smaller than x. Note that for $\lambda \notin \{-1, 0, 1\}$ the equivalence $\mathcal{B}_k^{(\lambda)} \simeq \mathcal{A}_0^{(\lambda)}$ is given by sending the quiver (B_*, p_*, q_*) to (B_0, ζ_0). We will not describe explicitly the functors $\chi_k^{(1)}$ and $\chi_k^{(-1)}$ referring to the proof of Proposition 4.4 in [7]. We need only to know how they change the vector spaces in the quiver. One has

$$(\chi_k^{(-1)}(B))_j = B_{2j} \tag{2}$$

$$\text{if } n = 2m \text{ is even then } (\chi_k^{(1)}(B))_j = B_{2j} \tag{3}$$

$$\text{if } n = 2m - 1 \text{ is odd then } (\chi_k^{(1)}(B))_j = B_{2j}$$

$$\text{for } j < m \text{ and } (\chi_k^{(1)}(B))_m = B_n. \tag{4}$$

Let us define functors $F_B : \mathcal{B}_k \to \mathcal{A}_0$ and $F_A : \mathcal{A}_k \to \mathcal{A}_0$ by

$$F_B((B_*, p_*, q_*)) = (B_0, \bar\zeta_0)$$

$$F_A((A_*, p_*, q_*)) = (A_0, \zeta_0).$$

Then it immediately follows from (2)–(4) that

$$(\chi_k^{(1)}(B))_0 = B_0 \text{ and } (\chi_k^{(-1)}(B)) = B_0$$

for any object B from the corresponding category. It is also clear that $F_B(\mathcal{B}_k^-) = 0$.

One of the main results of [7] says that the category of perverse sheaves on H_k constructible with respect to the stratification $\{S_j\}$ is equivalent to the category $\mathcal{B}_k$. Let us present some more details on this equivalence. the vector spaces B_j corresponds to stalks of Morse local systems. We will not define them here. We need only to define B_0. This is the stalk at a fixed point on the open stratum S_0 of our perverse sheaf (which is just a local system on S_0). Moreover the operator $\bar{\zeta}_0$ is equal to the monodromy of this local system along the generator of $\pi_1(S_0)$. Let us denote by $j : S_0 \hookrightarrow H_k$ the open imbedding. Then the functor j^* from the category of perverse sheaves on H_k to perverse sheaves on S_0 in the language of quivers coincides with the functor F_B. However for our purposes we need to describe in the language of quivers the right adjoint functor j_*. This description was privately explained to us by T. Braden (private communication).

Since $j^*(\mathcal{B}_k^+) = 0$ then by adjointness $j_*(A_0) \subset \mathcal{B}_k^-$. Similarly $j_*(A_0^{(\lambda)}) \subset \mathcal{B}_k^{(\lambda)}$. Moreover by Proposition 4.8 for $\lambda \notin \{-1, 0, 1\}$ the functor $j_* : A_0^{(\lambda)} \tilde{\to} \mathcal{B}_k^{(\lambda)}$ is an equivalence of categories. This implies the following claim.

Claim 4.9. *If the monodromy $\mu = e^{i\pi(\alpha+\kappa)} \neq \pm 1$ then $j_* \mathcal{L}_0$ is irreducible.*

If $\lambda = 1$ then by Proposition 4.8 we can replace category $\mathcal{B}_k^{(1)}$ by the category $A_{\lceil k/2 \rceil}^{(1)}$, and for $\lambda = -1$ we can replace category $\mathcal{B}_k^{(-1)}$ by the category $A_{\lfloor k/2 \rfloor}^{(1)}$. Thus let us study the functor $j_* : A_0^{(1)} \to A_l^{(1)}$ (where we denote by j_* the functor right adjoint to F_A). In fact let us describe the functor $j_* : A_0 \to A_l$.

Claim 4.10.

$$j_*((A_0, \zeta_0)) = (A_*, p_*, q_*)$$

where $A_j = A_0$ for all $j = 0, \ldots, l$, $p_j = \zeta_0 - 1$, $q_j = 1$.

Proof. Let $X = (X_*, p_*, q_*) \in A_l$. We have to show that $Hom(X, j_*(A_0, \zeta_0)) = Hom((X_0, \zeta_0), (A_0, \zeta_0))$. We have to show that if we are given a map $f_0 : X_0 \to A_0$ commuting with ζ_0 then we can extend it (in a functorial way) to a morphism $X \to j_*(A_0, \zeta_0)$. Let us define the maps $f_j : X_j \to A_0$ inductively by $f_j := f_{j-1}q_j$. It is easy to check all the properties. $\square$

Now let us come back to our perverse sheaf $j_* \mathcal{L}_0$ on H_k. Let us assume that the monodromy $\mu = e^{i\pi(\alpha+\kappa)} = \pm 1$. Set $l := \lceil k/2 \rceil$ if $\mu = 1$ and $l := \lfloor k/2 \rfloor$ if $\mu = -1$. Then the corresponding object in $A_0^{(1)}$ is equal to $(\mathbb{C}, 1)$. From the proof of Claim 4.10 we obtain that $j_*((\mathbb{C}, 1)) = (A, p_*, q_*) \in A_l^{(1)}$ where $A_j = \mathbb{C}$, $p_j = 0$, $q_j = 1$ for all $j = 0, \ldots, l$. It is easy to see that every subquiver of $j_*((\mathbb{C}, 1))$ has the form $Y_k = (Y_*, p'_*, q'_*)$ where $Y_j = A_j$ for $j \leq k$ and $Y_j = 0$

for $j > k$, $p'_j = p_j$, $q'_j = q_j$ for $j \leq k$ and $p'_j = 0 = q'_j$ for $j > k$. It particular it follows that $j_*((\mathbb{C}, 1))$ has a unique composition series and length $l + 1$.

Thus the above argument shows that the representations $\mathcal{L}^{\alpha,\pm}$ have unique composition series for $\alpha \neq 1, 2, \ldots, n - 1$. Let us show that if in addition $\alpha < n/2$ then the Gelfand-Kirillov dimension of consecutive subquotients is strictly increasing. Let us remind a general method how to compute the Gelfand-Kirillov dimension of a finitely generated g-module M. It is well known that the Gelfand-Kirillov dimension is equal to the dimension of the associated variety (or the Bernstein variety) $V_g(M)$ of M which is an algebraic subvariety of g^*. Let us remind its definition. One can choose a filtration of M by finite dimensional subspaces $M_0 \subset M_1 \subset M_2 \subset \cdots \subset M$ such that $gM_i = M_{i+1}$ for large i. Such a filtration is called *good*. The associated graded module grM is a module over the algebra $grU(g) = S(g)$. In other words grM is a coherent sheaf on $g^* = \operatorname{Spec} S(g)$. The support of this sheaf is called the associated variety of M and is denoted by $V_g(M)$. It is well known (and can be easily checked) that it does not depend on the choice of a good filtration. Moreover if we identify g^* with g using an invariant bilinear form on g then $V_g(M) \subset g$ is contained in the nilpotent cone of g.

Next let us assume that $X = GL(n, \mathbb{C})/{}^C P_k$ is the complex Grassmannian of $(n-k)$-dimensional subspaces $\in \mathbb{C}^n$ with the standard action of ${}^C K = O(n, \mathbb{C})$. Let $\mathcal{D}'$ be a ${}^C K$-equivariant sheaf of twisted differential operators on X which satisfies the assumptions of the Beilinson–Bernstein localization theorem. Let $\mathcal{M}$ be an ${}^C K$-equivariant $\mathcal{D}'$-module on X. Let us denote by $Ch(\mathcal{M})$ the singular support of $\mathcal{M}$. Thus $Ch(\mathcal{M}) \subset T^*X$. Let $\pi : T^*X \to g$ denote the moment map.

Lemma 4.11. *Let the sheaf $\mathcal{D}'$ satisfies the assumptions of the Beilinson–Bernstein theorem. Let $\mathcal{M}$ be a coherent $\mathcal{D}'$-module. Let $M := \Gamma(X, \mathcal{M})$. Then $V_g(M) = \pi(Ch(\mathcal{M}))$.*

This lemma is proved in Proposition 4.2 in [1] where the argument is taken from [6].

Let $l = \min\{k, n - k\}$. Passing to the orthogonal complement if necessary we can replace the Grassmannian $Gr_{n-k}(\mathbb{C}^n)$ by $Gr_l(\mathbb{C}^n)$. Let us denote by U_j, $j = 0, \ldots, l$, the locally closed subvariety of X consisting of l-subspaces such that the restriction of our quadratic form onto them has rank $l - j$. Then the singular support on any ${}^C K$-equivariant coherent $\mathcal{D}'$-module is contained in the union of closures of conormal bundles $T^*_{U_j}X$.

Lemma 4.12. *The image under the moment map $\pi(T^*_{U_j}X)$ is equal to the variety of complex symmetric $(n \times n)$-matrices T of rank at most $l - j$ such that there exists a subspace $F \in Gr_l(\mathbb{C})$ so that $T(\mathbb{C}^n) \subset F$ and $T(F) = 0$.(Let us denote this variety of matrices by $\mathcal{R}_j$.)*

Proof. Let us fix $E \in U_j$. Then the space of conormal vectors to U_j at E coincides with the space of operators (again using the identification of $g = gl_n(\mathbb{C})$ with its dual by the Killing form) $A : \mathbb{C}^n \to \mathbb{C}^n$ such that

(1) A is symmetric;
(2) $A(\mathbb{C}^n) \subset E$;
(3) $A(E) = 0$.

First let us show that rk $A \leq l - j$. We have Im $A \subset E \subset$ Ker $A = (\text{Im } A)^{\perp}$. Hence $l - \dim(\text{Im } A) \geq j$. Hence rk $A \leq l - j$. Next let us show that T^*U_j is mapped onto $\mathcal{R}_j$. It is easy to see that the variety of such matrices is a closure of matrices of the above form of rank $l - j$ which form an $O(n, \mathbb{C})$-orbit. Since the moment map is $GL_n(\mathbb{C})$-equivariant it is enough to check that the image of T^*U_j contains at least one matrix of rank $l - j$. Indeed we may choose linearly independent vectors $e_1, \ldots, e_{l-j} \in$ Ker $B|_E$. Consider the operator $Ax = \sum_{i=1}^{l-j} < x, e_i > e_i$. Clearly A satisfies (1)–(3) and rk $A = l - j$. $\qquad\square$

In order to compute the singular support of a $\mathcal{D}'$-module let us remind that it is equal to the union of those closures of $T^*_{U_j} X$ where the corresponding Morse groups of the corresponding $\mathcal{D}'$-module do not vanish. We may replace X by the transversal to the minimal orbit identified with H_k as previously. Then the strata U_j will be replaced by S_j.

First let us state few more results from [7]. For a local system $\mathcal{N}$ on a stratum S_j let us denote by $IC(S_j, \mathcal{N})$ the Goresky-Macpherson extension of $\mathcal{N}$ to H_k. The fundamental group of S_j for $1 \leq j \leq n - 1$ is isomorphic to $\mathbb{Z}/2\mathbb{Z}$ (see [7]). Thus on S_j there are just two irreducible local systems: the constant one and the rank one local system corresponding to to the non-trivial representation of $\pi_1(S_j)$. This non-trivial local system on S_j will be denoted by $\mathcal{N}_j$. The following statement was proved in [7], Corollary 4.11.

Proposition 4.13. *For* $1 \leq j \leq n - 1$ *the quiver B corresponding to* $IC(S_j, \mathcal{L}^{\otimes j})$ *is given by* $B_j = \mathbb{C}$, *and by all other* $B_i = 0$. *It is an object in* $\mathcal{B}_n^+$.

For $1 \leq j \leq n - 1$ *the quiver corresponding to* $IC(S_j, \mathcal{N}^{\otimes(j+1)})$ *is given by* $B_j = B_{j+1} = \mathbb{C}$, *and by all other* $B_i = 0$. *It is an object in* $\mathcal{B}_n^{(\lambda)}$ *where* $\lambda = (-1)^{j+1}$.

The quiver B corresponding to $IC(S_n, \mathbb{C})$ *is given by* $B_n = \mathbb{C}$ *and by all other* $B_i = 0$. *It is an object of* $\mathcal{B}_n^{(\lambda)}$ *where* $\lambda = (-1)^{n+1}$.

We always have $j_*\mathcal{L}_0 \in \mathcal{B}_n^-$. Let us first consider the case $\mu = 1$. Then $\mathcal{L}_0 = \mathbb{C}$. Thus $j_*\mathbb{C} \in \mathcal{B}_m^{(1)}$, and all its irreducible subquotients also belong to $\mathcal{B}_n^{(1)}$. Hence by Proposition 4.13 all irreducible subquotients of $j_*\mathbb{C}$ must have the form $IC(S_j, \mathcal{N}_j^{\otimes(j+1)})$ where j is odd and $1 \leq j \leq n-1$, or $IC(n, \mathbb{C})$ if n is odd. From this we see that the singular support of $IC(S_{2k-1}, \mathcal{N}_{2k-1}^{\otimes(2k)})$ is contained and not equal to the singular support of $IC(S_{2k+1}, \mathcal{N}_{2k+1}^{\otimes(2k+2)})$. This implies the statement about the Gelfand-Kirillov dimension in the case $\mu = 1$ using Lemmas 4.11 and 4.12.

Let us consider the case $\mu = -1$. Then $j_*\mathcal{L}_0 \in \mathcal{B}_n^{(-1)}$. Hence all irreducible subquotients also belong to $\mathcal{B}_n^{(-1)}$. Again Proposition 4.13 implies that all of them must be of the form $IC(S_j, \mathcal{N}_j^{(j+1)})$ with j is even and $1 \leq j \leq n - 1$, or

$IC(S_n, \mathbb{C})$ if n is even. Similarly to the previous case, the statement follows. Thus Theorem 4.2 is proved.   □

Let us discuss now the K-type structure of the range of the α-cosine transform T_{ji}^{α}. First let us recall some standard facts on the representations of the special orthogonal group $SO(n)$ (see e.g. [37]).

Lemma 4.14. *The isomorphism classes of irreducible representations of $SO(n)$, $n > 2$ are parameterized by their highest weights, namely sequences of integers $(m_1, m_2, \ldots, m_{[n/2]})$ which satisfy:*

(i) If n is odd then $m_1 \geq m_2 \geq \cdots \geq m_{[n/2]} \geq 0$:
(ii) If $n > 2$ is even then $m_1 \geq m_2 \geq \cdots \geq m_{n/2-1} \geq |m_{n/2}|$.

Recall also that for $n = 2$ the representations of $SO(2)$ are parameterized by a single integer m_1. We will use the following notation. Let us denote by Λ^+ the set of all highest weights of $SO(n)$, and by Λ_k^+ the set of all highest weights $\lambda = (m_1, m_2, \ldots, m_{[n/2]})$ with $m_i = 0$ for $i > k$ and all m_i are *even*.

Let us recall the decomposition of the space of functions on the Grassmannian $Gr_{k,n}$ under the action of $SO(n)$ referring for the proofs to [32, 33]. Since $Gr_{k,n}$ is a symmetric space, each irreducible representation enters with multiplicity at most one. The representations which do appear have highest weights precisely from $\Lambda_k^+ \cap \Lambda_{n-k}^+$.

Theorem 4.15. *Let $\alpha \in \mathbb{C}$, $\alpha \neq -n, -(n-1), \ldots, -1$. The range of the α-cosine transform $T_{j,i}^{\alpha} : C^{\infty}(Gr_{i,n}) \to C^{\infty}(Gr_{j,n})$ is a closed subspace and is decomposed under the action of $SO(n)$ as follows.*

(1) If $\alpha \notin \mathbb{Z}$ then the range consists of the representations of $SO(n)$ with highest weights precisely from the set $\Lambda_i^+ \cap \Lambda_{n-i}^+ \cap \Lambda_j^+ \cap \Lambda_{n-j}^+$.
(2) If $\alpha \in \mathbb{Z}$ and $\alpha > -\frac{n}{2}$ (hence $\alpha \geq 0$) then the range of T_{ji}^{α} consists of the representations with highest weights $\lambda = (m_1, \ldots, m_{[n/2]})$ such that $\lambda \in \Lambda_i^+ \cap \Lambda_{n-i}^+ \cap \Lambda_j^+ \cap \Lambda_{n-j}^+$ with the restriction $|m_2| \leq 1 + \alpha$.

Proof of Theorem 4.15. Let us prove first the statement about the K-type structure. The case (1) follows from Propositions 2.2, 2.5, and 4.1. Let us consider case (2). First let us consider the case $i = j$. It follows from our assumptions that $\alpha \in \mathbb{Z}$ is non-negative. Then the constant function on $Gr_{i,n}$ is mapped into a non-zero constant function. Thus the range of T_{ii}^{α} is an unramified $GL_n(\mathbb{R})$-module. By Theorems 3.4.2 and 3.4.4 of [15] the K-type structure of the unramified subquotient can be described as follows. Set $l := \min\{k, n-k\}$. Then the set of highest weights is $\{(m_1 \geq m_2 \geq \ldots) \in \Lambda_l^+ \mid |m_2| \leq 1 + \alpha\}$. Thus for $i = j$ the result follows. The case of general case follows again from this, Propositions 2.2, and 2.5.

Now let us prove that the image of T_{ij}^{α} is closed in the C^{∞}-topology. The argument is exactly the same as in [2], but we partly reproduce it here for the sake of completeness. It remains to prove that if f is a C^{∞}- function on $Gr_{i,n}$ belonging to the closure of the sum of $SO(n)$- irreducible subspaces satisfying conditions in the statement of Theorem 4.15, then f is an image under T_{ji}^{α} of some C^{∞}- function on $Gr_{j,n}$. We may assume that $j < i$. By Proposition 2.2, $T_{ji}^\alpha = c \cdot T_{jj}^\alpha R_{ji}$. We will need the following fact due to Casselman and Wallach (see [8]).

Proposition 4.16. *Let G be a real reductive group. Let K be its maximal compact subgroup. Let $\xi : X \to Y$ be a morphism of two admissible Banach G-modules of finite length which has a dense image. Then ξ induces an epimorphism on the spaces of smooth vectors.*

In our situation we will need the following more precise form of Proposition 4.16 which was proved in [2] (Lemma 1.10).

Lemma 4.17. *Let $G = GL(n, \mathbb{R})$. Let $K = O(n)$ be the maximal compact subgroup. Let X and Y be G- modules of continuous sections of some finite dimensional G- equivariant vector bundles over the Grassmannians (or any other partial flag manifolds). Let $\xi : X \to Y$ be a morphism of these G- modules. Then if $f \in Y$ is a smooth vector then there exists a smooth vector $g \in X$ such that $\xi(g) = f$ and the K- types entering into the decomposition of g are the same as those of f.*

Now let us continue the proof of Theorem 4.15. We have given an interpretation of T_{jj}^α as an intertwining operator of two $GL(n, \mathbb{R})$- modules; they satisfy the assumptions of Proposition 4.16, since they are induced from characters of parabolic subgroups (see [35, 36]). Hence by Proposition 4.16 there exists a C^∞- smooth function g on the Grassmannian $Gr_{j,n}$ such that $f = T_{jj}^\alpha(g)$ and with the same K- types as f. Next there exists an interpretation of the Radon transform as an intertwining operator of some admissible $GL(n, \mathbb{R})$- modules of finite length (it was given in [10]). Hence Proposition 4.16 implies the statement. □

References

1. S. Alesker, Description of translation invariant valuations on convex sets with solution of P. McMullen's conjecture. Geom. Funct. Anal. **11**(2), 244–272 (2001)
2. A. Alesker, J. Bernstein, Range characterization of the cosine transform on higher Grassmannians. Adv. Math. **184**(2), 367–379 (2004)
3. D. Barbasch, S. Sahi, B. Speh, Degenerate series representations for GL(2n, R) and Fourier analysis. Symposia Mathematica, Rome, 1988. Sympos. Math., vol. XXXI (Academic, London, 1990), pp. 45–69
4. A. Beilinson, J. Bernstein, Localisation de g-modules (French). C. R. Acad. Sci. Paris Sèr. I Math. **292**(1), 15–18 (1981)
5. F. Bien, *D-Modules and Spherical Representations*. Mathematical Notes, vol. 39 (Princeton University Press, Princeton, 1990)
6. W. Borho, J.-L. Brylinski, Differential operators on homogeneous spaces, III, Characteristic varieties of Harish-Chandra modules and primitive ideals. Invent. Math. **80**(1), 1–68 (1985)
7. T. Braden, M. Grinberg, Perverse sheaves on rank stratifications. Duke Math. J. **96**(2), 317–362 (1999)
8. W. Casselman, Canonical extensions of Harish-Chandra modules to representations of G. Canad. J. Math. **41**(3), 385–438 (1989)

9. T. Fujimura, On some degenerate principal series representations of $O(p, 2)$. J. Lie Theor. **11**(1), 23–55 (2001)

10. I.M. Gelfand, M.I. Graev, R.I. Roşu, The problem of integral geometry and intertwining operators for a pair of real Grassmannian manifolds. J. Operat. Theor. **12**(2), 359–383 (1984)

11. P. Goodey, R. Howard, Processes of flats induced by higher-dimensional processes. Adv. Math. **80**(1), 92–109 (1990)

12. P. Goodey, R. Howard, in *Processes of Flats Induced by Higher-Dimensional Processes. II*. Integral Geometry and Tomography (Arcata, CA, 1989), Contemp. Math., vol. 113 (Am. Math. Soc., Providence, 1990), pp. 111–119

13. P. Goodey, R. Howard, M. Reeder, Processes of flats induced by higher-dimensional processes. III. Geom. Dedicata **61**(3), 257–269 (1996)

14. E.L. Grinberg, Radon transforms on higher Grassmannians. J. Diff. Geom. **24**(1), 53–68 (1986)

15. R. Howe, S.T. Lee, Degenerate principal series representations of $GL_n(\mathbf{C})$ and $GL_n(\mathbf{R})$. J. Funct. Anal. **166**(2), 244–309 (1999)

16. R. Howe, E.C. Tan, Homogeneous functions on light cones: The infinitesimal structure of some degenerate principal series representations. Bull. Am. Math. Soc. (N.S.) **28**(1), 1–74 (1993)

17. K.D. Johnson, Degenerate principal series and compact groups. Math. Ann. **287**(4), 703–718 (1990)

18. M. Kashiwara, Representation theory and D-modules on flag varieties. Orbites unipotentes et représentations, III. Astérisque **173–174**(9), 55–109 (1989)

19. A. Koldobsky, Inverse formula for the Blaschke-Levy representation. Houston J. Math. **23**, 95–108 (1997)

20. H. Kraft, C. Procesi, On the geometry of conjugacy classes in classical groups. Comment. Math. Helv. **57**(4), 539–602 (1982)

21. S.S. Kudla, S. Rallis, Degenerate principal series and invariant distributions. Israel J. Math. **69**(1), 25–45 (1990)

22. S.T. Lee, Degenerate principal series representations of $Sp(2n, R)$. Compos. Math. **103**(2), 123–151 (1996)

23. S.T. Lee, H.Y. Loke, Degenerate principal series representations of $U(p, q)$ and $Spin_0(p, q)$. Compos. Math. **132**(3), 311–348 (2002)

24. G. Matheron, Un théorème d'unicité pour les hyperplans poissoniens (French). J. Appl. Probab. **11**, 184–189 (1974)

25. G. Matheron, in *Random Sets and Integral Geometry*. Wiley Series in Probability and Mathematical Statistics (Wiley, New York, 1975)

26. S.T. Lee, On some degenerate principal series representations of $U(n, n)$. J. Funct. Anal. **126**(2), 305–366 (1994)

27. B. Rubin, The Calderón reproducing formula, windowed X-ray transforms and Radon transforms in L_p-spaces. J. Fourier Anal. Appl. **4**, 175–197 (1998)

28. B. Rubin, Inversion formulas for the spherical Radon transform and the generalized cosine transform. Adv. Appl. Math. **29**, 471–497 (2002)

29. S. Sahi, Jordan algebras and degenerate principal series. J. Reine Angew. Math. **462**, 1–18 (1995)

30. R. Schneider, Über eine Integralgleichung in der Theorie der konvexen Körper (German). Math. Nachr. **44**, 55–75 (1970)

31. E. Spodarev, Cauchy-Kubote type integral formula for generalized cosine transforms. IZV. Nats. Akad. Nauk Armenii Mat. **37**(1), 52–69 (2002) translation in J. Contemp. Math. Anal. **37**(1), 47–63 (2002)

32. M. Sugiura, Representations of compact groups realized by spherical functions on symmetric spaces. Proc. Jpn. Acad. **38**, 111–113 (1962)

33. M. Takeuchi, *Modern Spherical Functions*. Translated from the 1975 Japanese original by Toshinobu Nagura. Translations of Mathematical Monographs, vol. 135 (American Mathematical Society, Providence, 1994)

34. D. Vogan, Gelfand–Kirillov dimension of Harish-Chandra modules. Inv. Math. **48**, 75–98 (1978)

35. N.R. Wallach, in *Real Reductive Groups. I*. Pure and Applied Mathematics, vol. 132 (Academic, Boston, 1988)
36. N.R. Wallach, in *Real Reductive Groups. II*. Pure and Applied Mathematics, vol. 132-II (Academic, Boston, 1992)
37. D.P. Želobenko, *Compact Lie Groups and Their Representations*. Translated from the Russian by Israel Program for Scientific Translations. Translations of Mathematical Monographs, vol. 40 (American Mathematical Society, Providence, 1973)
38. G.K. Zhang, Jordan algebras and generalized principal series representations. Math. Ann. **302**(4), 773–786 (1995)

On Modules Over Valuations

Semyon Alesker

Abstract To any smooth manifold X an algebra of smooth valuations $V^\infty(X)$ was associated in [Alesker, Israel J. Math. **156**, 311–339 (2006); Adv. Math. **207**(1), 420–454 (2006); Theory of Valuations on Manifolds, IV. New Properties of the Multiplicative Structure (2007); Alesker, Fu, Trans. Am. Math. Soc. **360**(4), 1951–1981 (2008)]. In this note we initiate a study of $V^\infty(X)$-modules. More specifically we study finitely generated projective modules in analogy to the study of vector bundles on a manifold. In particular it is shown that for a compact manifold X there exists a canonical isomorphism between the K-ring constructed out of finitely generated projective $V^\infty(X)$-modules and the classical topological K^0-ring constructed out of vector bundles over X.

1 Introduction

Let X be a smooth manifold of dimension n.[1] In [1–3, 5] the notion sof a smooth valuation on X was introduced. Roughly put, a smooth valuation is a $\mathbb{C}$-valued finitely additive measure on compact submanifolds of X with corners, which satisfies in addition some extra conditions. We omit here the precise description of the conditions due to their technical nature. Let us notice that basic examples of smooth valuations include any smooth measure on X and the Euler characteristic. There are many other natural examples of valuations coming from convexity,

---

[1]All manifolds are assumed to be countable at infinity, i.e. presentable as a union of countably many compact subsets. In particular they are paracompact.

S. Alesker (✉)
Sackler Faculty of Exact Sciences, Department of Mathematics, Tel Aviv University,
Ramat Aviv, 69978 Tel Aviv, Israel
e-mail: semyon@post.tau.ac.il

B. Klartag et al. (eds.), *Geometric Aspects of Functional Analysis*, Lecture Notes in Mathematics 2050, DOI 10.1007/978-3-642-29849-3_2,
© Springer-Verlag Berlin Heidelberg 2012

integral, and differential geometry. We refer to recent lecture notes [4, 7, 8] for an overview of the subject, examples, and applications.

The space $V^\infty(X)$ of all smooth valuations is a Fréchet space. It has a canonical product making $V^\infty(X)$ a commutative associative algebra over $\mathbb{C}$ with a unit element (which is the Euler characteristic).

In this note we initiate a study of modules over $V^\infty(X)$. Our starting point is the analogy to the following well known fact due to Serre and Swan [11, 12]: if X is compact, then the category of smooth vector bundles of finite rank over X is equivalent to the category of finitely generated projective modules over the algebra $C^\infty(X)$ of smooth functions (the functor in one direction is given by taking global smooth sections of a vector bundle).

In order to state our main results we need to remind a few general facts about valuations on manifolds. We have a canonical homomorphism of algebras

$$V^\infty(X) \to C^\infty(X) \tag{1}$$

given by the evaluation on points, i.e. $\phi \mapsto [x \mapsto \phi(\{x\})]$. This is an epimorphism. The kernel, denoted by W_1, is a nilpotent ideal of $V^\infty(X)$:

$$(W_1)^{n+1} = 0.$$

Next, smooth valuations form a sheaf of algebras which is denoted by $\mathcal{V}_X^\infty$: for an open subset $U \subset X$,

$$\mathcal{V}_X^\infty(U) = V^\infty(U),$$

where the restriction maps are obvious. We denote by $\mathcal{O}_X$ the sheaf of C^∞-smooth functions on X. Then the map (1) gives rise to the epimorphism of sheaves

$$\mathcal{V}_X^\infty \twoheadrightarrow \mathcal{O}_X. \tag{2}$$

Recall now the notion of a projective module. Let A be a commutative associative algebra with a unit. An A-module M is called *projective* if M is a direct summand of a free A-module, i.e. there exists an A-module N such that $M \oplus N$ is a free A-module (not necessarily of finite rank). It is easy to see that if M is in addition finitely generated then M is a direct summand of a free A-module of finite rank.

Let $\mathcal{A}$ be a sheaf of algebras on a topological space X. A sheaf $\mathcal{M}$ of $\mathcal{A}$-modules is called a *locally projective* $\mathcal{A}$-module if any point $x \in X$ has an open neighborhood U such that $\mathcal{M}(U)$ is a projective $\mathcal{A}(U)$-module.

Let us denote by $\mathrm{Proj}_f \, \mathcal{V}_X^\infty - mod$ the full subcategory of $\mathcal{V}_X^\infty$-modules consisting of locally projective $\mathcal{V}_X^\infty$-modules of finite rank. Let us denote by $\mathrm{Proj}_f \, V^\infty(X) - mod$ the full subcategory of the category of $V^\infty(X)$-modules consisting of projective $V^\infty(X)$-modules of finite rank. In Sect. 2 we prove the following result.

Theorem 1.1. *Let X be a smooth manifold.*

(1) Any locally projective $\mathcal{V}_X^\infty$-module of finite rank is locally free.

(2) Assume in addition that X is compact. Let $\mathcal{E}$ be a locally free $\mathcal{V}_X^\infty$-module of finite rank. Then there exists another locally free $\mathcal{V}_X^\infty$-module $\mathcal{H}$ of finite rank such that $\mathcal{E} \oplus \mathcal{H}$ is isomorphic to $(\mathcal{V}_X^\infty)^N$ for some natural number N.

(3) Assume again that X is compact. Then the functor of global sections

$$\Gamma: \mathrm{Proj}_f \, \mathcal{V}_X^\infty - mod \;\to\; \mathrm{Proj}_f \, V^\infty(X) - mod$$

is an equivalence of categories.

Notice that all the statements of the theorem are completely analogous to the classical situation of vector bundles (whose spaces of sections are projective finitely generated $C^\infty(X)$-modules). For example a version of (2) for vector bundles says that any vector bundle is a direct summand of a free bundle. A classical version of (3) is the above mentioned theorem of Serre-Swan. The method of proof of Theorem 1.1 is a minor modification of the proof for the analogous statement for vector bundles.

To formulate our next main result observe that to any $\mathcal{V}_X^\infty$-module we can associate an $\mathcal{O}_X$-module via

$$\mathcal{M} \mapsto \mathcal{M} \otimes_{\mathcal{V}_X^\infty} \mathcal{O}_X, \tag{3}$$

where $\mathcal{O}_X$ is considered as $\mathcal{V}_X^\infty$-module via the epimorphism (2). Clearly under this correspondence locally free $\mathcal{V}_X^\infty$-modules of finite rank are mapped to locally free $\mathcal{O}_X$-modules of equal rank, i.e. to vector bundles.

Theorem 1.2. *Assume that X is a compact manifold. Let N be a natural number. The map (3) induces a bijection between the isomorphism classes of locally free $\mathcal{V}_X^\infty$-modules of rank N and isomorphism classes of vector bundles of rank N.*

Theorem 1.2 is proved in Sect. 3. The proof is an application of general results of Grothendieck [9] on non-abelian cohomology of topological spaces and the existence of a finite decreasing filtration on $\mathcal{V}_X^\infty$ such that the associated graded sheaf is a sheaf of $\mathcal{O}_X$-modules.

Acknowledgements I thank M. Borovoi for useful discussions on non-abelian cohomology, and F. Schuster for numerous remarks on the first version of the paper. Partially supported by ISF grant 701/08.

2 Locally Free Sheaves Over Valuations

A sheaf $\mathcal{E}$ of $\mathcal{V}_X^\infty$-modules is called locally projective of finite rank if every point $x \in X$ has a neighborhood U such that there exists a sheaf of $\mathcal{V}_U^\infty$-modules $\mathcal{F}$ with the property that $\mathcal{E}|_U \oplus \mathcal{F}$ is isomorphic to $(\mathcal{V}_X^\infty)^N$ for some natural number N.

The technique used in the proofs of most of the results of this section is rather standard and is a simple modification of that from [6].

Proposition 2.1. *On a manifold X any locally projective V_X^∞-module of finite rank is locally free.*

Proof. Fix a point $x_0 \in X$. Let us denote for brevity $V_{x_0} := V_{X,x_0}^\infty$ (resp. $\mathcal{O}_{X,x_0}$) the stalk at x_0 of the sheaf V_X^∞ (resp. $\mathcal{O}_X$). Let $\mathcal{E}$ be a locally projective V_X^∞-module of finite rank. Consider its stalk $\mathcal{E}_{x_0}$ as V_{x_0}-module. Then there exists a V_{x_0}-module $\mathcal{F}$ such that

$$\mathcal{E}_{x_0} \oplus \mathcal{F} \simeq V_{x_0}^N$$

for some natural number N. Consider the idempotent endomorphism of the V_{x_0}-module

$$e : V_{x_0}^N \to V_{x_0}^N$$

given by the projection onto $\mathcal{E}_{x_0}$. Thus $e^2 = e$. Notice that V_{x_0} is a local ring with the maximal ideal

$$m := \{\phi \in V_{x_0} \mid \phi(\{x_0\}) = 0\}.$$

Clearly $V_{x_0}/m = \mathbb{C}$.

We have

$$\mathbb{C}^N = V_{x_0}^N \otimes_{V_{x_0}} (V_{x_0}/m) = (\mathcal{E}_{x_0} \otimes_{V_{x_0}} V_{x_0}/m) \oplus (\mathcal{F} \otimes_{V_{x_0}} V_{x_0}/m). \qquad (4)$$

Let us choose a basis

$$\xi_1', \ldots, \xi_k', f_1', \ldots, f_{N-k}'$$

of $\mathbb{C}^N$ such that the ξ_i''s form a basis of the first summand in the right hand side of (4), and the f_j''s form a basis of the second summand.

Let $\tilde{\xi}_i \in V_{x_0}^N$, $\tilde{f}_j \in V_{x_0}^N$ be their lifts. Define finally

$$\xi_i := e(\tilde{\xi}_i) \in \mathcal{E}_{x_0},$$
$$f_j := (1-e)(\tilde{f}_j) \in \mathcal{F}.$$

It is clear that

$$\xi_i \equiv \tilde{\xi}_i \ mod(m),$$
$$f_j \equiv \tilde{f}_j \ mod(m).$$

Consider the morphism of V_{x_0}-modules $\theta : V_{x_0}^k \to \mathcal{E}_{x_0}$ given by

$$\theta(\phi_1, \ldots, \phi_k) := \sum_{i=1}^k \phi_i \xi_i.$$

Consider also another morphism $\tau \colon \mathcal{V}_{x_0}^{N-K} \to \mathcal{F}$ given by

$$\tau(\psi_1, \ldots, \psi_{N-k}) = \sum_{j=1}^{N-k} \psi_j f_j.$$

Now define a morphism of $\mathcal{V}_{x_0}$-modules by

$$\sigma := \theta \oplus \tau \colon \mathcal{V}_{x_0}^N = \mathcal{V}_{x_0}^k \oplus \mathcal{V}_{x_0}^{N-k} \to \mathcal{E}_{x_0} \oplus \mathcal{F} \simeq \mathcal{V}_{x_0}^N.$$

We claim that σ is an isomorphism. It is equivalent to the property that $\det(\sigma) \in \mathcal{V}_{x_0}$ is invertible. In order to see this it suffices to show that $(\det \sigma)(\{x_0\}) \neq 0$. But the last condition is satisfied since

$$\sigma \otimes \mathrm{Id}_{\mathcal{V}_{x_0}/m} \colon \mathcal{V}_{x_0}^N \otimes_{\mathcal{V}_{x_0}} \mathcal{V}_{x_0}/m \to \mathcal{V}_{x_0}^N \otimes_{\mathcal{V}_{x_0}} \mathcal{V}_{x_0}/m$$

is an isomorphism $\mathbb{C}^N \to \mathbb{C}^N$ since by construction

$$\xi_1' = \xi_1(\{x_0\}), \ldots, \xi_k' = \xi_k(\{x_0\}), f_1' = f_1(\{x_0\}), \ldots, f_{N-k}' = f_{N-K}(\{x_0\})$$

form a basis of $\mathbb{C}^N$.

Since $\sigma \colon \mathcal{V}_{x_0}^N \to \mathcal{V}_{x_0}^N$ is an isomorphism, it follows that there exists an open neighborhood U of x_0 and an isomorphism of $\mathcal{V}_U^\infty$-modules

$$\tilde{\sigma} \colon (\mathcal{V}_U^\infty)^N \to (\mathcal{V}_U^\infty)^N$$

which extends σ, i.e. σ is the stalk of $\tilde{\sigma}$ at x_0. It follows that θ extends to an isomorphism

$$\tilde{\theta} \colon (\mathcal{V}_U^\infty)^k \overset{\sim}{\to} \mathcal{E}|_U$$

of $\mathcal{V}_U^\infty$-modules (and similarly for τ). $\qquad \square$

Lemma 2.2. *Let $\mathcal{E}$ be a locally free $\mathcal{V}_X^\infty$-module of finite rank N. Let $\xi_1, \ldots, \xi_k \in H^0(X, \mathcal{E})$ be chosen such that for every point $x_0 \in X$ their images $\bar{\xi}_1, \ldots, \bar{\xi}_k$ in $\mathcal{E} \otimes_{\mathcal{V}_X^\infty} \mathcal{V}_X^\infty/m_{x_0} \simeq \mathbb{C}^N$ form a linearly independent sequence. Consider the morphism of $\mathcal{V}_X^\infty$-modules*

$$f \colon (\mathcal{V}_X^\infty)^k \to \mathcal{E},$$

given by $f(\phi_1, \ldots, \phi_k) = \sum_{i=1}^k \phi_i \xi_i$.

Then $f \colon (\mathcal{V}_X^\infty)^k \to Im(f)$ is an isomorphism, and $\mathcal{E}/Im(f)$ is a locally free $\mathcal{V}_X^\infty$-module of rank $N - k$.

Proof. Let us denote for brevity $\mathcal{V} := \mathcal{V}_X^\infty$. The statement is local on X. Fix $x_0 \in X$. We can choose $\eta_1, \ldots, \eta_{N-k} \in H^0(\mathcal{E})$ such that their images

$$\bar{\xi}_1, \ldots, \bar{\xi}_k, \bar{\eta}_1, \ldots, \bar{\eta}_{N-k} \in \mathcal{E} \otimes_{\mathcal{V}} \mathcal{V}/m_{x_0}$$

form a basis. Consider a morphism of $\mathcal{V}$-modules $g: \mathcal{V}^k \oplus \mathcal{V}^{N-k} \to \mathcal{E}$ given by

$$g(\phi_1, \ldots, \phi_k; \psi_1, \ldots, \psi_{N-k}) = \sum_{i=1}^{k} \phi_i \xi_i + \sum_{j=1}^{N-k} \psi_j \eta_j.$$

Clearly $g|_{\mathcal{V}^k} \equiv f$. In a neighborhood of x_0 we may and will identify $\mathcal{E} \simeq \mathcal{V}^N$. Then

$$g: \mathcal{V}^k \oplus \mathcal{V}^{N-K} = \mathcal{V}^N \to \mathcal{V}^N.$$

It is easy to see that the map

$$g \otimes \mathrm{Id}_{\mathcal{V}/m_{x_0}}: \mathbb{C}^N \simeq (\mathcal{V}/m_{x_0})^N \to \mathbb{C}^N \simeq (\mathcal{V}/m_{x_0})^N$$

is an isomorphism. Hence $(\det g)(\{x_0\}) \neq 0$. It follows that $\det g \in \mathcal{V}$ is invertible in a neighborhood of x_0. Hence g is an isomorphism in a neighborhood of x_0. This implies the lemma immediately. $\qquad\square$

Lemma 2.3. *Let P be a locally free $\mathcal{V}_X^\infty$-module of finite rank. Then for any $\mathcal{V}_X^\infty$-module A,*

$$\mathrm{Ext}^i_{\mathcal{V}_X^\infty - mod}(P, A) = 0 \text{ for } i > 0.$$

Proof. We abbreviate again $\mathcal{V} := \mathcal{V}_X^\infty$. First notice that in the category of $\mathcal{V}$-modules the following two functors

$$F, G: \mathcal{V} - mod \to Vect$$

are naturally isomorphic:

$$F(A) = \mathrm{Hom}_{\mathcal{V}-mod}(P, A),$$
$$G(A) = H^0(X, P^* \otimes_{\mathcal{V}} A),$$

where $P^* := \underline{\mathrm{Hom}}_{\mathcal{V}-mod}(P, \mathcal{V})$ is the inner Hom as usual. Indeed the natural morphism

$$P^* \otimes_{\mathcal{V}} A = \underline{\mathrm{Hom}}_{\mathcal{V}-mod}(P, \mathcal{V}) \otimes_{\mathcal{V}} A \to \underline{\mathrm{Hom}}_{\mathcal{V}-mod}(P, A)$$

is an isomorphism of sheaves. Taking global sections, we get an isomorphism

$$H^0(X, P^* \otimes_{\mathcal{V}} A) \xrightarrow{\sim} H^0(X, \underline{\mathrm{Hom}}_{\mathcal{V}-mod}(P, A)).$$

But the last space is equal to $\mathrm{Hom}_{\mathcal{V}-mod}(P, A)$ (see [10], Chap. II, Sect. 1, Exc. 1.15).

Consequently F and G have isomorphic derived functors. Hence

$$\text{Ext}^i_{\mathcal{V}-mod}(P, A) \simeq H^i(X, P^* \otimes_{\mathcal{V}} A).$$

But the last group vanishes for $i > 0$ by [3], Lemma 5.1.2.   □

Corollary 2.4. *Let*

$$0 \to A \to B \to C \to 0$$

be a short exact sequence of $\mathcal{V}^\infty_X$-modules. If C is locally free of finite rank, then this exact sequence splits.

Proof. Indeed $\text{Ext}^1_{\mathcal{V}-mod}(C, A) = 0$ by Lemma 2.3.   □

Proposition 2.5. *Let X be a compact manifold. Let $\mathcal{E}$ be a locally free $\mathcal{V}^\infty_X$-module of finite rank. Then there exists another locally free $\mathcal{V}^\infty_X$-module $\mathcal{H}$ of finite rank such that*

$$\mathcal{E} \oplus \mathcal{H} \simeq (\mathcal{V}^\infty_X)^N.$$

Proof. Let us choose a finite open covering $X = \cup_\alpha U_\alpha$ such that the sheaf $\mathcal{E}|_{U_\alpha}$ is free for each α. Let $\{\phi_\alpha\}$ be a partition of unity in the algebra of valuations subordinate to this covering (it exist by [3], Proposition 6.2.1). We can find a finite dimensional subspace $L_\alpha \subset H^0(U_\alpha, \mathcal{E})$ which generates $\mathcal{E}|_{U_\alpha}$ as $\mathcal{V}^\infty_{U_\alpha}$-module. Consider $\phi_\alpha \cdot L_\alpha \subset H^0(X, \mathcal{E})$ (where all sections are extended by zero outside of U_α). Then the finite dimensional subspace

$$L := \sum_\alpha \phi_\alpha \cdot L_\alpha \subset H^0(X, \mathcal{E})$$

generates $\mathcal{E}$ as $\mathcal{V}$-module (indeed at every $x \in X$ there exists an α such that ϕ_α is invertible in a neighborhood of x).

Let us choose a basis $\xi_1, \ldots, \xi_s$ of L. Consider the morphism of $\mathcal{V}$-modules $F: \mathcal{V}^s \to \mathcal{E}$ given by

$$F(\phi_1, \ldots, \phi_s) = \sum_{i=1}^{s} \phi_i \xi_i.$$

Clearly F is an epimorphism of $\mathcal{V}$-modules. Let $\mathcal{A} := Ker(F)$. By Corollary 2.4 the short exact sequence

$$0 \to \mathcal{A} \to \mathcal{V}^s \to \mathcal{E} \to 0$$

splits. Thus $\mathcal{E} \oplus \mathcal{A} \simeq \mathcal{V}^s$. Hence $\mathcal{A}$ is locally projective. Hence $\mathcal{A}$ is locally free by Proposition 2.1.   □

Let us denote by $\text{Proj}_f \mathcal{V}^\infty_X - mod$ (or just $\text{Proj}_f \mathcal{V} - mod$) the full subcategory of $\mathcal{V} - mod$ consisting of locally free $\mathcal{V}$-modules of finite rank. Let us denote by $\text{Proj}_f V^\infty(X) - mod$ the category of projective $V^\infty(X)$-modules of finite rank.

Theorem 2.6. *Let X be a compact manifold. Then the functor of global sections*

$$\Gamma: \text{Proj}_f \, \mathcal{V}_X^\infty - mod \to \text{Proj}_f \, V^\infty(X) - mod$$

is an equivalence of categories.

Proof. We denote again by $\mathcal{V} := \mathcal{V}_X^\infty$. Let $\mathcal{A}, \mathcal{B} \in \text{Proj}_f \, \mathcal{V} - mod$. First let us show that

$$\text{Hom}_{\mathcal{V}-mod}(\mathcal{A}, \mathcal{B}) = \text{Hom}_{V^\infty(X)-mod}(\Gamma(\mathcal{A}), \Gamma(\mathcal{B})).$$

Both Hom functors respect finite direct sums with respect to both arguments. Since $\mathcal{A}, \mathcal{B}$ are direct summands of free $\mathcal{V}$-modules by Proposition 2.5 we may assume that $\mathcal{A} = \mathcal{B} = \mathcal{V}$. But clearly

$$\text{Hom}_{\mathcal{V}-mod}(\mathcal{V}, \mathcal{V}) = V^\infty(X),$$

$$\text{Hom}_{V^\infty(X)}(V^\infty(X), V^\infty(X)) = V^\infty(X).$$

Thus Γ is fully faithful.

Let us define a functor in the opposite direction (the localization functor),

$$G: \text{Proj}_f \, V^\infty(X) - mod \to \text{Proj}_f \, \mathcal{V} - mod,$$

by $G(A) := A \otimes_{V^\infty(X)} \mathcal{V}$. G is also fully faithful: it commutes with direct sums, and for trivial $V^\infty(X)$-modules the statement is obvious.

The functors $F \circ G$ and $G \circ F$ are naturally isomorphic to the identity functors. $\square$

3 Isomorphism Classes of Bundles Over Valuations

Recall that the sheaf of smooth valuations $\mathcal{V}_X^\infty$, which we will denote for brevity by $\mathcal{V}$, has a canonical filtration by subsheaves,

$$\mathcal{V} = \mathcal{W}_0 \supset \mathcal{W}_1 \supset \cdots \supset \mathcal{W}_n.$$

This filtration is compatible with the product, and $\mathcal{V}/\mathcal{W}_1 \simeq \mathcal{O}_X$ canonically [3]. Let us fix a natural number N. Let us denote by $\underline{GL_N(\mathcal{V})}$ (resp. $\underline{GL_N(\mathcal{O}_X)}$) the sheaf on X of invertible $N \times N$ matrices with entries in $\mathcal{V}$ (resp. $\mathcal{O}_X$). We have a natural homomorphism of sheaves of groups

$$\underline{GL_N(\mathcal{V})} \to \underline{GL_N(\mathcal{O}_X)}. \tag{5}$$

It is well known that isomorphism classes of usual vector bundles are in bijective correspondence with the (Cech) cohomology set $H^1(X, \underline{GL_N(\mathcal{O}_X)})$. Similarly it is clear that locally free $\mathcal{V}$-modules of rank N are in bijective correspondence with the set $H^1(X, \underline{GL_N}(\mathcal{V}))$. The main result of this section is

Theorem 3.1. *Let X be a compact manifold. The natural map*

$$H^1(X, \underline{GL_N}(\mathcal{V})) \to H^1(X, \underline{GL_N}(\mathcal{O}_X))$$

induced by (5) is a bijection. Thus, if X is compact, the isomorphism classes of rank N locally free $\mathcal{V}$-modules are in natural bijective correspondence with isomorphism classes of rank N vector bundles.

We will need some preparations before the proof of the theorem. First we observe that $\underline{GL_N}(\mathcal{V})$ has a natural filtration by subsheaves of normal subgroups

$$\underline{GL_N}(\mathcal{V}) =: \mathcal{K}_0 \supset \mathcal{K}_1 \supset \cdots \supset \mathcal{K}_n,$$

where for any $i > 0$, $\mathcal{K}_i(U) := \{\xi \in \underline{GL_N}(\mathcal{V})(U) | \xi \equiv I \bmod \mathcal{W}_i(U)\}$ for any open subset $U \subset X$. We have the canonical isomorphisms of sheaves of groups:

$$\mathcal{K}_0/\mathcal{K}_1 \simeq \underline{GL_N}(\mathcal{O}_X), \tag{6}$$

$$\mathcal{K}_i/\mathcal{K}_{i+1} \simeq (\mathcal{W}_i/\mathcal{W}_{i+1})^N \text{ for } i > 0. \tag{7}$$

We have to remind some general results due to Grothendieck [9]. Let $\underline{G}$ be a sheaf of groups (not necessarily abelian) on a topological space X. Let $\underline{F} \subset \underline{G}$ be a subsheaf of *normal* subgroups. Let $\underline{H} := \underline{G}/\underline{F}$ be the quotient sheaf, which is also a sheaf of groups. Notice that the sheaf of groups $\underline{G}$ acts on $\underline{F}$ by conjugations.

Let E' be a $\underline{G}$-torsor. Let $c' := [E'] \in H^1(X, \underline{G})$ be its class. Define a new sheaf $\underline{F}(E')$ to be the sheaf associated to the presheaf

$$U \mapsto \underline{F}(U) \times_{\underline{G}(U)} E'(U).$$

Since $\underline{G}$ acts on $\underline{F}$ by automorphisms, it follows that $\underline{F}(E')$ is a sheaf of groups. Grothendieck [9] has constructed a map

$$i_1 \colon H^1(X, \underline{F}(E')) \to H^1(X, \underline{G}),$$

and he has shown (see Corollary after Proposition 5.6.2 in [9]) that the set of classes $c \in H^1(X, \underline{G})$ which have the same image as c' under the natural map

$$H^1(X, \underline{G}) \to H^1(X, \underline{H})$$

is equal to the image of the map i_1. In particular we deduce immediately the following claim.

Claim 3.2. *If, for any $\underline{G}$-torsor E',*

$$H^1(X, \underline{F}(E')) = 0$$

then the natural map $H^1(X, \underline{G}) \to H^1(X, \underline{H})$ is injective.

We will need the following proposition.

Proposition 3.3. *For any $1 \leq i < j$ and for any $\mathcal{K}_0/\mathcal{K}_j$-torsor E', one has*

$$H^1(X, (\mathcal{K}_i/\mathcal{K}_j)(E')) = 0.$$

Proof. The proof is by induction in $j - i$. Assume first that $j - i = 1$. $(\mathcal{K}_i/\mathcal{K}_{i+1})(E')$ is a sheaf of $\mathcal{O}_X$-modules; this follows easily from (7) and the fact that $\mathcal{W}_i/\mathcal{W}_{i+1}$ is a sheaf of $\mathcal{O}_X$-modules. Hence $(\mathcal{K}_i/\mathcal{K}_{i+1})(E')$ is acyclic.

Assume now that $j - i > 1$. Then we have a short exact sequence of sheaves

$$1 \to (\mathcal{K}_j/\mathcal{K}_{j-1})(E') \to (\mathcal{K}_i/\mathcal{K}_j)(E') \to (\mathcal{K}_i/\mathcal{K}_{j-1})(E') \to 1.$$

Hence we have an exact sequence of pointed sets (see [9], Sect. 5.3),

$$H^1(X, (\mathcal{K}_j/\mathcal{K}_{j-1})(E')) \to H^1(X, (\mathcal{K}_i/\mathcal{K}_j)(E')) \to H^1(X, (\mathcal{K}_i/\mathcal{K}_{j-1})(E')).$$

The first and the third terms of the last sequence vanish by the induction assumption. Hence the middle term vanishes too. Proposition is proved. □

We easily deduce a corollary.

Corollary 3.4. *The natural map*

$$H^1(X, \underline{GL_N(\mathcal{V})}) \to H^1(X, \underline{GL_N(\mathcal{O}_X)})$$

is injective.

Proof. By Proposition 3.3 $H^1(X, \mathcal{K}_1(E')) = 0$ for any $\mathcal{K}_0$-torsor E'. Hence, by Claim 3.2, the map $H^1(X, \mathcal{K}_0) \to H^1(X, \mathcal{K}_0/\mathcal{K}_1)$ is injective. □

We will need a few more results from [9]. Assume X is a paracompact topological space (remind that all our manifolds are always assumed to be paracompact). Let $\underline{G}$ be a sheaf of groups on X as before. Let $\underline{F} \lhd \underline{G}$ be a subsheaf of *normal abelian* subgroups. Let $\underline{H} := \underline{G}/\underline{F}$ be the quotient sheaf as before. The action of $\underline{G}$ on $\underline{F}$ by conjugation induces in this case an action of $\underline{H}$ on $\underline{F}$. For any $\underline{H}$-torsor E'' one has the sheaf $\underline{F}(E'')$ defined similarly as before. This is a sheaf of abelian groups since $\underline{F}$ is, and $\underline{H}$ acts on $\underline{F}$ by automorphisms. Grothendieck ([9], Sect. 5.7) has constructed an element

$$\delta E'' \in H^2(X, \underline{F}(E''))$$

with the following property: $\delta E''$ vanishes if and only if the class $[E''] \in H^1(X, \underline{H})$ lies in the image of the canonical map $H^1(X, \underline{G}) \to H^1(X, \underline{H})$.

In order to apply this result in our situation we will need two lemmas.

Lemma 3.5. *For any $i > 0$ and any $\mathcal{K}_0/\mathcal{K}_i$-torsor E'',*

$$H^2(X, (\mathcal{K}_i/\mathcal{K}_{i+1})(E'')) = 0.$$

Proof. It is easy to see that $(\mathcal{K}_i/\mathcal{K}_{i+1})(E'')$ is a sheaf of $\mathcal{O}_X$-modules. Hence it is acyclic. □

Lemma 3.6. *For any $i \geq 0$ the natural map*

$$H^1(X, \mathcal{K}_0/\mathcal{K}_{i+1}) \to H^1(X, \mathcal{K}_0/\mathcal{K}_i)$$

is onto.

Proof. Let $c'' \in H^1(X, \mathcal{K}_0/\mathcal{K}_i)$ be an arbitrary element. Let E'' be a $\mathcal{K}_0/\mathcal{K}_i$-torsor representing c''. Consider the element $\delta E'' \in H^2(X, (\mathcal{K}_i/\mathcal{K}_{i+1})(E''))$. Since the last group vanishes by Lemma 3.5, by the above mentioned result of Grothendieck, c'' lies in the image of $H^1(X, \mathcal{K}_0/\mathcal{K}_{i+1})$. Lemma is proved. □

Corollary 3.7. *The natural map*

$$H^1(X, \underline{GL_N(V)}) \to H^1(X, \underline{GL_N(\mathcal{O}_X)})$$

is onto.

Proof. The map in the statement factorizes into the sequence of maps

$$H^1(X, \underline{GL_N(V)}) = H^1(X, \mathcal{K}_0) \to H^1(X, \mathcal{K}_0/\mathcal{K}_n) \to H^1(X, \mathcal{K}_0/\mathcal{K}_{n-1}) \to \cdots$$

$$\cdots \to H^1(X, \mathcal{K}_0/\mathcal{K}_1) = H^1(X, \underline{GL_N(\mathcal{O}_X)})$$

where all the maps are surjective by Lemma 3.6. Hence their composition is onto too. □

Now Theorem 3.1 follows immediately from Corollaries 3.4 and 3.7.

Remark 3.8. Theorem 3.1 has the following immediate consequence. For a compact manifold X we can construct a K-ring generated by finitely generated projective $V^\infty(X)$-modules in the standard way. Namely as a group it is equal to the quotient of the free abelian group generated by isomorphism classes of such modules by the relations

$$[M \oplus N] = [M] + [N].$$

The product is induced by the tensor product of such $V^\infty(X)$-modules. Then Theorem 3.1 implies that there is a canonical isomorphism of this K-ring with the classical topological K^0-ring (see [6]) constructed from vector bundles.

Remark 3.9. The main results of this paper are of general nature. It would be interesting to have concrete geometric examples of $\mathcal{V}_X^\infty$-modules; in the classical case of $\mathcal{O}_X$-modules we have the tangent bundle and its tensor powers.

As a first small step in this direction let us mention the following construction. Let $\mathcal{L}$ be a flat vector bundle over a manifold X. By an abuse of notation, we will also denote by $\mathcal{L}$ the sheaf of its locally constant sections. Let $\underline{\mathbb{C}}$ be the constant sheaf of $\mathbb{C}$-vector spaces. Consider the $\mathcal{V}_X^\infty$-module defined by

$$\tilde{\mathcal{L}} := \mathcal{L} \otimes_{\underline{\mathbb{C}}} \mathcal{V}_X^\infty$$

where we consider $\mathcal{V}_X^\infty$ as $\underline{\mathbb{C}}$-module via the imbedding $\underline{\mathbb{C}} \hookrightarrow \mathcal{V}_X^\infty$ where 1 goes to the Euler characteristic. It is easy to see that $\tilde{\mathcal{L}}$ is a locally free $\mathcal{V}_X^\infty$-module.

References

1. S. Alesker, Theory of valuations on manifolds, I. Linear spaces. Israel J. Math. **156**, 311–339 (2006). math.MG/0503397
2. S. Alesker, Theory of valuations on manifolds. II. Adv. Math. **207**(1), 420–454 (2006). math.MG/0503399
3. S. Alesker, *Theory of Valuations on Manifolds, IV. New Properties of the Multiplicative Structure.* Geometric Aspects of Functional Analysis, Lecture Notes in Math., vol. 1910 (Springer, Berlin, 2007), pp. 1–44. math.MG/0511171
4. S. Alesker, *New Structures on Valuations and Applications.* Lecture Notes of the Advanced Course on Integral Geometry and Valuation Theory at CRM, Barcelona, Preprint. arXiv:1008.0287
5. S. Alesker, J.H.G. Fu, Theory of valuations on manifolds, III. Multiplicative structure in the general case. Trans. Am. Math. Soc. **360**(4), 1951–1981 (2008); math.MG/0509512
6. M.F. Atiyah, in *K-Theory.* Lecture Notes by D.W. Anderson (W.A. Benjamin Inc., New York, 1967)
7. A. Bernig, Algebraic integral geometry. Global Differential Geometry, edited by C. Bär, J. Lohkamp and M. Schwarz, Springer 2012. arXiv:1004.3145
8. J.H.G. Fu, in *Algebraic Integral Geometry.* Lecture Notes of the Advanced Course on Integral Geometry and Valuation Theory at CRM, Barcelona. Preprint
9. A. Grothendieck, *A General Theory of Fibre Spaces with Structure Sheaf* (University of Kansas, KS, 1955) Preprint. http://www.math.jussieu.fr/leila/grothendieckcircle/GrothKansas.pdf
10. R. Hartshorne, in *Algebraic Geometry.* Graduate Texts in Mathematics, vol. 52 (Springer, New York, 1977)
11. J.-P. Serre, Modules projectifs et espaces fibrés à fibre vectorielle. *Séminaire Dubreil-Pisot: algèbre et théorie des nombres 11 (1957/58); Oeuvres I,* pp. 531–543
12. R.G. Swan, Vector bundles and projective modules. Trans. Am. Math. Soc. **105**, 264–277 (1962)

On Multiplicative Maps of Continuous and Smooth Functions

Shiri Artstein-Avidan, Dmitry Faifman, and Vitali Milman

Abstract In this note, we study the general form of a multiplicative bijection on several families of functions defined on manifolds, both real or complex valued. In the real case, we prove that it is essentially defined by a composition with a diffeomorphism of the underlying manifold (with a bit more freedom in families of continuous functions). Our results in the real case are mostly simple extensions of known theorems. We then show that in the complex case, the only additional freedom allowed is complex conjugation. Finally, we apply those results to characterize the Fourier transform between certain function spaces.

1 Introduction and Main Results

The following is the simplest form of a lemma regarding multiplicative maps. It is standard, and was used recently for example in the paper [2] where a characterization of the derivative transform as an essentially unique bijection (up to constant) from $C^1(\mathbb{R})$ to $C(\mathbb{R})$ which satisfies the chain rule was derived.

Lemma 1.1. *Assume that* $K : \mathbb{R} \to \mathbb{R}$ *is measurable, not identically zero and satisfies for all* $u, v \in \mathbb{R}$ *that* $K(uv) = K(u)K(v)$. *Then there exists some* $p > 0$ *such that*

$$K(u) = |u|^p \quad or \quad K(u) = |u|^p \operatorname{sgn}(u).$$

When instead of $\mathbb{R}$ we have a more complicated set with a multiplication operation, such as a class of functions, things become more involved. This already became apparent in the papers [1, 3], where characterizations of the Fourier transform were proved as a unique bijection between corresponding classes of

S. Artstein-Avidan (✉) · D. Faifman · V. Milman
Sackler Faculty of Exact Sciences, Department of Mathematics, Tel Aviv University,
Ramat Aviv, 69978 Tel Aviv, Israel
e-mail: shiri@post.tau.ac.il; dfaifmand@gmail.com; milman@post.tau.ac.il

B. Klartag et al. (eds.), *Geometric Aspects of Functional Analysis*, Lecture Notes
in Mathematics 2050, DOI 10.1007/978-3-642-29849-3_3,
© Springer-Verlag Berlin Heidelberg 2012

functions which maps products to convolutions. Let us recall a result from the paper [3]. Here $\mathcal{S} = \mathcal{S}_{\mathbb{C}}(n)$ denotes the Schwartz space of infinitely smooth *rapidly decreasing* functions $f : \mathbb{R}^n \to \mathbb{C}$, namely functions such that for any $l \in \mathbb{Z}_+$ and any multi-index $\alpha = (\alpha_1, \ldots, \alpha_n)$ of non-negative integers one has

$$\sup_{x \in \mathbb{R}^n} \left| \frac{\partial^\alpha f(x)}{\partial x^\alpha} (1 + |x|^l) \right| < \infty$$

where as usual $\frac{\partial^\alpha f(x)}{\partial x^\alpha} := \frac{\partial^{|\alpha|} f}{\partial x_1^{\alpha_1} \ldots \partial x_n^{\alpha_n}}$, $|\alpha| := \sum_{i=1}^n \alpha_i$.

Let $\mathcal{S}_{\mathbb{C}}'(n)$ be the topological dual of $\mathcal{S}_{\mathbb{C}}(n)$.

Theorem 1.2 (Alesker–Artstein–Faifman–Milman). *Assume we are given a bijective map $T : \mathcal{S}_{\mathbb{C}}(n) \to \mathcal{S}_{\mathbb{C}}(n)$ which admits an extension $T' : \mathcal{S}_{\mathbb{C}}'(n) \to \mathcal{S}_{\mathbb{C}}'(n)$ and such that for every $f \in \mathcal{S}_{\mathbb{C}}(n)$ and $g \in \mathcal{S}_1'\mathbb{C}(n)$ we have $T_u(f \cdot g) = (Tf) \cdot (T_u g)$. Then there exists a C^∞-diffeomorphism $u : \mathbb{R}^n \to \mathbb{R}^n$ such that*

$$\text{either } T(f) = f \circ u \text{ for all } f \in \mathcal{S}_{\mathbb{C}}(n),$$

$$\text{or } T(f) = \overline{f \circ u} \text{ for all } f \in \mathcal{S}_{\mathbb{C}}(n).$$

Thus, multiplicativity is valid only for transforms which are essentially a "change of variables". One of the elements in the proof was a lemma similar to those appearing in Appendix A below.

An obvious corollary of Theorem 1.2, which appeared in [3], was a theorem characterizing Fourier transform which is denoted by $\mathcal{F}$ and defined by

$$(\mathcal{F}f)(t) = \int_{\mathbb{R}} f(x) e^{-2\pi i x t} \, dx.$$

It is well known that Fourier transform exchanges pointwise product on $\mathbb{C}$ with usual convolution, which is denoted by $f * g$; that is, $\mathcal{F}(f \cdot g) = \mathcal{F}f * \mathcal{F}g$ and vice versa. The corollary of Theorem 1.2 is that the Fourier transform is, up to conjugation and up to a diffeomorphism, the only one which maps product to convolution among bijections $\mathcal{F} : \mathcal{S} \to \mathcal{S}$ which have an extension $\mathcal{F}' : \mathcal{S}' \to \mathcal{S}'$. It is not hard to check that if convolution is also mapped back to product then the diffeomorphism u above must be the identity mapping, for details see [3].

A similar characterization of the derivative through a functional equation was undertaken in [7, 8]. The functional equation was taken to be the chain rule.

One of the main theorems in the present note is that the assumption of the existence of $\mathcal{F}' : \mathcal{S}' \to \mathcal{S}'$ may be omitted in this theorem (and the corresponding extension of $\mathcal{T}$ in Theorem 1.2). This is presented in Theorem 1.10, one instance of which is $\mathcal{B}$ being Schwartz space. A direct corollary of the theorem is

Theorem 1.3. *Let $T : \mathcal{S}_{\mathbb{C}}(n) \to \mathcal{S}_{\mathbb{C}}(n)$ be a bijection.*

1. Assume T satisfies

$$T(f * g) = Tf \cdot Tg.$$

Then there exists a C^∞-diffeomorphism u : $\mathbb{R}^n \to \mathbb{R}^n$ such that either $Tf(u(x)) = \mathcal{F}f(x)$ or $Tf(u(x)) = \overline{\mathcal{F}f(x)}$.

2. *Assume* T *satisfies*

$$T(f \cdot g) = Tf * Tg.$$

Then there exists a C^∞-diffeomorphism u : $\mathbb{R}^n \to \mathbb{R}^n$ such that either $Tf = \mathcal{F}(f \circ u)$ or $Tf = \overline{\mathcal{F}(f \circ u)}$.

Remark 1.4. Similarly, Theorem 1.10 may also be used to characterize bijections $T : \mathcal{S}_\mathbb{C}(n) \to \mathcal{S}_\mathbb{C}(n)$ which satisfy $T(f * g) = Tf * Tg$.

Let us quote one more application of the method to Fourier theory. We denote by $C_c^\infty(\mathbb{R}, \mathbb{C})$ the smooth complex valued function on $\mathbb{R}$ which have compact support. It is well known, and referred to as a Paley-Wiener type theorem, that the class $C_c^\infty(\mathbb{R}, \mathbb{C})$ is the image under Fourier transform of the class $PW(\mathbb{R})$ consisting of functions F which decay on the real axis faster than any power of $|x|$, and have an analytic continuation on the complex plane satisfying the estimate $|F(z)| < A \exp(B|z|)$ for some constants A, B, see for example [6]. A similar characterization holds for functions of several variables, and we denote this class $PW(\mathbb{R}^n) = \mathcal{F}(C_c^\infty(\mathbb{R}^n, \mathbb{C}))$. The following will be an immediate corollary of Theorem 1.10.

Theorem 1.5. *Let $T : C_c^\infty(\mathbb{R}^n, \mathbb{C}) \to PW(\mathbb{R}^n)$ be a bijection which satisfies*

$$T(f * g) = Tf \cdot Tg.$$

Then there exists a C^∞-diffeomorphism $u : \mathbb{R}^n \to \mathbb{R}^n$ such that either $Tf(u(x)) = \mathcal{F}f(x)$ or $Tf(u(x)) = \overline{\mathcal{F}f(x)}$.

The setting of Schwartz space, and of its dual, in previous results, was very specific, and from the point of view of merely multiplicative mappings—not very natural. It was discussed mainly for its application to Fourier transform. However, in other characterization problems we found that similar tools were used in their proofs, and it turned out that in most of the natural situations in which we encounter multiplicative transforms, it is possible to characterize their form. One example was already given above in the form of compactly supported infinitely smooth functions. Below are several other such examples, and these are the main theorems to be proven in this note. The exposition is intended to make these tools available to the reader, more than to demonstrate the specific results, most of which we later discovered have already been proved in the literature (some more than 60 years ago, and some very recently). We view our method as very straightforward and natural, and believe it can be applied in many different situations.

In the following, a map $T : \mathcal{B} \to \mathcal{B}$ between some class $\mathcal{B}$ of real- or complex-valued functions, is called multiplicative if $T(fg) = Tf \cdot Tg$ pointwise for all $f, g \in \mathcal{B}$. Throughout the paper, we will address several families of C^k functions, defined on a C^k manifold M. When discussing Schwartz functions, it should always be understood that $M = \mathbb{R}^n$, and $k = \infty$.

Out first theorem regards multiplicative maps on continuous real valued functions, and it goes back to Milgram [9]. We also extend it to the class of continuous compactly supported functions.

Theorem 1.6. *Let M be a real topological manifold, and $\mathcal{B}$ is either $C(M, \mathbb{R})$ or $C_c(M, \mathbb{R})$. Let $T : \mathcal{B} \to \mathcal{B}$ be a multiplicative bijection. Then there exists some continuous $p : M \to \mathbb{R}_+$, and a homeomorphism $u : M \to M$ such that*

$$(Tf)(u(x)) = |f(x)|^{p(x)} \operatorname{sgn}(f(x)). \tag{1}$$

Remark 1.7. Without some non-degeneracy (and above we assume bijectivity, which is very strong non-degeneracy) there is a simple counterexample: let $Tf = f$ on $x \leq 0$, let $Tf(x) = f(x - 1)$ on $x \geq 1$ and let $Tf = f(0)$ on $[0, 1]$. However, this counterexample may actually hint that in a more general situation the map u may be a set valued map.

Next we move to the classes of C^k functions, where similar theorems hold, and moreover, no extra power is allowed, so that the mapping is automatically linear. This theorem is also known, but much more recent—it appears in [11] for $k < \infty$. The C^∞ case remained open in [11], and our method is able to clarify it as well. However, it also was already settled (by a considerably different method altogether) in [13]. We also obtain the same results for some subspaces of C^k, namely the compactly supported functions C_c^k, and the Schwartz functions $\mathcal{S}(n)$.

Theorem 1.8. *Let M be a C^k real manifold, $1 \leq k \leq \infty$, and $\mathcal{B}$ is one of the following function spaces: $C^k(M, \mathbb{R})$, $C_c^k(M, \mathbb{R})$ or $\mathcal{S}_\mathbb{R}(n)$. Let $T : \mathcal{B} \to \mathcal{B}$ be a multiplicative bijection. Then there exists some C^k-diffeomorphism $u : M \to M$ such that*

$$(Tf)(u(x)) = f(x), \tag{2}$$

In particular, T is linear.

We also address the case of complex-valued functions, which seems not to have been treated in previous works.

Theorem 1.9. *Let M be a topological real manifold, and $\mathcal{B}$ is either $C(M, \mathbb{C})$ or $C_c(M, \mathbb{C})$. Let $T : \mathcal{B} \to \mathcal{B}$ be a multiplicative bijection. Then there exists some homeomorphism $u : M \to M$ and a function $p \in C(M, \mathbb{C})$, $\operatorname{Re}(p) > 0$ such that either*

$$T(re^{i\theta})(u(x)) = |r(x)|^{p(x)} e^{i\theta(x)}$$

or

$$T(re^{i\theta})(u(x)) = |r(x)|^{p(x)} e^{-i\theta(x)}$$

Theorem 1.10. *Let M be a C^k real manifold, $1 \leq k \leq \infty$, and $\mathcal{B}$ is one of the following function spaces: $C^k(M, \mathbb{C})$, $C_c^k(M, \mathbb{C})$ or $\mathcal{S}_\mathbb{C}(n)$. Let $T : \mathcal{B} \to \mathcal{B}$ be a multiplicative bijection. Then there exists some C^k-diffeomorphism $u : M \to M$ such that either $Tf(u(x)) = f(x)$ or $Tf(u(x)) = \overline{f(x)}$. In particular, T is $\mathbb{R}$-linear.*

We give other variants of these theorems, and applications to Fourier transform, in Sect. 5.

Acknowledgements The authors would like to thank Mikhail Sodin for several useful discussions, Bo'az Klartag for explaining to us the failure of our original method of zero sets in the C^∞ n-dimensional setting, and the referee for numerous useful remarks and references, which allowed us to improve our results, and helped to provide better structure and context for the article.

2   Zero Sets

In the following section, $0 \le k \le \infty$, and M is a C^k real manifold. We use for $f \in C^k(M)$ the notation $Z(f)$ for the zero-set of the function, namely $Z(f) = \{x \in M : f(x) = 0\}$. We will also need (though in a very mild manner) the notion of the "jet" of a function at a point; in fact, we will only need here a function ρ whose k-jet at a point x_0, denoted $J\rho(x_0)$, is vanishing. This roughly means that all its derivatives at the point vanish. For the precise definition of a jet, see Appendix A. Finally, we fix a field $\mathbb{F}$ which is either $\mathbb{R}$ or $\mathbb{C}$. For the remainder of the section, all functions will have values in $\mathbb{F}$, and it will be often omitted from the notation.

The goal of this section is to establish the following

Proposition 2.1. *Let $0 \le k \le \infty$ be an integer, and let M be a C^k manifold. Let $\mathcal{B}$ be one of the following function families: $C^k(M, \mathbb{F})$, $C_c^k(M, \mathbb{F})$, $\mathcal{S}_\mathbb{F}(n)$. Assume $T : \mathcal{B} \to \mathcal{B}$ is a multiplicative bijection. Then there exists a homeomorphism $u : M \to M$ such that $Z(Tf) = u(Z(f))$ for all $f \in \mathcal{B}$.*

We present two different proofs. The first only applies to $\mathcal{B} = C^k(M)$ with $k < \infty$, and also in several 1-dimensional cases for the other function families, which will be specified later. The second proof is due to Mrcun [10], which applies in all cases, with slight modifications for the cases $\mathcal{B} = C_c^k(M, \mathbb{F})$ and $\mathcal{B} = \mathcal{S}_\mathbb{F}(n)$.

2.1   The Case of $\mathcal{B} = C^k(M, \mathbb{F})$, $k < \infty$

Lemma 2.2. *Let $f \in C^k(M)$. If $f(x_0) = 0$, then there exists $h \in C^k(M)$ s.t. $Z(h) = \{x_0\}$, and f^{4k+4} is divisible by h in $C^k(M)$.*

Proof. Fix $\rho \in C^k(M)$ which is non-negative, with $J\rho(x_0) = 0$, and $\rho(x) > 0$ for $x \ne x_0$. Take $h = |f|^2 + \rho$. It is then easy to see (Say, by induction) that $f^{4k+4}/h \in C^k(M)$, and we are done.   □

Remark 2.3. If $M = \mathbb{R}$ or $M = S^1$, this also holds for $k = \infty$ (with f instead of f^{4k+4}): simply take $h(x) = x$ for $M = \mathbb{R}$, $x_0 = 0$. Note that the statement is local, so it applies to S^1 as well. If in addition all functions are required to belong to $\mathcal{S}(1)$, one can construct $h \in \mathcal{S}(1)$ with the required property. Since those constructions only apply in the 1-dimensional case, while the corresponding C^∞ results hold in all dimensions and will be proven differently, we omit the details.

Corollary 2.4. *Let* $f \in C^k(M)$ *and* $x, y \in M$, $x \neq y$ *s.t.* $f(x) = f(y) = 0$. *One can then represent* $f^{4k+4} = f_1 f_2 f_3$ *with* $f_j \in C^k(M)$ *s.t.* $f_1(x) = f_3(y) = 0$, *and* $Z(f_1) \cap Z(f_3) = \emptyset$.

Proof. Take, using Lemma 2.2 f_1 and f_3 such that $Z(f_1) = \{x\}$, $Z(f_3) = \{y\}$, and f^{4k+4} is divisible by both f_1 and f_3 in $C^k(M)$. It is then obvious that $(f_1 f_3)$ divides f^{4k+4} in $C^k(M)$, since for all $z \in M$ either $f_1(z) \neq 0$ or $f_3(z) \neq 0$. Thus, $f_2 = \frac{f^{4k+4}}{f_1 f_3} \in C^k(M)$ is well-defined. $\square$

Lemma 2.5. *Let* $f, g \in C^k(M)$ *s.t.* $f(x) = g(x) = 0$. *Then one can find* $h \in C^k(M)$ *with* $h(x) = 0$ *s.t. both* f^{4k+4} *and* g^{4k+4} *are divisible by* h.

Proof. Take $h = f^2 + g^2$, and verify divisibility by induction. $\square$

We denote by $gcd(f^{4k+4}, g^{4k+4})$ the family of all such functions h.

Remark 2.6. Again, if $M = \mathbb{R}$ or $M = S^1$, this also holds for $k = \infty$, with f, g instead of f^{4k+4}, g^{4k+4}, since if the zero is assumed to be, say, at the point 0, then $h(x) = x$ is a common divisor.

We next prove that there is a function $u : M \to M$ which governs the behavior of zero-sets of functions under the transform $T : C^k(M) \to C^k(M)$.

Proposition 2.7. *Let* $0 \leq k < \infty$ *and let* M *be a* C^k *manifold. Assume* $T : C^k(M) \to C^k(M)$ *is a multiplicative bijection. Then there exists a homeomorphism* $u : M \to M$ *such that* $Z(Tf) = u(Z(f))$ *for all* $f \in C^k(M)$.

Proof. Step 1. For $f \in C^k(M)$, $Z(f) = \emptyset$ if and only if $Z(Tf) = \emptyset$.

    Simply note that if $Z(f) = \emptyset$ then $g = 1/f \in C^k(M)$. Thus $(Tf)(Tg) = T(fg) = T(1)$, and obviously $T(1) = 1$, so that $Z(Tf) = \emptyset$. For the reverse implication, consider T^{-1}, which is multiplicative as well.

Step 2. For $f, g \in C^k(M)$ we have that $Z(f) \cap Z(g) \neq \emptyset$, if and only if $Z(Tf) \cap Z(Tg) \neq \emptyset$.

    Take by Lemma 2.5 $h \in gcd(f^{4k+4}, g^{4k+4})$. Denote $f^{4k+4} = vh$ and $g^{4k+4} = wh$. Therefore $Tf^{4k+4} = TvTh$ and $Tg^{4k+4} = TwTh$. By assumption, $Z(h) \neq \emptyset$, and therefore $\emptyset \neq Z(Th) \subset Z(Tf) \cap Z(Tg)$. For the reverse implication, consider T^{-1}.

Step 3. There exists an invertible map $u : M \to M$ such that $Z(f) = \{x\}$ implies $Z(Tf) = \{u(x)\}$.

    Assume $Z(f) = \{x\}$. By (1), $Z(Tf) \neq \emptyset$. If $y, z \in Z(Tf)$ and $y \neq z$, apply the previous lemma: write $Tf^{4k+4} = g_1 g_2 g_3$ with $g_1(x) = g_3(z) = 0$, $Z(g_1) \cap Z(g_3) = \emptyset$. By bijectivity of T, $g_j = T(f_j)$. Thus $f^{4k+4} = f_1 f_2 f_3$. By step 2, f_1 and f_3 have no common zeros, while by step 1 both f_1 and f_3 have zeros. Thus f has at least two zeros, a contradiction. We conclude that Tf has a unique zero. If $Z(f) = Z(g) = \{x\}$, by step 2 Tf and Tg have a common (and by above unique) zero, thus $Z(Tf) = Z(Tg)$, i.e. $Z(Tf) = \{u(x)\}$ for some $u : M \to M$. Note that u must be invertible by observing T^{-1}.

Step 4. For $f \in C^k(M)$, $x \in M$, one has $f(x) = 0$ if and only if $Tf(u(x)) = 0$, i.e. $Z(Tf) = u(Z(f))$.

Take f such that $f(x) = 0$, and using Lemma 2.2 take h dividing f^{4k+4} with $Z(h) = \{x\}$. Denote $f^{4k+4} = hv$. Then $Tf^{4k+4} = ThTv$, and since $Th(u(x)) = 0$ by step 3, one has $Tf(u(x)) = 0$. For the reverse implication, consider T^{-1}.

Step 5. The map $u : M \to M$ is a homeomorphism.

For a chart $\mathbb{R}^n \simeq U \subset M$, take a C^k function f with $Z(f) = M \setminus U$. Then $u(B) = M \setminus Z(T(f))$ is an open set. Similarly, the preimage of a chart is open. Since the charts form a basis of the topology, images and preimages by u of open sets are open. Therefore, u is a homeomorphism.

$\square$

Remark 2.8. The construction of a homeomorphism with the property as in Lemma 2.7 does not extend to the general $C^\infty(M)$ case. Nevertheless, it can be carried out when $M = \mathbb{R}$ or $M = S^1$ by Remarks 2.3 and 2.6.

2.2 The Cases of $\mathcal{B} = C^k(M, \mathbb{F})$, $C_c^k(M, \mathbb{F})$ and $\mathcal{S}_{\mathbb{F}}(n)$, $0 \leq k \leq \infty$

We use the construction of u from [10] that is used in [11]. Some attention should be paid when repeating it for the different families of functions, and we do it in full detail for the convenience of the reader.

Proposition 2.9. *Let $\mathcal{B}$ be a multiplicatively closed family of functions s.t. $C_c^k(M) \subset \mathcal{B} \subset C^k(M)$. Let $T : \mathcal{B} \to \mathcal{B}$ be a multiplicative bijection, which restricts to a bijection of $C_c^k(M)$. Then there exists a homeomorphism $u : M \to M$ s.t. $u(Z(f)) = Z(Tf)$ for all $f \in \mathcal{B}$.*

Proof. Step 1. Recall the notion of a characteristic sequence of functions f_j at a point $x \in M$: this a sequence $f_j \in C_c^k(M)$ s.t. $f_j f_{j+1} = f_{j+1}$, and $\bigcap \text{supp}(f_j) = \{x\}$. Fix some $x \in M$ and a characteristic sequence of functions for it, f_j. By our assumptions, $g_j = T(f_j)$ satisfy $g_j g_{j+1} = g_{j+1}$, so $\text{supp}(g_{j+1}) \subset \text{supp}(g_j)$, and those are also compact sets, so $K = \bigcap \text{supp}(g_j) \neq \emptyset$. We want to show that this intersection is in fact a single point. Fix $y \in K$, a neighborhood U of y, and choose a characteristic sequence β_j at y with $\text{supp}(\beta_1) \subset U$. Take $\alpha_j = T^{-1}\beta_j$. Then $\gamma_j = f_j \alpha_j$ has compact support and satisfies $\gamma_{j+1} = \gamma_j \gamma_{j+1}$, so $\bigcap \text{supp}(\gamma_j) \neq \emptyset$, but $\text{supp}(\gamma_j) \subset \text{supp}(f_j)$, so $\bigcap \text{supp}(\gamma_j) = \{x\}$ and γ_j is a characteristic family at x. In particular, $\gamma_1 f_j = f_j$ for large j, and applying T, $\beta_1 g_1 g_j = g_j$. So $\text{supp}(g_j) \subset \text{supp}(\beta_1) \subset U$ for large j. This holds for every U, implying $\bigcap \text{supp}(g_j) = \{y\}$. We claim that y depends only on x and not on the choice of characteristic sequence f_j. Assuming $\tilde{f}_j$ is another such sequence, $f_j \tilde{f}_j$ is also a characteristic sequence at x, so if $\bigcap \text{supp}(Tf_j) = \{y\}$, $\bigcap \text{supp}(T\tilde{f}_j) = \{z\}$ and $y \neq z$

then $\bigcap \mathrm{supp}(Tf_j T \tilde{f}_j) \subset \{y\} \cap \{z\} = \emptyset$, a contradiction. We thus define the map $u : M \to M$ by $u(x) = y$. By bijectivity of T, it is obvious that u is also bijective.

Step 2.   We next claim that for all $f \in \mathcal{B}, x \in M, T(f)(u(x))$ depends only on the germ f_x. Indeed, assume $f_x = g_x$. Take a characteristic sequence ϕ_j at x. Then for some large j, $\phi_j f = \phi_j g$, so $T(\phi_j)T(f) = T(\phi_j)T(g)$. By construction of u, $T(\phi_j) \equiv 1$ in a neighborhood of $u(x)$, so $T(f)(u(x)) = T(g)(u(x))$.

Step 3.   Take $f \in \mathcal{B}$ s.t. $f(x) \neq 0$. Choose $g \in \mathcal{B}$ s.t. $(fg)_x = 1_x$ (which can be done since $C_c^k(M, \mathbb{F}) \subset \mathcal{B}$). Then $T(f)(u(x))T(g)(u(x)) = T(1_x)(u(x))$. Now $T(f_x)(u(x))T(1_x)(u(x)) = T(f_x)(u(x))$ by multiplicativity for any germ f_x, implying by surjectivity of T that $T(1_x)(u(x)) \neq 0$ (in fact, since $1_x^2 = 1_x$, we immediately conclude that $T(1_x)(u(x)) = 1$). Thus $T(f)(u(x)) \neq 0$. By considering T^{-1}, we get $f(x) \neq 0 \iff Tf(u(x)) \neq 0$, as required. Finally, u is a homeomorphism by Step 5 of the proof of Proposition 2.7.   □

Remark 2.10. It follows from step 2 that in fact T maps germs of functions at x to germs of functions at $u(x)$. This is also an immediate consequence of Proposition 2.1, as will be seen in the next section.

Corollary 2.11. *Let $\mathcal{B}$ be a multiplicatively closed family of functions s.t. $C_c^k(M) \subset \mathcal{B} \subset C^k(M)$ and for all $f \in \mathcal{B}$, $\{x \in M : f(x) = 1\} \subset M$ is compact. Let $T : \mathcal{B} \to \mathcal{B}$ be a multiplicative bijection. Then there exists a homeomorphism $u : M \to M$ s.t. $u(Z(f)) = Z(Tf)$ for all $f \in \mathcal{B}$.*

Proof. By the proposition above, it only remains to verify that T restricts to a bijection of $C_c^k(M)$. Observe that

$$f \in C_c^\infty(M) \iff fg = f \text{ for some } g \in \mathcal{B}$$

Indeed, if $f \in C_c^\infty(M)$, just choose any $g \in C_c^k(M)$ with $g \equiv 1$ on $\mathrm{supp}(f)$. In the other direction, if $fg = f$ with $g \in \mathcal{B}$, then $g \neq 1$ outside some compact set K, implying $\mathrm{supp}(f) \subset K$. Since T is multiplicative and bijective, $fg = f \iff TfTg = Tf$, so T restricts to a bijection of $C_c^k(M)$ as required.   □

Corollary 2.12. *For both $\mathcal{B} = C_c^k(M, \mathbb{F})$ and $\mathcal{B} = \mathcal{S}_{\mathbb{F}}(n)$, a multiplicative bijection $T : \mathcal{B} \to \mathcal{B}$ defines a homeomorphism $u : M \to M$ s.t. $u(Z(f)) = Z(Tf)$ for all $f \in \mathcal{B}$.*

Proof. Simply apply the Corollary above. In fact, for $\mathcal{B} = C_c^k(M, \mathbb{F})$ Proposition 2.9 applies immediately.   □

Remark 2.13. The family $\mathcal{B} = C^k(M, \mathbb{F})$ does not satisfy all assumptions automatically: it is not immediate that $T : C^k(M) \to C^k(M)$ preserves the subspace of compactly supported functions, which makes the construction in [10] slightly more involved. We do not repeat here the proof in this case, since no details of the original proof should be modified.

3 Real Valued Functions

In the following section, we describe the general form of $T : \mathcal{B} \to \mathcal{B}$ for the real-valued function families $\mathcal{B}$ from Proposition 2.1. We then separately treat the cases of $k = 0$ and $0 < k \leq \infty$. The reader might want to review the notion of the jet of a function before proceeding (see Appendix A). We will write $\mathcal{B}^k$ instead of $\mathcal{B}$ for any of the families $C^k(M)$, $C_c^k(M)$, and also $\mathcal{S}(n)$ if $k = \infty$. One always has $C_c^k(M) \subset \mathcal{B}^k$. In the following, f_x denotes the germ of f at x.

Proposition 3.1. *Given a multiplicative bijection $T : \mathcal{B}^k \to \mathcal{B}^k$ (with $0 \leq k \leq \infty$) there exists a homeomorphism u, given by Lemma 2.1, such that letting T_u be defined by $T_u(f) = T(f) \circ u$, the new map $T_u : \mathcal{B}^k \to C(M, \mathbb{R})$ has the following properties:*

(1) It is multiplicative, that is, $T_u(fg) = T_u(f)T_u(g)$

(2) It is local, namely $(T_u f)_x = (T_u g)_x$ when $f_x = g_x$. Moreover, $(T_u f)(x) = F(x, J^k f(x))$.

(3) It is determined by its action on non-negative functions, namely

$$T_u(f)(x) = \begin{cases} 0, & f(x) = 0 \\ T_u(|f|)(x)\,\mathrm{sgn}(f(x)), & f(x) \neq 0 \end{cases}$$

Proof. Part (1) is obvious. For part (2), observe:

Step 1. $T(0) = 0$. Immediate since $T(f)T(0) = T(0)$ for all f, and T is bijective.

Step 2. For any open set V, $f = g$ on V implies $Tf = Tg$ in $cl(u(V))$.

Indeed, take any open ball $B \subset V$, and take a function h with $Z(h) = M \setminus B$ (a bump function over B). We have $f \cdot h = g \cdot h$ in M, and so $Tf \cdot Th = Tg \cdot Th$ in M and by Proposition 2.7 $Z(Th) = M \setminus u(B)$, implying $Tf = Tg$ in $u(B)$. This holds for all $u(B) \subset u(V)$; since u is a homeomorphism, $Tf = Tg$ in $u(V)$, and by continuity in $cl(u(V))$.

Put another way, we proved that the germ $(Tf)_{u(x)}$ only depends on the germ f_x of f at x. We may write $(Tf)_{u(x)} = T(f_x)_{u(x)}$, and $T(f)(u(x)) = T_u(f)(x) = T_u(f_x)(x)$. Thus we may compute $T_u(C_x)$ for the constant germ C at x, even if the constant function $C \notin \mathcal{B}$, by completing C to a compactly supported function away from x.

Step 3. $(Tf)(u(x)) = F(x, J^k f(x))$ for some $F : J^k \to \mathbb{R}$.

Indeed, fix $x_0 \in M$. Choose an open ball U around x_0, and two open sectors $V_1, V_2 \subset U$, having x_0 as a common vertex, and $cl(V_1) \cap cl(V_2) = \{x_0\}$. Given two functions $f_1, f_2 \in C(M)$, assume that $J^k f_1(x_0) = J^k f_2(x_0)$. By Whitney's extension theorem, one can choose a $C_c^k(M)$ function f_3 that equals f_j on V_j for $j = 1, 2$. Then $T_u(f_3)$ and $T_u(f_j)$ coincide on $cl(V_j)$, and in particular, $Tf_1(u(x_0)) = Tf_3(u(x_0)) = Tf_2(u(x_0))$. Therefore, $(T_u f)(x) = (Tf)(u(x)) = F(x, J^k f(x))$. This completes the proof of (2).

Step 4. $T_u(-1_x) = ((-1)^{\delta(x)})_x$ with $\delta(x) \in \{0, 1\}$ a locally constant function. Indeed, for any germ f_x, $T_u(f_x) = T_u(f_x)T_u(1_x)$ so $T_u(1_x) \equiv 1_x$. Again by multiplicativity, $T_u(-1_x)^2 = T_u(1_x) = 1_x$, so $T_u(-1_x) = ((-1)^{\delta(x)})_x$ Finally, note that $T_u(-1_x)$ is the germ of a continuous function to conclude $\delta(x)$ is locally constant.

Step 5.

$$(T_u f)(x) = \begin{cases} 0, & f(x) = 0 \\ T_u(|f|)(x) T_u(\mathrm{sgn}(f(x))), & f(x) \neq 0 \end{cases}$$

Indeed, since by continuity, for any $x \in M$ such that $f(x) \neq 0$, sgn f is locally constant at x, and by step 2 $T_u(f_x)(x) = T_u(\mathrm{sgn}\, f(x)|f_x|)(x)$.

Step 6. The function $\delta(x)$ from step 4 satisfies $\delta(x) = 1$ for all x. Indeed, $f(x_0) > 0$ implies $(T_u f)(x_0) > 0$: One can choose $g \in C_c^k(M)$ with $f(x) = g(x)^2$ for x near x_0. Then $T_u(f)(x_0) = T_u(g)^2(x_0) > 0$. If $T_u(-1)(x) = +1$ for some x, it implies that $(Tf)(u(x)) = T(|f|)(u(x))$ is always non-negative on the connected component of $u(x)$ in M, thus contradicting surjectivity of T. Therefore, $T_u(-1_x) = -1_x$. $\qquad\square$

From now on we work with T_u instead of T, and only return to the original T when we show that u is a C^k-diffeomorphism. Thus, for now we cannot assume that the image of T_u is C^k, but only that it is continuous.

By part (3) of Proposition 3.1, we need to study our transform only on non-negative functions.

Lemma 3.2. *Let T_u satisfy the conclusion of Proposition 3.1. Then there exists a global section of $(J^k)^*$, c_k, such that for $f(x_0) > 0$,*

$$T_u(f)(x_0) = \exp(\langle c_k(x_0), J^k(\log f)(x_0)\rangle)$$

If $k < \infty$, c_k is a continuous global section. If $k = \infty$, it is locally finite dimensional and continuous, i.e. every $x_0 \in M$ has an open neighborhood U such that $c_\infty = Q_n(c_n)$ in U for some finite n, and c_n is a continuous section of $(J^n)^*$ over U.

Proof. As in the proof above, T_u clearly maps positive functions to positive functions. Define $A : C_c^k(M) \to C(M)$ by $A(f) = \log T_u(\exp(f))$. Then A is an additive transformation, with the additional property that $A(f)(x) = B(x, J^k f(x))$, where $B(x, \cdot) : J_x^k \to \mathbb{R}$ is an additive functional for every $x \in M$. Apply Lemma A.2 from Appendix A to conclude the stated result. $\qquad\square$

We are ready to conclude the proof of Theorem 1.6.

Proof of Theorem 1.6. Recall that $T_u = T \circ u$, and observe that it is surjective, as T is. We already know by Lemma 3.2 with $k = 0$ that

$$(Tf)(u(x)) = \begin{cases} 0, & f(x) = 0 \\ |f(x)|^{c_0(x)} \mathrm{sgn}(f(x)), & f(x) \neq 0 \end{cases}$$

with $c_0(x)$ continuous. We are left to show that $c_0(x) > 0$ everywhere. Indeed, if $c_0(x) = 0$ then $T_u f(x)$ is either 0 or ± 1 for every f, contradicting surjectivity of T_u; while if $c_0(x) < 0$, we could take a positive function f with an isolated zero at x_0, and then $\lim_{x \to x_0} Tf(x) = \infty$, contradicting continuity of Tf.

Next we assume $k \geq 1$ and prove Theorem 1.8, i.e. that $(Tf)(u(x)) = f(x)$. We will denote $v = u^{-1}$.

Fix some $x_0 \in M$, and choose a relatively compact neighborhood U of x_0 as in Lemma 3.2. Thus $(Tf)(u(x)) = \exp(\langle c_k(x), J^k(\log f)(x) \rangle)$ for positive $f \in C^k(U)$, and $c_k = Q(n,k)(c_n)$ for some finite n, c_n a continuous section of J^n over U (i.e., c_k only depends on the n-jet of the function). We claim that one can take $n = 0$. We may assume that U is a coordinate chart with x_0 at the origin. Then, if c_n at x_0 depends on terms of the jet other than the constant term, one has

$$\langle c_k(x), J^k(\log f)(x) \rangle = a_0(x) \log f(x) + \sum_{1 \leq |\alpha| \leq n} a_\alpha(x) \frac{\partial^{|\alpha|} \log f}{\partial x^\alpha}$$

where $a_\alpha \in C(U)$, and $a_{\alpha_0}(0) \neq 0$ for some $\alpha_0 \neq 0$. Fix such α of maximal modulus $m = |\alpha|$, and take $f(x) = \lambda_1 x_1 + \ldots + \lambda_d x_d$ where $d = \dim M$. Then

$$\frac{\partial^{|\alpha|} \log f}{\partial x^\alpha} = \frac{\pm \prod \lambda_j^{\alpha_j} (|\alpha| - 1)!}{(\sum \lambda_j x_j)^{|\alpha|}}$$

so an appropriate choice of λ_j (not all zero) will guarantee that

$$\sum_{|\alpha| = m} a_\alpha(x) \frac{\partial^{|\alpha|} \log f}{\partial x^\alpha} = \frac{C(x)}{(\sum \lambda_j x_j)^m}$$

with $C(x)$ continuous and non-vanishing near 0. The same would hold, with a different $C(x)$, also if we sum up all the α-derivatives for $1 \leq |\alpha| \leq n$. It follows that

$$T_u(\sum \lambda_j x_j) = |\sum \lambda_j x_j|^{a_0(x)} e^{\sum_{1 \leq |\alpha| \leq n} a_\alpha(x)(\log \sum \lambda_j x_j)^{(\alpha)}}$$

cannot be continuous at 0, a contradiction.

Thus $(Tf)(x) = f(v(x))^{a_0(v(x))}$ for positive f. Taking $f_x \equiv 2_x$, we conclude that $a_0(v(x)) \in C^k(M)$. As in the case $k = 0$ we see that $a_0 > 0$, so $f(v(x)) \in C^k(M)$ for all positive f, which implies $v \in C^k(M, M)$. The same reasoning applied to T^{-1}, we conclude that u is a C^k-diffeomorphism. Now it is obvious that $C_c^k(M)$ is invariant under T, so $T : C_c^k(M) \to C_c^k(M)$ is a bijection. Since $k \geq 1$, we must have $a_0 \equiv 1$, so $(Tf)(u(x)) = f(x)$, as claimed. $\qquad \square$

4   Complex Valued Functions

In this section, we describe the general form of $T : \mathcal{B} \to \mathcal{B}$ for the complex-valued function families $\mathcal{B}$ from Proposition 2.1. We again treat the cases of $k = 0$ and $0 < k \leq \infty$ separately. We write $\mathcal{B}^k$ instead of $\mathcal{B}$ for any of the families $C^k(M)$, $C_c^k(M)$, and also $\mathcal{S}(n)$ if $k = \infty$. One always has $C_c^k(M) \subset \mathcal{B}^k$.

Proposition 4.1. *Given a multiplicative bijection* $T : \mathcal{B}^k \to \mathcal{B}^k$ *(with* $0 \le k \le \infty$*) there exists a homeomorphism u, given by Lemma 2.1, such that letting T_u be defined by* $T_u(f) = T(f) \circ u$, *the new map* $T_u : \mathcal{B}^k \to C(M, \mathbb{C})$ *has the following properties:*

(1) It is multiplicative: $T_u(fg) = T_u(f)T_u(g)$
(2) It is local, namely $(T_u f)_x = (T_u g)_x$ *when* $f_x = g_x$. *Moreover,* $(T_u f)(x) = F(x, J^k f(x))$.
(3) $T_u(f)(x) = 0$ *if and only if* $f(x) = 0$.

The proof of this statement is as in the real case, and is omitted. We denote $v = u^{-1}$. It is then obvious that $C_c^k(M, \mathbb{C})$ is an invariant subspace of T, on which T is bijective. Also by part (2), T_u extends naturally to a map $T_u : C^k(M, \mathbb{C}) \to C(M, \mathbb{C})$ which retains properties (1)–(3).

We next make some helpful decompositions, which enable us to treat the various parts of the transform separately. A complex valued function $f \in C^k(M, \mathbb{C})$ can be written as $r(x)e^{i\theta(x)}$ where $r \ge 0$ continuous s.t. $r \in C^k(\{x : r(x) > 0\}$, and $\theta \in C^k(\{x : r(x) > 0\}, S^1)$ and

Proposition 4.2. *There exists a function* $g_0 \in C(M, \mathbb{R}_+)$, *and global sections* $d_k \in (J^k)^*(M \times S^1, \mathbb{R})$, $h_k \in (J^k)^*(M \times \mathbb{R})$ *and* $e_k \in (J^k)^*(M \times S^1, S^1)$ *such that*

$$T_u(re^{i\theta})(x) = \begin{cases} 0, & r(x) = 0 \\ r(x)^{g_0(x)} e^{i\langle h_k(x), J^k \log r(x)\rangle} e^{\langle d_k(x), J^k \theta(x)\rangle} e^{i\langle e_k(x), J^k \theta(x)\rangle}, & r(x) \ne 0 \end{cases}$$

The sections h_k, d_k, e_k *are continuous when* $k < \infty$, *and locally finite dimensional and continuous when* $k = \infty$.

Proof. We may, using multiplicativity of T_u, write

$$T_u(r(x) \exp(i\theta(x))) = T_u(r(x)) S(\theta(x))$$

where $S(\theta) = T_u(\exp(i\theta))$. Since we already know that zeros are mapped to zeros, $T_u : C^k(M, \mathbb{R}_+) \to C(M, \mathbb{C}^*)$ and $S : C^k(M, S^1) \to C(M, \mathbb{C}^*)$ are group homomorphisms. Denote further $T_u(r) = G(r)H(r)$ and $S(\theta) = D(\theta)E(\theta)$, where $D : C^k(M, S^1) \to C(M, \mathbb{R}_+)$, $E : C^k(M, S^1) \to C(M, S^1)$, $G : C^k(M, \mathbb{R}_+) \to C(M, \mathbb{R}_+)$, $H : C^k(M, \mathbb{R}_+) \to C(M, S^1)$ are homomorphisms of groups. Furthermore, Property (2) immediately implies that D, E, G, H are all local, namely depend only on the jets of the functions.

As in the real case, we apply lemma A.2 of Appendix A to conclude that $G(r)(x) = r(x)^{g_0(x)}$ with $g_0 \in C(M, \mathbb{R}_+)$ (if $k \ge 1$, G cannot depend on higher derivatives of r as in the proof of Theorem 1.8, while $g_0 > 0$ as in the proof of Theorem 1.6: $g_0 \ge 0$ to guarantee continuity of $T_u(r(x))$ at a zero point of r, and $g_0(x) = 0$ would immediately contradict surjectivity of T). Then by Lemmas A.3–A.5 of Appendix A, $D(\theta)(x) = \exp(\langle d_k(x), J^k\theta(x)\rangle)$; $H(r)(x) = \exp(i\langle h_k(x), J^k \log r(x)\rangle)$; and $E(\theta) = \exp(i\langle e_k(x), J^k\theta(x)\rangle)$ where d_k, h_k, e_k are as stated. $\square$

We now can complete the proof of Theorem 1.9.

Proof of Theorem 1.9. Recall that $k = 0$. Then $d_k(x) = 0$, $e_k(x) = m$ for some fixed $m \in \mathbb{Z}$ and $h_k(x) = h_0$ for some $h_0 \in C(M)$. Thus $T_u(re^{i\theta}) = r^{g_0}e^{i(m\theta + h_0 \log r)}$, $g_0, h_0 \in C(M)$. From injectivity of T_u on $\mathcal{B}$, $m = \pm 1$: otherwise, either $m = 0$ and T_u does not depend on θ; or $|m| \geq 2$, so θ can be replaced with $\theta + 2\pi/m$ without affecting $T_u(re^{i\theta})$ (for any compactly supported $r(x)$). Thus

$$T(re^{i\theta})(u(x)) = r(x)^{p(x)}e^{\pm i\theta}$$

where $p(x) = g_0(x) + ih_0(x)$, as claimed. □

Next we proceed to prove Theorem 1.10.

Proof of Theorem 1.10.

Step 1. We prove that u is a C^k-diffeomorphism of M. Taking $r \equiv 2$, $\theta \equiv 0$ we see that $|T(2)(x)| = 2^{g_0(v(x))} \in C^k(M)$, in particular $g_0(v(x)) \in C^k(M)$. Thus for any $r(x) \in C^k(M, \mathbb{R}_+)$ (again $\theta(x) \equiv 0$)

$$\log|T(r)(x)| = g_0(v(x))\log r(v(x)) \in C^k(M)$$

so also $r(v(x)) \in C^k(M, \mathbb{R}_+)$. Thus $v \in C^k(M, M)$. By considering T^{-1}, u is also C^k, implying u is a C^k-diffeomorphism of M. From now on we only consider $T_u f = Tf \circ u$, and prove that $T_u f = f$ or $T_u f = \overline{f}$. Note that $T_u : C^k(M) \to C^k(M)$, and its restriction $T_u : C_c^k(M) \to C_c^k(M)$ is a bijection.

Step 2. We show here that $d_k = 0$. If $d_k \neq 0$, one could choose for any $x_0 \in M$ a function $\theta \in C^k(U \setminus \{x_0\})$ where U is a small neighborhood of x_0 contained in a coordinate chart with its origin at x_0, for which d_k only depends on the m-jet, $m < \infty$, s.t. $\langle d_k(x_n), J^k\theta(x_n)\rangle \geq \frac{g_0(x_n)}{|x_n|^2}$ for some $U \setminus \{x_0\} \ni x_n \to x_0$, while $|J^j\theta(x)| \leq C_j|x|^{-N_j}$ for all $j \leq k$ and $x \in U \setminus \{x_0\}$ (simply take $\theta = C|x|^{-2}$ with appropriate C in a contractible neighborhood of the sequence (x_n), and extend it to $U \setminus x_0$ arbitrarily). Then, taking $r(x) = \exp(-1/|x|^2)$, one has $f(x) = r(x)e^{i\theta(x)} \in C^k(U)$, $f(x_0) = 0$ while

$$|T_u f(x_0)| = \lim_{n \to \infty} r(x_n)^{g_0(x_n)} \exp(\langle d_k(x_n), J^k\theta(x_n)\rangle)$$

$$\geq \lim_{n \to \infty} \exp(-g_0(x_n)/|x_n|^2)\exp(g_0(x_n)/|x_n|^2) = 1$$

a contradiction. Thus, $d_k \equiv 0$.

Step 3. We show that $g_0 \equiv 1$. Indeed, since $|T_u(r(x)e^{i\theta(x)})| = r(x)^{g_0(x)} \in C^k(M, \mathbb{R}_+)$ for all $r \in C^k(M, \mathbb{R}_+)$, and since T_u is surjective on $C_c^k(M, \mathbb{C})$, we must have $g_0 \equiv 1$

Step 4. We next claim that $h_k \equiv 0$. Note that for all $r(x) \in C^k(M, \mathbb{R}_+)$,

$$T_u(r(x)) = r(x)e^{i\langle h_k(x), J^k \log r(x)\rangle} \in C^k(M)$$

so h_k is a C^k section of $(J^k)^*(M \times \mathbb{R})$. First, h_k only depends on the constant term of the jet, or else $T_u(r(x))$ would not be in C^k for all $r \in C^k(M, \mathbb{R}_+)$: this is obvious when $k < \infty$; and if $k = \infty$, we proceed as was done in the real case. Fix a coordinate chart U s.t. $h_k(x) = (a_\alpha(x))_{|\alpha| \le m}$ where $a_\alpha \in C^k(U)$, $m \ge 1$ and $a_\alpha(0) \ne 0$ for some α with $|\alpha| = m$. Take $f(x) = \lambda_1 x_1 + \ldots + \lambda_d x_d$ with coefficients λ_j s.t.

$$\sum_{|\alpha|=m} a_\alpha(x) \frac{\partial^{|\alpha|} \log f}{\partial x^\alpha} = \frac{C(x)}{(\sum \lambda_j x_j)^m}$$

with $C(x) \in C^k(U)$ and non-vanishing near 0. Then, assuming $\lambda_1 \ne 0$ and considering only points x where $f(x) > 0$,

$$\frac{\partial}{\partial x_1}(T_u f) = \lambda_1 e^{i \langle h_k(x), J^k \log f(x) \rangle}$$

$$+ \left(\sum \lambda_j x_j \right) \left(\left(\frac{\partial C(x)}{\partial x_1} \frac{1}{(\sum \lambda_j x_j)^m} - \frac{m \lambda_1 C(x)}{(\sum \lambda_j x_j)^{m+1}} \right) \right.$$

$$\left. + \sum_{|\alpha| \le m} \frac{C_\alpha(x)}{(\sum \lambda_j x_j)^{|\alpha|}} \right) e^{i \langle h_k(x), J^k \log f(x) \rangle}$$

where all C_α are continuous. Thus there is no limit to $\frac{\partial}{\partial x_1}(T_u f)$ as $x \to 0$ (along points of positivity for f), a contradiction. So $h_k = h_0 \in C(M, \mathbb{R})$. Then, if $h_0(x_0) \ne 0$, take a chart with x_0 at the origin, and consider $f(x) = x_1 \in C^k(M)$—the first coordinate function. Then

$$\frac{\partial}{\partial x_1}(T_u f)(0) = \lim_{x_1 \to 0^+} \frac{x_1 e^{i h_0(x) \log x_1}}{x_1} = \lim_{x_1 \to 0^+} e^{i h_0(x) \log x_1}$$

which diverges since $h_0(x) \log x_1$ is continuous when $x_1 \in (0, \infty)$, and $\lim_{x_1 \to 0^+} |h_0(x) \log x_1| = \infty$. This is a contradiction.

Step 5. Finally, we want to show that $\langle e_k(x), J^k \theta(x) \rangle = \pm \theta(x)$. First, by considering a coordinate chart and polynomial functions θ, we see that the components of $e_k(x)$ are in fact C^k. We now treat separately the cases $k < \infty$ and $k = \infty$.

Case 1: $k < \infty$. Then $T_u(e^{i \theta(x)}) = \exp(i \langle e_k(x), J^k \theta(x) \rangle) \in C^k(M, S^1)$ for all $e^{i\theta} \in C^k(M, S^1)$, which is impossible unless e_k only depends on $J^0 \theta$, i.e. $\langle e_k(x), J^k \theta(x) \rangle = m \theta(x)$. From injectivity of T_u on $C_c^k(M, \mathbb{C})$, $m = \pm 1$.

Case 2: $k = \infty$. Let us prove the following

Proposition. $P : C^\infty(M, \mathbb{R}) \to C^\infty(M, \mathbb{R})$, $\theta \mapsto \langle e_k(x), J^\infty \theta(x) \rangle$ *induces an isomorphism of stalks of smooth functions at every $p \in M$.*

Proof. Indeed, fix $p \in M$, and a germ u_p represented by $u \in C^\infty(U, \mathbb{R})$ with some small neighborhood $U \ni p$.

To see that $\text{Ker}(P_p) = 0$, assume $(Pu)_p = 0$, so Pu vanishes identically in some neighborhood V of p. Take any $r \in C_c^\infty(M, \mathbb{R}_+)$ with $\text{supp}(r) \subset V$ and $r(p) > 0$. Then $T_u(r \exp(iu)) = r \exp(iPu) \equiv r$ and also $T_u(r) = r$, so by injectivity of T_u, u must be a multiple of 2π whenever $r \neq 0$, in particular $u_p \equiv (2\pi l)_p$—a constant germ with $l \in \mathbb{Z}$. Since $P(c) = mc$ for constant functions c, and $m \neq 0$ from injectivity of T_u on $C_c^k(M, \mathbb{C})$, we may conclude that $c = 0$ and so $u_p = 0$.

For surjectivity of P_p, let us find v_p s.t. $(Pv)_p = u_p$. Choose any smooth continuation of u to M, and some $r(x) \in C_c^\infty(M, \mathbb{R}_+)$ with $r(p) = 1$. By surjectivity of T_u, one can find $v \in C^\infty(M, \mathbb{R})$ s.t. $r \exp(iu) = T_u(r \exp(iv)) = r \exp(iPv)$. Thus $Pv \equiv u$ modulo 2π in some neighborhood $V = \{r(x) \neq 0\}$ of p. We then may replace v by $v + 2\pi l$ if necessary, and replace V by a connected neighborhood of p so that $(Pv)_p \equiv u_p$, as required. □

Step 6. Now fix a small neighborhood W in M, s.t. $\langle e_k(x), J^k \theta(x) \rangle$ only depends on a finite jet, i.e. P is a differential operator in W. We apply a consequence of Peetre's theorem (see Lemma A.7 below) to conclude that P is of order 0, implying $J^k \theta = m\theta$. From injectivity of T_u on $C_c^k(M, \mathbb{C})$, $m = \pm 1$. This concludes the proof of Theorem 1.10. □

5 Various Generalizations and Applications to Fourier Transform

One of our main motivations for studying multiplicative transforms is that Fourier transform can be, in certain settings, characterized by the property that it carries product to convolution, as explained in the introduction. We already stated two corollaries of Theorem 1.10 following from this point of view, namely Theorems 1.3 and 1.5. Of course, in general the Fourier transform is not defined on all continuous functions, and even when it is, its image is usually not as well understood as in the case of Schwartz and compactly supported function. Let us first state a formal corollary of Theorems 1.6 and 1.8, in the case where $M = S^1$, that is, of 2π-periodic real-valued functions.

Denote the subclass of $\ell_2(\mathbb{Z})$ consisting of those sequences which are the coefficients of the fourier series of 2π-periodic real valued C^k functions by E_k. Fourier series is a bijection $\hat{} : C^k(S^1, \mathbb{R}) \to E_k$, given by

$$\hat{f}(n) = \int_0^{2\pi} f(x) e^{-2\pi i n x} dx.$$

It satisfies that $\widehat{f \cdot g} = \hat{f} * \hat{g}$ where here we use $*$ for two series to mean their convolution (or Cauchy product), that is,

$$\{a_n\} * \{b_n\} = \{c_n\} \qquad \text{where } c_k = \sum_{j \in \mathbb{Z}} a_j b_{k-j}.$$

Corollary 5.1. *Let $F : C^k(S^1, \mathbb{R}) \to E_k$ be a bijection which satisfies*

$$F(f \cdot g) = (Ff) * (Fg). \tag{3}$$

Then there exists some continuous $p : \mathbb{R}_+ \to \mathbb{R}_+$ and a C^k-diffeomorphism $u : S^1 \to S^1$ such that

$$Ff = \hat{g} \text{ with } g(x) = |f(u(x))|^{p(x)} \operatorname{sgn}(f(u(x))). \tag{4}$$

and if $k \geq 1$ then $p \equiv 1$, i.e. $g = f \circ u$.

Another interesting class to work with is $L_2(\mathbb{R}) \cap C^k(\mathbb{R}) \cap L_\infty(\mathbb{R})$, and although our theorems do not formally apply to this class, it is not hard to check that their corresponding variants are valid as well, as the interested reader may care to verify.

One may thus apply the Fourier transform $\mathcal{F}$ in this case, and conclude that the only bijections from this class to its images under $\mathcal{F}$ which map product to convolution are the standard Fourier transform composed with the additional terms coming from out main theorems (a diffeomorphism u only, if $k \geq 1$, and some power $p(x)$ and sign if $k = 0$). Clearly not every choice of u and p will give a bijection, but the statement is only on the existence of such functions. The "permissible" u and p are to be determined by the class in question.

We next briefly present an observation regarding possible generalizations of the main theorems. We state them in the simplest case of continuous functions on M.

Theorem 5.2. *Let M be a real topological manifold, let $V, W, U : C(M) \to C(M)$ satisfy that V is a bijection and that for all $f, g \in C(M)$*

$$V(f \cdot g) = (Wf) \cdot (Ug). \tag{5}$$

Then there exist continuous $a, b : M \to \mathbb{R}^+$, $p : M \to \mathbb{R}_+$, and a homeomorphism $u : M \to M$ such that

$$Vf(u(x)) = a(x)|f(x)|^{p(x)} \operatorname{sgn}(f(x)),$$

$$Wf(u(x)) = b(x)|f(x)|^{p(x)} \operatorname{sgn}(f(x)),$$

$$Uf(u(x)) = c(x)|f(x)|^{p(x)} \operatorname{sgn}(f(x)),$$

with $c(x) = \frac{a(x)}{b(x)}$.

Proof. Take $g \equiv 1$, and denote $Wg(x) = c(x) \in C(\mathbb{R})$ and $Ug(x) = d(x) \in C(\mathbb{R})$. Then $V(f) = U(f) \cdot c(x) = d(x) \cdot W(f)$. From bijectivity of V we see that c and d can never vanish, and that

$$V(fg) = \frac{1}{c(x)d(x)} V(f)V(g).$$

Define $Tf = Vf/(cd)$ we have

$$T(fg) = \frac{1}{cd} V(fg) = \frac{1}{cd} Vf \cdot \frac{1}{cd} Vg = Tf \cdot Tg.$$

Since V is a bijection, so is T, and we may apply Theorem 1.6 to conclude that there exists some continuous $p : M \to \mathbb{R}_+$, and a homeomorphism $u : M \to M$ such that

$$(Tf)(u(x)) = |f(x)|^{p(x)} \operatorname{sgn}(f(x)). \tag{6}$$

Therefore, letting $a(x) = c(u(x))d(u(x))$ and $b(x) = d(u(x))$, the proof is complete. $\qquad\square$

Remark 5.3. The version of the above theorem in which Fourier transform can be applied (say, the $L_2(\mathbb{R}) \cap C(\mathbb{R}) \cap L_\infty(\mathbb{R})$ case) has, as usual, a direct consequence regarding the exchange of product and convolution. Assume W, U, V satisfy

$$V(f \cdot g) = W(f) * U(g).$$

Apply $\mathcal{F}$ and get that $V' = \mathcal{F}V$, $W' = \mathcal{F}W$ and $U' = \mathcal{F}U$ satisfy

$$V'(f \cdot g) = W'(f) \cdot U'(g),$$

which is equation (5). Of course, one needs some assumption on the range of V to get that V' is a bijetcion, and apply a modification of Theorem 5.2. The conclusion would be of the form

$$(Vf)(u(x)) = \hat{a}(x) * \mathcal{F}\left(|f(x)|^{p(x)} \operatorname{sgn}(f(x))\right).$$

Appendix A: Additive Local Operators on Jet Bundles

A.1 A Review of Jet Bundles

We briefly outline the basic definitions concerning jet bundles. For more details, see [14].

Let M be a C^n manifold ($0 \le n \le \infty$), and fix a C^n smooth real vector bundle E over M. We denote by $\mathcal{O}_k(E, U)$ the C^k sections on $U \subset M$, and $\Gamma(\mathcal{O}^k(E))$

$= \mathcal{O}_k(E, M)$ the global C^k section of E. Also, $\Gamma^k_c(E)$ will denote the compactly supported global C^k sections of E. For our purposes, we really only need two cases: $E = M \times \mathbb{R}$ and $E = M \times \mathbb{C}$. The C^k sections are then simply C^k functions on M with values in $\mathbb{R}$ or $\mathbb{C}$. Let $J^k = J^k(E)$ (with $k \le n$) denote the associated k-jet bundle for which the fiber over $x \in M$ is denoted J^k_x. We give two equivalent definitions of jet bundles, and consider first the case $k < \infty$.

1. Consider the sheaf of modules $\mathcal{O}_k(E)$ of C^k sections of E over the sheaf of rings $\mathcal{O}_k(M) := \mathcal{O}_k(M \times \mathbb{R})$. The stalk $\mathcal{O}_{k,x}$ at x of $\mathcal{O}_k(M)$ is a local ring, with the maximal ideal $n_x = \{f_x(x) = 0\}$. We then define the space $J^k_x(E)$ of k-jets at x as the quotient $\mathcal{O}_{k,x}(E)/n_x^{k+1}\mathcal{O}_{k,x}(E)$. We denote the projection

$$J^k(x) : \mathcal{O}_{k,x}(E) \to J^k_x(E)$$

and for $k < n$ one also has the natural projections

$$P(k)(x) : J^{k+1}_x \to J^k_x$$

We then topologize the disjoint union $J^k = \bigcup_{x \in M} J^k_x$ by requiring all set-theoretic sections of the form $\sum_{j=1}^N a_j(x)J^k f_j(x)$ to be continuous, for $a_j \in C(M)$ and $f_j \in \Gamma(\mathcal{O}_k(E))$.

Informally, $J^k f(x)$ is the k-th Taylor polynomial of f, in a coordinate-free notation. For instance, $J^0(E) = E$, and $J^1(M \times \mathbb{R}) = (M \times \mathbb{R}) \oplus T^*M$.

2. It is well known that if $x \in M$ is a critical point for a function $f \in C^2(M)$, i.e. $d_x f = 0$, then the Hessian of f is well defined. For a general bundle E, the same happens already in the first order: the 1-jet of a germ $s_x \in \mathcal{O}_{k,x}(E)$ is well defined at $x \in M$ (it is an element of $T^*_x M \otimes E_x$), given that $s(x) = 0$. Proceeding by induction, one can show that the notion $J^k_x s_x = 0$ is well defined for all finite k, as well as for $k = \infty$. The space of k-jets at x is then defined as the quotient $\mathcal{O}_{k,x}(E)/\{s_x : J^{k+1}_x s_x = 0\}$.

For $k = \infty$ (assuming M is C^∞) only the second definition applies. Alternatively, one could generalize definition 1 as follows: define the space $J^\infty_x(E)$ of ∞-jets at x as the n_x-adic completion of $\mathcal{O}_{k,x}(E)$. This is just the inverse limit of J^k_x through the projections $P(k)$. Thus,

$$J^\infty_x(E) = \{(v_j)_{j=0}^\infty : P(j)(v_{j+1}) = v_j\}$$

The equivalence of the two definition is known as Borel's lemma. The set-theoretic bundle of jets $J = J^\infty$ will be the disjoint union of all fibers $J^\infty_x(E)$. We denote $Jf = J^\infty f$ the section of J defined by a C^∞ section f of E. For $0 \le j < k \le \infty$, denote $P(k, j) : J^k \to J^j$ the natural projection map, and $Q(j, k) = P(k, j)^* : (J^j)^* \to (J^k)^*$. Also, denote $P_k = P(\infty, k)$, $Q_k = P_k^*$. We will assume that some inner product is chosen on J^n for $n < \infty$.

In general, there is no canonic map $J^k_x \to J^n_x$ when $k < n$. However, when $k = 0$ and $E = M \times V$ for some vector space V, such a map does exist: We have $J_x^0 \simeq E_0 x \simeq V$ and all isomorphisms are canonical, so one can choose a constant function with the given value and consider its n-jet. Such jets are called constant jets.

We list a few more basic properties of jets, which will not be used in the paper. Assume that E is trivial over M, with fiber either $\mathbb{R}$ or $\mathbb{C}$. The fiber J_x^n with $0 \leq n \leq \infty$ is then naturally a local ring, with maximal ideal $m_x = \{J^n f : f(x) = 0\} = \operatorname{Ker} P(n, 0)$, $m_x^k = \operatorname{Ker} P(n, k)$. Thus we get an m_x-filtration of J_x^n. The induced m_x-adic topology on J_x^n is Hausdorff by definition, and complete.

A.2 S^1-Bundles

We would like to discuss separately the case $E = M \times S^1$, where S^1 is considered as a Lie group. We make the following definition: For $f = e^{i\theta} \in C^k(M, S^1)$, $J^k f := J^k \theta$. Note that $J^0 f$ is only defined up to 2π. Further, note that a functional $c : J_x^k(M \times S^1) \to \mathbb{R}$ is induced from a functional $\tilde{c} : J_x^k(M \times \mathbb{R}) \to \mathbb{R}$ which vanishes on constant jets. Denote the bundle of such functionals by $(J^k)^*(M \times S^1, \mathbb{R})$. We also define a linear character $\chi : J_x^k(M \times S^1) \to S^1$ by $\chi = \exp(ic)$ where $c : J_x^k(M \times \mathbb{R}) \to \mathbb{R}$ is a linear functional s.t. $c(v) = mv$ for constant jets v, for some fixed $m \in \mathbb{Z}$. The bundle of linear characters χ (as well as that of the corresponding functionals c) is denoted $(J^k)^*(M \times S^1, S^1)$.

A.3 Main Lemmas

Particular cases of the following Lemma have appeared before, i.e. in [5].

Lemma A.1. *Fix a C^s manifold M, $0 \leq s \leq \infty$, and a C^s real vector bundle E over M. Assume $B : J^s(E) \to \mathbb{R}$ satisfies the following conditions:*

(1) $B(x, u + v) = B(x, u) + B(x, v)$ for all $x \in M$, $u, v \in J_x^s$.
(2) For all $f \in \Gamma_c^s(E)$, one has $B(J^s f) \in C(M)$.
    Then B is linear in every fiber.

Proof. Let $A = \{x \in M \,|\, B(x, \cdot) : J_x^s \to \mathbb{R} \text{ is non-linear}\}$. First we prove that A has no accumulation points in M.

Assume the contrary, i.e. $A \ni x_k \to x_\infty$. We can assume that $x_k \neq x_l$ for $1 \leq k < l \leq \infty$. By the assumption, there exists a sequence $v_k \in J_{x_k}^s$ such that the functions $B(x_k, t v_k)$ are additive and non-linear in t. We will construct in A.4 a sequence of sections $f_k \in \Gamma^s(E)$ such that:

(a) f_k is supported in a small neighborhood of x_k, such that $\operatorname{supp}(f_k) \cap \operatorname{supp}(f_l) = \emptyset$ for $k \neq l$, and $x_\infty \notin \operatorname{supp}(f_k)$, for all $k < \infty$.

(b) $J^s f_k(x_k) = \epsilon_k v_k$ for some $0 < \epsilon_k < 1$.

(c) $|J^{\min(k,s)} f_k(x)| < 2^{-k}$ for all $x \in M$ and $k < \infty$.

Given that, choose $0 < t_k \to 0$ such that $|B(x_k, t_k \epsilon_k v_k)| > 1$, using that non-linear additive functions are not locally bounded, and consider $f = \sum_{k=1}^{\infty} t_k f_k$. By condition (c), $f \in \Gamma^s(E)$, and by (a), (b), $J^s f(x_k) = t_k J^s f_k(x_k) = t_k \epsilon_k v_k$ and $J^s f(x_\infty) = 0$. Thus $B(x_k, t_k J^s f(x_k)) \to B(x_\infty, J^s f(x_\infty))$ by condition (2) on B, but $|B(x_k, t_k J^s f(x_k))| > 1$ while $B(x_\infty, 0) = 0$, a contradiction.

Next, we show that A is empty. Indeed, take any sequence $x_k \notin A$ converging to arbitrary $x_\infty \in M$. Fix any $v \in J^s_{x_\infty}$, and choose $f \in \Gamma^s(E)$ with $J^s f(x_\infty) = v$. Then for all $t \in \mathbb{R}$, $B(x_k, t J^s f(x_k)) \to B(x_\infty, t J^s f(x_\infty))$, i.e. $B(x_\infty, tv)$ is a pointwise limit of continuous functions (of $t \in \mathbb{R}$), thus a measurable function. Next use a theorem of Banach and Sierpinski, see [4, 15] which states that if $B(x_\infty, tv)$ is a measurable function of t and $B(x_\infty, \cdot)$ is additive, then it must be linear. $\square$

Lemma A.2. *Under the conditions of the lemma above*

(1) *If $s < \infty$ then B is a continuous section of $(J^s)^*$.*

(2) *If $s = \infty$ then B is locally finite dimensional and continuous in the following sense: Denoting $F_n = \{x : B(x, \cdot) \in Image(Q_n)_x\}$, one has $F_n \subset F_{n+1}$ are closed sets, and for all $K \subset M$ compact, there exists an n such that $K \subset F_n$ and $B = Q_n(c_n)$ where c_n is a continuous section of $(J^n)^*$ over K.*

Proof. (1) Fix some coordinate chart $U \subset M$, with a trivialization of E over M. Take $c_s(x) \in (J^s_x)^*$ s.t. $B(x, v) = c_s(x)(P_s(v)) = \sum_{|\alpha| \leq s} a_{\alpha, \beta}(x) v_\beta^{(\alpha)}$ (here α parametrizes the order of the derivative, and β the coordinate in E_x) for $x \in U$, $v \in J^s_x$. Take $x_k \to x_\infty$ within U, with x_∞ at the origin. We claim that $a_{\alpha, \beta}(x_k) \to a_{\alpha, \beta}(x_\infty)$: this is straightforward by condition (2) if one considers polynomial sections. Thus c_s is continuous.

(2) *Step 1.* Let $C = \{x \in M | B(x, \cdot) \notin Image(Q_n)$ for all $n < \infty\}$. We prove that C has no accumulation points. Like before, assume the contrary, i.e. $C \ni x_k \to x_\infty$, and all x_k are different for $k \leq \infty$. Choose a sequence $v_k \in J^\infty_{x_k}$ with $P_k(v_k) = 0$ and $B(x_k, v_k) = 1$. Again, we construct in A.4 a sequence of functions $g_k \in \Gamma^\infty(E)$ such that:

    (a) g_k is supported in a small neighborhood of x_k, such that supp$(g_k) \cap$ supp$(g_l) = \emptyset$ for $k \neq l$, and $x_\infty \notin$ supp(g_k), for all k.

    (b) $J g_k(x_k) = v_k$.

    (c) $|J^k g_k(x)| < 2^{-(k-1)}$ for all $x \in M$.

Now consider $g = \sum_{k=1}^{\infty} g_k$. By condition (c), $g \in \Gamma^\infty(E)$, $J g(x_k) = J g_k(x_k) = v_k$ and $J g(x_\infty) = 0$. Thus $1 = B(x_k, J g(x_k)) \to B(x_\infty, J g(x_\infty)) = 0$ by condition (2) on B, a contradiction.

Step 2. Fix some coordinate chart $U \subset M$, with a trivialization of E over M. Then for $x \in F_n$ one has $c_n(x) \in (J^n_x)^*$ s.t. $B(x, v) = c_n(x)(P_n(v)) = \sum_{|\alpha| \leq n} a_{\alpha, \beta}(x) v_\beta^{(\alpha)}$ (here α parametrizes the order of the derivative, and β the coordinate in E_x) for $x \in F_n \cap U$, $v \in J^\infty_x$. Take $x_k \to x_\infty$ within

U, with $x_k \in F_n$. We will show that $a_{\alpha,\beta}(x_k)$ converges as $k \to \infty$: this is straightforward by condition (2) if one considers polynomial sections. And so c_n extends to a continuous section of $(J^n)^*$ on the closure of F_n. This implies that for all $v \in J_x^\infty$ one can choose any f with $Jf(x_\infty) = v$, and then by condition (2)

$$B(x_\infty, v) = B(x_\infty, Jf(x_\infty)) = \lim_{k \to \infty} B(x_k, Jf(x_k))$$

$$= \lim_{k \to \infty} c_n(x_k)(J^n f(x_k))$$

$$= c_n(x_\infty)(J^n f(x_\infty)) = c_n(x_\infty)(v)$$

i.e. $x_\infty \in F_n$. Thus, F_n is closed.

Step 3. We now show that C is in fact empty. Assume otherwise, and take $M \setminus C \ni x_k \to x_\infty$, $x_\infty \in C$. Since F_k are all closed sets, we may assume $x_k \in M \setminus F_k$. We may further assume that all x_k are distinct. Now we simply repeat the construction of step 1: Choose a sequence $v_k \in J_{x_k}^\infty$ with $P_k(v_k) = 0$ and $B(x_k, v_k) = 1$. Choose the functions g_k, g as before. Thus $Jg(x_k) = v_k$ and $Jg(x_\infty) = 0$, so $1 = B(x_k, Jg(x_k)) \to B(x_\infty, Jg(x_\infty)) = 0$ by condition (2) on B, a contradiction.

Step 4. Finally, if $K \subset M$ is compact, and $F_k \cap K$ is a strictly increasing sequence of subsets, one can choose a converging sequence $K \setminus F_k \ni x_k \to x_\infty$, which is impossible by Step 3. Thus $K \subset F_n$ for large n. $\square$

A.4 The Construction of the Function Families

We finally construct the functions f_k, g_k which we used in the lemmas above, for some fixed k. We do this by mimicking the proof of Borel's lemma. Since the construction is local, we can assume E is trivial. For simplicity, we further assume $M = \mathbb{R}$, and $E = M \times \mathbb{R}$. We will denote $v = v_k$. For convenience, assume $x_k = 0$. Take

$$\delta = \min\left(\frac{1}{2}\min\{|x_m - x_k| : 1 \le m \le \infty, m \ne k.\}\right)$$

Fix a smooth, non-negative function $h \in C^\infty(\mathbb{R})$ such that $h(x) = 0$ for $|x| > \delta$, and $h(x) = 1$ for $|x| \le \delta/2$. Let $\psi_n(x) = x^n h(x)$, then $\psi_n^{(j)}(0) = n!\delta_n^j$.

Define

$$\lambda_n = \max\{1, |v^{(j)}|, \sup|\psi_n|, \sup|\psi_n'|, \ldots, \sup|\psi_n^{(n)}|\}$$

$$\phi_n = \frac{v^{(n)}}{n!\lambda_n^n}\psi_n(\lambda_n x)$$

Then

$$\phi_n^{(j)} = \frac{v^{(n)}}{n!\lambda_n^{n-j}}\psi_n^{(j)}(\lambda_n x)$$

and $\phi_n^{(j)}(0) = v^{(j)}\delta_n^j$. Take $H = \sum_{n=0}^{\infty} \phi_n(x)$. This is a C^{∞} function, since

$$\sum |\phi_n^{(j)}| \le \sum_{n=0}^{j+1} \frac{|v^{(n)}|}{n! \lambda_n^{n-j}} \sup |\psi_n^{(j)}| + \sum_{n=j+2}^{\infty} \frac{1}{n!} \frac{1}{\lambda_n^{n-j-2}} \frac{|v^{(n)}|}{\lambda_n} \frac{\sup |\psi_n^{(j)}|}{\lambda_n}$$

and $JH(0) = v$.

Now we consider our two cases separately: For an arbitrary $v_k \in \mathbb{R}^{\infty}$, take

$$\epsilon_k = \frac{1}{2^k(1 + \max\{\sup |H|, \sup |H'|, \dots, \sup |H^{(k)}|\})}$$

and $f_k = \epsilon_k H$ satisfies (a)–(c).

For $v_k \in \text{Ker}(P_k)$, we have for all $j < k$

$$|H^{(j)}| \le \sum_{n=k+1}^{\infty} \frac{1}{n!} \le \frac{1}{2^{k-1}}$$

so $g_k = H$ satisfies (a)-(c)   □

A.5   Replacing $\mathbb{R}$ with S^1

Next we state without proof several variants of the lemmas above. To see how the proofs above adapt to those S^1 cases, see [3].

Lemma A.3. *Fix a C^{∞} manifold M, and $E = M \times \mathbb{R}$. Assume $B : J^{\infty} \to S^1$ (here S^1 is a Lie group) satisfies the following conditions:*

(1) *$B(x, u + v) = B(x, u)B(x, v)$ for all $x \in M$, $u, v \in J_x^{\infty}$.*
(2) *For all $f \in C_c^{\infty}(M)$, one has $B(J^{\infty}f) \in C(M)$.*
     Then $B(x, u) = \exp(i \langle c(x), u \rangle)$ where $c \in J^*$. Moreover, $M = \bigcup_{n=0}^{\infty} F_n$ where $F_n = \{x : c(x) \in Image(Q_n)_x\}$. The sets $F_n \subset F_{n+1}$ are closed, and for all $K \subset M$ compact, one has $K \subset F_n$ for some n, and B is a continuous section of $(J^n)^*$ over K.

Lemma A.4. *Fix a C^{∞} manifold M, and $E = M \times S^1$. Assume $B : J^{\infty}(E) \to \mathbb{R}$ satisfies the following conditions:*

(1) *$B(x, uv) = B(x, u) + B(x, v)$ for all $x \in M$, $u, v \in J_x^{\infty}$.*
(2) *For all $f \in C_c^{\infty}(M, S^1)$, one has $B(J^{\infty}f) \in C(M)$.*
     Then $B(x, v) = \langle c(x), v \rangle$ where $c \in J^*$ vanishes on constant jets. Moreover, $M = \bigcup_{n=0}^{\infty} F_n$ where $F_n = \{x : c(x) \in Image(Q_n)_x\}$. The sets $F_n \subset F_{n+1}$ are closed, and for all $K \subset M$ compact, one has $K \subset F_n$ for some n, and B is a continuous section of $(J^n)^*$ over K.

Lemma A.5. *Fix a C^∞ manifold M, and $E = M \times S^1$. Assume $B : J^\infty(E) \to S^1$ satisfies the following conditions:*

(1) $B(x, uv) = B(x, u)B(x, v)$ for all $x \in M$, $u, v \in J_x^\infty$.
(2) For all $f \in C_c^\infty(M, S^1)$, one has $B(J^\infty f) \in C(M, S^1)$.
  Then $B(x, v) = \exp(i \langle c(x), v \rangle)$ where $c \in J^*$. Moreover, $M = \bigcup_{n=0}^\infty F_n$ where $F_n = \{x : c(x) \in Image(Q_n)_x\}$. The sets $F_n \subset F_{n+1}$ are closed, and for all $K \subset M$ compact, one has $K \subset F_n$ for some n, and B is a continuous section of $(J^n)^*$ over K. Also, $c(x)(v) = mv$ for constant jets v, with $m \in \mathbb{Z}$.

Remark A.6. One also has the C^s simpler versions of those lemmas, with $s < \infty$.

A.6 A Consequence of Peetre's Theorem

In the following, M is a real smooth manifold, and E, F are smooth vector bundles over M. First, recall Peetre's Theorem [12]:

Theorem. *Let $Q : \Gamma^\infty(E) \to \Gamma^\infty(F)$ be a linear operator which decreases support: $\mathrm{supp}(Qs) \subset \mathrm{supp}(s)$. Then for every $x \in M$ there is an integer $k \geq 0$ and an open neighborhood $U \ni x$ s.t. P restricts to a differential operator of order k on $\Gamma^\infty(U, E)$.*

Lemma A.7. *Let $P : \Gamma^\infty(E) \to \Gamma^\infty(E)$ be an invertible differential operator, which for all $x \in M$ induces an isomorphism on the stalk $\mathcal{O}_x(E)$. Then P is of order 0.*

Proof. Denote $Q = P^{-1}$. We claim that Q is a local operator (i.e. $Qf(x)$ only depends on the germ f_x of $f \in \Gamma^\infty(E)$). Indeed, assume $f_x = 0$ and take $g = Qf$. Then $(Pg)_x = f_x = 0$, implying $g_x = 0$, as required. In particular, Q does not increase supports: $\mathrm{supp}(Qf) \subset \mathrm{supp}(f)$. By Peetre's theorem, Q is locally a differential operator. Thus in small neighborhoods $U \subset M$, P and Q are two differential operators that are inverse to each other, and we conclude they have order 0.   □

Appendix B: Two Characterizations of Continuous Functions

In this appendix we give two lemmas in the same spirit, which may be of use in other settings. One concerns bijections of open sets preserving intersection. The other is almost trivial but nevertheless may be useful.

Denote by $U(\mathbb{R})$ the open subsets of $\mathbb{R}$.

Lemma B.1. *Assume $F : U(\mathbb{R}) \to U(\mathbb{R})$ is a bijection which satisfies*

$$F(U_1 \cap U_2) = F(U_1) \cap F(U_2).$$

Then F is given by an open point map $u : \mathbb{R} \to \mathbb{R}$,

$$F(U) = \{u(x) : x \in U\}.$$

Proof. First we claim that F is an order isomorphism. Indeed, (using injectivity)

$$U_1 \subset U_2 \Leftrightarrow U_1 \cap U_2 = U_1 \Leftrightarrow F(U_1) \cap F(U_2) = F(U_1) \Leftrightarrow F(U_1) \subset F(U_2).$$

Next, note that this implies that

$$F(U_1 \cup U_2) = F(U_1) \cup F(U_2).$$

Indeed, denote (using surjectivity) $F(U_3) = F(U_1) \cup F(U_2)$, then $F(U_1) \subset F(U_1 \cup U_2)$ and $F(U_2) \subset F(U_1 \cup U_2)$ and thus $F(U_3) \subset F(U_1 \cup U_2)$ and so $U_3 \subset U_1 \cup U_2$. On the other hand $U_1 \subset U_3$ and $U_2 \subset U_3$, so that $U_1 \cup U_2 \subset U_3$. We get equality, thus F preserves union too.

Next we claim that the set of special open sets $\{A_x : x \in \mathbb{R}\}$ given by $A_x = \{y \in \mathbb{R} : y \neq x\}$ is invariant under F. Indeed, start by noticing that $F(\mathbb{R}) = \mathbb{R}$, as it satisfies that $\mathbb{R} \cap A = A$ for every A. (Also that $\mathbb{R} \cup B = \mathbb{R}$ for every B) and these properties are preserved under F and unique to this subset.

Next, notice that a set A_x satisfies $A_x \cup B = A_x$ or $\mathbb{R}$ for every B. Whereas every other open set A has at least two points which are not elements in it, and thus one may construct at least three different sets as combinations $A \cup B$ for various open B's. Therefore, $F(A_x) = A_{u(x)}$. Notice that F^{-1} satisfies the same conditions as F and thus u is a bijection.

Finally, for a set A we know that for any $x \notin A$

$$F(A) = F(A \cap A_x) = F(A) \cap A_{u(x)}$$

and thus $u(x) \notin F(A)$, so that $F(A) \subset u(A)$. However, using the same argument for u^{-1} we get that $F^{-1}(A) \subset u^{-1}(A)$, and applying this to the set $F(A)$ we get $A \subset u^{-1}(F(A))$ which means $u(A) \subset F(A)$ and the proof is complete.   □

Remark B.2. Clearly u is a continuous bijection. It is also clear that the setting of $\mathbb{R}$ can be vastly generalized.

Lemma B.3. *Let $f : \mathbb{R}^n \times \mathbb{R} \to \mathbb{R}$ be a function such that for any $g \in C(\mathbb{R}^n)$, also $f(x, g(x)) \in C(\mathbb{R}^n)$. Then f is continuous.*

Proof. Without loss of generality it suffices to show that f is continuous at $(0, 0) \in \mathbb{R}^n \times \mathbb{R}$. Choose $\mathbb{R}^n \times \mathbb{R} \ni (x_j, y_j) \to 0$, and show that $f(x_j, y_j) \to f(0, 0)$. Assume otherwise, namely, that one can choose such a sequence and $\epsilon > 0$ such that $|f(x_j, y_j) - f(0, 0)| > \epsilon$. Since by assumption $f(\cdot, y_j)$ is continuous for all j, we may choose $x'_j \in \mathbb{R}^n$ such that $|x'_j - x_j| < \frac{1}{j}$ and $|f(x'_j, y_j) - f(x_j, y_j)| < \frac{\epsilon}{2}$. Moreover, we can do it in such a way that all x'_j are different. Then one can choose a continuous $g : \mathbb{R}^n \to \mathbb{R}$ such that $g(x'_j) = y_j$. Thus $f(x'_j, y_j) = f(x'_j, g(x'_j)) \to f(0, 0)$. But also $|f(x'_j, y_j) - f(0, 0)| > \frac{\epsilon}{2}$, a contradiction.   □

References

1. S. Alesker, S. Artstein-Avidan, V. Milman, A characterization of the Fourier transform and related topics (A special volume in honor of Prof. V.l Havin). Am. Math. Soc. Transl. (2) **226**, 11–26 (2009)
2. S. Artstein-Avidan, H. König, V. Milman, The chain rule as a functional equation. J. Funct. Anal. **259**, 2999–3024 (2010)
3. S. Alesker, S. Artstein-Avidan, D. Faifman, V. Milman, A characterization of product preserving maps with applications to a characterization of the Fourier transform. Illinois J. Math. **54**(3), 1115–1132 (2010)
4. S. Banach, Sur l'équation fonctionelle $f(x + y) = f(x) + f(y)$. Fund. Math. **1**, 123–124 (1920)
5. H. König, V. Milman, Characterizing the derivative and the entropy function by the Leibniz rule. With appendix by D. Faifman. J. Funct. Anal. **261**(5), 1325–1344 (2011)
6. I.M. Gel'fand, G.E. Shilov, *Generalized Functions*, vol. 1. Properties and Operations (Academic, New York, 1964 [1958])
7. H. König, V. Milman, A functional equation characterizing the second derivative. J. Funct. Anal. **261**(4), 876–896 (2011)
8. H. König, V. Milman, The chain rule functional equation on $\mathbb{R}^n$. J. Funct. Anal. **261**(4), 861–875 (2011)
9. A.N. Milgram, Multiplicative semigroups of continuous functions. Duke Math. J. **16**, 377–383 (1949)
10. J. Mrcun, On isomorphisms of algebras of smooth functions. Proc. Am. Math. Soc. **133**, 3109–3113 (2005)
11. J. Mrcun, P. Semrl, Multiplicative bijections between algebras of differentiable functions. Annales Academae Scientiarum Fennicae Mathematica **32**, 471–480 (2007)
12. J. Peetre, Une caractérisation abstraite des opérateurs différentiels. Math. Scand. **7**, 211–218 (1959)
13. F.C. Sánchez, J.C. Sánchez, Some preserver problems on algebras of smooth functions. Ark. Mat. **48**, 289–300 (2010)
14. D.J. Saunders, *The Geometry of Jet Bundles* (Cambridge University Press, London, 1989)
15. W. Sierpinski, Sur l'équation fonctionelle $f(x + y) = f(x) + f(y)$. Fundamenta Math. **1**, 116–122 (1920)

Order Isomorphisms on Convex Functions in Windows

Shiri Artstein-Avidan, Dan Florentin, and Vitali Milman

Abstract In this paper we give a characterization of all order isomorphisms on some classes of convex functions. We deal with the class $Cvx(K)$ consisting of lower-semi-continuous convex functions defined on a convex set K, and its subclass $Cvx_0(K)$ of non negative functions attaining the value zero at the origin. We show that any order isomorphism on these classes must be induced by a point map on the epi-graphs of the functions, and determine the exact form of this map. To this end we study convexity preserving maps on subsets of $\mathbb{R}^n$, and also in this area we have some new interpretations, and proofs.

1 Introduction

In recent years, a big research project initiated by the first and third named authors has been carried out, in which a characterization of various transforms by their most simple and basic properties has been found. Among these are the Fourier transform (see [1,2]), the Legendre transform (see [6]), polarity for convex sets (see [5,10,16]), the derivative (see [9]) and various other transforms. This paper is part of this effort.

One example for such characterization is the understanding of bijective order isomorphisms for certain partially ordered sets (see Sect. 3 for definitions and details). In this category one includes the Legendre transform, which is, up to linear terms, the unique bijective order reversing map on the set of all convex (lower-semi-continuous) functions on $\mathbb{R}^n$, with respect to the pointwise order. In this paper we

S. Artstein-Avidan (✉) · D. Florentin · V. Milman
Sackler Faculty of Exact Sciences, Department of Mathematics, Tel Aviv University,
Ramat Aviv, 69978 Tel Aviv, Israel
e-mail: shiri@post.tau.ac.il; danflorentin@gmail.com; milman@post.tau.ac.il

B. Klartag et al. (eds.), *Geometric Aspects of Functional Analysis*, Lecture Notes
in Mathematics 2050, DOI 10.1007/978-3-642-29849-3_4,
© Springer-Verlag Berlin Heidelberg 2012

discuss analogous results where the convex functions are defined on a "window", that is on a convex set $K \subseteq \mathbb{R}^n$, and several other variants as well.

One main tool in the proof of such results is the fundamental theorem of affine geometry, which states that an injective mapping on $\mathbb{R}^n$ which sends lines to lines, and whose image is not contained in a line, must be affine linear. When working with "windows", one immediately encounters a need for a similar theorem for maps defined on a subset of $\mathbb{R}^n$. Such theorems exist in the literature, and a mapping which maps intervals to intervals on a subset of $\mathbb{R}^n$ must be of a very specific form, which we call here "fractional linear", discussed in Sect. 2. A remark about this name is in need: the mappings are of the form: $\frac{Ax+b}{\langle c,x \rangle + d}$ for $A \in L_n(\mathbb{R}), b, c \in \mathbb{R}^n$ and $d \in \mathbb{R}$, with some extra restriction. In the literature, the name "fractional linear maps" sometimes refers to Möbius transformations, which is not the case here (note that a Möbius transformation on a subset of $\mathbb{C} = \mathbb{R}^2$ does not preserve intervals, but these notions are indeed connected—see Example 2.7). One could name them "permissable projective transformation" but we prefer to think about them exclusively in $\mathbb{R}^n$ and not on the projective space. Another option was to call them "convexity preserving maps", which describes their action rather than their functional form, but this hides the fact that they are of a very simple form.

The classification of interval preserving transforms of a convex subset of $\mathbb{R}^n$ is known (see [15]). However, since this is an essential part for our study of the order isomorphisms on functions in windows, we dedicate the whole of Sect. 2 to this topic. We also provide some new insights and results, and give a seemingly new geometric proof of the main fact which is that such maps are fractional linear. Some parts of this section are elementary and may be known to the reader, but we include them as they too serve as intuition for the way these maps behave.

In Sects. 3–5 we turn to the main topic of this paper, namely characterization of order preserving (and reversing) isomorphisms on classes of convex functions defined on windows. Let $T \subseteq K$ be two closed convex sets. The class of all lower-semi-continuous convex functions $\{f : K \to \mathbb{R} \cup \{\infty\}\}$ is denoted $Cvx(K)$, and its subclass of non negative functions satisfying $f(T) = 0$ is denoted $Cvx_T(K)$. In Sect. 3 we give background on general order isomorphisms. In Sects. 4 and 5 we deal with characterization of such transforms on $Cvx(K)$ and $Cvx_0(K)$ respectively. In both cases, the proof is based on finding a subset of convex functions which are extremal, in some sense. The extremal elements are relatively simple functions, which can be described by a point in $\mathbb{R}^{n+1}$. We show that an order isomorphism is determined by its action on the extremal family, and that its restriction to this family must be a bijection. Therefore the transform induces a bijective point map on a subset of $\mathbb{R}^{n+1}$. By applying our uniqueness theorem, we show that this point map must be fractional linear. We then discuss some generalizations of these theorems to other classes of non-negative convex functions.

Acknowledgements The authors would like to thank Leonid Polterovich for helpful references and comments. They also wish to thank the anonymous referee for useful remarks. Supported in part by Israel Science Foundation: first and second named authors by grant No. 865/07, second and third named authors by grant No. 491/04. All authors were partially supported by BSF grant No. 2006079.

2 Interval Preserving Maps

We start this section with a simple but curious fact which is stated again and proved as Theorem 2.27 below. This fact demonstrates the idea that fractional linear maps should be a key ingredient in convexity theory. Consider a convex body K (actually any closed convex set will do) which includes the origin and is included in the half-space $H_1 = \{x_1 < 1\} \subseteq \mathbb{R}^n$. One may take its polar, defined by

$$K^\circ = \{y : \sup_{x \in K}\langle x, y\rangle \leq 1\}.$$

Since polarity reverses the partial order of inclusion on closed convex sets including the origin, K° includes the set $[0, e_1]$ which is the polar of H_1. Translate it by $-e_1$ (so now it includes $[-e_1, 0]$, and in particular includes the origin) and then take its polar again. In other words, we constructed a map mapping certain convex sets (which include 0 and are included in the half-space H_1) to convex sets, given by

$$F(K) = (K^\circ - e_1)^\circ.$$

Clearly this mapping is order preserving.

While polarity is a "global" operation, it turns out that this mapping is actually induced by a point map on H_1, $\tilde{F} : H_1 \to \mathbb{R}^n$, which preserves intervals, and can be explicitly written as $\tilde{F}(x) = \frac{x}{1-x_1}$. (This is a simple calculation, and for completeness we provide it in the proof of Theorem 2.27 below.)

This map is a special case of the so called "fractional linear maps" which are the main topic of this section.

2.1 Definition and Simple Observations

Definition 2.1. Let $D \subseteq \mathbb{R}^n$. A function $f : D \to \mathbb{R}^n$ is called an interval preserving map, if f maps every interval $[x, y] \subseteq D$ to an interval $[z, w]$.

Lemma 2.2. *Let $D \subseteq \mathbb{R}^n$, and let $f : D \to \mathbb{R}^n$ be an injective interval preserving map. Then for every $x, y \in D$ with $[x, y] \subseteq D$ we have that $f([x, y]) = [f(x), f(y)]$.*

Proof. Indeed, assume that $f([x, y]) = [w, z]$, and that, say, $f(y) \in (w, z)$. Pick a point $b \in (w, f(y))$, and a point $b' \in (f(y), z)$. Then for some $a, a' \in (x, y)$, $b = f(a)$ and $b' = f(a')$. Consider $f([a, a'])$. It is an interval that includes the points b and b', and therefore it includes $f(y)$, whereas $y \notin [a, a']$—in contradiction to the injectivity of f. ☐

Lemma 2.3. *Let $D \subseteq \mathbb{R}^n$, and let $f : D \to \mathbb{R}^n$ be an injective interval preserving map. Then the inverse $f^{-1} : f(D) \to D$ is interval preserving.*

Proof. Let $I = [f(a), f(b)]$ be an interval in the image, and $f(c) \in I$. From Lemma 2.2 $f([a, b]) = I$, so by injectivity, $c \in [a, b]$.   □

Remark 2.4. Clearly, an interval preserving map f must be convexity preserving, i.e. f must map every convex set K to a convex set $f(K)$. We will actually need the opposite direction, given in the following lemma.

Lemma 2.5. *Let $D \subseteq \mathbb{R}^n$ be a convex set, and let $f : D \to \mathbb{R}^n$ be an injective interval preserving map. Then the inverse image of a convex set in $f(D)$ is convex.*

Proof. Let $K \subseteq f(D)$ be a convex set, and let $x, y \in f^{-1}(K)$. We wish to show that $[x, y] \subseteq f^{-1}(K)$. The interval $[x, y]$ is contained in the convex domain D, and by Lemma 2.2 we know that

$$f([x, y]) = [f(x), f(y)] \subseteq K,$$

where the last inclusion is due to convexity of K. This implies $[x, y] \subseteq f^{-1}(K)$.

□

Lemma 2.6. *Let $D \subseteq \mathbb{R}^n$ be an open domain and $f : D \to \mathbb{R}^n$ an injective interval preserving map, then f is continuous.*

Proof. Let us prove that f is continuous at a point $x \in D$. We may assume that D is convex (restrict f to an open convex neighborhood of x). Let $y = f(x) \in f(D)$ and B_y an open ball containing y. If $x \in int(f^{-1}(B_y))$ we are finished (we have a neighborhood of x that is mapped into B_y). We claim this must be the case. Indeed, assume otherwise, then x is on the boundary of the set $f^{-1}(B_y)$, which by Lemma 2.5, is convex. Let $x_{out} \in D$ such that $[x, x_{out}] \cap f^{-1}(B_y) = \{x\}$. Then $f([x, x_{out}])$ is an interval $I \subseteq f(D)$ such that $I \cap B_y = \{y\}$, but no such interval exists, since B_y is open.   □

2.2   Fractional Linear Maps

Clearly, linear maps are interval preserving. It turns out that when the domain of the map is contained in a half-space of $\mathbb{R}^n$, there is a larger family of (injective) interval preserving maps. Indeed, fix a scalar product $\langle \cdot, \cdot \rangle$ on $\mathbb{R}^n$, let $A \in L_n(\mathbb{R})$ be a linear map, $b, c \in \mathbb{R}^n$ two vectors and $d \in \mathbb{R}$ some constant, then the map

$$v \mapsto \frac{1}{\langle c, v \rangle + d}(Av + b)$$

is defined on the open half space $\langle c, v \rangle < -d$ and is interval preserving. One can check interval preservation directly, or deduce it from the projective description in Sect. 2.2.1, as well as an injectivity argument. A necessary and sufficient condition for this map to be injective is that the associated matrix $\hat{A}$ (defined below) is invertible. The matrix A itself need not be invertible, for an example see Remark 2.9. We call these maps "fractional linear maps", see the introduction for a remark about this name.

2.2.1 Projective Description

Consider the projective space $\mathbb{R}P^n = P(\mathbb{R}^{n+1})$, the set of 1-dimensional subspaces of $\mathbb{R}^{n+1}$. It is easily seen that

$$\mathbb{R}P^n \simeq \mathbb{R}^n \cup \mathbb{R}P^{n-1},$$

where one can geometrically think of $\mathbb{R}^n$ as $\mathbb{R}^n \times \{1\} \subseteq \mathbb{R}^{n+1}$, so that each line which is not on the hyperplane $e_{n+1}^{\perp}(\approx \mathbb{R}^n)$, intersects the shifted copy of $\mathbb{R}^n$ at exactly one point. The lines which lie on $e_{n+1}^{\perp}$ are thus lines in an n-dimensional subspace, and are identified with $\mathbb{R}P^{n-1}$.

Any invertible linear transformation on $\mathbb{R}^{n+1}$ induces a transformation on $\mathbb{R}P^n$, mapping lines to lines. Thus it induces in particular a map $F : \mathbb{R}^n \to \mathbb{R}^n \cup \mathbb{R}P^{n-1}$. The part mapped to $\mathbb{R}P^{n-1}$ is either empty—in which case the induced transformation on $\mathbb{R}^n$ is linear, or an affine hyperplane H—in which case the induced transformation $F : \mathbb{R}^n \setminus H \to \mathbb{R}^n$ is fractional linear.

Indeed, if the matrix associated with the original transformation in $\mathcal{L}(\mathbb{R}^{n+1}, \mathbb{R}^{n+1})$ is $\hat{A} \in GL_{n+1}$, then the hyperplane in $\mathbb{R}^n$ mapped to $e_{n+1}^{\perp}$ is simply $\{x \in \mathbb{R}^n : (\hat{A}(x, 1))_{n+1} = 0\}$. If $\hat{A}$ is given by

$$\hat{A} = \begin{pmatrix} A & b \\ c^T & d \end{pmatrix}$$

with $A \in M_{n \times n}$, $b, c \in \mathbb{R}^n$ and $d \in \mathbb{R}$, then the set $\{x : \langle c, x \rangle + d \neq 0\} \subset \mathbb{R}^n$ is exactly the pre image of $\mathbb{R}P^{n-1}$ under F. It is an $n-1$ dimensional subspace if $c \neq 0$, and empty if $c = 0$ ($\hat{A} \in GL_{n+1}$ implies $d \neq 0$ in that case).

Pick any vector $x \in \mathbb{R}^n$ which is not in this hyperplane, then it is mapped to $y = \hat{A}(x, 1) \in \mathbb{R}^{n+1}$ which has a non-vanishing $(n+1)$th coordinate, $y_{n+1} = \langle c, x \rangle + d$. Normalize $y \to y/y_{n+1}$, so that the last coordinate is 1, and consider only the first n coordinates of this vector (we denote the projection to the first n coordinates by P_n). Thus, under the above map, x is mapped to

$$F(x) = P_n\left(\hat{A}(x, 1)/(\hat{A}(x, 1))_{n+1}\right) = \frac{Ax + b}{\langle c, x \rangle + d}. \tag{1}$$

Denote the domain of the map by $D \subseteq \mathbb{R}^n$. Clearly, if $D = \mathbb{R}^n$, the condition $\phi(x) = (\hat{A}(x, 1))_{n+1} \neq 0$ for all $x \in D$ implies that this (affine linear) function $\phi(x)$ is constant, which means that our induced map is affine linear. However, when D is contained in a half space (for example, if D is a convex set strictly contained in $\mathbb{R}^n$), there are many choices of $\hat{A}$ which satisfy this condition. Indeed, $(\hat{A}(x, 1))_{n+1} = \langle c, x \rangle + d$ for some $c \in \mathbb{R}^n$ and $d \in \mathbb{R}$ ((c, d) is the $(n + 1)$th row of $\hat{A}$), and the condition is that

$$\forall x \in D \qquad \langle c, x \rangle \neq -d,$$

which can be satisfied for appropriate chosen c and d; for every direction c in which D is bounded, there exists a critical d such that from this value onwards the condition is satisfied (the critical d may or may not be chosen, depending on the boundary of D). Other ways of describing these maps will be given in Sect. 2.4.

Notation. For future reference we denote the map F associated with a matrix $\hat{A}$ by F_A and the matrix $\hat{A}$ associated with a map F by A_F. Note that $\hat{A}$ is defined uniquely up to a multiplicative constant. We say that A_F induces F, and may also write $F \sim A_F$.

Example 2.7. We are, in fact, very much familiar with one class of projective transformations: Möbius transformations of the extended complex plane. These are just projective transformations of the complex projective line $P^1(\mathbb{C})$ to itself. We describe points in $P^1(\mathbb{C})$ by homogeneous coordinates $[z_0, z_1]$, and then a projective transformation τ is given by $\tau([z_0, z_1]) = [az_0 + bz_1, cz_0 + dz_1]$ where $ad - bc \neq 0$. This corresponds to the invertible linear transformation

$$T = \begin{pmatrix} a & b \\ c & d \end{pmatrix}.$$

It is convenient to write $P^1(\mathbb{C}) = \mathbb{C} \cup \{+\infty\}$ where the point $\{+\infty\}$ is now the 1-dimensional space $z_1 = 0$. Then if $z_1 \neq 0$, $[z_0, z_1] = [z, 1]$ and $\tau([z, 1]) = [az + b, cz + d]$ and if $cz + d \neq 0$ we can write $\tau([z, 1]) = [(az + b)/(cz + d), 1]$ which is the usual form of a Möbius transformation, i.e.

$$z \mapsto \frac{az + b}{cz + d}.$$

The advantage of projective geometry is that the point $1 = [1, 0]$ plays no special role. If $cz + d = 0$ we can still write $\tau([z, 1]) = [az + b, cz + d] = [az + b, 0] = [1, 0]$ and if $z = 1$ (i.e. $[z_0, z_1] = [1, 0]$) then we have $\tau([1, 0]) = [a, c]$. In this note, however, we work over $\mathbb{R}$ and these transformations when considered as acting over $\mathbb{R}^2$ do not preserve intervals.

Example 2.8. If we view the real projective plane $P^2(\mathbb{R})$ in the same way, we get some less familiar transformations. Write $P^2(\mathbb{R}) = \mathbb{R}^2 \cup P^1(\mathbb{R})$ where the projective line at infinity is $\{(x, y, 0), x \in \mathbb{R}, y \in \mathbb{R}\}$. A linear transformation $T : \mathbb{R}^3 \to \mathbb{R}^3$ can then be written as the matrix

$$T = \begin{pmatrix} a_{11} & a_{12} & b_1 \\ a_{21} & a_{22} & b_2 \\ c_1 & c_2 & d \end{pmatrix},$$

and its action on $[x, y, 1]$ can be expressed, with $v = (x, y) \in \mathbb{R}^2$, as

$$v \rightarrow \frac{1}{\langle c, v \rangle + d}(Av + b)$$

where A is the matrix $\{a_{ij}\}$ and b, c are the vectors (b_1, b_2), (c_1, c_2). Each such transformation can be considered as a composition of an invertible linear transformation, a translation and an inversion $v \rightarrow v/(\langle c, v \rangle + d)$. Clearly it is easier here to consider projective transformations defined by 3×3 matrices.

Remark 2.9. Consider the matrix

$$\hat{A} = \begin{pmatrix} 1 & 0 & 0 \\ 0 & 0 & 1 \\ 0 & 1 & 0 \end{pmatrix}.$$

It gives rise to the transformation

$$(x, y) \mapsto \left(\frac{x}{y}, \frac{1}{y} \right),$$

which is injective (where it is defined). The upper 2×2 block (or $n \times n$, in the general case) is not invertible, though. We will get back to this transformation in later sections.

2.2.2 Basic Properties

1. *Preservation of intervals.* It is very easy to check that the map F defined above in (1) preserves intervals. Indeed, an interval in $\mathbb{R}^n$ is a subset of a line, which corresponds to a two dimensional plane in $\mathbb{R}^{n+1}$. The latter is mapped by A_F to a two dimensional plane, and after the radial projection to the level $x_{n+1} = 1$ we again get a line.
2. *Maximal domain.* A non affine fractional linear map F can be extended to a half space. The only restriction is that for $x \in D$ one has $\langle c, x \rangle \neq -d$, that is, D cannot intersect some given affine hyperplane H. Since we are interested in a convex domain, we must choose one side, which means the domain can be extended to a half space. It is not immediately clear why it cannot be extended further. To see why it cannot be extended further *while preserving intervals*, consider a point $x_0 \in H$. We shall see that there is no way to define F on x_0. Indeed, take two rays emanating from x_0 into the domain of F, say H^+ (and not

on H); $\{x_0 + \lambda y : \lambda > 0\}, \{x_0 + \lambda y' : \lambda > 0\}$. The fact that the rays are not on H means that $\langle c, y \rangle \neq 0$, likewise for y'. Moreover, $\langle c, y \rangle$ and $\langle c, y' \rangle$ have the same sign (say, positive, if we are in H^+). Assume $F(x) = \frac{Ax+b}{\langle c,x \rangle + d}$. Remember $\langle c, x_0 \rangle = -d$. Then $F(x_0 + \lambda y) = \frac{A(x_0 + \lambda y) + b}{\langle c, x_0 + \lambda y \rangle + d} = \frac{Ay}{\langle c,y \rangle} + \frac{1}{\lambda} \cdot \left(\frac{Ax_0 + b}{\langle c,y \rangle} \right)$, and similarly $F(x_0 + \lambda y') = \frac{Ay'}{\langle c,y' \rangle} + \frac{1}{\lambda} \cdot \left(\frac{Ax_0 + b}{\langle c,y' \rangle} \right)$. We see the two rays are mapped to two parallel half lines, which by injectivity of F are not identical, and therefore $F(x_0)$ cannot be chosen so that it lies on both these lines. This means F cannot be extended to a domain which intersects H, and still preserve intervals.

3. *The image.* The image of a (non-affine) fractional linear map F, whose domain is maximal (meaning it is an open half space) is an open half space. Indeed, let $\hat{A} = A_F$ be the associated matrix, and let $\hat{A}(\{(x, 1) : x \in \mathbb{R}^n\}) = E \subseteq \mathbb{R}^{n+1}$. Then the image of F is the radial projection into $\{(x, 1) : x \in \mathbb{R}^n\}$ of the part of E with positive $(n + 1)$th coordinate. It is easily checked that this is a half space in $\{(x, 1) : x \in \mathbb{R}^n\} \sim \mathbb{R}^n$, whose boundary is the hyperplane

$$\partial(Im(F)) = \{Ax : \langle c, x \rangle = 1\}, \quad \text{where} \quad \hat{A} = \begin{pmatrix} A & b \\ c^T & d \end{pmatrix}.$$

Note that it does *not* depend on b and d.

4. *Composition.* It is easily checked that $A_{F \circ G} = A_F \cdot A_G$. In particular, the composition of two fractional linear maps is again a fractional linear map. As for the domains: The maximal domain of each of the maps is a half space, and so is the image, thus the map is formally defined only on $G^{-1}(Im(G) \cap dom(F))$, and by the previous remarks it can be extended to be defined on some half space.

5. *The inverse map.* It is easily checked that $A_{F^{-1}} = A_F^{-1}$ and in particular, every fractional linear map has an inverse, which is also fractional linear. The domain of F^{-1} is the image of F, which by previous remarks is exactly the radial projection into $\{(x, 1) : x \in \mathbb{R}^n\}$ of the part of $E = A(\{(x, 1) : x \in \mathbb{R}^n\})$ with, say, positive $(n + 1)$th coordinate.

2.2.3 More Properties

To continue we first need two properties of fractional linear maps, given in Lemmas 2.10 and 2.15. The first is a transitivity result.

Lemma 2.10. *Fix a point p in the interior of the simplex $\Delta = \{z = \sum z_i e_i : 0 \leq z_i, \sum_{i=1}^n z_i \leq 1\}$, where $\{e_i\}_{i=1}^n$ is the standard basis of $\mathbb{R}^n$. Given $n + 2$ points, $x_0, x_1, \ldots, x_n, y$ in $\mathbb{R}^n$ such that y is in the interior of $conv(x_i)_{i=0}^n$, there exists an open convex domain D which contains the points and a fractional linear map $F : D \to \mathbb{R}^n$ such that for $1 \leq i \leq n$, $F(x_i) = e_i$, $F(x_0) = 0$ and $F(y) = p$.*

Remark 2.11. By invertibility (see Item (5) above) an equivalent formulation is as follows: there exists a fractional linear map $F : \Delta \to \mathbb{R}^n$ such that for $1 \leq i \leq n$, $F(e_i) = x_i$, $F(0) = x_0$ and $F(p) = y$.

Remark 2.12. From Lemma 2.14, it will follow that the map in Lemma 2.10 is unique.

Remark 2.13. Let us compare Lemma 2.10 with the more standard transitivity result of projective geometry, which can be found for example in [13] (Theorem 2, p. 59):

Let $A_1, \ldots, A_{n+2}$ and $B_1, \ldots, B_{n+2}$ be two sets of points in general position in $\mathbb{R}P^n$. Then there exists a unique projective transformation $f : \mathbb{R}P^n \to \mathbb{R}P^n$ such that $f(A_i) = B_i$ for $i = 1, \ldots, n + 2$. Indeed, they have the same flavor, however we demand more (in both sets, one point is in the convex hull of all the others) and get more; the whole convex hull is in the domain of the fractional linear map (i.e. it is mapped, within $\mathbb{R}P^n = \mathbb{R}^n \cup \mathbb{R}P^{n-1}$, to the part not in $\mathbb{R}P^{n-1}$).

Proof of Lemma 2.10. First let us build an affine linear map which maps x_i to e_i for $i = 1, \ldots, n$ and x_0 to 0. This is clearly possible by linear algebra. So we are left with the following task: given z in the interior of the simplex, build a fractional linear map F whose domain contains the simplex, such that $F(e_i) = e_i$, $F(0) = 0$ and $F(z) = p$.

To describe this map F, consider its associated matrix $\hat{A}$ in GL_{n+1}. Let us give the matrix elements which produce the desired map. Let the matrix be given by

$$\hat{A} = \begin{pmatrix} & & 0 \\ A & & \vdots \\ & & 0 \\ c^T & & d \end{pmatrix},$$

where A is an $n \times n$ matrix, $c \in \mathbb{R}^n$, and $d \in \mathbb{R}^+$. Let A be the diagonal matrix with diagonal entries $A_{i,i} = \frac{p_i}{z_i}$, let $d = \frac{1 - \sum p_i}{1 - \sum z_i}$, and let the vector c be given by $c_i = \frac{p_i}{z_i} - d$. The matrix induces a fractional linear map on the domain $\{x : \langle c, x \rangle > -d\}$. We must verify that the points $0, \{e_i\}_{i=1}^n$ are in this domain. Indeed, $d > 0$ since the points are in the simplex, and also $c_i > -d$. Finally, it is easily checked that the associated fractional linear map satisfies the desired conditions. $\square$

Once we know that the map from Lemma 2.10 exists, it follows that it is unique. Indeed, by the Theorem quoted in Remark 2.13, there exists only one fractional linear map which maps $n + 2$ given points to another $n + 2$ given points. We formulate it below, and for completeness provide the proof.

Lemma 2.14. *Let $F_1 : D_1 \to \mathbb{R}^n$ and $F_2 : D_2 \to \mathbb{R}^n$ be two fractional linear maps, where $D_i \subseteq \mathbb{R}^n$. Let $\{x_i\}_{i=0}^{n+1}$ be $(n + 2)$ points in $D_1 \cap D_2$ such that one is in the interior of the convex hull of the others. If $F_1(x_i) = F_2(x_i)$ for every $0 \leq i \leq n + 1$, then the two maps coincide on all of $D_1 \cap D_2$ and moreover, are induced by the same matrix in GL_{n+1} (up to multiplication by a non-zero scalar).*

Proof of Lemma 2.14. Without loss of generality, by Lemma 2.10, we can assume that $x_0 = 0$, $x_i = e_i$ for $i = 1, \ldots, n$, and that $x_{n+1} = p$ is any point we desire in the interior of the convex hull of $\{x_i\}_{i=1}^n$. Furthermore, by the same lemma, we may assume that $F_1(x_i) = F_2(x_i) = x_i$ for all i, and therefore we may simply compare, say, F_1 to Id (and then also F_2). Consider the matrix which induces F_1, given by some

$$\begin{pmatrix} A & b \\ c^T & d \end{pmatrix},$$

where A is an $n \times n$ matrix, b and c are vectors in $\mathbb{R}^n$ and $d \in \mathbb{R}$. In fact, $d \neq 0$ since 0 is in the domain of F_1, which is $\{x : \langle c, x \rangle > -d\}$. Without loss of generality we let $d = 1$. From the condition $F_1(0) = 0$ we see that $b = 0$. From $F_1(e_i) = e_i$ we see that A is diagonal, let us denote $A_{i,i} = a_i$, so that $c_i = a_i - 1$. Finally, for $F_1(p) = p$ we see that

$$p_i = \frac{a_i p_i}{1 + \sum (a_j - 1) p_j}.$$

This implies that for all i, $a_i = 1 + \sum (a_j - 1) p_j$, and in particular $a_1 = \ldots = a_n$. This means that $(a_1 - 1)(1 - \sum p_j) = 0$, and since p is not on the hyperplane passing through $\{e_i\}_{i=1}^n$, it implies $a_i = 1$ for all i, that is, F_1 is the identity mapping. The same holds for F_2. $\qquad\qquad\qquad\qquad\qquad\qquad\qquad\qquad\qquad\qquad\qquad\qquad\square$

As a consequence we get the following useful fact:

Corollary 2.15. *Let $F_1 : D_1 \to \mathbb{R}^n$ and $F_2 : D_2 \to \mathbb{R}^n$ be two fractional linear maps, where $D_i \subseteq \mathbb{R}^n$. Let $D \subseteq D_1 \cap D_2$ be some open domain in $\mathbb{R}^n$ such that $F_1|_D = F_2|_D$. Then the two maps coincide, i.e. they are induced by the same matrix, and their maximal extension is the same function, with the same maximal domain.*

2.3 Uniqueness

When the domain of an interval preserving map is assumed to be all of $\mathbb{R}^n$, it is a well known classical theorem that the map must be affine linear, as stated in the fundamental theorem of affine geometry, quoted below as Theorem 2.16. As a reference see, for example, [13], or [3] for the projective counterpart. More generally, interval preservation can be replaced by "collineation". More far reaching generalizations also exist, and we refer the reader to the forthcoming [8] where an elaborate account of these is given.

Theorem 2.16 (The Fundamental Theorem of Affine Geometry). *Let $m \geq 2$ and $f : \mathbb{R}^n \to \mathbb{R}^m$ be a bijective interval preserving map. Then f must be an affine transformation.*

In this section we discuss the fact that when the domain is a convex set (or, more generally, a connected open domain), the only interval preserving maps are fractional linear. This result was obtained by Shiffman in [15]. His method of proof is different from ours, and works in the projective setting, where he shows that any such map can be extended (using Desargues' theorem) to a mapping of the whole projective space, and then, from the fundamental theorem of projective geometry he concludes that it must be projective linear. We work in a more elementary way, never leaving $\mathbb{R}^n$. However, Shiffman's result is in a more general setting where not *all* intervals are assumed to be mapped to intervals, but only a subfamily which is large enough. This is important in some applications, in particular in the proof of Theorem 5.23.

The main theorem discussed in this section is the following.

Theorem 2.17. *Let $n \geq 2$ and let $K \subseteq \mathbb{R}^n$ be a convex set with non empty interior. If $F : K \to \mathbb{R}^n$ is an injective interval preserving map, then F is a fractional linear map.*

The proof of the theorem relies on the following lemma:

Lemma 2.18. *Assume $n \geq 2$. Let $\Delta \subseteq U \subseteq \mathbb{R}^n$, where U is an open set, and Δ is a non-degenerate simplex with vertices $x_0, \ldots, x_n$. Let p belong to the interior of Δ. If $F : U \to \mathbb{R}^n$ is an injective interval preserving map that fixes all $(n + 1)$ vertices of Δ and the interior point, that is $F(x_i) = x_i$ for every $0 \leq i \leq n$, and $F(p) = p$, then $F|_{\Delta} = Id|_{\Delta}$.*

Proof of Lemma 2.18. The proof goes by induction on the dimension n. Begin with $n = 2$. Consider a two dimensional simplex Δ, that is, a triangle in $\mathbb{R}^2$, with vertices a, b, c, and a point $p \in int(\Delta)$. Since F is injective and interval preserving, by Lemma 2.6 it is continuous, which implies that the set $D = \{x \in \Delta : F(x) = x\}$ is closed.

Let us check that all the edges are contained in D. Assume the contrary, namely that there is a point $e \in [a, b]$, $e \notin D$. Since D is closed, there exists an interval $[a', b'] \subseteq [a, b]$, such that $a', b' \in D$, but $(a', b') \cap D = \emptyset$. Now we will find a point $e' \in (a', b') \cap D$ in contradiction, thus concluding that no such e exists. Let us find two points $a'' \in [a', c]$ and $b'' \in [b', c]$, such that $a'', b'' \in D$. To this end, consider the intervals $[a', c]$, $[b', c]$. They are both mapped to themselves by F, and both intersect the line L containing a and p, for which we have $F(L) \subseteq L$. Let $a'' \in [a', c]$ and $b'' \in [b', c]$ be the points of intersection with L. Then $F(a'') = a''$ since this is the only point in $[a', c]$ and in L, and similarly $F(b'') = b''$.

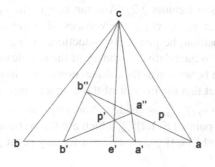

Now we look at the intersection of $[a'', b']$ with $[b'', a']$. This is a point p' in the interior of the triangle $a'b'c$. The line between c and p' intersects with $[a', b']$ at some point $e' \in (a', b')$, and by the same argument as before, $e' \in D$. We get a contradiction which proves that F is the identity map on the edges of Δ.

Next, for every point y in the interior, we draw two intervals containing y— each connecting a vertex with an edge, and get that the two intervals must be mapped to themselves (since the end points are on the edges and are thus mapped to themselves). This implies, as before, $F(y) = y$, which completes the proof for $n = 2$.

For the inductive step, we assume that the proposition is true for dimension $n - 1$, and prove it for dimension n. Let Δ be an n dimensional simplex. Denote by $\Delta_i := Conv\{x_0, \ldots, x_{i-1}, x_{i+1}, \ldots, x_n\}$ the face of Δ opposite to x_i. First we claim that $F(\Delta_i) = \Delta_i$. Indeed, this is due to interval preservation, together with the fact that the vertices are mapped to themselves. Denote by $y \in relint(\Delta_i)$ the unique point in the intersection of Δ_i, with the line connecting x_i and $p \in int(\Delta)$. Interval preservation implies that $F(y)$ remains on this line, and since it must remain on the face, we get $F(y) = y$. By applying the induction hypothesis to the $(n - 1)$ dimensional simplex Δ_i, we conclude that $F|_{\Delta_i} = Id|_{\Delta_i}$. The fact that the restriction of F to each of the faces is the identity, combined with interval preservation, implies that $F|_\Delta = Id|_\Delta$ simply by representing a point in the interior as the intersection of two intervals with endpoints on faces. $\qquad\square$

By the transitivity result from Lemma 2.10, we may state a corollary of the above lemma for general maps on the simplex.

Corollary 2.19. *Assume $n \geq 2$. Let $\Delta \subseteq U \subseteq \mathbb{R}^n$, where U is an open set, and Δ is a non-degenerate simplex with vertices $x_0, \ldots, x_n$. If $F : U \to \mathbb{R}^n$ is an injective interval preserving map then there exists a fractional linear map F_A such that $F|_\Delta = F_A|_\Delta$.*

Proof of Corollary 2.19. Let p belong to the interior of Δ. The main step is to show that the mapping F maps the point p to a point in the interior of $conv\{F(x_i)\}_{i=0}^n$, so that we may invoke transitivity and Lemma 2.18. To this end we shall use induction and prove the following claim: an injective interval preserving map must map simplices of dimension k, for any $k \geq 1$, to simplices of the same dimension, whose vertices are the images of the original vertices. Once this is done, an interior point must be mapped to an interior point by injectivity of F. The case $k = 1$ is almost by definition (see Lemma 2.2). Assume this is the case for simplices of dimension $\leq k$ and let $y_0, \ldots y_{k+1}$, the vertices of some $(k + 1)$ dimensional simplex, in general position, be given. By induction, the relative boundary of the convex hull is mapped to the relative boundary of the simplex $\{F(y_j)\}_{j=0}^{k+1}$. Since a point in the interior can be written as the intersection of two intervals with endpoints on the boundary, we get that the interior of the simplex $conv\{y_j\}_{j=0}^{k+1}$ is mapped to the interior of $conv\{F(y_j)\}_{j=0}^{k+1}$, as needed. Applying this, we have that the points $\{x_i\}_{i=0}^n$ are mapped to points $\{F(x_i)\}_{i=0}^n$ which are the vertices of a non degenerate simplex, $F(\Delta) = conv\{F(x_i)\}_{i=0}^n =: \Delta'$, and for any point $p \in int(\Delta)$ we have that $F(p) \in int(\Delta')$. To prove the corollary, chose any $p \in \Delta$, and compose F with some fractional linear G so that $(G \circ F)(x_i) = x_i$ for $i = 0, \ldots n$ and $(G \circ F)(p) = p$. Using Lemma 2.18 we have that $G \circ F = Id$ on Δ, and therefore $F|_{\Delta} = G^{-1}|_{\Delta}$, which is fractional linear, as claimed. □

Proof of Theorem 2.17. First we prove the theorem under the assumption that K is open and convex, and at the end of the proof we remark on the extension to general convex K (with non empty interior).

First we note that for every simplex Δ inside K the statement holds: consider $n + 2$ points $x_0, \ldots, x_n, p \in \mathbb{R}^n$, arranged as a simplex Δ and a point in its interior, as in Corollary 2.19. Since F is injective and interval preserving, by Corollary 2.19 $F|_{\Delta}$ is fractional linear.

Next, consider the union of two simplices Δ_1 and Δ_2 such that the intersection has a non empty interior. $F|_{\Delta_i}$ is fractional linear on each simplex Δ_1 and Δ_2, and these mappings coincide on the intersection, so they must be induced by the same matrix, by Corollary 2.15.

Finally, by covering the domain K with simplices so that each two are connected by a chain of simplices $\{\Delta_i\}_{i=0}^{N}$, with the property that the intersection of Δ_i and Δ_{i+1} has a non empty interior, we get that there is one map which induces all of the maps $F|_{\Delta}$ for all these simplices, meaning that F itself is a fractional linear map. Such a covering exists, for example an infinite family $\{\Delta_{x,y} : x, y \in K\}$, where $\Delta_{x,y}$ is some simplex which contains x and y in the interior will do (such a simplex exists for every x and y). This completes the proof in the case where K is open.

For a general convex K with non empty interior we must deal with the boundary of K. We know there exists a fractional linear map $G : U \to \mathbb{R}^n$ s.t. $F|_{int(K)} = G|_{int(K)}$, where U is the maximal domain of G (an open half space), and of course $int(K) \subseteq U$. We wish to show that $K \subseteq U$, and that $F = G$ also on $K \cap \partial K$. Take $x \in K \cap \partial K$. We first claim that $x \in U$, for which we need only show that $x \notin H = \partial U$. However, we have shown in item (2) of Sect. 2.2.2 that G cannot be extended to be defined on any point of H so that it is still interval preserving, from which we conclude $K \subseteq U$. Indeed, this was shown by considering two points a, b in the interior of K, to which correspond intervals $[a, x)$ and $[b, x)$ which are mapped to intervals, by G. Were x on the boundary, these intervals would have been parallel, and no way to define $F(x)$ would have existed. When $x \notin H$, the intervals $[G(a), G(x)]$ and $[G(b), G(x)]$ have a unique point of intersection $G(x)$, and we conclude that $F(x) = G(x)$. □

Remark 2.20. Theorem 2.17 can be proved for a general open connected set K; we only used convexity of K when arguing that K can be covered by simplices to get the wanted chains. This argument holds also whenever K is open and connected. Indeed, to get this covering we took between every two points $x, y \in K$ a simplex $\Delta_{x,y}$. This simplex is now replaced by a chain of simplices connecting x and y, constructed using an ϵ neighborhood of the path between x and y.

To complete the picture let us also attend to the case $n = 1$, although this will not be used in the sequel. Obviously, a similar theorem cannot be proved in

$\mathbb{R}$, since, for example, all continuous functions are interval preserving. The next theorem, Theorem 2.23, gives a characterization of one dimensional fractional linear maps. The theorem is a local version of the more well-known fact from projective geometry, stating that maps preserving cross ratio are linear when the domain and range are lines, and projective when the domain and range are extended lines.

We recall that the *cross ratio* of four numbers (thought of as coordinates of points on a line) is defined to be

$$[a, b, c, d] := \left(\frac{c - a}{c - b}\right) / \left(\frac{d - a}{d - b}\right).$$

For details and discussion see, for example, [13].

Remark 2.21. Note that $[a, b, c, x] = [a', b', c', x']$ implies $x' = \frac{\alpha x + \beta}{\gamma x + \delta}$, where $\alpha, \beta, \gamma, \delta$ are some function of a, b, c, a', b', c'. Conversely, every fractional linear map on $\mathbb{R}$ preserves the cross ratio of any four points in its domain.

Remark 2.22. Regarding permutations of a, b, c, d, we have the following:

$$[A, B, c, d] = [B, A, c, d]^{-1},$$

$$[a, b, C, D] = [a, b, D, C]^{-1},$$

$$[a, B, C, d] = 1 - [a, C, B, d],$$

and using the rule for these three transpositions, the cross ratio of any permutation of a, b, c, d can be derived from $[a, b, c, d]$. Moreover, as a consequence, we see that if we have $[a, b, c, d] = [x, y, z, w]$, then for every permutation σ we also have that $[\sigma(a), \sigma(b), \sigma(c), \sigma(d)] = [\sigma(x), \sigma(y), \sigma(z), \sigma(w)]$.

A basic notion when dealing with one dimensional fractional linear maps is the projection of one line to another line, through a so called "focus point" situated outside the two lines. See [13] for more details on the relation between fractional linear maps, preservation of cross ratio, and projection.

Theorem 2.23. *Let $I \subseteq \mathbb{R}$ be a convex set, either bounded or not, and $f : I \to \mathbb{R}$. Assume further that f preserves cross ratio on I, so for every four distinct points $a < b < c < d \in I$*

$$[f(a), f(b), f(c), f(d)] = [a, b, c, d].$$

Then f is fractional linear on I. In fact, it is true also if $a, b, c \in I$ are three (distinct) fixed points, and we assume only that f preserves cross ratio of a, b, c, d for any $d \in I \setminus \{a, b, c\}$.

Proof. Let $a, b, c \in I$ such that $a < b < c$, and f preserves cross ratio of a, b, c, x for any $x \in I \setminus \{a, b, c\}$. Let $x \in I$. We consider four cases; $x < a$, $a < x < b$, $b < x < c$, and $c < x$. For each case, the preservation of cross ratio yields a different equation;

$$x < a \Rightarrow [f(x), f(a), f(b), f(c)] = [x, a, b, c],$$

$$a < x < b \Rightarrow [f(a), f(x), f(b), f(c)] = [a, x, b, c],$$

$$b < x < c \Rightarrow [f(a), f(b), f(x), f(c)] = [a, b, x, c],$$

$$c < x \Rightarrow [f(a), f(b), f(c), f(x)] = [a, b, c, x].$$

By Remark 2.22, each of these equations implies $[f(a), f(b), f(c), f(x)] = [a, b, c, x]$, and thus by Remark 2.21, we get $f(x) = \frac{\alpha x + \beta}{\gamma x + \delta}$ for some $\alpha, \beta, \gamma, \delta$ which depend only on $a, b, c, f(a), f(b), f(c)$. Therefore f is a fractional linear map on I. □

2.4 Other Representations and Properties

2.4.1 Canonical Form

In what follows, we denote by $x = (x_1, \dots, x_n)$ the coordinates of a point x with respect to the standard basis $\{e_i\}$.

Definition 2.24. Let H^+ be the half space $\{x_1 > 1\}$. The mapping $F_0 : H^+ \to H^+$ given by

$$F_0(x) = \frac{x}{x_1 - 1}$$

will be called the *canonical* fractional linear map.

It is useful to note that the group of fractional linear maps is generated by its subgroup of affine linear maps, and the above map.

Theorem 2.25. *Let F be an injective non-affine fractional linear map with $F(x_0) = y_0$. Then there exist $B, C \in GL_n$ such that $B(F(Cx + x_0) - y_0) = F_0(x)$.*

Proof of Theorem 2.25. Define $G(x) := F(x + x_0) - y_0$, then $G(0) = 0$. G is an injective non-affine fractional linear map, with an inducing matrix of the form:

$$\begin{pmatrix} A' & b \\ c^T & d \end{pmatrix}.$$

From $0 \in Dom(G)$ it follows $d \neq 0$, so (using the multiplicative degree of freedom) we let $d = -1$. Also, $G(0) = 0$ implies $b = 0$. Since G is injective, the inducing matrix is invertible, and by $b = 0$ this implies that $A' \in GL_n$. Non-linearity of G implies $c \neq 0$. Therefore we can write for some $A' \in GL_n$, $0 \neq c \in \mathbb{R}^n$, that

$$G(x) = \frac{A'x}{\langle c, x \rangle - 1}.$$

Pick $C \in GL_n$ such that $C^t c = e_1$. We get $\langle c, Cx \rangle = \langle e_1, x \rangle = x_1$. Therefore

$$G(Cx) = \frac{A'Cx}{x_1 - 1}.$$

Finally, by letting $B = (A'C)^{-1}$, we get $(B \circ G \circ C)(x) = \frac{x}{x_1 - 1}$, and so

$$B(F(Cx + x_0) - y_0) = \frac{x}{x_1 - 1},$$

as required. $\square$

Remark 2.26. For simplicity, assume below $x_0 = y_0 = 0$. The representation in Theorem 2.25 is clearly not unique, as C can be chosen in any way satisfying just one linear condition, and B depends on C. Another form which can be given is:

$$C^{-1}A'^{-1}FC = \frac{x}{x_1 - 1},$$

where A' is uniquely determined, and C as before. Yet a third way to view this representation is:

$$F(x) = \frac{A'x}{\langle c, x \rangle - 1},$$

as was shown in the proof. This form has the advantage of emphasizing the degrees of freedom of a fractional linear map, since both the point c and the matrix A' are determined uniquely.

2.4.2   Geometric Structure

The mapping $F_0(x) = \frac{x}{x_1 - 1}$ is defined on $H^+ = \{x_1 > 1\}$, and satisfies $F_0(H^+) = H^+$. It is an involution on H^+ (and on $H^- = \{x_1 < 1\}$ as well). Denote the boundary of H^+ by H.

For every affine hyperplane parallel to H, namely $H_t = \{x : x_1 = t\}$ (for $t \neq 1$), we have $F_0(H_t) = H_{f(t)}$, where $f(t) = \frac{t}{t-1}$. The restriction $F_0 : H_t \to H_{f(t)}$, thought of as a map on $\mathbb{R}^{n-1}$, is a linear map—in fact, it is simply a scalar map; $x \mapsto \frac{1}{t-1}x$. In particular we see that in this family of parallel hyperplanes (shifts of H), parallel hyperplanes are mapped to parallel hyperplanes. This behavior is unique to shifts of H. Indeed, take $v \in \mathbb{R}^n$, then $(F_0^{-1} = F_0)$:

$$F_0(\{x : \langle x, v \rangle = c\}) = \{F_0(x) : \langle x, v \rangle = c\} = \{x : \langle F_0(x), v \rangle = c\}$$

$$= \{x : \langle x, v \rangle = c(x_1 - 1)\} = \{x : \langle x, v \rangle = c\langle x, e_1 \rangle - c\}$$

$$= \{x : \langle x, v - ce_1 \rangle = -c\}.$$

And so we see that if $v \neq \lambda e_1$, hyperplanes parallel to $v^\perp$ are mapped to hyperplanes which are *not* parallel; $(v - ce_1)^\perp \neq (v - c'e_1)^\perp$ for $c \neq c'$.

These considerations, by Theorem 2.25, may be applied to a general fractional linear mapping F. There are two hyperplanes, the first of which, say H_1, is the boundary of the maximal domain of F, and the second, H_2, is the boundary of the image of F, such that any translate of H_1 (which is in the domain) is mapped to a translate of H_2, and moreover, the map F restricted to each translate of H_1 is linear. In any other direction, however, two parallel hyperplanes are mapped to two hyperplanes which are not parallel.

As for a linear subspace V of $\mathbb{R}^n$ of dimension $0 \leq k \leq n$, we have $F_0(V) = V$ (by this we mean $F_0(V \cap H^-) = V \cap H^-$ and $F_0(V \cap H^+) = V \cap H^+$, since F_0 is not defined on the intersection with H). For $n - 1$ dimensional subspaces, we have seen it in the formula given above for the image of hyperplanes under F_0; substituting $c = 0$ yields $F_0(v^\perp) = v^\perp$ for every $v \in \mathbb{R}^n$. But in fact it is true trivially for subspaces of any dimension; simply note that $F_0(x)$ is in the direction of x. In fact, this is a particular case of the more general phenomenon; lines (more precisely: their intersection with the domain) through a fixed point in the domain of F, $x_0 \in dom(F)$, are mapped into lines through $F(x_0)$. This is due to interval preservation of F. Since F is smooth, this mapping of lines (but not of points along the lines) is the linear map given by the differential of F, $dF(x_0)$.

We can say even more about the geometric structure of F. For a point y_0 on the boundary of the maximal domain of F, the family of all the rays emanating from the point y_0 (into the domain) is mapped to the family of all half lines in the image of F which are parallel to some vector y_0', and vice versa. Again, by Theorem 2.25 it is enough to show this for the specific map $F_0(x) = \frac{x}{x_1 - 1}$. Consider a point $\hat{y} = (1, y)$ on H; a ray emanating from $\hat{y}$ *into the domain* can be written, for some $(1, u) \in H$ as

$$R = \{(1, y) + t(1, u) : t \in \mathbb{R}^+\}.$$

It is mapped to the half line

$$l' = \{F((1, y) + t(1, u)) : t \in \mathbb{R}^+\} = \{(1, u) + \frac{1}{t}(1, y) : t \in \mathbb{R}^+\}$$

$$= \{(1, u) + s(1, y) : s \in \mathbb{R}^+\}.$$

So we have seen that for a and b on H, the ray $a + b\mathbb{R}^+$ is mapped under F_0 to $b + a\mathbb{R}^+$ and vice versa. For example all rays emanating from the point $e_1 \in H$

are mapped to all lines perpendicular to H. Note that the part of l which is close to the point $\hat{y}$ (small t) is mapped to the part of l' which is far from the hyperplane H (large s). In a sense, the point $\hat{y}$ is mapped to "infinity" in direction opposite to H.

This also shows that fractional linear maps act as a lens on straight lines intersecting the defining hyperplane. Indeed, a cone of rays with base B, emanating from the point a in H, is mapped to a half infinite cylinder with base B, in the direction a. If $a \in B$, the corresponding line is the only one in the cone which is mapped to itself. When considering a general non-affine fractional linear map, we get that an infinite cone with base B is mapped to a half infinite cylinder with base $T(B)$ for some linear T, and vice versa. Of course, if the fractional linear map is affine it also does this, but by mapping cones to themselves and cylinders to themselves.

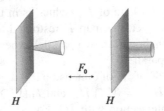

$$H \qquad\qquad H$$

2.5 Additional Results

2.5.1 Fractional Linear Maps and Polarity

For a closed convex set T containing 0, denote its polar set as before by T°. We claim that in a sense, the "root" of a fractional linear map is the polar map. The following theorem states that the so called "distortion" of fractional linear maps corresponds to two actions of polarity, each with respect to a different point of origin.

Theorem 2.27. *Let* $0 \in K \subseteq \{x_1 < 1\} \subseteq \mathbb{R}^n$ *be a closed convex set. Then for the canonical form of a fractional linear map,* $F_0(x) = \frac{x}{x_1 - 1}$*, the following holds:*

$$F_0(K) = (e_1 - K^\circ)^\circ.$$

In [10], [16] the authors prove uniqueness theorems for order isomorphisms on various families of convex sets. Here we see new such maps, on the family of closed convex bodies which are contained in a half space. Uniqueness of these maps in some weak sense (among point maps) follows immediately from the uniqueness Theorem 2.17. Applying techniques from those papers one can get uniqueness of these maps among all order isomorphisms on this class of convex bodies.

Proof of Theorem 2.27. Let T be a closed convex set. Clearly

$$[0, e_1] \subseteq T \quad \Leftrightarrow \quad [-e_1, 0] \subseteq T - e_1 \quad \Leftrightarrow \quad (T - e_1)^\circ \subseteq \{x_1 > -1\},$$

and therefore under our assumptions for every $x \in (T - e_1)^\circ$ we have $0 < 1 + x_1$. We define $G(-x) = -F_0(x)$, or explicitly $G(x) = \frac{-x}{x_1 + 1}$. Note that F_0 is an involution on $\{x_1 \neq 1\}$, and hence G is an involution on $\{x_1 \neq -1\}$. Compute

$$
\begin{aligned}
(T - e_1)^\circ &= \{x \in \mathbb{R}^n : \langle x, y - e_1 \rangle \leq 1 \quad \forall y \in T\} \\
&= \{x \in \mathbb{R}^n : \langle x, y \rangle \leq 1 + x_1 \quad \forall y \in T\} \\
&= \left\{x \in \mathbb{R}^n : \left\langle \frac{-x}{1 + x_1}, -y \right\rangle \leq 1 \quad \forall y \in T\right\} \\
&= \{x \in \mathbb{R}^n : \langle G(x), -y \rangle \leq 1 \quad \forall y \in T\} \\
&= \{G^{-1}(x) \in \mathbb{R}^n : \langle x, y \rangle \leq 1 \quad \forall y \in (-T)\} \\
&= G^{-1}(\{x \in \mathbb{R}^n : \langle x, y \rangle \leq 1 \quad \forall y \in (-T)\}) \\
&= G^{-1}((-T)^\circ) = G(-T^\circ) = -F_0(T^\circ),
\end{aligned}
$$

which in turn implies

$$F_0(T^\circ) = (e_1 - T)^\circ,$$

for sets T which contain the interval $[0, e_1]$, or conversely, such that $T^\circ \subseteq \{x_1 < 1\}$. Therefore we can formulate it in the following way, for a closed convex $K \subseteq \{x_1 < 1\}$ such that $0 \in K$ we have

$$F_0(K) = (e_1 - K^\circ)^\circ.$$

□

Remark 2.28. Recall that $\{x_1 = 1\}$ is the defining hyperplane of F_0, so we cannot hope to get that result for K which intersects this hyperplane. In the other side of this hyperplane, however, we do not have 0, and again cannot work with K°.

Remark 2.29. By Theorem 2.25, once we understand the action of F_0 on convex bodies, we understand the action of all (non-affine) fractional linear maps on convex bodies, and the only difference is in some linear maps and translations.

2.5.2 Sets That Can be Preserved

The fractional linear maps clearly have a non-linear "distortion" of the image. As we saw above, when approaching the defining hyperplane, the map diverges. However, fractional linear maps preserve some structure, for example, they preserve combinatorial structure of polytopes (number of vertices, faces of every dimension, intersection between faces, etc.). We will investigate which sets can be preserved by fractional linear maps.

We present some examples of simple convex sets K for which there exist fractional linear maps F with $F(K) = K$. This will also shed some light on the question: "given sets K_1, K_2, does there exist a fractional linear map F such that $F(K_1) = K_2$?". This question will have consequences in the next section, where we deal with classes of functions supported on convex sets ("windows"), and see that the existence of any order isomorphism between two such classes depends on the existence of a fractional linear map between the corresponding windows (more precisely; between the corresponding *cylinders*, either $K_i \times \mathbb{R}^+$ or $K_i \times \mathbb{R}$).

Let us start with an explicit two dimensional example: A non-affine fractional linear map which preserves the Euclidean disk.

Example 2.30 (Euclidean ball, 2 dimensions). Define $T : D \to \mathbb{R}^2$, where $D = \{(x, y) \in \mathbb{R}^2 : x < 2\}$, in the following way:

$$\begin{pmatrix} x \\ y \end{pmatrix} \mapsto \begin{pmatrix} T_1(x) \\ T_2(x, y) \end{pmatrix} = \begin{pmatrix} \frac{2x-1}{2-x} \\ \frac{\sqrt{3}y}{2-x} \end{pmatrix}.$$

Note that $x^2 + y^2 = 1$ implies $T_1(x)^2 + T_2(x, y)^2 = 1$, that is, S^1 is mapped to itself by T. It is easy to check that T maps S^1 onto itself. By the interval preservation property of T, this implies that the unit ball is mapped to itself. Note that $T(0) \neq 0$, with correspondence to Theorem 2.37.

Example 2.31 (Ellipsoids in n dimensions). The above explicit example can be extended easily to the Euclidean ball in $\mathbb{R}^n$. However, let us discuss this case, or more generally, the case of an ellipsoid in $\mathbb{R}^n$, in a slightly more abstract way. Note that a conic section is always mapped by a fractional linear map to a conic section. Indeed, a conic section in $\mathbb{R}^n$ is given as a section of the cone $C = \{x_{n+1}^2 = \sum_{i=1}^n x_i^2\}$ by a hyperplane (identified with $\mathbb{R}^n$). Equivalently, we may take the section of a linear image of the cone, $A(C)$ (for $A \in GL_{n+1}(\mathbb{R})$) by the hyperplane $\{x_{n+1} = 1\} \subseteq \mathbb{R}^{n+1}$. Viewing fractional linear maps as traces of linear maps on $\mathbb{R}^{n+1}$ (say, given by a matrix B), we immediately get that the image of the conic section corresponding to $A(C)$ is the conic section corresponding to $BA(C)$. Next, letting $\mathcal{E}$ be some closed ellipsoid in the domain of a non-affine fractional linear map (so, it is bounded away from the defining hyperplane), it is mapped to a conic section, but since F is continuous, this must be a compact conic section, and in particular a *bounded* one. Thus, $F(\mathcal{E})$ is an ellipsoid $\mathcal{E}'$. Finally, since any two ellipsoids can be mapped to one another via an affine linear map, we can find an invertible affine transformation A such that $AF(\mathcal{E}) = \mathcal{E}$, and AF is a non-affine fractional linear map.

Before moving on to the next convex set, we mention that for Euclidean balls (and hence ellipsoids) we also have a transitivity result, in the flavor of Lemma 2.10 for simplices. It is given in the following proposition.

Proposition 2.32. *Let B_n denote the open unit ball in $\mathbb{R}^n$, and $\mathcal{E}$ be some open ellipsoid, with $p \in \mathcal{E}$. Then there exists a bijective fractional linear map $F : \mathcal{E} \to B_n$ with $F(p) = 0$.*

Proof. There exists an *affine linear* map that maps $\mathcal{E}$ to B_n and p to p', and an orthogonal transformation which maps p' to λe_1 for $0 \leq \lambda < 1$. If $\lambda = 0$ we are done, with F being an affine map.

Assume otherwise; then by invertibility of f.l. maps, our task is to find a bijective fractional linear map $G : B_n \to B_n$ such that $G(0) = \lambda e_1$, for a given $0 < \lambda < 1$. Let $a := 1/\lambda$, $c := \sqrt{a^2 - 1}$ (so, $1 < a, 0 < c$). One possible choice of G is induced by the $(n+1) \times (n+1)$ matrix

$$
A_G = \begin{pmatrix} D & e_1 \\ e_1^T & a \end{pmatrix},
$$

where D is diagonal with eigenvalues $\{a, c, \ldots, c\}$. The direct formula corresponding to that choice of G is:

$$
G \begin{pmatrix} x_1 \\ \vdots \\ x_n \end{pmatrix} = \frac{1}{x_1 + a} \begin{pmatrix} ax_1 + 1 \\ cx_2 \\ \vdots \\ cx_n \end{pmatrix}.
$$

$\square$

We turn to the second example which is again in $\mathbb{R}^2$, a trapezoid.

Example 2.33 (Trapezoid). Let $\alpha > 0$, and $D = \{(x, y) \in \mathbb{R}^2 : x < 1 + \alpha^{-1}\}$. Define $T : D \to \mathbb{R}^2$ and $A : \mathbb{R}^2 \to \mathbb{R}^2$ in the following way:

$$
T \begin{pmatrix} x \\ y \end{pmatrix} = \begin{pmatrix} T_1(x) \\ T_2(x, y) \end{pmatrix} = \begin{pmatrix} \frac{x}{1+\alpha-\alpha x} \\ \frac{(1+\alpha)y}{1+\alpha-\alpha x} \end{pmatrix}, \quad A \begin{pmatrix} x \\ y \end{pmatrix} = \begin{pmatrix} 1 - y \\ x \end{pmatrix}.
$$

The affine linear map A is the $\pi/2$ rotation around $(1/2, 1/2)$, and so it maps the four points $(0,0), (1,0), (0,1), (1,1)$ to themselves in a cyclic manner i.e. to $(1,0), (0,1), (1,1), (0,0)$ respectively. The fractional linear map T fixes the three points $(0,0), (1,0), (0,1)$, and maps $(1,1)$ to $(1, 1 + \alpha)$.

Denote by K the trapezoid with vertices $(0,0), (1,0), (0,1), (1, 1 + \alpha)$, and consider $F : K \to K$ defined by $F := T \circ A \circ T^{-1}$. It is obvious that F is not affine, and that it maps the four vertices of K to themselves cyclically, thus by interval preservation, $F(K) = K$. These two facts can also be verified from the direct formula of F:

$$
\begin{pmatrix} x \\ y \end{pmatrix} \mapsto \begin{pmatrix} \frac{\alpha x - y + 1}{\alpha x + \alpha y + 1} \\ \frac{(\alpha+1)^2 x}{\alpha x + \alpha y + 1} \end{pmatrix}.
$$

Note, had we chosen $\alpha = 0$, our trapezoid K would be a square, and we would get that T, therefore F, are both affine maps, and thus we see that at least with this construction, we did not get a non-affine fractional linear map that preserves the cube in $\mathbb{R}^2$. This is, in fact, a general result in $\mathbb{R}^n$.

We denote by Q^n the unit ball of the l_∞ norm in $\mathbb{R}^n$, and by B_1^n the unit ball of the l_1 norm in $\mathbb{R}^n$:

$$Q^n := \{x \in \mathbb{R}^n : -1 \le x_i \le 1, \quad i = 1, \dots, n\},$$

$$B_1^n := \{x \in \mathbb{R}^n : \sum_{i=1}^n |x_i| \le 1, \quad i = 1, \dots, n\}.$$

Theorem 2.34. *Any bijective fractional linear map $F : Q^n \to Q^n$ is affine.*

Theorem 2.35. *Any bijective fractional linear map $F : B_1^n \to B_1^n$ is affine.*

We use the following lemma:

Lemma 2.36. *Let $K \subset \mathbb{R}^n$ be a non-degenerate closed polytope, and $f : K \to \mathbb{R}^n$ a fractional linear map. If two pairs of opposite and parallel facets are mapped to such pairs, the map must be affine.*

Proof. By Sect. 2.4.2, if f is not affine, there is only one direction in which f maps parallel hyperplanes to parallel hyperplanes. Therefore, if two $n - 1$ dimensional subsets are parallel (but are not contained in the same hyperplane), and mapped to parallel sets, they must lie on a translate of the defining hyperplane of f. Assume that F_1, F_2 are two parallel facets of K, and likewise F_3, F_4. There is no hyperplane whose shifts contain all four facets, since K is a polytope of full dimension (there are no more than two parallel facets). Therefore, the fact that the pair F_1, F_2 is mapped to a similar pair, and likewise F_3, F_4, implies that f is affine.   □

Proof of Theorems 2.34, 2.35. Both the facets of Q^n and of B_1^n have the property that every two non-opposite facets intersect. Therefore, every pair of opposite facets is mapped to such a pair. In particular, we have two such pairs, and by the previous lemma this implies that f is affine.   □

Next, we prove that if K is a centrally symmetric convex body, the only fractional linear maps which may preserve both K and $\{0\}$ are affine.

Theorem 2.37. *Let $K \subseteq \mathbb{R}^n$ be a closed, convex, centrally symmetric body, and let $F : K \to K$ be a bijective fractional linear map. If $F(0) = 0$, then F is linear.*

Proof of Theorem 2.37. As usual, since $F(0) = 0$ we assume that the inducing matrix of F has the form:

$$F \sim \begin{pmatrix} A & 0 \\ v^T & -1 \end{pmatrix},$$

where $A \in GL_n$, and $0, v \in \mathbb{R}^n$. Therefore $F(x) = \frac{Ax}{\langle v, x \rangle - 1}$.

We need to show that $v = 0$. Otherwise, let $x \in \partial K$ be such that $\langle v, x \rangle \neq 0$ (for example, take x in the direction of v). The interval $[x, -x]$ is mapped by F to the interval $[F(x), F(-x)]$. Since F is surjective, $F(x)$ and $F(-x)$ are also on the boundary of K, and by the formula they are in opposite direction, which means that $F(-x) = -F(x)$, by symmetry of K. By $\|F(x)\| = \|F(-x)\|$ we get $|\langle v, x \rangle + 1| = |\langle v, x \rangle - 1|$, meaning $\langle v, x \rangle = 0$, in contradiction to our choice of x. Thus we conclude $v = 0$, which means that F is linear. □

Remark 2.38. The theorem remains correct also when the condition "closed" is omitted. If the closure of K is contained in the maximal domain of F (the half space parallel to the defining hyperplane), then by continuity of F we get that the same conditions hold for the closure of K, apply the theorem, and conclude that F is linear. In the other case, i.e. when the closure of K intersects the defining hyperplane, one must be more careful, and we omit the details completing the proof.

Remark 2.39. The condition $F(0) = 0$ cannot be omitted. Indeed, we have seen examples of symmetric bodies preserved by non-affine fractional linear maps, for instance in Example 2.30.

Theorem 2.40. *Let $\Delta \subseteq \mathbb{R}^n$ be a closed, non-degenerate simplex, and $p \in \Delta$ its center of mass. If $F : \Delta \to \Delta$ is a bijective fractional linear map with $F(p) = p$, then F is affine linear.*

Proof of Theorem 2.40. Denote by $x_0, \ldots, x_n$ the vertices of Δ. Let $A : \Delta \to \Delta$ be the affine map defined by the conditions $A(x_i) = F(x_i)$, $i = 0, \ldots, n$. Such a map obviously exists, moreover it is unique, and it is invertible. Note that $A(\Delta) = \Delta$ implies $A(p) = p$, since the center of mass is a linear invariant. By Lemma 2.14, this implies $F = A$, meaning that F is affine linear. □

Remark 2.41. As in the case of symmetric bodies, the condition $F(p) = p$ cannot be omitted. In fact we have seen in Lemma 2.10 a transitivity result, stating that fractional linear maps can map any simplex to itself, with an arbitrary permutation on the vertices, and in addition map a given point inside—say, the center of mass—to an arbitrary point inside. In the last theorem we have seen that among these maps, the affine maps are the only ones which map the center of mass to itself.

However, the choice of a different point inside will not give the same result. Meaning, for any point p' in the interior of Δ which is not the center of mass, there exists a non-affine fractional linear map F such that $F(\Delta) = \Delta$ and $F(p') = p'$. The construction is quite simple—find a linear map $A : \Delta \to \Delta$ which permutes the vertices and does *not* fix the point p' (such a map is easily seen to exist), and then compose it with a fractional linear which fixes the vertices but "restores" $A(p')$ to p' (that map will be non-affine, since the only affine map which fixes all the vertices is the identity map). This composition is the wanted map.

3 Background on Order Isomorphisms

Our main interest in what follows is order preserving and order reversing transforms on convex functions, when the functions are restricted to being defined on a convex body in $\mathbb{R}^n$ rather than the whole space. It turns out that this restriction changes the picture entirely, and a new family of transformations appears. These transformations are based on fractional linear maps, which we studied in detail in Sect. 2.

3.1 General Order Isomorphisms

Definition 3.1. If S_1, S_2 are partially ordered sets, and $\mathcal{T} : S_1 \to S_2$ is a bijective transform, such that for every $f, g \in S_1$: $f \leq g \iff \mathcal{T}f \leq \mathcal{T}g$, we say that $\mathcal{T}$ is an *order preserving isomorphism*.

Definition 3.2. If S_1, S_2 are partially ordered sets, and $\mathcal{T} : S_1 \to S_2$ is a bijective transform, such that for every $f, g \in S_1$: $f \leq g \iff \mathcal{T}f \geq \mathcal{T}g$, we say that $\mathcal{T}$ is an *order reversing isomorphism*.

Definition 3.3. A partially ordered set S is said to be *closed under supremum*, if for every $\{f_\alpha\} \subseteq S$, there exists a unique element in S, denoted $\sup\{f_\alpha\}$, with the following two properties:

1. For every α, $f_\alpha \leq \sup\{f_\alpha\}$ (bounding from above).
2. If $g \in S$ also bounds $\{f_\alpha\}$ from above, then $\sup\{f_\alpha\} \leq g$ (minimality).

Definition 3.4. A partially ordered set S is said to be *closed under infimum*, if for every $\{f_\alpha\} \subseteq S$, there exists a unique element $f \in S$, with the following two properties:

1. For every α, $f \leq f_\alpha$ (bounding from below).
2. If $g \in S$ also bounds $\{f_\alpha\}$ from below, then $g \leq f$ (maximality).

Consider the case where S is a partially ordered set which contains a minimal element, and is closed under supremum. When S is one of the classes of convex functions we deal with, $\sup\{f_\alpha\}$ may be given by the pointwise supremum. However, the corresponding pointwise $\inf\{f_\alpha\}$ operation may not give a convex function. To obtain an infimum operation (denoted $\widehat{\inf}$), we use the supremum operation in the following way:

$$\widehat{\inf_{\alpha \in A}}\{f_\alpha\} := \sup\{g \in S : \forall \alpha \in A \quad g \leq f_\alpha\}.$$

That is, $\widehat{\inf}\{f_\alpha\}$ is the largest element which is below the family $\{f_\alpha\}$. Using $\widehat{\inf}$, we see that these classes are also closed under infimum; the first property is due to the minimality of sup, and the second holds since sup is a bound from above. Dealing with convex functions, we have:

1. $\hat{\inf}\{f_\alpha\} \leq \inf\{f_\alpha\}$.
2. When $\inf\{f_\alpha\}$ is already a convex function, $\inf\{f_\alpha\} = \hat{\inf}\{f_\alpha\}$.

   For example, if f is a convex function, then $f = \inf\{\delta_{x,f(x)}\}$ (recall that $\delta_{x,c}(y) = +\infty$ for $y \neq x$, and $\delta_{x,c}(x) = c$). Thus $\inf\{\delta_{x,f(x)}\} = \hat{\inf}\{\delta_{x,f(x)}\} = f$.

Next we follow Proposition 2.2 from [4], which states that an order preserving isomorphism $\mathcal{T}$ must satisfy $\mathcal{T}(\sup\{f_\alpha\}) = \sup\{\mathcal{T} f_\alpha\}$ and $\mathcal{T}(\hat{\inf}\{f_\alpha\}) = \hat{\inf}\{\mathcal{T} f_\alpha\}$, that is, sup and $\hat{\inf}$ are *preserved* by $\mathcal{T}$. Similarly, an order reversing isomorphism satisfies $\mathcal{T}(\sup\{f_\alpha\}) = \hat{\inf}\{\mathcal{T} f_\alpha\}$ and $\mathcal{T}(\hat{\inf}\{f_\alpha\}) = \sup\{\mathcal{T} f_\alpha\}$, that is, sup and $\hat{\inf}$ are *interchanged* by $\mathcal{T}$. We will prove this lemma for the case of order isomorphisms and order reversing isomorphisms between two *possibly different* partially ordered sets.

Proposition 3.5. *Let $\mathcal{S}_1, \mathcal{S}_2$ be partially ordered sets closed under supremum and infimum, and let $\mathcal{T} : \mathcal{S}_1 \to \mathcal{S}_2$ be an order preserving isomorphism. Then for any family $f_\alpha \in \mathcal{S}_1$ we have*

$$\mathcal{T}(\hat{\inf}\{f_\alpha\}) = \hat{\inf}\{\mathcal{T} f_\alpha\},$$

$$\mathcal{T}(\sup\{f_\alpha\}) = \sup\{\mathcal{T} f_\alpha\}.$$

Proposition 3.6. *Let $\mathcal{S}_1, \mathcal{S}_2$ be partially ordered sets closed under supremum and infimum, and let $\mathcal{T} : \mathcal{S}_1 \to \mathcal{S}_2$ be an order reversing isomorphism. Then for any family $f_\alpha \in \mathcal{S}_1$ we have*

$$\mathcal{T}(\hat{\inf}\{f_\alpha\}) = \sup\{\mathcal{T} f_\alpha\},$$

$$\mathcal{T}(\sup\{f_\alpha\}) = \hat{\inf}\{\mathcal{T} f_\alpha\}.$$

Both proofs are almost identical to the proof of Proposition 2.2 in [4], but we cannot apply it directly, since here the domain and image of $\mathcal{T}$ may be different sets. Therefore we prove below only Proposition 3.5 (the proof of Proposition 3.6 follows the exact same lines).

Proof of Proposition 3.5. Let $\{f_\alpha\}_{\alpha \in A} \subseteq \mathcal{S}_1$. Denote $f = \sup\{f_\alpha\}$, and g such that $\mathcal{T} g = \sup\{\mathcal{T} f_\alpha\}$—such g exists due to surjectivity of $\mathcal{T}$. We wish to show that $\mathcal{T} f = \mathcal{T} g$, i.e. $f = g$. Since $f \geq f_\alpha$ for all α, we get $\mathcal{T} f \geq \mathcal{T} f_\alpha$ for all α, thus $\mathcal{T} f \geq \sup\{\mathcal{T} f_\alpha\} = \mathcal{T} g$, which implies $f \geq g$. On the other hand, since $\mathcal{T} g \geq \mathcal{T} f_\alpha$ for all α, we have $g \geq f_\alpha$ for all α, thus $g \geq \sup\{f_\alpha\} = f$. We have seen $f \geq g$ and $g \geq f$, therefore $f = g$.

For $\inf$, denote $f = \hat{\inf}\{f_\alpha\}$, and g such that $\mathcal{T} g = \hat{\inf}\{\mathcal{T} f_\alpha\}$. We wish to show that $\mathcal{T} f = \mathcal{T} g$, i.e. $f = g$. Since $f \leq f_\alpha$ for all α, we get $\mathcal{T} f \leq \mathcal{T} f_\alpha$ for all α, thus $\mathcal{T} f \leq \hat{\inf}\{\mathcal{T} f_\alpha\} = \mathcal{T} g$, which implies $f \leq g$. On the other hand, since $\mathcal{T} g \leq \mathcal{T} f_\alpha$ for all α, we get $g \leq f_\alpha$ for all α, thus $g \leq \hat{\inf}\{f_\alpha\} = f$. We have seen $f \geq g$ and $g \geq f$, therefore $f = g$. $\square$

3.2   Order Isomorphisms of Convex Functions

In a recent series of papers, the first and third named authors have crystallized the concept of duality and investigated order reversing isomorphisms (called there "abstract duality") for various classes of objects and functions, see [5, 6]. The main theorem in [6] can be stated in two equivalent forms which we quote here for future reference.

Recall the Legendre transform $\mathcal{L}$ for a function $\phi : \mathbb{R}^n \to \mathbb{R} \cup \{\infty\}$; one first fixes a scalar product $\langle \cdot, \cdot \rangle$ on $\mathbb{R}^n$ (that is, a pairing between the space and the dual space). The Legendre transform $\mathcal{L}$ is then defined by

$$(\mathcal{L}\phi)(x) = \sup_y \{\langle x, y \rangle - \phi(y)\}. \tag{2}$$

It is an involution on the class of all lower-semi-continuous convex functions on $\mathbb{R}^n$, denoted $Cvx(\mathbb{R}^n)$. More precisely, $Cvx(\mathbb{R}^n)$ consists of all convex l.s.c. functions $f : \mathbb{R}^n \to \mathbb{R} \cup \{+\infty\}$, together with the constant $-\infty$ function.

Theorem 3.7. *Let $\mathcal{T} : Cvx(\mathbb{R}^n) \to Cvx(\mathbb{R}^n)$ be an order reversing involution. Then there exist $C_0 \in \mathbb{R}$, $v_0 \in \mathbb{R}^n$ and a symmetric transformation $B \in GL_n$, such that*
$$(\mathcal{T}\phi)(x) = (\mathcal{L}\phi)(Bx + v_0) + \langle x, v_0 \rangle + C_0.$$

We call these two properties "abstract duality", and so we say that on the class $Cvx(\mathbb{R}^n)$ there is, up to linear terms, only one duality transform, $\mathcal{L}$. More generally we have:

Theorem 3.8. *Let $\mathcal{T} : Cvx(\mathbb{R}^n) \to Cvx(\mathbb{R}^n)$ be an order reversing isomorphism. Then, there exist $C_0 \in \mathbb{R}, C_1 \in \mathbb{R}^+, v_0, v_1 \in \mathbb{R}^n$ and $B \in GL_n$, such that*
$$(\mathcal{T}\phi)(x) = C_0 + \langle v_1, x \rangle + C_1(\mathcal{L}\phi)(B(x + v_0)).$$

As usual, this is equivalent to the following

Theorem 3.9. *Let $\mathcal{T} : Cvx(\mathbb{R}^n) \to Cvx(\mathbb{R}^n)$ be an order preserving isomorphism. Then there exist $C_0 \in \mathbb{R}, C_1 \in \mathbb{R}^+, v_0, v_1 \in \mathbb{R}^n$ and $B \in GL_n$, such that*
$$(\mathcal{T}\phi)(x) = C_1\phi(Bx + v_0) + \langle v_1, x \rangle + C_0.$$

3.3   Order Isomorphisms of Geometric Convex Functions

The subclass of $Cvx(\mathbb{R}^n)$ consisting of non negative functions with $f(0) = 0$ is denoted by $Cvx_0(\mathbb{R}^n)$. Next we follow [7] to define two transforms $\mathcal{J}$ and $\mathcal{A}$ on this class. Consider the following transform, defined on $Cvx_0(\mathbb{R}^n)$:

$$(\mathcal{A}f)(x) = \begin{cases} \sup_{\{y \in \mathbb{R}^n : f(y) > 0\}} \frac{\langle x, y \rangle - 1}{f(y)} & \text{if } x \in \{f^{-1}(0)\}^\circ \\ +\infty & \text{if } x \notin \{f^{-1}(0)\}^\circ \end{cases}. \tag{3}$$

(with the convention $\sup \emptyset = 0$). One may check that it is order reversing. This transform (with its counterpart $\mathcal{J}$ defined below) first appeared in the classical monograph [14], but remained practically unnoticed until recently. For details, a geometric description, and more, see [7]. Next define:

$$\mathcal{J} = \mathcal{LA} = \mathcal{AL}.$$

Clearly, as a composition of two order reversing isomorphisms, it is an order preserving isomorphism. The formula for $\mathcal{J}$ can be computed (again, see [7] for details), and has the form:

$$(\mathcal{J}f)(x) = \inf\{r > 0 : f(x/r) \le 1/r\},$$

with the convention $\inf \emptyset = +\infty$. It turns out that, apart from the identity transform, up to linear variants, this is the only order preserving transform on the class $Cvx_0(\mathbb{R}^n)$. It was shown in [7] that the following uniqueness theorems for $\mathcal{J}$ hold.

Theorem 3.10. *If* $\mathcal{T} : Cvx_0(\mathbb{R}^+) \to Cvx_0(\mathbb{R}^+)$ *is an order isomorphism, then there exist two constants* $\alpha > 0$ *and* $\beta > 0$ *such that either (a-la-i) for every* $\phi \in Cvx_0(\mathbb{R}^+)$,

$$(\mathcal{T}\phi)(x) = \beta\phi(x/\alpha),$$

or (a-la-$\mathcal{J}$), for every $\phi \in Cvx_0(\mathbb{R}^+)$,

$$(\mathcal{T}\phi)(x) = \beta(\mathcal{J}\phi)(x/\alpha).$$

In higher dimensions, it was shown that

Theorem 3.11. *Let* $n \ge 2$. *Any order isomorphism* $\mathcal{T} : Cvx_0(\mathbb{R}^n) \to Cvx_0(\mathbb{R}^n)$ *is either of the form* $\mathcal{T}f = C_0 f \circ B$ *or of the form* $\mathcal{T}f = C_0(\mathcal{J}f) \circ B$ *for some* $B \in GL_n$ *and* $C_0 > 0$.

It is interesting to notice, and will be quite important in the sequel, that the map (on functions) $\mathcal{J}$ is actually induced by a point map on the epi-graphs of those functions. Indeed, one can check that for every $f \in Cvx_0(\mathbb{R}^n)$, the bijective map $F : \mathbb{R}^n \times \mathbb{R}^+ \to \mathbb{R}^n \times \mathbb{R}^+$ given by

$$F(x, y) = \left(\frac{x}{y}, \frac{1}{y}\right),$$

satisfies

$$epi(\mathcal{J}f) = F(epi(f)),$$

where
$$epi(f) = \{(x, y) \in \mathbb{R}^n \times \mathbb{R}^+ : f(x) < y\}.$$

See [7] for details. Moreover, we see that F is actually a fractional linear map. We will get back to this issue frequently in the next two sections.

Clearly, if we have a point map which preserves the set "epi-graphs of (a certain subset of) convex functions" then it induces an order preserving transform on this subset. It is not clear that, in some cases, any order preserving transform is induced by such a point map. However, this turns out to be the case both in the theorems described above, and in all theorems in the next two sections. Let us emphasize that this is also, usually, the idea behind the proof. First one shows that the transform must be induced by some point map, and moreover, one which preserves intervals. Next one uses some theorem which classifies all interval preserving maps (for example, the fundamental theorem of affine geometry, or Theorem 2.17), and finally one checks which of these maps really induces a transform on the right class, by this getting a full classification of order preserving transforms.

3.4   Order Reversing Isomorphisms

Considering order reversing transforms, the situation is slightly different, since there are two different cases. The first case is when one is given a set on which there is a known order reversing transform, such as $\mathcal{L}$ on $Cvx(\mathbb{R}^n)$ or on $Cvx_0(\mathbb{R}^n)$, for example. In that case the classification of order reversing transforms is completely equivalent to the classification of order preserving ones, by composing each of them with the known transform. For example, the theorems above give the following:

Theorem 3.12. *Let $n \geq 2$. If $\mathcal{T} : Cvx(\mathbb{R}^n) \to Cvx(\mathbb{R}^n)$ is an order reversing involution, then $\mathcal{T}$ is of the form $\mathcal{T}f = (\mathcal{L}f) \circ B + C_0$, for some symmetric $B \in GL_n$ and $C_0 \in \mathbb{R}$.*

Theorem 3.13. *Let $n \geq 2$. If $\mathcal{T} : Cvx_0(\mathbb{R}^n) \to Cvx_0(\mathbb{R}^n)$ is an order reversing involution, then $\mathcal{T}$ is either of the form $\mathcal{T}f = (\mathcal{L}f) \circ B$, or of the form $\mathcal{T}f = C_0(\mathcal{A}f) \circ B$, for some symmetric $B \in GL_n$ and $C_0 > 0$.*

However, there exists a second case in which there is no order reversing transform and this requires a different treatment, since one cannot use the above mentioned strategy, and is forced to find the real obstruction for the existence of such a transform (see [4] for examples). In Sect. 4.5 we deal with order reversing isomorphisms on $Cvx(K)$, and show that when $K \neq \mathbb{R}^n$, there are no such transforms.

4 The Cone of Convex Functions on a Window

4.1 Introduction

We investigate the question of characterizing order isomorphisms on convex functions, when the domain of the functions is not the whole of $\mathbb{R}^n$ but a convex subset. One such example which has already been studied (see [7]) is the case of geometric convex functions on $\mathbb{R}^+$. Since this example is central also for our setting, we describe it in detail below. First, let us recall the following definition:

Definition 4.1. The class of all lower-semi-continuous convex functions $f : K \to \mathbb{R} \cup \{\infty\}$ together with the constant $-\infty$ function on K will be denoted $Cvx(K)$. It can be naturally embedded into $Cvx(\mathbb{R}^n)$ by assigning to f the value $+\infty$ outside K.

We often call K a window, on which we observe the functions of $Cvx(\mathbb{R}^n)$. Our first results regard a description of order isomorphisms on the class of convex functions defined on a window. We state two versions, one of which does not assume surjectivity, but in which the order preservation condition is replaced by a slightly stronger condition of preservation of supremum and generalized infimum.

Theorem 4.2. *Let $n \geq 1$, and let $K_1, K_2 \subseteq \mathbb{R}^n$ be convex sets with non empty interior. If $\mathcal{T} : Cvx(K_1) \to Cvx(K_2)$ is an order preserving isomorphism, then there exists a bijective fractional linear map $F : K_1 \times \mathbb{R} \to K_2 \times \mathbb{R}$, such that $\mathcal{T}$ is given by*

$$epi(\mathcal{T}f) = F(epi(f)).$$

In particular, K_2 is a fractional linear image of K_1.

Theorem 4.3. *Let $n \geq 1$, and let $K_1, K_2 \subseteq \mathbb{R}^n$ be convex sets with non empty interior. If $\mathcal{T} : Cvx(K_1) \to Cvx(K_2)$ is an injective transform satisfying:*

1. $\mathcal{T}(\sup_\alpha f_\alpha) = \sup_\alpha \mathcal{T} f_\alpha$.
2. $\mathcal{T}(\hat{\inf}_\alpha f_\alpha) = \hat{\inf}_\alpha \mathcal{T} f_\alpha$.

for any family $\{f_\alpha\} \subseteq Cvx(K_1)$, then there exist $K_2' \subseteq K_2$, and a bijective fractional linear map $F : K_1 \times \mathbb{R} \to K_2' \times \mathbb{R}$, such that $\mathcal{T}$ is given by

$$epi(\mathcal{T}f) = F(epi(f)).$$

Note that for $x \notin K_2'$ we get $(\mathcal{T} f)(x) = +\infty$.

Note that by Proposition 3.5, an order isomorphism respects the actions of sup and inf. Therefore Theorem 4.3 is stronger, and implies Theorem 4.2. However, in the bijective case some of the reasoning is much simpler, and therefore below we prove both theorems independently, for clarity.

Remark 4.4. Let us elaborate on the meaning of the equation $epi(\mathcal{T}f) = F(epi(f))$. When F induces a transform on $Cvx(K)$, it is shown in Sect. 4.4 that up to some affine linear functional L_1, F is of the form

$$F(x, y) = \left(\frac{Ax + u}{\langle v, x \rangle + d}, \frac{y}{\langle v, x \rangle + d} \right),$$

where $A \in L_n(\mathbb{R})$, $u, v \in \mathbb{R}^n$, and $d \in \mathbb{R}$. Denoting $L_0 = \langle v, \cdot \rangle + d$ for the affine linear functional in the denominator, and $F_b(x) = \frac{Ax+u}{\langle v,x \rangle +d}$ for the base-map (projection of F to the first n coordinates), we conclude that

$$(\mathcal{T}f) = \left(\frac{f}{L_0} \right) \circ F_b^{-1} + L_1, \tag{4}$$

where L_1 is some affine linear functional and $F_b^{-1} : K_2 \to K_1$ is bijective. Note that L_0 and F_b are not independent, since L_0 must vanish on the defining hyperplane of F_b (where F_b is not defined). Moreover, note that for a general f, the function $\frac{f}{L_0}$ may not be convex, but the composition with F_b^{-1} exactly compensates this problem, and the result is again a convex function. In the special case of $A = I, u = 0, L_0(x) = x_1 + 1, L_1(x) \equiv 0$ we get $F(x, y) = (\frac{x}{x_1+1}, \frac{y}{x_1+1})$, and $(\mathcal{T}f)(x) = (1 - x_1) f(\frac{x}{1-x_1})$. This simpler form of the transform is not general, but if one allows linear actions on the epi-graphs, before and after F acts on them, it suffices to consider this form. There is another important, *different,* instance of the equation $epi(\mathcal{T}f) = F(epi(f))$, which may occur when the transform is defined on the subset of $Cvx(K)$ consisting of non-negative functions vanishing at the origin. We state it now for comparison and elaborate below (Theorem 5.2). A transform of this second, essentially different, type (a-la-$\mathcal{J}$, see [7]), corresponds to the inducing fractional linear map:

$$F_{\mathcal{J}}(x, y) = \left(\frac{x}{y}, \frac{1}{y} \right),$$

and to the explicit formula:

$$(\mathcal{J}f)(x) = \inf \left\{ r > 0 : rf \left(\frac{x}{r} \right) \leq 1 \right\}.$$

4.2 The Bijective Case

Proof of Theorem 4.2. The proof is composed of several steps.

Extremality of delta functions. As in [7], we define the following family P of extremal functions: $f \in P$ if every two functions above f are comparable, that is:

$$f \leq g, h \quad \Rightarrow \quad g \leq h \quad \text{or} \quad h \leq g.$$

This implies that the support of f (the set on which f is finite) consists of only one point. We call these functions **delta functions**, and denote by $\delta_{x,c}$ the function which equals c at the point x, and $+\infty$ elsewhere.

$\mathcal{T}$ is a bijection between the family P in $Cvx(K_1)$ and the family P in $Cvx(K_2)$, since this property is defined only using the "$\leq$" relation, which $\mathcal{T}$ preserves in both directions. Thus $\mathcal{T}(\delta_{x,c}) = \delta_{y,d}$, and this map between delta functions is bijective. This allows us to define a bijection $F : K_1 \times \mathbb{R} \to K_2 \times \mathbb{R}$; $F(x,c) = (y(x,c), d(x,c))$, such that $\mathcal{T}(\delta_{x,c}) = \delta_{F(x,c)}$. In fact, we get that $y = y(x)$ and $d = d(x,c)$ because two functions $\delta_{x,c}$ and $\delta_{x,c'}$ are comparable, and so must be mapped to comparable functions. Note that also $y(x)$ is bijective. Indeed, it is injective since the images of two functions are comparable if, and *only* if, the original functions are comparable, and it is surjective since all delta functions are in the image of $\mathcal{T}$.

Preservation of intervals. The "projection" of F to the first n coordinates, i.e. the mapping $x \mapsto y(x)$, is a bijective interval preserving map. Indeed, assume $y(x_1) = y_1$, $y(x_2) = y_2$, and $x_3 \in [x_1, x_2]$. Since $\delta_{x_3,0} \geq \inf\{\delta_{x_1,0}, \delta_{x_2,0}\}$, the function $\delta_{x_3,0}$ must be mapped to a function δ_{y_3,d_3} which is above $\inf\{\delta_{y_1,d_1}, \delta_{y_2,d_2}\}$. Since $\inf\{\delta_{y_1,d_1}, \delta_{y_2,d_2}\}$ is $+\infty$ outside $[y_1, y_2]$, this implics $y_3 \in [y_1, y_2]$. For $n \geq 2$, it implies that $y(x)$ is fractional linear, by Theorem 2.17. In fact this is true also when $n = 1$, but for $n = 1$ it follows from interval preservation of F itself. To see that F is interval preserving, consider (x_3, c_3) on the interval between (x_1, c_1) and (x_2, c_2). We know it is mapped to (y_3, d_3) with $y_3 \in [y_1, y_2]$ and moreover, letting $y_3 = \lambda y_1 + (1 - \lambda)y_2$, we know $d_3 \geq \lambda d_1 + (1 - \lambda)d_2$. Using surjectivity, we deduce that $F(x_3, c_3) = \delta_{y_3, \lambda d_1 + (1-\lambda)d_2}$, since $\delta_{y_3, \lambda d_1 + (1-\lambda)d_2}$ is above the function $\inf\{\delta_{y_1,d_1}, \delta_{y_2,d_2}\}$ and for all $c < c_3$, $\delta_{x_3,c}$ is *not* above the function $\inf\{\delta_{x_1,c_1}, \delta_{x_2,c_2}\}$.

Since F is an injective interval preserving map, we may apply Theorem 2.17, to conclude that F is a fractional linear map.

To complete the proof of Theorem 4.2, let $f \in Cvx(K_1)$, and write it as

$$f = \inf\{\delta_{x,y} : (x, y) \in epi(f)\}.$$

$$\Rightarrow \mathcal{T}f = \inf\{\mathcal{T}(\delta_{x,y}) : (x, y) \in epi(f)\}$$

$$= \inf\{\delta_{F(x,y)} : (x, y) \in epi(f)\}$$

$$= \inf\{\delta_{x,y} : (x, y) \in F(epi(f))\}.$$

On the other hand:

$$\mathcal{T}f = \inf\{\delta_{x,y} : (x, y) \in epi(\mathcal{T}f)\}.$$

Therefore we get
$$epi(\mathcal{T}f) = F(epi(f)),$$
as desired. This completes the proof.   □

Of course, there are restrictions on the structure of F for it to induce such a transform. This is elaborated in Sect. 4.4.

4.3 The Injective Case

We next move to the case of injective transforms. Let us first remark why in Theorem 4.3 we had to change the conditions from mere order preservation to preservation of sup and inf.

Remark 4.5. In the bijective case, order preservation (in both directions) is equivalent to preservation of sup and inf. One direction is given in Proposition 3.5, and the other is given here:

$$f \le g \quad \Rightarrow \quad \mathcal{T}(g) = \mathcal{T}(\sup\{f, g\}) = \sup\{\mathcal{T}(f), \mathcal{T}(g)\} \quad \Rightarrow \quad \mathcal{T}(f) \le \mathcal{T}(g),$$

$$\mathcal{T}(f) \le \mathcal{T}(g) \Rightarrow \mathcal{T}(g) = \sup\{\mathcal{T}(f), \mathcal{T}(g)\} = \mathcal{T}(\sup\{f, g\}) \Rightarrow g = \sup\{f, g\} \Rightarrow f \le g.$$

This direction is true also in the injective case (preservation of sup and inf implies order preservation), but the opposite (order preservation in both directions implies preservation of sup and inf) is not, as shown in the following example. The following $\mathcal{T} : Cvx(\mathbb{R}^n) \to Cvx(\mathbb{R}^n)$ is injective and $f \le g$ if and only if $\mathcal{T}f \le \mathcal{T}g$:

$$(\mathcal{T}f)(x) = f(x) + x_1^2.$$

But $\mathcal{T}$ does *not* map inf to inf. The reason behind this fact is that $\mathcal{T}$ is not surjective. Moreover, there exist f, g, such that $\inf\{\mathcal{T}(f), \mathcal{T}(g)\}$ is not in the image of $\mathcal{T}$, and in particular it is not equal to $\mathcal{T}(\inf\{f, g\})$; for example take $f(x) = x_1, g(x) = -x_1$.

For the proof of the more general Theorem 4.3, we need the following known geometric lemma. The dimension of a set K denotes the minimal dimension of an affine subspace which contains the set.

Lemma 4.6. *In an m-dimensional affine space, let M be a closed convex set. Let $\mathcal{F}$ be a family of m-dimensional closed convex sets such that $K \ne M$ for all $K \in \mathcal{F}$, and $K_1 \cap K_2 = M$ whenever $K_1 \ne K_2$ and $K_1, K_2 \in \mathcal{F}$. Then $\mathcal{F}$ is at most countable.*

We reformulate it, to better suit our need:

Lemma 4.7. *Let $M \subseteq \mathbb{R}^n$ be a fixed closed convex set of dimension m. Let $\mathcal{F}$ be an uncountable family of closed convex sets such that $K \ne M$ for all $K \in \mathcal{F}$, and $K_1 \cap K_2 = M$ whenever $K_1 \ne K_2$ and $K_1, K_2 \in \mathcal{F}$. Then for at least one set $K \in \mathcal{F}$, $dim(K) \ge m + 1$. In particular, $m \le n - 1$.*

Lemma 4.7 follows from Lemma 4.6, where the minimal subspace which contains M is taken to be the m-dimensional affine space of Lemma 4.6. Our application of this lemma requires a little more, so we prove:

Lemma 4.8. *Let $M \subseteq \mathbb{R}^n$ be a fixed closed convex set of dimension m. Let $\mathcal{F}$ be an uncountable family of closed convex sets such that $K \neq M$ for all $K \in \mathcal{F}$, and $K_1 \cap K_2 = M$ whenever $K_1 \neq K_2$ and $K_1, K_2 \in \mathcal{F}$. Then for at least one set $K \in \mathcal{F}$, $\dim(K) \geq m + 1$. Moreover, $m \leq n - 2$.*

Proof. We wish to prove that $m \neq n - 1$; the rest follows from Lemma 4.7. Assume otherwise, then let $H = \{\langle x, u \rangle = c\}$ be the affine subspace of dimension $n - 1$ which contains M. Our assumption is that the relative interior of M in H is not empty. The set $\{K \in \mathcal{F} : K \subseteq \mathcal{H}\}$ is at most countable, by Lemma 4.6. Since $\mathcal{F}$ is not countable, there are at least three sets which are not contained in H, and therefore (without loss of generality) we have $A, B \in \mathcal{F}$ such that $A \cap H^+ \neq \emptyset$, $B \cap H^+ \neq \emptyset$, where $H^+ := \{\langle x, u \rangle > c\}$. Let $a \in A, b \in B$ such that $a, b \in H^+$, and let $x \in M$ be a point in the relative interior of M. Since $conv\{M, a\} \subseteq A$, we conclude that there is some open half ball of the form $B_{(x,r)} \cap H^+$ contained in A, and likewise for B. The two half balls have non empty intersection, in contradiction to $A \cap B = M$. $\qquad \square$

We will use this lemma for epi-graphs of functions. Noting that

$$epi(\max\{f, g\}) = epi(f) \cap epi(g),$$

we get the following lemma for convex functions:

Lemma 4.9. *Let $M : \mathbb{R}^n \to \mathbb{R}$ be a fixed convex function, such that $epi(M) \subseteq \mathbb{R}^{n+1}$ is of dimension m. Let $\mathcal{F}$ be an uncountable family of convex functions such that $f < M$ for all $f \in \mathcal{F}$, and $\max\{f_1, f_2\} = M$ whenever $f_1, f_2 \in \mathcal{F}$ and $f_1 \neq f_2$. Then for at least one function $f \in \mathcal{F}$, $\dim(epi(f)) \geq m + 1$. Moreover, $m \leq n - 1$.*

Proof of Theorem 4.3. We start by checking where the constant function $+\infty$ is mapped to. Let us call its image f_∞. Consider the family $\{\delta_x\}_{x \in K_1}$, and its image $\{\mathcal{T}\delta_x\}_{x \in K_1}$. It is uncountable, and every two functions in the second family satisfy $\max\{g_1, g_2\} = f_\infty$.

This means, by Lemma 4.9, that there exists $x_1 \in K_1$ such that the dimension of the epi-graph of $\mathcal{T}\delta_{x_1}$ must be higher by at least 1 than the dimension of the epi-graph of f_∞. Similarly, for x_1 we construct an uncountable family of functions $\{\delta_{[x_1,y]}\}_{y \in K_1}$ such that the maximum of every two is δ_{x_1}, and by applying Lemma 4.9 again we get that there exists at least one such function, the image of which has an epi-graph with dimension higher by at least 1 than the dimension of the epi-graph of $\mathcal{T}\delta_{x_1}$. After repeating this construction an overall of $n - 1$ times, we conclude that there exist $x_1, \ldots, x_{n-1} \in K_1$ such that the epi-graph of the function $\mathcal{T}\delta_{conv\{x_1,\ldots,x_{n-1}\}}$ is of dimension higher by at least $n - 1$ than the dimension of the epi-graph of f_∞. Applying Lemma 4.9 one last time, we get that the dimension of the epi-graph of $\mathcal{T}\delta_{conv\{x_1,...,x_{n-1}\}}$ is at most $(n + 1) - 2 = n - 1$. This means that the epi-graph of f_∞ is of dimension 0, that is, $f_\infty = +\infty$.

This also shows that $\mathcal{T}(\delta_{x,c}) = \delta_{y,d}$. Indeed, since the only epi-graph with dimension 0 has already been designated to f_∞, the dimension of the epi-graph of $\mathcal{T}(\delta_{x,c})$ is at least 1; but we may construct a chain as above which implies that it is also at most 1. We define the injective map $F : K_1 \times \mathbb{R} \to K_2 \times \mathbb{R}$ by the relation $\mathcal{T}(\delta_{x,c}) = \delta_{F(x,c)}$, and denote $F(x, c) = (y(x, c), d(x, c))$.

In fact, we get that $y = y(x)$ and $d = d(x, c)$ because the two functions $\delta_{x,c}$ and $\delta_{x,c'}$ are comparable, and so must be mapped to comparable functions (by Remark 4.5). Note that $y(x)$ is injective because the images of two functions are comparable if and *only if* the original functions are comparable. In addition, $y(x)$ is interval preserving. Indeed, assume $y(x_1) = y_1$, $y(x_2) = y_2$, and $x_3 \in [x_1, x_2]$. Since $\delta_{x_3} \geq \hat{\inf}\{\delta_{x_1}, \delta_{x_2}\}$, the function δ_{x_3} must be mapped to a function $\delta_{y_3,c}$ which is above $\hat{\inf}\{\delta_{y_1,c_1}, \delta_{y_2,c_2}\}$, which implies $y_3 \in [y_1, y_2]$. For $n \geq 2$, the fact that $\hat{y}(x)$ is an injective interval preserving map implies that it is fractional linear, by Theorem 2.17. Actually this is true also for $n = 1$, but it only follows from the fact that $(x, c) \mapsto (y, d)$ is also interval preserving, which we will next show.

Remark. We note that until this point in the proof (for $n \geq 2$) we only use the max/min condition, and not the stronger assumed condition for sup/inf; we already get that the map F is very restricted: it is a fractional linear map on the base, and some one dimensional map $d_x(c)$ on each fiber, and all these maps d_x must join together to preserve convexity of epi-graphs. This seems to restrict $d(x, c)$ enough to determine its form, but we chose to continue using a different argument, which works also for $n = 1$, but requires the preservation of sup/inf.

To see that F is interval preserving consider the function $\hat{\min}\{\delta_{x_1,c_1}, \delta_{x_2,c_2}\}$, which is $+\infty$ outside the interval $[x_1, x_2]$ and linear in it, with $f(x_1) = c_1$ and $f(x_2) = c_2$. By assumption, it is mapped to $\hat{\min}\{\delta_{y_1,d_1}, \delta_{y_2,d_2}\}$. Taking $(x_3, c_3) \in [(x_1, c_1), (x_2, c_2)]$ we have that $\delta_{x_3,c_3} \geq \hat{\min}\{\delta_{y_1,d_1}, \delta_{y_2,d_2}\}$ and so the point (y_3, d_3) lies above or on the segment $[(y_1, d_1), (y_2, d_2)]$.

On the other hand, look at $x_3 = \lambda x_1 + (1 - \lambda)x_2$ and $c_3' < \lambda c_1 + (1 - \lambda)c_2$. That is, we take a point (x_3, c_3') which is under the segment $[(x_1, c_1), (x_2, c_2)]$. From the "only if" condition, we have that $\mathcal{T}(\delta_{x_3,c_3'}) \not\geq \hat{\min}\{\delta_{x_1,c_1}, \delta_{x_2,c_2}\}$. So (y_3, d_3') is under the segment $[(y_1, d_1), (y_2, d_2)]$, since $y_3 \in [y_1, y_2]$ and it cannot be above or on it. Since $\delta_{x_3,c_3} = \sup_{c_3' < c_3}\{\delta_{x_3,c_3'}\}$, we may use the condition of supremum to get $d_3 = \sup\{d_3'\}$, and thus (y_3, d_3) is below or on the segment $[(y_1, d_1), (y_2, d_2)]$. Together with what we saw before, this implies $(y_3, d_3) \in [(y_1, d_1), (y_2, d_2)]$.

So, we have shown that $F : K_1 \times \mathbb{R} \to K_2 \times \mathbb{R}$ is an injective interval preserving map, and we may apply Theorem 2.17 to conclude that it is fractional linear.

To complete the proof of Theorem 4.3, we proceed in exactly the same way as in the proof of Theorem 4.2, to conclude that

$$\mathcal{T}f = \hat{\inf}\{\delta_{x,y} : (x, y) \in F(epi(f))\}$$

$$= \hat{\inf}\{\delta_{x,y} : (x, y) \in epi(\mathcal{T}f)\},$$

and thus

$$epi(\mathcal{T}f) = F(epi(f)),$$

which completes the proof.   □

Both proofs generalize without any complication to various other settings in which one considers different classes, such as the class of all non negative functions in $Cvx(\mathbb{R}^n)$, or in $Cvx(K)$, or more generally:

$$S_{f_0} = Cvx(\mathbb{R}^n) \cap \{f : f_0 \le f\},$$

for some fixed $f_0 \in Cvx(\mathbb{R}^n)$. We get:

Theorem 4.10. *Let $n \ge 1$, and let $f_1, f_2 \in Cvx(\mathbb{R}^n)$ be convex functions with support of full dimension. If $\mathcal{T} : S_{f_1} \to S_{f_2}$ is an order isomorphism, then there exists a bijective fractional linear map $F : epi(f_1) \to epi(f_2)$, such that $\mathcal{T}$ is given by*

$$epi(\mathcal{T}f) = F(epi(f)).$$

Theorem 4.11. *Let $n \ge 1$, and let $f_1, f_2 \in Cvx(\mathbb{R}^n)$ be convex functions with support of full dimension. If $\mathcal{T} : S_{f_1} \to S_{f_2}$ is an injective transform satisfying:*

1. $\mathcal{T}(\sup_\alpha f_\alpha) = \sup_\alpha \mathcal{T} f_\alpha$.
2. $\mathcal{T}(\hat{\inf}_\alpha f_\alpha) = \hat{\inf}_\alpha \mathcal{T} f_\alpha$.

for any family $\{f_\alpha\} \subseteq S_{f_1}$, then there exist $f_2' \in S_{f_2}$, and a bijective fractional linear map $F : epi(f_1) \to epi(f_2')$, such that $\mathcal{T}$ is given by

$$epi(\mathcal{T}f) = F(epi(f)).$$

It is tempting to consider Theorems 4.2 and 4.3 as manifestations of Theorems 4.10 and 4.11, where f_i is the function which attains only the values $-\infty$ on K_i and $+\infty$ outside K_i. The only problem is that these functions are not elements of $Cvx(\mathbb{R}^n)$, but in fact Theorems 4.10 and 4.11 can be further generalized without any effort. Instead of considering only classes of the form $S_{f_0} = \{f \in Cvx(\mathbb{R}^n) : epi(f) \subseteq epi(f_0)\}$, consider also $\{f \in Cvx(\mathbb{R}^n) : epi(f) \subseteq K\}$, where K is some convex set (in the case of Theorems 4.2 and 4.3, K is the infinite cylinder $K_i \times \mathbb{R}$).

4.4   Classification of Admissible Fractional Linear Maps

Since fractional linear maps send intervals to intervals, it is clear (a-posteriori, once we know the transform is induced by a fractional linear map) that a delta function $\delta_{x,c}$ is mapped to a delta function $\delta_{y,d}$; since these are the only functions with epi-graphs that are half-lines. Moreover, by order preservation, we see that y is a function only of x. Observations of this kind allow us to classify the type of fractional linear maps that induce transforms as in Theorem 4.2.

Let the inducing matrix $A_F \in GL_{n+2}$ be given by

$$
A_F = \begin{pmatrix} & & & u'_1 & u_1 \\ & A & & \vdots & \vdots \\ & & & u'_n & u_n \\ v'_1 & \cdots & v'_n & a & b \\ v_1 & \cdots & v_n & c & d \end{pmatrix},
$$

where A is an $n \times n$ matrix, $v, v', u, u' \in \mathbb{R}^n$, and $a, b, c, d \in \mathbb{R}$.

The infinite cylinder $K_1 \times \mathbb{R}$ is contained in the domain of F, so it must not intersect the defining hyperplane $H = \{\langle v, x \rangle + cy = -d\}$, which implies $c = 0$. In particular, $K_1 \subseteq \{\langle v, x \rangle > -d\}$ (the sign of the denominator is constant on $dom(F)$, and we choose it to be positive; we may do so due to the multiplicative degree of freedom in the choice of A_F).

Since F must map fibers $\{(x, y) : y \in \mathbb{R}\}$ to fibers, we see that for $i = 1, \ldots, n$, $F((x, y))_i = \left(\frac{Ax + yu' + u}{\langle v, x \rangle + d} \right)_i$ does not depend on y, which implies $u' = 0$.

Let $f \in Cvx(K_1)$. The image of $epi(f)$ must be the epi-graph of some $g \in Cvx(K_2)$. Since we have chosen a positive sign for the denominator, this simply means that $a > 0$, and we choose $a = 1$, thus exhausting the multiplicative degree of freedom in the choice of A_F.

Finally, let F' be the map corresponding to the following $(n+1) \times (n+1)$ matrix, having removed the next to last row and column from A_F:

$$
A_{F'} = \begin{pmatrix} A & u \\ v^T & d \end{pmatrix}.
$$

The map $F' : K_1 \to K_2$ is fractional linear, and corresponds to the action of F on fibers (the "projection" of F to $\mathbb{R}^n$). Thus $A_{F'}$ must be invertible. We note that this condition always holds; we have $A_F \in GL_{n+2}$, and since the $(n + 1)$th column of A_F is e_{n+1}, $det(A_{F'}) = \pm det(A_F) \neq 0$.

We claim that these restrictions are not only necessary but also sufficient:

Proposition 4.12. *Let $K_1 \subseteq \mathbb{R}^n$ be a convex set with interior, for $n \geq 1$. Let A be an $n \times n$ matrix, $u, v, v' \in \mathbb{R}^n$, $b, d \in \mathbb{R}$, and let F, F' be the fractional linear maps defined by the following matrices:*

$$
A_F = \begin{pmatrix} & & 0 \\ & A & \vdots & u \\ & & 0 \\ v'^T & & 1 & b \\ v^T & & 0 & d \end{pmatrix}, \qquad A_{F'} = \begin{pmatrix} A & u \\ v^T & d \end{pmatrix}.
$$

If the following two conditions are satisfied:

1. $K_1 \subseteq \{\langle v, x \rangle > -d\}$.
2. $A_{F'} \in GL_{n+1}$, or equivalently $A_F \in GL_{n+2}$.

then F induces an order isomorphism from $Cvx(K_1)$ to $Cvx(K_2)$ by its action on epi-graphs, where $K_2 = F'(K_1)$.

Proof. The following four conditions must be checked: that epi-graphs are mapped to epi-graphs, that convexity of the functions is preserved under the transform, that it is bijective, and that it is order preserving. Bijectivity and convexity preservation follow easily by the bijectivity and interval preservation properties of fractional linear maps, and order preservation is immediate for transforms induced by a point map. The fact that epi-graphs are mapped to epi-graphs follows from the zeros in the $(n + 1)$th (next to last) column of A_F. ☐

Denote the map from the fiber above x_1 to the fiber above $F'(x_1) = x_2$ by $F_{x_1} : \mathbb{R} \to \mathbb{R}$. It is an affine linear map, given by

$$F_{x_1}(y) = \frac{\langle v', x_1 \rangle + y + b}{\langle v, x_1 \rangle + d}.$$

Remark 4.13. Letting $x_2 = F'(x_1)$ we get

$$(\mathcal{T}f)(x_2) = F_{x_1}(f(x_1)).$$

Note that there is a sort of coupling between the "projected" map F', which determines the $x \in \mathbb{R}^n$ dependency, and F_{x_1}, which determines the y dependency. More precisely: given F', the transform induced by F is determined, up to multiplication by a positive scalar, and addition of an affine linear function. We next show that the linear part is determined by v' and b. Consider a transform $\mathcal{T}$ induced by a map F, where

$$A_F = \begin{pmatrix} & & 0 & \\ A & & \vdots & u \\ & & 0 & \\ 0 \cdots 0 & 1 & 0 \\ v^T & & 0 & d \end{pmatrix}.$$

Next, consider the transform: $(\tilde{\mathcal{T}}f)(x) = (\mathcal{T}f)(x) + \langle x, w \rangle + e$, induced by a map $\tilde{F}$, where $w \in \mathbb{R}^n$ and $e \in \mathbb{R}$. As before, denote

$$A_{\tilde{F}} = \begin{pmatrix} & & 0 \\ \tilde{A} & \vdots & \tilde{u} \\ & & 0 \\ \tilde{v}'^T & 1 & \tilde{b} \\ \tilde{v}^T & 0 & \tilde{d} \end{pmatrix}.$$

Then $A = \tilde{A}$, $u = \tilde{u}$, $v = \tilde{v}$, and $d = \tilde{d}$. The only difference is in the next to last row, namely v' and b, and a simple calculation shows that

$$\begin{pmatrix} \tilde{v} \\ \tilde{b} \end{pmatrix} = \begin{pmatrix} A^T & v \\ u^T & d \end{pmatrix} \begin{pmatrix} w \\ e \end{pmatrix}.$$

The matrix appearing above is exactly $A_{F'}^T$, so it is invertible, and therefore, the set of all v, b corresponds exactly to the set of all affine linear additions to $\mathcal{T}$ (clearly these affine additions do not harm the properties of order preservation, bijectivity, etc.).

4.5 Order Reversing Isomorphisms

The Legendre transform $\mathcal{L} : Cvx(\mathbb{R}^n) \to Cvx(\mathbb{R}^n)$, is the unique order reversing isomorphism on $Cvx(\mathbb{R}^n)$. The corresponding question for windows is, given $K_1, K_2 \subseteq \mathbb{R}^n$, what are all the possible order reversing isomorphisms between $Cvx(K_1)$ and $Cvx(K_2)$? It turns out that there are no such order reversing isomorphisms, except in the aforementioned case where $K_1 = K_2 = \mathbb{R}^n$. This is due to the fact that the delta functions "have nowhere to be mapped to". We formulate this simple observation in the following Proposition 4.17. To this end we use the following two definitions.

Definition 4.14. Let $P_K \subset Cvx(K)$ denote the following subset of extremal functions:

$$P_K := \{f \in Cvx(K) : g, h \geq f \Rightarrow g, h \text{ are comparable}\}.$$

Definition 4.15. Let $Q_K \subset Cvx(K)$ denote the following subset of extremal functions (dual to P):

$$Q_K := \{f \in Cvx(K) : g, h \leq f \Rightarrow g, h \text{ are comparable}\}.$$

Recall that in this new notation, for any closed convex K (actually, for any $K \subseteq \mathbb{R}^n$), P_K consists exactly of the delta functions. In $Cvx(\mathbb{R}^n)$, it is clear that $Q_{\mathbb{R}^n}$

consists of linear functions; it follows from the fact that the only functions below $f = \langle c, x \rangle + d$ are of the form $g(x) = \langle c, x \rangle + d'$, for $d' < d$. In the next lemma we see that when $K \neq \mathbb{R}^n$ is a convex set with non empty interior, $Q_K = \emptyset$.

Lemma 4.16. *If $K \subsetneq \mathbb{R}^n$ is a convex set with non empty interior, then $Q_K = \emptyset$.*

Proof. Clearly, if f is a non linear convex function, $f \notin Q$ (take two hyperplanes supporting $epi(f)$ in different directions). For a linear function f, one may easily construct two non-parallel linear functions below it, which are not comparable (they will satisfy $g(x), h(x) \leq f(x)$ for every $x \in K$, not for every $x \in \mathbb{R}^n$). Note that the fact that K has non empty interior is essential, otherwise there is no guarantee that the functions will differ on K, as demonstrated by the example of K being a subspace. $\square$

We have shown in the proof of Theorem 4.2 that an order preserving isomorphism $\mathcal{T} : Cvx(K_1) \to Cvx(K_2)$ defines a bijection from P_{K_1} to P_{K_2}. Similarly, an order reversing isomorphism defines a bijection from P_{K_1} to Q_{K_2} (and from Q_{K_1} to P_{K_2}, of course), which is why we say Q is "dual" to P.

Proposition 4.17. *Let $n \geq 1$, and let $K_1, K_2 \subseteq \mathbb{R}^n$ be convex sets with non empty interior, such that either $K_1 \neq \mathbb{R}^n$ or $K_2 \neq \mathbb{R}^n$. Then there does not exist any order reversing isomorphism $\mathcal{T} : Cvx(K_1) \to Cvx(K_2)$.*

Proof of Proposition 4.17. Without loss of generality, assume $K_2 \neq \mathbb{R}^n$ (otherwise consider $\mathcal{T}^{-1}$). Let $x \in K_1$, then $\delta_{x,0} \in P_{K_1}$. Therefore $\mathcal{T}(\delta_{x,0}) \in Q_{K_2}$, which contradicts the conclusion of Lemma 4.16. $\square$

5 Geometric Convex Functions on a Window

Recall the definition of geometric convex functions on a window:

Definition 5.1. For a convex set $K \subseteq \mathbb{R}^n$ with $0 \in K$, the subclass of $Cvx(K)$ containing non negative functions satisfying $f(0) = 0$ is called the class of *geometric convex functions*, and denoted by $Cvx_0(K)$, i.e.

$$Cvx_0(K) = \{f \in Cvx(K) : f \geq 0, f(0) = 0\}.$$

It is naturally embedded in $Cvx_0(\mathbb{R}^n)$ by assigning to f the value $+\infty$ outside K. Therefore an equivalent definition is

$$Cvx_0(K) = \{f \in Cvx(\mathbb{R}^n) : 1_K \leq f \leq 1_{\{0\}}\}$$

where 1_K denotes the *convex* indicator function of K, which is zero on K and $+\infty$ elsewhere, and similarly $1_{\{0\}}$. Note that these functions are usually denoted by 1_K^∞, however, we never use in this paper the standard characteristic functions, so this notation can not lead to a misunderstanding.

In this section we deal with order isomorphisms from $Cvx_0(K_1)$ to $Cvx_0(K_2)$, where K_i are convex sets (containing 0, of course), and some generalizations of these classes.

As the example of $\mathcal{J}$ in $Cvx_0(\mathbb{R}^n)$ (which was discussed in Sect. 3.3) shows us, the case of $Cvx_0(K)$ is more involved than $Cvx(K)$, and a transform can be more complicated than a mere fractional linear change in the domain with the corresponding change in the fiber. Indeed, here we know already of an example where an indicator function is not mapped to such.

However, for the cases of $K = \mathbb{R}^+$ and $K = \mathbb{R}^n$ we do have theorems of the sort, see Theorems 3.10 and 3.11. There, the transform *is* given by a fractional linear point map on the epi-graphs. In each of these cases we observe two different types of behavior; one where fibers *are* mapped to fibers (a-la-i), and one when they are not (a-la-$\mathcal{J}$).

In this section we generalize these theorems to apply to an order isomorphism $\mathcal{T} : Cvx_0(K_1) \to Cvx_0(K_2)$, for convex domains K_1, K_2.

Theorem 5.2. *Let $n \geq 2$, and let $K_1, K_2 \subseteq \mathbb{R}^n$ be convex sets with non empty interior. If $\mathcal{T} : Cvx_0(K_1) \to Cvx_0(K_2)$ is an order preserving isomorphism, then there exists a bijective fractional linear map $F : K_1 \times \mathbb{R}^+ \to K_2 \times \mathbb{R}^+$, such that $\mathcal{T}$ is given by*

$$epi(\mathcal{T}f) = F(epi(f)).$$

The case $n = 1$ is slightly different since the two domains $\mathbb{R}^+$ and $\mathbb{R}^-$ do not interact. Other than that, the result is the same, for example see Theorem 5.7.

Remark 5.3. Of course, it is not true that every fractional linear map on $K_1 \times \mathbb{R}^+$ induces such a transform. A discussion of which fractional linear maps do induce such a transform (similar to that in Sect. 4.4) is given in Sect. 5.3.2.

Remark 5.4. In Sect. 5.3.2 we will also see that there is a difference between the cases $0 \in \partial K$ and $0 \in int(K)$, where in the former a "$\mathcal{J}$-type" transform does exist, and in the latter it does not (except in the case $K_1 = K_2 = \mathbb{R}^n$).

First, we will prove the one-dimensional theorem. We will do this in two ways. The first (in Sect. 5.1) is by using the known uniqueness Theorem 3.10 for $\mathcal{J}$ and i. The second is a direct proof, which we postpone to Sect. 5.3.1. We add this second proof for two purposes; to make the paper self contained, and also to clarify the case of a transform $\mathcal{T} : Cvx_0([0, x_1]) \to Cvx_0([0, x_2])$, that is when the domain of all functions is bounded.

Second, we will prove the multi-dimension theorem, in the following stages: we show that the transform must act "ray-wise". Then, on each ray, we *could* already apply the one-dimensional conclusion, but in fact we need much less—thus we continue directly and show that two extremal families of functions, namely linear functions and indicator functions, determine the full shape of $\mathcal{T}$. The extremality property forces the transform to act bijectively on these two families, and in a monotone way. Here, we do not need to discover the exact rule of this monotone mapping (even though we have it, since we've solved the one dimensional case).

Instead, we prove that there is some point map on the epi-graphs, controlling the rule of the transform for a third family, namely triangle functions. We show that this point map is interval preserving, and then apply Theorem 2.17 to show that it is fractional linear. Finally, we show that the rule of the transform for triangles determines the whole transform, thus completing the proof. This plan follows the proof from [7] of the case $K_1 = K_2 = \mathbb{R}^n$.

5.1 Dimension One

In [7], the first and third named authors showed that essentially, any order isomorphism $\mathcal{T} : Cvx_0(\mathbb{R}^+) \to Cvx_0(\mathbb{R}^+)$ is either i or $\mathcal{J}$, see Theorem 3.10. We note that in this case, indeed, for each of these two families of transforms, the transform is induced by a point map on the epi-graphs which is fractional linear. The first family of transforms (a-la-i) is given by

$$(\mathcal{T}\phi)(x) = \beta\phi(x/\alpha),$$

for positive α and β, and the inducing maps are $F^{i}_{\alpha,\beta}(x, y) = (\alpha x, \beta y)$. The second family of transforms (a-la-$\mathcal{J}$) is given by

$$(\mathcal{T}\phi)(x) = \beta(\mathcal{J}\phi)(x/\alpha),$$

for positive α and β, and the inducing maps are $F^{\mathcal{J}}_{\alpha,\beta}(x, y) = \left(\frac{\alpha x}{y}, \frac{\beta}{y}\right)$.

We introduce a third transform, with a parameter $z > 0$, to be able to switch between the bounded and non bounded cases;

Definition 5.5. Let $z > 0$, and $F_z : [0, z) \times \mathbb{R}^+ \to \mathbb{R}^+ \times \mathbb{R}^+$ be the bijective fractional linear map defined by $F_z(x, y) = \left(\frac{x}{z-x}, \frac{y}{z-x}\right)$.

Lemma 5.6. F_z induces an order isomorphism $\mathcal{T}_z : Cvx_0([0, z)) \to Cvx_0(\mathbb{R}^+)$ by its action on epi-graphs, that is

$$epi(\mathcal{T}_z(f)) = F_z(epi(f)).$$

Proof. To see that a transform defined using a point map on the epi-graphs, is an order isomorphism, three things need to be checked; that it is well defined, that it is bijective, and that it preserves order in both directions. For $\mathcal{T}_z$ to be well defined, F_z must map epi-graphs of geometric convex functions to epi-graphs of geometric convex functions. Since F_z is fractional linear, it is interval preserving, thus a convex epi-graph is mapped to some convex set. Among all convex sets, epi-graphs of geometric convex functions are characterized by two inclusions;

$$\{(0, y) : y > 0\} = epi(1_{\{0\}}) \subseteq epi(f) \subseteq epi(1_K) = \{(x, y) : x \in K, y > 0\}.$$

Note that F_z maps the half line $\{(0, y) : y > 0\}$ onto itself, and the entire domain $[0, z) \times \mathbb{R}^+$ onto the image $\mathbb{R}^+ \times \mathbb{R}^+$. Therefore also $F_z(epi(f))$ is between these two sets, which means it is the epi-graph of some geometric convex function. Bijectivity of F_z implies bijectivity of $\mathcal{T}_z$. Since $f \leq g \Leftrightarrow epi(g) \subseteq epi(f)$, a transform induced by a bijective point map on the epi-graphs, automatically preserves order in both directions. $\square$

We are ready to prove the one dimensional theorem, dealing with $I_1, I_2 \subseteq \mathbb{R}$ which may be either bounded intervals or half lines.

Theorem 5.7. *Let $I_1 \subseteq \mathbb{R}$ be either of the form $I_1 = [0, x_1)$ for some positive x_1, or $I_1 = [0, \infty)$, and likewise I_2. If $\mathcal{T} : Cvx_0(I_1) \to Cvx_0(I_2)$ is an order isomorphism, then there exists a bijective fractional linear map $F : I_1 \times \mathbb{R}^+ \to I_2 \times \mathbb{R}^+$, such that $\mathcal{T}$ is given by*

$$epi(\mathcal{T}f) = F(epi(f)).$$

Proof of Theorem 5.7. Define $\tilde{\mathcal{T}} : Cvx_0(\mathbb{R}^+) \to Cvx_0(\mathbb{R}^+)$ in the following way:

If $I_1 = [0, x_1)$ and $I_2 = [0, \infty)$, then: $\tilde{\mathcal{T}} := \mathcal{T} \circ \mathcal{T}_{x_1}^{-1}$

If $I_1 = [0, x_1)$ and $I_2 = [0, x_2)$, then: $\tilde{\mathcal{T}} := \mathcal{T}_{x_2} \circ \mathcal{T} \circ \mathcal{T}_{x_1}^{-1}$

If $I_1 = [0, \infty)$ and $I_2 = [0, \infty)$, then: $\tilde{\mathcal{T}} := \mathcal{T}$

If $I_1 = [0, \infty)$ and $I_2 = [0, x_2)$, then: $\tilde{\mathcal{T}} := \mathcal{T}_{x_2} \circ \mathcal{T}$

$\tilde{\mathcal{T}}$ is clearly an order isomorphism. Next, by simply applying Theorem 3.10, we get that our original $\mathcal{T}$ is some composition of the transforms i, $\mathcal{J}$, $\mathcal{T}_z$, and $\mathcal{T}_z^{-1}$, which are all induced by fractional linear point maps on the epi-graphs. Thus we conclude that $\mathcal{T}$ is also induced by such a map. $\square$

Remark 5.8. For transforms on (or to) $Cvx_0([0, z])$ simply note that all elements of $Cvx_0([0, z))$ are non decreasing and lower-semi-continuous functions, and thus have a unique extension to $[0, z]$, which preserves order in both directions. Therefore, by embedding $Cvx_0([0, z)) = Cvx_0([0, z])$ (where f is mapped to its unique extension) we get an order isomorphism of the form described in Theorem 5.7, and thus have the same result for closed intervals $[0, z]$, where epi-graphs are taken *without* the point z. In particular we see that there exist order isomorphisms between $Cvx_0([0, z])$ and $Cvx_0(\mathbb{R}^+)$.

5.1.1 Table of One Dimension Transforms

Straightforward computation of the transform in each of the cases gives, in each of the four scenarios, two types of transforms; a-la-identity and a-la-$\mathcal{J}$. We list them here, indicated by the fractional linear maps which induce them, namely $F_{a,b} : I_1 \times \mathbb{R}^+ \to I_2 \times \mathbb{R}^+$. Each family is two-parametric, for convenience we choose the parameters a, b such that $a, b > 0$ gives exactly all the functions in the family:

I_1	I_2	a-la-i; $F_{a,b}(x,y)$	a-la-$\mathcal{J}$; $F_{a,b}(x,y)$
$[0,x_1)$	$[0,x_2)$	$\frac{x_2}{x(1-a)+x_1 a}\cdot\begin{pmatrix}x\\by\end{pmatrix}$	$\frac{bx_2}{bx+y}\cdot\begin{pmatrix}x\\a(x_1-x)\end{pmatrix}$
$[0,x_1)$	$[0,\infty)$	$\frac{a}{x_1-x}\cdot\begin{pmatrix}x\\by\end{pmatrix}$	$\frac{b}{y}\cdot\begin{pmatrix}x\\a(x_1-x)\end{pmatrix}$
$[0,\infty)$	$[0,x_2)$	$\frac{ax_2}{ax+1}\cdot\begin{pmatrix}x\\by\end{pmatrix}$	$\frac{bx_2}{bx+y}\cdot\begin{pmatrix}x\\a\end{pmatrix}$
$[0,\infty)$	$[0,\infty)$	$a\cdot\begin{pmatrix}x\\by\end{pmatrix}$	$\frac{b}{y}\cdot\begin{pmatrix}x\\a\end{pmatrix}$

There is an essential difference between the i-type and $\mathcal{J}$-type transforms; they handle differently the extremal elements of $Cvx_0(I)$, which are indicators and linear functions (see Sect. 5.2.2 for exact definitions). The i-type transforms map indicators to themselves (bijectively), and likewise linear functions. The $\mathcal{J}$-type transforms, however, interchange between the two sub-families, mapping indicators to linear functions (bijectively) and vice versa. In the inducing maps, we also have a natural distinction between the i-type and $\mathcal{J}$-type maps. In both cases the determinant of the Jacobian of the inducing map never vanishes; it is positive for i-type maps, and negative for $\mathcal{J}$-type maps.

5.2 Multi Dimension

5.2.1 Acting on Rays

We next prove that in the n-dimensional case, one merely deals with many copies of the one dimensional problem (in fact, the case of functions on $\mathbb{R}^+$).

The next lemma states that an order isomorphism basically works in the following way: first, there is a permutation on the rays, and then on each ray, the transform acts independently of the functions' values on other rays.

There are two nuances here; first, if $K \neq \mathbb{R}^n$, then in some directions it does not contain a *full* ray. Since this does not affect the argumentation in any way, we don't distinguish between a full ray ($\mathbb{R}^+z$) and a restricted ray ($\mathbb{R}^+z \cap K$), which may be a bounded interval, and use "ray" to describe both. Second, if $0 \in int(K)$, then the set of all relevant rays can be described by S^{n-1}, but if $0 \in \partial K$, then there are less relevant rays (in some directions z, $\mathbb{R}^+z \cap K = \{0\}$). Therefore we are again forced to add another definition, for the set of all relevant rays—$S(K) \subseteq S^{n-1}$. $S(K) := \{z \in S^{n-1} : \mathbb{R}^+z \cap K \neq \{0\}\}$. In what follows, the support of a function is defined to be (the closure of) the set on which it is finite; $\overline{\{x : f(x) < \infty\}}$

Lemma 5.9. *Let $n \geq 2$, and let $K_1, K_2 \subseteq \mathbb{R}^n$ be convex sets with non empty interior. If $\mathcal{T} : Cvx_0(K_1) \to Cvx_0(K_2)$ is an order preserving isomorphism, then there exists a bijection $\Phi : S(K_1) \to S(K_2)$, such that any function supported on*

$\mathbb{R}^+ y$ is mapped to a function supported on $\mathbb{R}^+ z$, for $z = \Phi(y)$. Moreover, $\mathcal{T}$ acts ray-wise, namely $(\mathcal{T} f)|_{\mathbb{R}^+ z}$ depends only on $f|_{\mathbb{R}^+ y}$, for $z = \Phi(y)$.

We remark that if we were to prove the theorem directly for order reversing transformations then we would not encounter this ray-wise behavior, and get a transform $\mathcal{A}$ (or $\mathcal{L}$) which, miraculously, when combined with $\mathcal{L}$ acts ray-wise. Later on, it will follow that Φ must be induced by a linear map.

The proof uses the following simple observation: if $x, y \in \mathcal{S}(K_1)$ are two different points, and $f_x, f_y \in Cvx_0(K_1)$ are two functions supported on $\mathbb{R}^+ x$ and $\mathbb{R}^+ y$ respectively, then $\max\{f_x, f_y\} = 1_{\{0\}}$, and thus also $\max\{\mathcal{T} f_x, \mathcal{T} f_y\} = 1_{\{0\}}$, which means that $\mathcal{T} f_x$ and $\mathcal{T} f_y$ are supported on different sets.

Proof of Lemma 5.9. For two functions f, g to have $\max\{f, g\} = 1_{\{0\}}$ they must be supported on two sets whose intersection equals $\{0\}$. A function with support in a line cannot be mapped to one whose support includes two positively-linearly-independent points because then $\mathcal{T}^{-1}$ would map two functions whose support intersects at $\{0\}$ only, to functions supported on the same ray—impossible. Thus functions supported on a given ray are all mapped to functions supported on another fixed ray. By invertibility, we get that this defines a mapping $\Phi : \mathcal{S}(K_1) \twoheadrightarrow \mathcal{S}(K_2)$ which is bijective.

As for the ray-wise action of $\mathcal{T}$, the values of $\mathcal{T} f$ on $\mathbb{R}^+ z$ are the same as the values of $\max\{\mathcal{T} f, R_z\}$, where R_z denotes the function which is 0 on $\mathbb{R}^+ z \cap K_2$ and $+\infty$ elsewhere. This maximum is the image of the function $\max\{f, R_y\}$, because $\mathcal{T} R_y = R_z$ (each being the smallest function supported on the corresponding ray). Since $\max\{f, R_y\}$ does not depend on the values f attains outside $\mathbb{R}^+ y$, our claim follows.   □

5.2.2 Extremal Elements and Monotonicity

Restricted to a ray I, we consider two families of extremal functions in $Cvx_0(I)$; indicator functions, and linear functions.

(a) $1_{[0,z]}$ which equals to 0 on $[0, z]$ and $+\infty$ elsewhere (indicator).
(b) $l_c(t) = \max\{ct, 1_I(t)\}$ (linear).

Formally, the function l_c is defined on the whole of $\mathbb{R}^n$, therefore it is not really linear, but we will use this name in short. All the $\mathcal{J}$-type transforms switch (a) and (b)—bijectively, and all the i-type transforms fix (a) and fix (b)—again, bijectively. We will show that this is no coincidence—a general order isomorphism $\mathcal{T}$ must act in one of these two ways. We derive this from two properties of these families—the extremality property, and the non-comparability relation between these two families.

Definition 5.10. A function $f \in Cvx_0(I)$ is called *extremal* if there exist no two functions $g, h \in Cvx_0(I)$ such that $g \ngeq f$ and $h \ngeq f$ but $\max\{g, h\} \geq f$.

In the language of epi-graphs, this means that for $epi(f)$ to contain $A \cap B$, it must contain either A or B—whenever A, B are also epi-graphs of geometric convex functions.

We claim that extremality *characterizes* indicator and linear functions in $Cvx_0(I)$:

Lemma 5.11. *The only extremal functions in $Cvx_0(I)$ are either of the form $1_{[0,z]}$ for some $z \in I$ or of the form l_c for some $c \in \mathbb{R}^+$.*

Proof of Lemma 5.11. It is easy to check that both families are extremal. To show that any extremal function $f \in Cvx_0(I)$, must be of one of the two forms, we first show that if it assumes some value $0 < c \neq \infty$, it must be linear. Indeed, let $f(x) = c$. Without loss of generality we may assume $x \in int(I)$, since f is lower-semi-continuous. Consider the function $1_{[0,x]}$ assuming 0 in the interval $[0, x]$ and $+\infty$ elsewhere; $f \nleq 1_{[0,x]}$, since $1_{[0,x]}(x) = 0 < f(x)$. Consider the function $L_x(y) = \frac{c}{x}y$. By convexity of f, on the interval $[0, x]$, $f \leq L_x$. Since outside $[0, x]$ we have $f \leq 1_{[0,x]}$, this implies $f \leq \max\{1_{[0,x]}, L_x\}$, and so by extremality it must be that $f \leq L_x$. Since x is in the interior of I, this means that $f = L_x$, and therefore f is linear. The only other option is that f assumes only the values 0 and $+\infty$, which implies it is an indicator function, by convexity. $\square$

Lemma 5.12. *If $\mathcal{T} : Cvx_0(I_1) \to Cvx_0(I_2)$ is an order isomorphism then either:*

$\mathcal{T}$ is a bijection from linear functions to indicators, and a bijection from indicators to linear functions, or:
$\mathcal{T}$ is a bijection from linear functions to themselves, and a bijection from indicator functions to themselves.

Proof. Extremality is preserved under $\mathcal{T}$. Indeed, if there exist two functions $g, h \in Cvx_0(I_2)$ such that $g \ngeq \mathcal{T}f$ and $h \ngeq \mathcal{T}f$ but $\max\{g, h\} \geq \mathcal{T}f$, then the functions $\mathcal{T}^{-1}g$ and $\mathcal{T}^{-1}h$ contradict extremality for f. So, we see that the family of all extremal functions is mapped to itself, and by Lemma 5.11 this family is exactly the union of linear and indicator functions. Since $\mathcal{T}^{-1}$ shares the same properties as $\mathcal{T}$, we see that the map is surjective.

Secondly, all linear functions are comparable to one another and all indicator functions are comparable to one another (by f and g comparable we mean that either $f \leq g$ or $g \leq f$). However, no indicator function is comparable to a linear function—except for the trivial examples of $1_{\{0\}}$ and 0, whose behavior is obvious—since in $Cvx_0(I)$, these are the maximal and minimal elements (they are also the only mutual elements in both families). Hence, once we know that one linear function is mapped to a linear function then all of them must be, and then all indicator functions are mapped to indicators. The alternative is of course that all linear functions are mapped to indicators, and then all indicators are mapped to linear functions. $\square$

In this last lemma, a dichotomy, not apparent at first sight, appears. We have two very different possibilities, one corresponding to i, the identity transform (which clearly maps linear functions to themselves, likewise for indicator functions), and the other possibility corresponds to the transform $\mathcal{J}$, which—as can be checked—maps linear functions to indicator functions and vice-versa. Despite this dichotomy, in the statement of the next lemma we do not need to separate the two cases.

Next we claim that $\mathcal{T}$ is a *monotone* bijection on each of the extremal families. Monotonicity has a meaning here since both families are fully ordered subsets of $Cvx_0(I)$—"chains"—bounded together by the minimal and maximal elements $f_0 \equiv 0$ and $f_\infty = 1_{\{0\}}$.

If $\mathcal{T}$ maps linear functions to themselves (and likewise indicator functions), we define $S : I_1 \to I_2$ to be the function for which $\mathcal{T}1_{[0,x]} = 1_{[0,S(x)]}$, and $A : \mathbb{R}^+ \to \mathbb{R}^+$, for which $\mathcal{T}(l_c) = l_{A(c)}$. If $\mathcal{T}$ interchanges between the two families, we define $S : I_1 \to \mathbb{R}^+$ to be the function for which $\mathcal{T}1_{[0,x]} = l_{S(x)}$, and $A : \mathbb{R}^+ \to I_2$, for which $\mathcal{T}(l_c) = 1_{[0,A(c)]}$. In this next simple lemma we formulate the monotonicity property:

Lemma 5.13. *Assume $\mathcal{T} : Cvx_0(I_1) \to Cvx_0(I_2)$ is an order isomorphism.*

If $\mathcal{T}$ maps linear functions to themselves, then S and A are increasing bijections.

If $\mathcal{T}$ interchanges between the two families, then S and A are decreasing bijections.

Proof. S and A are bijections, since $\mathcal{T}$ is a bijection. Note that $1_{[0,x]} \leq 1_{[0,y]} \Leftrightarrow x \geq y$ and $l_c \leq l_d \Leftrightarrow c \leq d$. Therefore, if $\mathcal{T}$ fixes each of the families, S and A are increasing, and if $\mathcal{T}$ switches between the families, S and A are decreasing. $\square$

5.2.3 Triangles Functions: Completing the Proof

Next, we handle another family of functions, "triangle" functions. We show it is preserved under $\mathcal{T}$, and that the rule of the transform for it is monotone. We show that when leaving the one-dimensional perspective, the rule of the transform for triangles is controlled by an interval preserving bijection; and thus we apply our uniqueness theorem for such maps, Theorem 2.17. Finally we show that the transform is determined by its behavior on triangles, which proves Theorem 5.2.

For $z \in K$ and $c \in \mathbb{R}^+$, we introduce the "triangle" functions, denoted $\triangleleft_{z,c} \in Cvx_0(K)$:

$$\triangleleft_{z,c}(x) = \begin{cases} c|x|, & \text{if } x \in [0,z] \\ +\infty, & \text{otherwise.} \end{cases}$$

Note that they are one-dimensional (i.e. supported on a ray), so they can be thought of as elements of $Cvx_0(I)$ where I is a ray, and then $\triangleleft_{z,c} = \max\{1_{[0,z]}, l_c\}$.

Lemma 5.14. *If $\mathcal{T} : Cvx_0(I_1) \to Cvx_0(I_2)$ is an order isomorphism then a triangle function $\triangleleft_{z,c}$ is mapped under $\mathcal{T}$ to a triangle function $\triangleleft_{z',c'}$, where (z',c') is a function of (z,c).*

Proof of Lemma 5.14. A triangle is the maximum of an indicator and a linear function. By Proposition 3.5 $\mathcal{T}$ respects sup and inf, and thus in both cases of Lemma 5.12, a triangle is mapped to the maximum of an indicator and a linear function; that is, to a triangle.   □

Remark. Since in Lemma 5.13 we showed that $\mathcal{T}$ maps indicator and linear functions in a monotone way, it is obvious that this is the case also for triangles, meaning either $\mathcal{T}(\lhd_{z,c}) = \lhd_{S(z),A(c)}$, or $\mathcal{T}(\lhd_{z,c}) = \lhd_{A(c),S(z)}$, and in both cases, fixing any of the parameters z or c and changing the other monotonously, changes also the triangle in the image monotonously. Since we already know the exact shape of 1D transforms, we could have concluded this immediately. However, in what follows, we only use the fact that $\mathcal{T}(\lhd_{x,c}) = \lhd_{y,d}$, and that this map is monotone, meaning that on a fixed ray, either $y = y(x)$, $d = d(c)$, and both functions are bijective and increasing, or $y = y(c)$, $d = d(x)$, and both functions are bijective and decreasing.

We return to the n-dimensional picture, the first time since we reduced the discussion to ray-wise action. We wish to see how the different mappings of triangles on different rays all fit together. To this end, we replace the "parametrization" of triangles, from the point z (indicating the support of the function) and the slope c, to the point z and the *value of the function at that point* $h = c|z|$. To avoid abuse of notation, for $h = c|z|$ we will denote $\lhd_{z,c}$ by $\lhd^{z,h}$. With this notation, we denote by $F : (K_1 \setminus \{0\}) \times \mathbb{R}^+ \to (K_2 \setminus \{0\}) \times \mathbb{R}^+$ the bijective map for which $\mathcal{T} \lhd^{z,h} = \lhd^{F(z,h)}$.

Proposition 5.15. *Let $n \geq 2$, $K_1, K_2 \subseteq \mathbb{R}^n$ convex sets with non empty interior, and $\mathcal{T} : Cvx_0(K_1) \to Cvx_0(K_2)$ an order preserving isomorphism. Assume $F : (K_1 \setminus \{0\}) \times \mathbb{R}^+ \to (K_2 \setminus \{0\}) \times \mathbb{R}^+$ is the bijection satisfying $\mathcal{T}(\lhd^{x,h}) = \lhd^{F(x,h)}$ for every $(x,h) \in (K_1 \setminus \{0\}) \times \mathbb{R}^+$. Then F is a fractional linear map.*

Proof of Proposition 5.15. First we show that the restriction of F to any domain for which $(0,0)$ is an extreme point, is fractional linear. Let $(x_1, h_1), (x_2, h_2) \in K_1 \times \mathbb{R}^+$ such that $0 \notin [x_1, x_2]$. This merely means that our argument does not hold if x_1 and x_2 are on opposite rays. Letting $(x_3, h_3) \in [(x_1, h_1), (x_2, h_2)]$, and denoting $F(x_i, h_i) = (y_i, l_i)$, we need to prove that $(y_3, l_3) \in [(y_1, l_1), (y_2, l_2)]$. If x_i are on the same ray, then it follows from the one dimensional case, handled in Sect. 5.3.1, that the restriction of F to this line is fractional linear, and in particular it maps intervals to intervals, that is $F([(x_1, h_1), (x_2, h_2)]) = [(y_1, l_1), (y_2, l_2)]$. Assume otherwise, that x_i are linearly independent. Note that $\lhd^{x_3,h_3} \geq \inf\{\lhd^{x_1,h_1}, \lhd^{x_2,h_2}\}$, and that in this inequality x_3 is maximal, and h_3 is minimal. Therefore $\lhd^{y_3,l_3} \geq \inf\{\lhd^{y_1,l_1}, \lhd^{y_2,l_2}\}$, and in *this* inequality—due to the monotonicity of $\mathcal{T}$ on triangles—again y_3 is maximal, and l_3 is minimal (recall that in Lemma 5.13 we saw that if indicators and linear functions are exchanged, S and A are decreasing, and if they are preserved, S and A are increasing—thus in any case maximality of x_3 and minimality of h_3 coincides with maximality of y_3 and minimality of l_3). Therefore y_3, which lies on a different ray than those of y_1, y_2 (Φ is bijective), is in the triangle with vertices $0, y_1, y_2$, and due to its maximality—$y_3 \in [y_1, y_2]$. Moreover, the point (y_3, l_3) is above or on the interval $[(y_1, l_1), (y_2, l_2)]$, and due to its minimality, it is *on* this line. Therefore $(y_3, l_3) \in [(y_1, l_1), (y_2, l_2)]$, which means that F preserves intervals which do not intersect the positive h-axis; $\{(0, h) : h \geq 0\}$. In other words, the restriction of F to any domain for which $(0, 0)$ is an extreme point, is interval preserving. By applying Theorem 2.17, we conclude that F is fractional linear on each such domain, and thus, since $n \geq 2$, we may use Corollary 2.15 to conclude that F is a fractional linear map on the whole of $(K_1 \setminus \{0\}) \times \mathbb{R}^+$.   □

Remark. The proof of Proposition 5.15 does *not* work in one dimension, since the only two rays; $\mathbb{R}^+, \mathbb{R}^-$ cannot interact—they have 0 in their convex hull, and therefore a direct proof is needed in this case, to show that the transform is given by a fractional linear map on the epi-graphs. In fact, while it is true for transforms on a ray, it is indeed *not* the case for transforms on $Cvx_0(\mathbb{R})$, or on $Cvx_0(I)$ where I is an interval containing 0 in the interior.

Remark. The function F which is defined formally only for $(x, h) \in (K \setminus \{0\}) \times \mathbb{R}^+$, can in fact be extended to $K \times \mathbb{R}^+$, since the defining hyperplane of F does not intersect $epi(1_{\{0\}}) = \{(0, h) : h > 0\}$. Indeed, it is obvious that if it intersects this ray in one point it must contain the whole ray. In such a case, it follows from the properties of fractional linear maps, that rays emanating from a point in the hyperplane are mapped to parallel rays emanating from the hyperplane. Such a point map does not induce a transform on $Cvx_0(K)$. Therefore F can be defined on the whole of $K \times \mathbb{R}^+$. Moreover, using the fact that the supremum of all triangles is $1_{\{0\}}$, we get that $F(epi(1_{\{0\}})) = epi(1_{\{0\}})$.

Finally, knowing that the transform rule for triangle functions is controlled by a fractional linear map F, we turn to see that this is also the case for the epi-graph of any function. We use the following simple equality $epi(f) = \{(x, h) \in (K \setminus \{0\}) \times \mathbb{R}^+ : \triangleleft^{x,h} > f\} \cup epi(1_{\{0\}})$ which holds for every $f \in Cvx_0(K)$.

Proof of Theorem 5.2. By the previous proposition, there exists a bijective fractional linear map $F : (K_1 \setminus \{0\}) \times \mathbb{R}^+ \to (K_2 \setminus \{0\}) \times \mathbb{R}^+$, and we need to show that $F(epi(f)) = epi(\mathcal{T}f)$.

$$
\begin{aligned}
epi(\mathcal{T}f) = \ & \{(y, l) \in (K_2 \setminus \{0\}) \times \mathbb{R}^+ : \triangleleft^{y,l} > \mathcal{T}f\} \quad \cup epi(1_{\{0\}}) \\
= \ & F(\{(x, h) \in (K_1 \setminus \{0\}) \times \mathbb{R}^+ : \triangleleft^{F(x,h)} > \mathcal{T}f\}) \cup F(epi(1_{\{0\}})) \\
= \ & F(\{(x, h) \in (K_1 \setminus \{0\}) \times \mathbb{R}^+ : \mathcal{T} \triangleleft^{x,h} > \mathcal{T}f\}) \ \cup F(epi(1_{\{0\}})) \\
= \ & F(\{(x, h) \in (K_1 \setminus \{0\}) \times \mathbb{R}^+ : \triangleleft^{x,h} > f\}) \quad \cup F(epi(1_{\{0\}})) \\
= \ & F(epi(f))
\end{aligned}
$$

□

5.3 Additional Results

5.3.1 Direct Uniqueness Proof in the One Dimensional Bounded Case

We focus on the possible transforms in the case where linear functions are mapped to themselves, likewise indicator functions. Clearly, the function $S : I_1 \to I_2$ for which we have that $\mathcal{T}1_{[0,x]} = 1_{[0,S(x)]}$ is bijective and increasing (so it is continuous as well). Similarly $A : \mathbb{R}^+ \to \mathbb{R}^+$, for which $\mathcal{T}(l_c) = l_{A(c)}$, is bijective, increasing, and continuous. Note that we deal now only with I_1 and I_2 which are bounded, which means that S maps an interval to an interval, and A maps a full ray to a full ray.

Lemma 5.16. *Let $I_1 = [0, x_1), I_2 = [0, x_2)$, where $x_i \in \mathbb{R}$ are two positive numbers. Let $\mathcal{T} : Cvx_0(I_1) \to Cvx_0(I_2)$ be an order preserving isomorphism. Assume further, that for some increasing bijective function $S : I_1 \to I_2$ we have $\mathcal{T}1_{[0,x]} = 1_{[0,S(x)]}$, and for another increasing bijective function $A : \mathbb{R}^+ \to \mathbb{R}^+$, we have that $\mathcal{T}l_c = l_{A(c)}$. Then there exist two constants $\alpha > 0$ and $d < 1$ such that $A(c) = \alpha c$ and $S(x) = \frac{x_2}{x_1} \cdot \frac{x}{d(x/x_1-1)+1}$.*

Proof of Lemma 5.16. Denote as before $\lhd_{x,c} = \max\{1_{[0,x]}, l_c\}$, and similarly $g_{x,c} = \hat{\inf}\{1_{[0,x]}, l_c\}$. We get (on I_2 replace x_1 by x_2):

$$
g_{x,c}(z) = \begin{cases} 0 & ; \text{if } z \in [0, x] \\ c(z-x)\frac{x_1}{x_1-x} & ; \text{if } z \in [x, x_1] \\ +\infty & ; \text{otherwise} \end{cases}, \quad \lhd_{x,c}(z) = \begin{cases} cz & ; \text{if } z \in [0, x] \\ +\infty & ; \text{otherwise} \end{cases}.
$$

By Proposition 3.5 we get $\mathcal{T}(\lhd_{x,c}) = \lhd_{S(x),A(c)}$, $\mathcal{T}(g_{x,c}) = g_{S(x),A(c)}$. Let $0 < t < 1$ and consider $g = g_{tx,(\frac{c}{1-t})(\frac{x_1-tx}{x_1})}$. It can be easily checked that $g \le \lhd_{x,c}$, and $g(x) = \lhd_{x,c} = cx$, so that $g \not\le \lhd_{x,c'}$ for any $c' < c$, and $g \not\le \lhd_{x',c}$ for any $x' > x$. In fact, when a g-type function and a $\lhd$-type function behave that way ($g \le \lhd_{x,c}$ with maximal x and minimal c) it must be that they are equal at the "breaking point of the triangle", i.e. at the point x. Since $\mathcal{T}$ preserves order in both directions, $\mathcal{T}(g)$ and $\mathcal{T}(\lhd_{x,c})$ behave in the same way, and therefore:

$$
g_{S(tx),A((\frac{c}{1-t})(\frac{x_1-tx}{x_1}))} \le \lhd_{S(x),A(c)}
$$

with equality between the two functions at the point $S(x)$, meaning:

$$
A\left(c\frac{x_1 - tx}{x_1 - tx_1}\right) \cdot (S(x) - S(tx)) \cdot \left(\frac{x_2}{x_2 - S(tx)}\right) = A(c)S(x)
$$

for every $0 < t < 1$, every $0 < x < x_1$, and every $0 < c$. By defining $u = \frac{x_1-tx}{x_1-tx_1}$ and rearranging the equation, we get:

$$\frac{A(cu)}{A(c)} = \left(\frac{S(x)}{S(x) - S(tx)}\right) \cdot \left(\frac{x_2 - S(tx)}{x_2}\right). \tag{5}$$

In particular, the ratio $\frac{A(cu)}{A(c)}$ does not depend on c—thus it is equal to $\frac{A(u)}{A(1)}$, and we may write

$$A(cu) = \frac{A(c)A(u)}{A(1)}, \tag{6}$$

which holds for all $0 < c$ and $1 < u$ (see the definition of u). For $u = 1$ it is true trivially. For $0 < u < 1$ we denote $u' := 1/u > 1$. Noticing the symmetry between u and c, we interchange their roles to see that $A(1) = \frac{A(u)A(u')}{A(1)}$, and write

$$\frac{A(cu)}{A(c)} = \frac{1}{\frac{A(cu \cdot u')}{A(cu)}} = \frac{1}{\frac{A(u')}{A(1)}} = \frac{1}{\frac{A(1)}{A(u)}} = \frac{A(u)}{A(1)}.$$

Equation (6), valid for all $c > 0, u > 0$, together with the continuity of A, implies that A is of the form

$$A(c) = \alpha c^\gamma$$

for some fixed $\alpha > 0$ and γ.

Therefore, $\frac{A(cu)}{A(u)} = u^\gamma$. Returning to equation (5) with this new information, and substituting $u = \frac{x_1 - tx}{x_1 - tx_1}$, we get

$$\left(\frac{x_1 - tx}{x_1 - tx_1}\right)^\gamma = \left(\frac{S(x)}{S(x) - S(tx)}\right) \cdot \left(\frac{x_2 - S(tx)}{x_2}\right). \tag{7}$$

This can be written also as

$$S(tx) = S(x) \cdot \left(\frac{x_2 \left(\frac{x_1 - tx}{x_1 - tx_1}\right)^\gamma - x_2}{x_2 \left(\frac{x_1 - tx}{x_1 - tx_1}\right)^\gamma - S(x)}\right),$$

to show that for a given $0 < x < x_1$, $f(t) := S(tx)$ is differentiable as a function of t, for all $0 < t < 1$. This means S is differentiable in $(0, x_1)$ (the interior of I_1).

Denote $D_{a,b} = \frac{S(b) - S(a)}{b - a}$ for $a, b \in [0, x_1]$, and similarly $D_{a,a} = S'(a)$ for $a \in (0, x_1)$, so that $D_{a,b} \to D_{a,a}$ when $b \to a$. Note that for $a \neq b, 0 < D_{a,b} < \infty$. Rearranging equation (7) yields:

$$\left(\frac{x_1 - tx}{x_1 - tx_1}\right)^{\gamma - 1} = \frac{D_{0,x} \cdot D_{tx,x_1}}{D_{0,x_1} \cdot D_{tx,x}}.$$

Choose $x < x_1$ such that $S'(x) \neq 0$ and let $t \to 1^-$, then the right hand side of the equation tends to a finite, strictly positive number, and since $\frac{x_1 - tx}{x_1 - tx_1} \to \infty$ when $t \to 1$, this implies $\gamma = 1$. Therefore for every $0 < t < 1, 0 < x < x_1$ we have:

$$D_{0,x_1} \cdot D_{tx,x} = D_{0,x} \cdot D_{tx,x_1}$$

or alternatively:

$$[0, tx, x, x_1] = [S(0), S(tx), S(x), S(x_1)]$$

which by Theorem 2.23 implies that S is fractional linear. Combined with $S(0) = 0$, $S(x_1) = x_2$, $S'(x) > 0$, and $\gamma = 1$, this implies that S and A each belongs to a one-parametric family of maps of the form

$$S(x) = x_2 \cdot \frac{x/x_1}{d(x/x_1 - 1) + 1} \qquad A(c) = \alpha c$$

where $d < 1$ and $\alpha < 0$. $\qquad\qquad\qquad\qquad\qquad\qquad\qquad\qquad\qquad\qquad$ □

5.3.2 Classification of Admissible Fractional Linear Maps

We wish to fully classify the type of fractional linear maps that induce transforms as in Theorem 5.2. (The one dimensional case was fully described in Sect. 5.1). Denote by $A_\infty = \{(0, y) : y > 0\}$ the epi-graph of $\delta_{0,0} = 1_{\{0\}}$; the maximal function in $Cvx_0(K)$, and by $A_0^1 = \{(x, y) : x \in K_1, y > 0\}$ the epi-graph of 1_{K_1}; the minimal function in $Cvx_0(K_1)$ (similarly $A_0^2 = \{(x, y) : x \in K_2, y > 0\}$ for $Cvx_0(K_2)$). Since $Cvx_0(K) = \{f \in Cvx(\mathbb{R}^n) : 1_K \leq f \leq 1_{\{0\}}\}$, it turns out that a necessary and sufficient condition for a bijection $F : K_1 \times \mathbb{R}^+ \to K_2 \times \mathbb{R}^+$ to induce an order isomorphism is that it maps the minimal and maximal elements in $Cvx_0(K_1)$ to the minimal and maximal elements in $Cvx_0(K_2)$, namely:

$$F(A_\infty) = A_\infty, \tag{8}$$

$$F(A_0^1) = A_0^2. \tag{9}$$

Indeed, since F is a bijection from the cylinder $K_1 \times \mathbb{R}^+$ to the cylinder $K_2 \times \mathbb{R}^+$, we see that the transform is bijective. Order preservation (in both directions) is automatic for point-map-induced transforms. One must check that $F(epi(f))$ is an epi-graph of some convex function, which follows from it being a convex set containing the fiber A_∞. Since $A_\infty \subseteq epi(f) \subseteq A_0^1$, we get $A_\infty \subseteq F(epi(f)) \subseteq A_0^2$, meaning that $F(epi(f))$ is an epi-graph of a function in $Cvx_0(K_2)$. Therefore, we give the description of a general fractional linear map F which satisfies (8) and (9). Let the matrix A_F be given by

$$A_F = \begin{pmatrix} & & & v_1' & u_1' \\ & A & & \vdots & \vdots \\ & & & v_n' & u_n' \\ v_1 & \cdots & v_n & a & b \\ u_1 & \cdots & u_n & c & d \end{pmatrix},$$

for $A \in L_n(\mathbb{R})$, $v, v', u, u' \in \mathbb{R}^n$, and $a, b, c, d \in \mathbb{R}$. Thus F is given by:

$$
\begin{pmatrix} x \\ y \end{pmatrix} \mapsto \begin{pmatrix} \frac{Ax + yv' + u'}{\langle u, x \rangle + cy + d} \\ \\ \frac{\langle v, x \rangle + ay + b}{\langle u, x \rangle + cy + d} \end{pmatrix}
$$

Condition (8) means that $\begin{pmatrix} 0 \\ y \end{pmatrix}$ is mapped to $\begin{pmatrix} 0 \\ g(y) \end{pmatrix}$, where $g : \mathbb{R}^+ \to \mathbb{R}^+$ is some bijection, and therefore

$$
\frac{yv' + u'}{cy + d} = 0 \qquad \text{for all } y > 0,
$$

which implies $v' = u' = 0$. For $g(y) = \frac{ay+b}{cy+d}$ to be a bijection there exist only two options, corresponding to the two types of transforms on $Cvx_0(K)$: either g is increasing, and then $g(y) = \frac{y}{d}$ for some $d > 0$, which is associated with the i-type transforms, or g is decreasing, and then $g(y) = \frac{b}{y}$ for some $b > 0$, which is associated with the $\mathcal{J}$-type transforms. We denote these two different cases by F^i and $F^{\mathcal{J}}$, and (using the multiplicative degree of freedom in A_F) get:

$$
A_{F^i} = \begin{pmatrix} & & 0 & 0 \\ A & \vdots & \vdots \\ & & 0 & 0 \\ v_1 \cdots v_n & 1 & 0 \\ u_1 \cdots u_n & 0 & d \end{pmatrix}, \qquad A_{F^{\mathcal{J}}} = \begin{pmatrix} & & 0 & 0 \\ A & \vdots & \vdots \\ & & 0 & 0 \\ v_1 \cdots v_n & 0 & b \\ u_1 \cdots u_n & 1 & 0 \end{pmatrix}
$$

Note that in both cases, $A_F \in GL_{n+2} \Leftrightarrow A \in GL_n$. Turning to condition (9), we separate the two cases, dealing first with the i-type.

This case is very similar to the $Cvx(K)$ case (see discussion in Sect. 4.4), where the preservation of infinite cylinders is replaced with preservation of a part of those cylinders. Since

$$
F^i \begin{pmatrix} x \\ y \end{pmatrix} = \begin{pmatrix} \frac{Ax}{\langle u, x \rangle + d} \\ \\ \frac{\langle v, x \rangle + y}{\langle u, x \rangle + d} \end{pmatrix}.
$$

we have that $\frac{\langle v, x \rangle + y}{\langle u, x \rangle + d} > 0$ for all $x \in K_1$ and $y > 0$, which implies $K_1 \subseteq \{\langle u, x \rangle + d > 0\}$. Since $\frac{\langle v, x \rangle + y}{\langle u, x \rangle + d}$ maps $\mathbb{R}^+$ to $\mathbb{R}^+$ (as a function of y), $\langle v, x \rangle = 0$ for all $x \in K_1$, which implies $v = 0$ (recall that K_1 has interior). The general form of an i-type inducing map, is thus given, for $A \in GL_n$, $u \in \mathbb{R}^n$, and $d > 0$, such that $K_1 \subseteq \{\langle u, x \rangle + d > 0\}$ by

$$F^{i}\begin{pmatrix} x \\ y \end{pmatrix} = \begin{pmatrix} \frac{Ax}{\langle u,x \rangle + d} \\ \frac{y}{\langle u,x \rangle + d} \end{pmatrix}.$$

For the $\mathcal{J}$-type case, we know that

$$F^{\mathcal{J}}\begin{pmatrix} x \\ y \end{pmatrix} = \begin{pmatrix} \frac{Ax}{\langle u,x \rangle + y} \\ \frac{\langle v,x \rangle + b}{\langle u,x \rangle + y} \end{pmatrix}.$$

Therefore $\frac{\langle v,x \rangle + b}{\langle u,x \rangle + y} > 0$ for all $x \in K_1$ and $y > 0$, which implies $K_1 \subseteq \{\langle u, x \rangle \geq 0\}$, and also $K_1 \subseteq \{\langle -v, x \rangle \leq b\}$.

In the image, we know that each fiber $\{(x_2, y)\}$ above a point $x_2 \in K_2$ must contain all positive y. The fiber above $\frac{Ax_0}{\langle u,x_0 \rangle + y_0}$ is given by $\frac{t\langle v,x_0 \rangle + b}{t(\langle u,x_0 \rangle + y_0)}$, and is the image of the ray (tx_0, ty_0) in $K_1 \times \mathbb{R}^+$, which may be bounded or not. If it is bounded, say of the form $[(0,0),(x_0,y_0)]$, we must have $\langle v, x_0 \rangle = -b$. If it is not bounded, we must have $\langle v, x_0 \rangle = 0$. Therefore we handle the following cases separately:

A cone K_1: In this case, all rays (tx_0, ty_0) in $K_1 \times \mathbb{R}^+$ are not bounded, therefore all directions x_0 in K_1 satisfy $\langle v, x_0 \rangle = 0$, and therefore $v = 0$, since K_1 has interior.

Bounded K_1: In this case, all rays (tx_0, ty_0) in $K_1 \times \mathbb{R}^+$ are bounded, therefore all rays in K_1 emanating from the origin have end points in the hyperplane $\{\langle v, \cdot \rangle = -b\}$ (in particular, $v \neq 0$). This means that K_1 is a truncated cone, i.e. $K_1 = \mathcal{K}_1 \cap S_1$, where $\mathcal{K}_1$ is the minimal cone containing K_1 and S_1 is the slab $\{0 \leq \langle -v, \cdot \rangle \leq b\}$.

General K_1: In this case, some rays (tx_0, ty_0) in $K_1 \times \mathbb{R}^+$ are bounded, which implies $v \neq 0$. All non bounded directions x_0 must satisfy as before $\langle v, x_0 \rangle = 0$, which implies that K_1 is bounded in directions $x_0 \notin v^{\perp}$, and $K_1 \cap v^{\perp}$ is a (degenerate) cone. As in the bounded case, we get $K_1 = \mathcal{K}_1 \cap S_1$.

We can sum up the above three options as follows. The set K_1 is the intersection of some cone, with the (possibly degenerate; if $v = 0$) slab $\{0 \leq \langle -v, \cdot \rangle \leq b\}$.

Similarly, since in every direction, K_2 is given by $\{tAx : 0 \leq t \leq \langle u, x \rangle^{-1}\}$ for some $x \in K_1$ (if $\langle u, x \rangle = 0$ we let $\langle u, x \rangle^{-1} = \infty$), it contains full rays in all directions $A(u^{\perp})$, and in other directions it contains intervals with end points x_0 which satisfy $\langle A^{-T}u, x_0 \rangle = 1$. This implies, as before, that K_2 is some cone, intersected with the (possibly degenerate) slab $\{0 \leq \langle A^{-T}u, \cdot \rangle \leq 1\}$.

One can further investigate the possible restrictions on v, u, A with respect to the bodies K_i, but it involves considering different cases for K_1 and K_2. We do not go into this in detail but instead give a few examples.

Remark 5.17. Under the condition $0 \in int(K_1)$, the only $\mathcal{J}$-type order isomorphism $\mathcal{T} : Cvx_0(K_1) \to Cvx_0(K_2)$ is possible when $K_1 = K_2 = \mathbb{R}^n$. Indeed, in that case $K_1 \subseteq \{\langle u, x \rangle \geq 0\}$ implies $u = 0$, which means that the projection of $F^{\mathcal{J}}$ to the first n coordinates is $P_n F^{\mathcal{J}} \begin{pmatrix} x \\ y \end{pmatrix} = \frac{1}{y} Ax$. Clearly, since A is invertible, and K_1 contains 0 in the interior, this means $K_2 = \mathbb{R}^n$, and by the exact same argument also $K_1 = \mathbb{R}^n$. In addition, $K_1 \subseteq \{\langle v, x \rangle + b \geq 0\}$ implies $v = 0$, thus the general form of $F^{\mathcal{J}}$ which induces a transform on $Cvx_0(\mathbb{R}^n)$ is

$$F^{\mathcal{J}} \begin{pmatrix} x \\ y \end{pmatrix} = \begin{pmatrix} \frac{Ax}{y} \\ \\ \frac{b}{y} \end{pmatrix}$$

for $A \in GL_n$ and $b > 0$, as stated in Theorem 3.11.

Example 5.18. When $u \neq 0$, the defining hyperplane of $F^{\mathcal{J}}$ intersects the cylinder $K_1 \times \mathbb{R}$, and the defining hyperplane of the image intersects the cylinder $K_2 \times \mathbb{R}$. This is a restriction on the bodies K_i; since K_2 is the intersection of a cylinder (which has its base on the defining hyperplane of the image) with a half space ($y > 0$), we have (see Sect. 2.4.2) that $K_1 \times \mathbb{R}^+$ is the intersection of a cone (emanating from the origin) with a half space. This is true also for $K_2 \times \mathbb{R}^+$, and so both our bodies are simultaneously the intersection of a half space with a cylinder and the intersection of a half space with a cone. For example, let $K_1 = (\mathbb{R}^+)^n$ and $K_2 = \text{conv}\{0, e_1, \ldots, e_n\}$, and let $F^{\mathcal{J}}$ be given by

$$A_{F^{\mathcal{J}}} = \begin{pmatrix} & & & 0\,0 \\ & I_n & & \vdots\,\vdots \\ & & & 0\,0 \\ 0 & \cdots & 0\,0 & b \\ 1 & \cdots & 1\,1 & 0 \end{pmatrix}.$$

Example 5.19. Let $K = K_1 = K_2$ be the slab $\{0 \leq x_1 \leq 1\}$. The following matrix induces a mapping $F^{\mathcal{J}} : K \times \mathbb{R}^+ \to K \times \mathbb{R}^+$:

$$A_{F^{\mathcal{J}}} = \begin{pmatrix} & & & 0\,0 \\ & I_n & & \vdots\,\vdots \\ & & & 0\,0 \\ -e_1^T & & 0 & 1 \\ e_1^T & & 1 & 0 \end{pmatrix},$$

which induces an order isomorphism on $Cvx_0(K)$.

5.4 Generalized Geometric Convex Functions

5.4.1 Introduction

Definition 5.20. Let $n \geq 2$, and let $T \subset K$ be two closed convex sets. The subclass of $Cvx(\mathbb{R}^n)$ consisting of functions above 1_K and below 1_T will be denoted $Cvx_T(K)$, that is

$$Cvx_T(K) := \{f \in Cvx(\mathbb{R}^n) : 1_K \leq f \leq 1_T\}.$$

Remarks. 1. In the case $n = 1$ this definition would still make sense, but it does not really generalize the case of $T = \{0\}$. Indeed, $Cvx_T(K)$ is isomorphic to $Cvx_0([0, 1])$ if $K \setminus T$ is connected, and to $Cvx_0([-1, 1])$ otherwise.
2. When $T = \emptyset$, this is the case of convex functions on a window, $Cvx(K)$.
3. When $T = \{0\}$, this is the case of geometric convex functions on a window, $Cvx_0(K)$.
4. Throughout this section we will assume that K is of dimension n, and that the interior of $K \setminus T$ is connected.

Definition 5.21. Let $\mathcal{T} : Cvx_{T_1}(K_1) \rightarrow Cvx_{T_2}(K_2)$ be an order preserving isomorphism, and $F : K_1 \times \mathbb{R}^+ \rightarrow K_2 \times \mathbb{R}^+$ a fractional linear map such that $epi(Tf) = F(epi(f))$ for every $f \in Cvx_{T_1}(K_1)$. The transform $\mathcal{T}$ and the map F are said to be of i-type in two cases: the first, if F is linear map, and the second, if F is a non affine fractional linear map with its defining hyperplane containing a ray in the $\mathbb{R}^+$ direction. Otherwise, $\mathcal{T}$ and F are said to be of $\mathcal{J}$-type.

Note that this definition coincides with that of the particular case $Cvx_0(K)$, given in the previous Sect. 5.3.2.

Definition 5.22. Let $\mathcal{T} : Cvx_{T_1}(K_1) \rightarrow Cvx_{T_2}(K_2)$ be an order reversing isomorphism. We say that $\mathcal{T}$ is of $\mathcal{A}$-type if the composition $\mathcal{T} \circ \mathcal{A}$ is an order preserving isomorphism of i-type, otherwise we say $\mathcal{T}$ is of $\mathcal{L}$-type.

We deal with order isomorphisms from $Cvx_{T_1}(K_1)$ to $Cvx_{T_2}(K_2)$. We show that order preserving isomorphisms are induced by fractional linear point maps on $K_1 \times \mathbb{R}^+$, which are always of i-type. We show that up to a composition with such transforms, the only order reversing isomorphism is the geometric duality $\mathcal{A}$. It may be formulated for order preserving or for order reversing transforms:

Theorem 5.23. *Let $n \geq 2$, and let $T_1 \subset K_1 \subset \mathbb{R}^n, T_2 \subset K_2 \subset \mathbb{R}^n$ be four non empty, convex, compact sets, and assume that $int(K_i) \neq \emptyset$, and that $int(K_1 \setminus T_1)$ is connected. If $\mathcal{T} : Cvx_{T_1}(K_1) \rightarrow Cvx_{T_2}(K_2)$ is an order preserving isomorphism, then there exists a fractional linear map $F : K_1 \times \mathbb{R}^+ \rightarrow K_2 \times \mathbb{R}^+$ such that for every $f \in Cvx_{T_1}(K_1)$, we have*

$$epi(\mathcal{T}f) = F(epi(f)).$$

Moreover, F is of i-type, and in particular T_2 is a fractional linear image of T_1, and K_2 is a fractional linear image of K_1.

Theorem 5.24. *Let $n \geq 2$, let $K \subset T \subseteq \mathbb{R}^n$ be two convex sets such that $0 \in int(K)$, and assume that K does not contain a full line, and that $int(T \setminus K)$ is connected. Let $T' \subset K' \subset \mathbb{R}^n$ be two non empty, convex, compact sets, and assume that $int(K') \neq \emptyset$. If $\mathcal{T} : Cvx_K(T) \to Cvx_{T'}(K')$ is an order reversing isomorphism, then $\mathcal{T}$ is of $\mathcal{A}$-type. In particular, T', K' are fractional linear images of $T°, K°$ respectively.*

Proof of Theorem 5.24. The composition $\tilde{\mathcal{T}} := \mathcal{T} \circ \mathcal{A} : Cvx_{T°}(K°) \to Cvx_{T'}(K')$ is an order preserving isomorphism, and the assumptions on T and K imply that T, K, T', K' satisfy the conditions of Theorem 5.23. Indeed, $T°, K°$ are non empty convex sets, and $0 \in int(K)$ implies that they are compact. Since K does not contain a full line, $K°$ is not contained in any hyperplane, thus it has non empty interior. It is easy to check that for two convex sets $A \subset B$, $int(B \setminus A)$ is connected if and only if $int(A° \setminus B°)$ is connected, thus we may apply Theorem 5.23. We get that $\tilde{\mathcal{T}}$ is induced by some fractional linear map $F : T° \times \mathbb{R}^+ \to T' \times \mathbb{R}^+$, which is of i-type, thus $\mathcal{T}$ is of $\mathcal{A}$-type, as desired. $\qquad\square$

For the proof of Theorem 5.23, we first need to define and characterize extremal elements in the class $Cvx_T(K)$. Then we show that extremal elements are mapped to such, which will imply that the transform induces a point map on a subset of $\mathbb{R}^{n+1}$. We show that this point map is interval preserving for a sufficiently large set of intervals, in order to use a theorem of Shiffman [15], which states that the map is fractional linear. Finally we show that under our assumptions, the transform is of i-type, thus completing the proof of Theorem 5.23. We will need the following notations throughout this section.

- Let $n \geq 2$, and let $A \subset B \subseteq \mathbb{R}^n$ be two closed convex sets. We denote

$$\mathcal{K}^n(A, B) = \{K \subseteq \mathbb{R}^n : K \text{ is closed, convex, and } A \subseteq K \subseteq B\}.$$

For $\mathcal{K}^n(\emptyset, \mathbb{R}^n)$ we simply write $\mathcal{K}^n$. Note that if $T \neq \emptyset$, any element in $\mathcal{K}^{n+1}(epi(1_T), epi(1_K))$ is an epi-graph of some function $f \in Cvx_T(K)$.
- For the convex hull of two sets A and B we write

$$A \vee B = \bigcap_{K \in \mathcal{K}^n, (A \cup B) \subseteq K} K.$$

5.4.2 Extremal Elements

Definition 5.25. A set $K \in \mathcal{K}^n(A, B)$ is called *extremal* if $\quad \forall T, P \in \mathcal{K}^n(A, B)$:

$$K = T \vee P \quad \Longrightarrow \quad K = T, \quad \text{or} \quad K = P.$$

Definition 5.26. A function $f \in Cvx_T(K)$ is called *extremal* if $\forall g, h \in Cvx_T(K)$:

$$f = \hat{\inf}\{h, g\} \quad \Longrightarrow \quad f = h, \quad \text{or} \quad f = g.$$

Another formulation of which is:

$$epi(f) = epi(h) \vee epi(g) \quad \Longrightarrow \quad f = h, \quad \text{or} \quad f = g,$$

which (in the case $T \neq \emptyset$), means that $epi(f)$ is extremal in $\mathcal{K}^{n+1}(epi(1_T), epi(1_K))$.

Recall that for bijective transforms, order-preservation in both directions is equivalent to preservation of the lattice operations $\hat{\inf}$ and sup (see Proposition 3.5 and Remark 4.5). Since the extremality property is defined by the $\hat{\inf}$ operation, all extremal elements in the domain are mapped to all extremal elements in the range.

In the next few lemmas we investigate extremal elements of $Cvx_T(K)$. We need the following simple observation.

Lemma 5.27. *Let $\varphi : \mathbb{R}^n \to \mathbb{R}$ be an affine linear functional and $K \subset \mathbb{R}^n$ a closed, convex set that does not contain a ray on which φ is constant. If $\varphi(K) > 0$, then there exists some $c \in \mathbb{R}$ such that $\varphi(K) \geq c > 0$.*

Proof. Consider the slab $S = \varphi^{-1}([0, 1])$. If the intersection $K \cap S$ is empty then we may take $c = 1$. Assume otherwise, then $K \cap S$ is a closed convex set, and moreover, it is bounded. Indeed, the slab S contains only rays on which φ is constant, and K contains no such rays, therefore $K \cap S$ contains no rays, and one can easily verify that for a convex set this is equivalent to boundedness. Since $K \cap S$ is compact and φ is continuous, there exists $x_0 \in K$ such that $\varphi(K) \geq \varphi(x_0) \equiv c > 0$. $\qquad\square$

Lemma 5.28. *Let $n \geq 2$, and let $T \subset K \subset \mathbb{R}^n$ be two non empty, compact, convex sets. Consider the subsets $A = T \times \mathbb{R}^+$, $B = K \times \mathbb{R}^+$ of $\mathbb{R}^n \times \mathbb{R} = \mathbb{R}^{n+1}$. If $K \in \mathcal{K}^{n+1}(A, B)$ is extremal, then $K = A \vee \{x\}$, for some $x \in B$.*

Proof of Lemma 5.28. Let $K \in \mathcal{K}^{n+1}(A, B)$ be extremal. By a Krein-Milman type theorem for non compact sets, see [11], K is the convex hull of its extreme points and extreme rays. Since the only rays in K are translates of $\{0\} \times \mathbb{R}^+$, and any extreme ray must emanate from an extreme point, K is the convex hull of A and its extreme points. Finally, since the set of exposed points is dense in the set of extreme points, see [17], if we denote by E the set of *exposed* extreme points of K which are not in A, we have $K = A \vee E$ (actually in [17] this is proved for compact convex sets, but the non compact case follows as an immediate consequence, and also appears in a more general setting of normed spaces in [12], as Theorem 2.3).

Let $x_1 \in E$, and let φ_1 be an affine functional such that $\varphi_1(K \setminus \{x_1\}) > 0$ and $\varphi_1(x_1) = 0$. Note that φ_1 cannot be constant on translates of $\{0\} \times \mathbb{R}^+$, since then it would be constant 0 on the translate of $\{0\} \times \mathbb{R}^+$ emanating from x_1, contradicting strict positivity on $K \setminus \{x_1\}$. If $E \subseteq A \vee \{x_1\}$, the proof is complete.

Assume otherwise; that there exists $x_2 \in E \setminus (A \vee \{x_1\})$. We may separate x_2 from the closed set $A \vee \{x_1\}$ by an affine functional φ_2 such that $\varphi_2(A \vee \{x_1\}) > 0$ and $\varphi_2(x_2) < 0$. Denote by H_2^- the (closed) half space on which $\varphi_2 \leq 0$ and by H_2^+ the (closed) half space on which $\varphi_2 \geq 0$. Consider the sets $K^+ = A \vee (E \cap H_2^+)$, $K^- = A \vee (E \cap H_2^-)$. Clearly $K^i \in \mathcal{K}^{n+1}(A, B)$ and $K = K^+ \vee K^-$, thus by extremality of K we must have either $K = K^+$ or $K = K^-$. Since both A and $E \cap H_2^+$ are contained in H_2^+, so is K^+, thus $x_2 \notin K^+$. This implies $K \neq K^+$, i.e. $K = K^-$. We next show that $x_1 \notin K^-$, which leads to the wanted contradiction. To this end we claim that $\varphi_1(K^-) > 0 = \varphi_1(x_1)$. Indeed, $\varphi_1(A) > 0$, and the only rays contained in A are translates of $\{0\} \times \mathbb{R}^+$, on which φ_1 is not constant. Thus, by Lemma 5.27, there exists some constant c such that $\varphi_1(A) \geq c > 0$. Similarly $\varphi_1(K \cap H_2^-) \geq c' > 0$. For the convex hull we get $\varphi_1(K^-) \geq \min\{c, c'\} > 0$, so $x_1 \notin K^-$.

The following is a simpler version of Lemma 5.28, which we do not use in this paper but add it to complete the picture.

Lemma 5.29. *Let $n \geq 2$, and let $A \subset B \subset \mathbb{R}^n$ be two compact convex sets. If $K \in \mathcal{K}^n(A, B)$ is extremal, then $K = A \vee \{x\}$, for some $x \in B$.*

We omit the proof, as it is contained in the proof of the previous lemma (the use of Lemma 5.27 is replaced by a straightforward compactness argument).

A reformulation of Lemma 5.28 is:

Lemma 5.30. *Let $n \geq 2$, and let $T \subset K \subset \mathbb{R}^n$ be two non empty, compact, convex sets. If $f \in Cvx_T(K)$ is extremal, then either:*

- $f = 1_T$, *or:*
- $f = \widehat{\inf}\{1_T, \delta_{k,h}\}$ *for some $k \in K \setminus T$ and $h \geq 0$.*

Proof of Lemma 5.30. By Lemma 5.28, $epi(f) = epi(1_T) \vee \{x\}$ for some $x \in epi(1_K)$. If $x \in epi(1_T)$, then $f = 1_T$. If $x \notin epi(1_T)$, then $f = \widehat{\inf}\{1_T, \delta_{k,h}\}$ for some k, h as stated above. $\square$

5.4.3 The Point Map

So far we have seen that an order isomorphism $\mathcal{T} : Cvx_{T_1}(K_1) \to Cvx_{T_2}(K_2)$ is in particular a bijection between the extremal families. Clearly $\mathcal{T}(1_{T_1}) = 1_{T_2}$. Aside of the maximal element 1_{T_1}, each extremal function in $Cvx_{T_1}(K_1)$ corresponds to a point in $\mathbb{R}^{n+1}$, thus $\mathcal{T}$ induces a bijective point map $F : (K_1 \setminus T_1) \times \mathbb{R}^+ \to (K_2 \setminus T_2) \times \mathbb{R}^+$.

Denote by E_1 the interior of the set $(K_1 \setminus T_1) \times \mathbb{R}^+$ (by our assumption, it is connected). The sets $(K_i \setminus T_i) \times \mathbb{R}^+$ inherit the partial order structure of $Cvx_{T_i}(K_i)$, after restriction to the set of extremal elements, and the bijective map $F : (K_1 \setminus T_1) \times \mathbb{R}^+ \to (K_2 \setminus T_2) \times \mathbb{R}^+$ is an order isomorphism. In the following lemmas we will use the fact that the injective map $F|_{E_1} : E_1 \to (K_2 \setminus T_2) \times \mathbb{R}^+$ is an order isomorphism on its image, to prove that for *some* intervals $[a, b] \subset E_1$,

$F([a, b])$ is again an interval (these can be characterized as the ones that, extended to a full line, do not intersect $epi(1_{T_1})$). Since the use of the uniqueness Theorem 2.17 requires the preservation of *all* intervals, we apply a result by Shiffman from [15], which roughly states that if a set of points is covered by an open set of intervals which are all mapped to intervals, then the inducing map is fractional linear. More precisely, denote by $\mathcal{L}(\mathbb{R}^n)$ the set of all lines in $\mathbb{R}^n$, not necessarily intersecting the origin. It may be seen as a subset of the Grassmannian $G_{n+1,2}$, therefore it is equipped with the usual inherited metric topology (for some details see Remark 4 below). Denoting by $\mathcal{L}(U) \subseteq \mathcal{L}(\mathbb{R}^n)$ the set of all such lines intersecting a given set $U \subseteq \mathbb{R}^n$, we have

Theorem 5.31 ([15]). *Let $n \geq 2$, let U be an open connected set in $\mathbb{R}^n$, and let $\mathcal{L}_0$ be an open subset of $\mathcal{L}(U)$, which covers U, i.e. $U \subseteq \bigcup_{l \in \mathcal{L}_0} l$. Assume that $F : U \to \mathbb{R}^n$ is a continuous injective map, and that $F(l \cap U)$ is contained in a line for all $l \in \mathcal{L}_0$. Then F is fractional linear.*

Remarks. 1. Theorem 5.31 is adjusted to the real, linear, setting (i.e. when U is a subset of $\mathbb{R}^n$, which is embedded in $\mathbb{R}P^n$), and is a particular case of the more general statement Shiffman proves in [15]. The general result applies for subsets of $\mathbb{R}P^n$ or $\mathbb{C}P^n$, and states that the map F is projective linear.
2. In [15], Theorem 5.31 is proved for $\mathbb{R}P^n$ and $\mathbb{C}P^n$ simultaneously. However, considering only the case of $\mathbb{R}P^n$, one may check (by following the proof in [15]), that in this case continuity is actually not required, and may be replaced by the following weaker condition; if $I \subset U$ is an interval and $I \subset l \in \mathcal{L}_0$, then $F(I)$ is again an interval. We will use this stronger version of Theorem 5.31.
3. In our setting, we have epi-graphs of functions in $Cvx_T(K)$, therefore we apply Theorem 5.31 to the function F defined on the set $U = E_1 \subset \mathbb{R}^{n+1}$.
4. A line in $\mathcal{L}(\mathbb{R}^n)$ is determined by its closest point to the origin and its direction. That is, for every $l \in \mathcal{L}(\mathbb{R}^n)$ let $x_l \in l$ be the unique point satisfying $|x_l| = \min\{|x| : x \in l\}$, and let $u_l \in \{x_l\}^\perp$ be one of the two points satisfying $l = \{x_l + tu_l, t \in \mathbb{R}\}$, $|u_l| = 1$ (the other being $-u_l$). Note that directions in $\{x_l\}^\perp$ correspond to S^{n-2} if $x_l \neq 0$, and to S^{n-1} if $x_l = 0$. Denoting the line l by the pair (x_l, u_l), we get a correspondence between $\mathcal{L}(\mathbb{R}^n) \subset G_{n+1,2}$ and $((\mathbb{R}^n \setminus \{0\}) \times S^{n-2}) \bigcup (\{0\} \times S^{n-1})$, which is 1–1, modulo the $\pm$ choice in the direction u. The metric d on $\mathcal{L}(\mathbb{R}^n)$ is inherited from that on $G_{n+1,2}$, and it follows that $d((x, u_1), (x, u_2)) = |u_1 - u_2|$, and that $d((x_1, u), (x_2, u)) = d((\hat{x_1}, 1), (\hat{x_2}, 1))$, where $\hat{x} = \frac{x}{|x|}$.
   A neighborhood of (x, u) is therefore constructed by perturbing simultaneously x and u. It can be checked that such a perturbation contains the following "cylinder" of lines; fix a point $z \in (x, u)$, let $M > 0$, let $a, b \in (x, u)$ satisfy $|a - z| = |b - z| = M$, and let A, B be open balls of radius $1/M$ and centers a, b respectively. We take our "cylinder" of lines to be $\mathcal{L}_{l,z,M} := \mathcal{L}(A) \cap \mathcal{L}(B)$. For every $z \in l$ and every $M > 0$, there exists a small perturbation of $l = (x, u)$ which is contained in $\mathcal{L}_{l,z,M}$. More precisely, there exists $\varepsilon > 0$ such that $G_{l,\varepsilon} = \{(y, v) : |y - x| + |v - u| < \varepsilon\} \subset \mathcal{L}_{l,z,M}$. This fact is useful in the proof of Lemma 5.32.

Let $\tilde{\mathcal{L}}_0 := \mathcal{L}(E_1) \setminus \mathcal{L}(T_1 \times \mathbb{R}^+)$, that is, $\tilde{\mathcal{L}}_0$ is the set of lines through E_1 (the domain of F), which do not intersect the inner half cylinder $T_1 \times \mathbb{R}^+$. In Lemma 5.32 we prove that the interior of $\tilde{\mathcal{L}}_0$, denoted $\mathcal{L}_0$, is an open subset of $\mathcal{L}(E_1)$ which covers E_1, and in Lemma 5.34 we prove that $F(l \cap E_1)$ is contained in a line for all $l \in \mathcal{L}_0$, and that intervals which are segments of lines in $\mathcal{L}_0$ are mapped to intervals.

Lemma 5.32. *The open set $\mathcal{L}_0 = int(\tilde{\mathcal{L}}_0)$ described above, covers E_1. That is,*

$$E_1 \subseteq \bigcup_{l \in \mathcal{L}_0} l.$$

Proof of Lemma 5.32. Let $x \in E_1$. We may separate x from the closed set $T_1 \times \mathbb{R}^+$ by a hyperplane. Denote by H the translate of this hyperplane containing x. We claim that if $l \subset H$ is a line containing x, which is not parallel to the ray $\{0\} \times \mathbb{R}^+$, then $l \in \mathcal{L}_0$. Indeed, it is clear that $l \in \tilde{\mathcal{L}}_0$. Consider the set of lines $\mathcal{L}_{l,x,M}$, for some $M > 0$ (see the last remark). It is an open neighborhood of l, and since T_1 is compact and l is not parallel to $\{0\} \times \mathbb{R}^+$, we have (for large enough M) that $\mathcal{L}_M \subset \tilde{\mathcal{L}}_0$, thus $l \in \mathcal{L}_0$. This implies $x \in \bigcup_{l \in \mathcal{L}_0} l$, and hence $\mathcal{L}_0$ covers E_1.   □

Next we prove that the set $\mathcal{L}_0$ consists exactly of all the lines in $\mathcal{L}(E_1)$, with the property that points along these lines are non comparable.

Lemma 5.33. *Let $a, b \in E_1$ be two different points, and let $l_{a,b}$ be the line containing a and b. Then $l_{a,b} \notin \mathcal{L}_0$ if and only if a and b are comparable.*

Proof of Lemma 5.33. The point a is "greater" than the point b, if and only if $a \in epi(1_{T_1}) \vee \{b\}$, therefore a and b are comparable if and only if $l_{a,b}$ is in the closure of $\mathcal{L}(\{b\}) \cap \mathcal{L}(T_1 \times \mathbb{R}^+) \subset \mathcal{L}(E_1) \setminus \tilde{\mathcal{L}}_0$ (in fact, the closure is only necessary if a and b are on the same translate of $\{0\} \times \mathbb{R}^+$). This closure does not intersect $\mathcal{L}_0$, the interior of $\tilde{\mathcal{L}}_0$, therefore we have shown:

$$a, b \text{ are comparable} \quad \Rightarrow \quad l_{a,b} \notin \mathcal{L}_0.$$

If $l_{a,b} \notin \mathcal{L}_0$ there are two cases. First assume $l_{a,b} \notin \tilde{\mathcal{L}}_0$ (that is, $l_{a,b}$ intersects $T_1 \times \mathbb{R}^+$). Thus a and b are comparable (one is in the convex hull of the other and $epi(1_{T_1})$). Otherwise, assume $l_{a,b} \in \tilde{\mathcal{L}}_0$. Since it is not in the interior $\mathcal{L}_0$, we get $l_{a,b} \in \partial \tilde{\mathcal{L}}_0$. Since $\mathcal{L}(E_1) \setminus \bigcup_{x \in K_1 \setminus T_1} \{x\} \times \mathbb{R}$ is open, and $\mathcal{L}(T_1 \times \mathbb{R}^+)$ is closed, $\mathcal{L}_0 = int(\tilde{\mathcal{L}}_0)$ must contain $\left(\mathcal{L}(E_1) \setminus \bigcup_{x \in K_1 \setminus T_1} \{\{x\} \times \mathbb{R}\} \right) \setminus \mathcal{L}(T_1 \times \mathbb{R}^+)$, and therefore $\mathcal{L}_0 = \tilde{\mathcal{L}}_0 \setminus \bigcup_{x \in K_1 \setminus T_1} \{x\} \times \mathbb{R}$. Thus $l_{a,b} \in \partial \tilde{\mathcal{L}}_0$ implies that $l_{a,b}$ is parallel to the ray $\{0\} \times \mathbb{R}$, and that a and b are comparable. Thus we have shown:

$$l_{a,b} \notin \mathcal{L}_0 \quad \Rightarrow \quad a, b \text{ are comparable}.$$

□

Lemma 5.34. *If $l \in \mathcal{L}_0$, then $F(l \cap E_1)$ is contained in a line. Moreover, if $I \subset E_1$ is an interval and $I \subset l \in \mathcal{L}_0$, then $F(I)$ is again an interval.*

Proof of Lemma 5.34. The intersection of every $l \in \mathcal{L}_0$ with the convex set $K_1 \times \mathbb{R}^+$ is either a ray or an interval. Since $l \in \mathcal{L}_0$, it does not intersect $T_1 \times \mathbb{R}^+$, and therefore also $l \cap E_1$ is either a ray or an interval. Thus, by Lemma 5.33, it is enough to show that for every two non comparable points $a, b \in E_1$, we have $F([a,b]) = [F(a), F(b)]$. Denote for every $x \in E_1$ by δ_x the function with epigraph $(T_1 \times \mathbb{R}^+) \vee \{x\}$. Of all the extremal functions δ_x, only those corresponding to $x \in [a,b]$ have the following minimality property: $\delta_x \geq \hat{\inf}\{\delta_a, \delta_b\}$, and for every y with $\delta_y \geq \hat{\inf}\{\delta_a, \delta_b\}$, we have $\delta_y \not< \delta_x$. This property is preserved by F, therefore the interval $[a,b]$ is mapped to the interval $[F(a), F(b)]$. $\square$

Proof of Theorem 5.23. The set E_1 is open and connected. Therefore, by Lemmas 5.32 and 5.34, we may apply Theorem 5.31 (see Remark 2 after Theorem 5.31) to the map $F|_{E_1}$, and conclude it is fractional linear. To see that $F : (K_1 \setminus T_1) \times \mathbb{R}^+ \to (K_2 \setminus T_2) \times \mathbb{R}^+$ is fractional linear, note that a point in the boundary of $(K_1 \setminus T_1) \times \mathbb{R}^+$ is the infimum of all the points below it which are in E_1. To see that F induces the transform $\mathcal{T} : Cvx_{T_1}(K_1) \to Cvx_{T_2}(K_2)$, note that the epi-graph of a function $f \in Cvx_{T_i}(K_i)$ corresponds to the set of extremal functions above it, and that f is given as the infimum of those extremal functions. Finally we need to show that F is of i-type, that is, assuming F is a non affine fractional linear map, we need to show that the defining hyperplane is parallel to the $\mathbb{R}^+$ direction. If it is not, then by Sect. 2.4.2, the half cylinder $K_1 \times \mathbb{R}^+$ is mapped to some cone, which must be $K_2 \times \mathbb{R}^+$. But since K_2 is compact, $K_2 \times \mathbb{R}^+$ is not a cone. Therefore the map F is either affine, or it is non affine, but with a defining hyperplane containing the direction of the epi-graphs (the ray $\{0\} \times \mathbb{R}^+$). $\square$

References

1. S. Alesker, S. Artstein-Avidan, V. Milman, in *A Characterization of the Fourier Transform and Related Topics*. Linear and Complex Analysis. Amer. Math. Soc. Transl. Ser. 2, vol. 226 (Amer. Math. Soc., Providence, 2009), pp. 11–26
2. S. Alesker, S. Artstein-Avidan, D. Faifman, V. Milman, A characterization of product preserving maps with applications to a characterization of the Fourier transform. Illinois J. Math. **54**, 1115–1132 (2010)
3. E. Artin, *Geometric Algebra* (Wiley-Interscience, NY, 1988)
4. S. Artstein-Avidan, V. Milman, A characterization of the concept of duality. Electronic Research Announcements in Mathematical Sciences, AIMS **14**, 48–65 (2007)
5. S. Artstein-Avidan, V. Milman, The concept of duality for measure projections of convex bodies. J. Funct. Anal. **254**, 2648–2666 (2008)
6. S. Artstein-Avidan, V. Milman, The concept of duality in convex analysis, and the characterization of the Legendre transform. Ann. Math. (2) **169**(2), 661–674 (2009)
7. S. Artstein-Avidan, V. Milman, Hidden structures in the class of convex functions and a new duality transform. J. Eur. Math. Soc. **13**(4), 975–1004 (2011)

8. S. Artstein-Avidan, B.A. Slomka, A new fundamental theorem of affine geometry and applications. Preprint
9. S. Artstein-Avidan, H. König, V. Milman, The chain rule as a functional equation. J. Funct. Anal. **259**, 2999–3024 (2010)
10. K. Böröczky, R. Schneider, A characterization of the duality mapping for convex bodies. Geom. Funct. Anal. **18**, 657–667 (2008)
11. V.L. Klee, Extremal structure of convex sets. Arch. Math. **8**, 234–240 (1957)
12. V.L. Klee, Extremal structure of convex sets II. Math. Zeitschr. **69**, 90–104 (1958)
13. V.V. Prasolov, V.M. Tikhomirov, *Geometry* (English summary). Translated from the 1997 Russian original by O.V. Sipacheva. Translations of Mathematical Monographs, vol. 200. (American Mathematical Society, Providence, 2001)
14. R.T. Rockafellar, *Convex Analysis* (Princeton University Press, Princeton, 1970)
15. B. Shiffman, Synthetic projective geometry and Poincaré's theorem on automorphisms of the ball. Enseign. Math. (2) **41**(3–4), 201–215 (1995)
16. B.A. Slomka, On duality and endomorphisms of lattices of closed convex sets. Adv. Geom. **11**(2), 225–239 (2011)
17. S. Straszewicz, Über exponierte Punkte abgeschlossener Punktmengen. Fund. Math. **24**, 139–143 (1935)

Finite Transitive Graph Embeddings into a Hyperbolic Metric Space Must Stretch or Squeeze

Itai Benjamini and Oded Schramm

Abstract The δ-hyperbolicity constant of a finite vertex transitive graph with more than two vertices is proportional to its diameter. This implies that any map from such a graph into a 1-Gromov hyperbolic metric space has to stretch or squeeze the metric.

Contrary to the situation in Euclidean space, there is a finite constant δ such that for every triangle T in hyperbolic space $\mathbb{H}^n$ there is a point $p \in \mathbb{H}^n$ whose distance from each of the three edges of T is at most δ. This observation led I. Rips to the following definition.

Definition 1 (δ-hyperbolic). Let X be a geodesic metric space. If there is a finite $\delta > 0$ such that for every geodesic triangle T in X there is a point $p \in X$ whose distance from the three edges is at most δ, then X is said to be δ-hyperbolic. The least such δ is denoted by $\delta(X)$. If $\delta(X) < \infty$, then we say that X is hyperbolic.

Recall that a geodesic segment in a metric space X is a subset that is isometric with an interval in $\mathbb{R}$ and that X is geodesic if for every pair of distinct points $x, y \in X$ is contained in at least one geodesic segment. A geodesic segment with endpoints x and y will be denoted by $[x, y]$ (with the understanding that this notation is sometimes ambiguous). A geodesic triangle T in X is a triple of geodesic segments of the form $[x, y], [y, z]$ and $[z, x]$.

Hyperbolic groups, also known as Gromov-hyperbolic groups [4], are groups whose Cayley graph is hyperbolic (hyperbolicity is independent of the choice of

I. Benjamini (✉)
Faculty of Mathematics and Computer Science, The Weizmann Institute of Science, P.O. Box 26, Rehovot 76100, Israel
e-mail: itai.benjamini@weizmann.ac.il

O. Schramm
Oded died on September 1, 2008, while solo climbing Guye Peak in Washington State

B. Klartag et al. (eds.), *Geometric Aspects of Functional Analysis*, Lecture Notes in Mathematics 2050, DOI 10.1007/978-3-642-29849-3_5,
© Springer-Verlag Berlin Heidelberg 2012

generators). There is much literature about hyperbolic metric spaces and groups, see, for instance [3].

Definition 2 (transitive graph). A graph G is *vertex transitive* if for any pair of vertices v and u of G, there is an automorphism of G mapping v to u.

Graphs may be viewed as metric spaces, with the graph distance. Our first observation is the following

Theorem 1. *There is a constant $c > 0$ such that for every finite vertex-transitive graph G with at least 3 vertices*

$$\delta(G) > c \operatorname{diam}(G).$$

The proof gives $c \geq 1/8$. The note [2] contains a related but weaker result for expanders. There are infinite vertex transitive hyperbolic graphs, e.g., regular trees.

The theorem will be used to show that any embedding of a transitive graph of large diameter into a hyperbolic metric space cannot preserve the metric. In order to state this result we define the *stretch* of a map.

Definition 3. The *distortion* of a map $f : X \to Y$ between metric spaces $(X, d_X), (Y, d_Y)$ is given by $D(f) = D_+(f)/D_-(f)$, where

$$D_+(f) = \max\left\{1, \sup_{a,b \in X} \frac{d_Y(f(a), f(b))}{1 + d_X(a, b)}\right\},$$

$$D_-(f) = \min\left\{1, \inf_{a,b \in X} \frac{1 + d_Y(f(a), f(b))}{d_X(a, b)}\right\}.$$

Such a choice of values for D_+ and D_- allows us to estimate the distortion of quasi-isometries, for example. In general such maps are not injective and not continuous and, moreover, every point of X can have many images in Y. An inverse map of a quasi-isometry is also a quasi-isometry.

Definition 4. A map $f : X \to Y$ of two metric spaces is called the (λ, c)-quasi-isometry if for any two points u_1 and u_2 of X we have

$$\frac{1}{\lambda}|u_1 - u_2| - c \leq |f(u_1) - f(u_2)| \leq \lambda|u_1 - u_2| + c.$$

The natural case of a quasi-isometry is when $c << \lambda$. The distortion of this map is $D \leq \lambda^2$. For a dilation the distortion is equal to the coefficient of dilation. As general hyperbolic graphs have non-degenerated triangles which are isomorphic to cycles, it is impossible to represent them injectively in a tree.

Corollary 2. *The metric distortion between finite vertex transitive graph and a tree is proportional to the diameter.*

By [2] similar result holds for expanders.

Gromov [4] proved that any n points in an hyperbolic metric space can be embedded in a tree with a distortion bounded by $O(\delta \log n)$.

Question 3. What bound can we get when the tree is replaced by a graph with some fixed finite δ-hyperbolicity?

1 Proofs

We start with Theorem 1.

A quasi-proof. Suppose that it does not have a fat triangles. Then we have a sequence of finite graphs G_n where the ratio of the diameter to the hyperbolicity constant goes to infinity. Rescale G_n so that the diameter is 1 and take a subsequential limit. The limit will be a transitive 0-hyperbolic metric space with diameter 1; necessarily a tree (a $\mathbb{R}$-tree). But bounded metric trees are not transitive, since they have endpoints and midpoints.

Proof. Suppose G is finite and transitive, and d is its diameter. Let A and D realize the diameter, i.e. $AD = d$. By transitivity there is a geodesic path BC that has D as its midpoint and length d. Let δ be the hyperbolicity constant. By its definition, there is a point X on BC such that the distance from X to AC is at most 2δ and the distance from X to AB is at most 2δ. Suppose, w.l.o.g. that X is closer to B than to C. We have $XB + XC = d$, $AX + XB < 2\delta + d$ (because X is within 2δ of AB), $AX + XC < 2\delta + d$. Add these latter two and subtract the previous equality, and get $AX < d/2 + 2\delta$. Since $AD = d$, this means that $DX > d/2 - 2\delta$. Since X is on BC and closer to B, this means that $BX < 2\delta$. Since X is within 2δ from AC, we have $AC > AX + XC - 2\delta$. Since $BX < 2\delta$ and $BC = d$ this gives $AC > AX + d - 4\delta$. Since AC is at most d, this implies $AX < 4\delta$. But DX is at most $d/2$. so $d = AD \leq AX + DX < 4\delta + d/2$ So $d < 8\delta$.

Question 4. What is the right constant in the theorem?

We turn to the corollary.

Proof. A geodesic triangle (as well as a geodesic path) is a structure that is roughly preserved by a map with a small distortion. The corollary follows from the following lemma applied to the fat triangle contained in the vertex transitive graph.

Lemma 5. If a map from G to H has distortion D, and G has a geodesic triangle C of length proportional to the diameter and hyperbolicity constant proportional to the diameter, then there is a simple self avoiding cycle $C' \subset H$ of length proportional to the diameter over D.

The lemma is a straight forward adaptation of lemma 5 in [1].

Regarding distortion of embedding into general hyperbolic spaces we have a weaker result. It is known [4] that in δ-hyperbolic spaces, quasi geodesics are within bounded distance, depending only on δ, of a true geodesic. This implies that the distortion of embedding geodesic triangles with hyperbolicity constant growing to ∞ into 1-hyperbolic space must grow to ∞. As a bounded distortion image of a geodesic triangle is a quasi geodesic triangle, which is within a bounded distance from a true geodesic triangle. A quantitative analysis of this will be useful in answering Question 2.

References

1. O. Angel, I. Benjamini, Phase transition for the metric distortion of percolation on the hypercube. Combinatorica **27**, 645–658 (2007)
2. I. Benjamini, Expanders are not hyperbolic. Isarel J. Math. **108**, 33–36 (1998)
3. E. Ghys, A. Haefliger, A. Verjovsky (eds.), *Groups Theory from a Geometrical Viewpoint* (World Scientific, Singapore, 1991)
4. M. Gromov, *Hyperbolic Groups, Essays in Group Theory*, ed. by S. Gersten. MSRI Publications, vol. 8 (Springer, Berlin, 1987), pp. 75–265

Tightness of Fluctuations of First Passage Percolation on Some Large Graphs

Itai Benjamini and Ofer Zeitouni

Abstract The theorem of Dekking and Host [Probab. Theor. Relat. Fields **90**, 403–426 (1991)] regarding tightness around the mean of first passage percolation on the binary tree, from the root to a boundary of a ball, is generalized to a class of graphs which includes all lattices in hyperbolic spaces and the lamplighter graph over $\mathbb{N}$. This class of graphs is closed under product with any bounded degree graph. Few open problems and conjectures are gathered at the end.

1 Introduction

In *First Passage Percolation* (FPP) random i.i.d lengths are assigned to the edges of a fixed graph. Among other questions one studies the distribution of the distance from a fixed vertex to another vertex or to a set, such as the boundary of a ball in the graph, see e.g. [9] for background. Formally, given a rooted, undirected graph $G = (V, E)$ with root o, let D_n denote the collection of vertices at (graph) distance n from the root. For $v \in D_n$, let $\mathcal{P}_v$ denote the collection of paths $(v_0 = o, v_1, v_2, v_3, \ldots, v_k = v)$ (with $(v_{i-1}, v_i) \in E$) from o to v. Given a collection of positive i.i.d. $\{X_e\}_{e \in E}$, define, for $v \in E$,

$$Z_v = \min_{p \in \mathcal{P}_v} \sum_{e \in p} X_e . \tag{1}$$

Because of the positivity assumption on the weights, we may and will assume that any path in $\mathcal{P}_v$ visits each vertex of G at most once.

I. Benjamini (✉) · O. Zeitouni
Faculty of Mathematics and Computer Science, The Weizmann Institute of Science, P.O. Box 26, Rehovot 76100, Israel
e-mail: itai.benjamini@weizmann.ac.il; ofer.zeitouni@weizmann.ac.il

Both authors were supported by their respective Israel Science Foundation grants.

B. Klartag et al. (eds.), *Geometric Aspects of Functional Analysis*, Lecture Notes in Mathematics 2050, DOI 10.1007/978-3-642-29849-3_6, © Springer-Verlag Berlin Heidelberg 2012

For n integer, let $Z_n = \min_{v \in D_n} Z_v$. Under a mild moment condition on the law of the random lengths X_e, Dekking and Host [6] proved that for any regular tree, $Z_n - EZ_n$, the random distance from the root to D_n minus its mean, is tight. (Recall that a sequence of real valued random variables $\{X_n\}_{n \geq 0}$ is tight iff for any $\epsilon > 0$, there is some $r_\epsilon \in \mathbb{R}$, so that for all n, $P(|X_n| > r_\epsilon) < \epsilon$.)

We formulate here a simple and general property of the underling graph G and prove that for graphs satisfying this property and a mild condition on the law of X_e, the collection $\{Z_n - EZ_n\}_{n \geq 0}$ is tight. Lattices in real hyperbolic spaces $\mathbb{H}^d$, the graph of the lamplighter over $\mathbb{N}$, as well as graphs of the form $G \times H$ where G satisfies the conditions we list below and H is any bounded degree graph, are shown to possess this property. (In passing, we mention that the Euclidean case is wide open; it is known that in two dimensions the fluctuations of the distance are not tight, see [12,13], however only very poor upper bounds are known [1]. For a special solved variant see [8].)

In the next section we formulate the geometric condition on the graph and the assumption on the distribution of the edge weights $\{X_e\}_{e \in V}$; we then state the tightness result, Theorem 2.1, which is proved in Sect. 3. We conclude with a few open problems.

2 A Recursive Structure in Graphs and Tightness

Throughout, let $dist_G$ denote the graph distance in G. The following are the properties of G and the law of X_e alluded to above.

(1) G contains two vertex-disjoint subgraphs G_1, G_2, which are isomorphic to G.
(2) There exists $K < \infty$ so that $EX_e < K$, and

$$dist_G(Root_G, Root_{G_1}) = dist_G(Root_G, Root_{G_2}).$$

One can replace Property (2) by the following.

(3) $X_e < K$ a.s., and every vertex at distance n from the root is connected to at least one vertex of distance $n + 1$.

Properties (1) and (2) imply that the binary tree embeds quasi-isometrically into G, thus G has exponential growth. Property (3) is called having "no dead ends" in geometric group theory terminology.

Theorem 2.1. *Assume Property (1) and either Property (2) or Property (3). Then the sequence $\{(Z_n - EZ_n)\}_{n \geq 1}$ is tight.*

Note that a hyperbolic lattice in $\mathbb{H}^d$, $d \geq 2$, intersected with a half space, admits the graph part of Properties (1) and (2) above (and probably (3) as well but we don't see a general proof). This is due to topological transitivity of the action on the space of geodesics, i.e. pairs of point of the boundary. There exist elements g in the authomorphism group of the hyperbolic space that map the half space into arbitrarily small open sets of the boundary and elements of this group map the lattice orbit to itself. Note also that by the Morse lemma of hyperbolic geometry (see, e.g., [4] p. 175), if one assumes in addition that $X_e \geq \delta > 0$ a.s. then a path with minimal FPP length will be within a bounded distance from a hyperbolic geodesic and will not wind around, thus tightness for half space for weights that are bounded below by a uniform positive constant implies tightness for the whole space. (Recall also that the regular tree is a lattice in $\mathbb{H}^2$; see [10] for some nice pictures of other planar hyperbolic lattices.)

An example satisfying Properties (1) and (2) is given by the semi group of the lamplighter over $\mathbb{N}$. Recall the graph of the lamplighter over $\mathbb{N}$: a vertex corresponds to a scenery of 0's and 1's over $\mathbb{N}$, with finitely many 1's with a position of a lamplighter in $\mathbb{N}$; edges either change the bit at the position of the lamplighter or move the lamplighter one step to the left or the right, see, e.g., [10]. If we fix the left most bit and restrict the lamplighter to integers strictly bigger than 1, we get the required G_0 and G_1.

It easy to see that if G satisfies the properties in the theorem then $G \times H$ will too. In particular the theorem applies to $T \times T'$ for two regular trees. Note also that if G satisfies the Property (1) in the theorem, then the lamplighter over G will admit it as well.

3   Proof of Theorem 2.1

The proof is based on a modification of an argument in [6]; a related modification was used in [3]. Note first that, by construction,

$$\text{(a) } EZ_{n+1} \geq EZ_n,$$

because to get to distance $n+1$ a path has to pass through distance n and the weights $\{X_e\}$ are positive.

Under Property (3), one has in addition

(a') Z_n and Z_{n+i} can be constructed on the same space so that

$$Z_{n+i} \geq Z_n \text{ while } Z_{n+i} \leq Z_n + Ki.$$

(The first inequality does not need Property (3), but the second does—one just goes forward from the minimum at distance n, i steps.)

On the other hand, from Property (1),

$$EZ_{n+1} \leq E(\min(Z_{n-R_1+1}, Z'_{n-R_2+1}) + KC,$$

where $R_i = dist_G(Root_G, Root_{G_i})$, $C = \max(R_1, R_2)$, and Z'_m denotes a identically distributed independent copy of Z_m.

Since $\min(a, b) = \frac{a+b}{2} - \frac{|a-b|}{2}$,

$$EZ_{n+1} \leq (1/2)[EZ_{n-R_1+1} + EZ'_{n-R_2+1} - E|Z_{n-R_1+1} - Z'_{n-R_2+1}|] + KC.$$

Therefore, with $n_i = n + 1 - R_i$,

$$E|Z_{n_1} - Z'_{n_2}| \leq [-2EZ_{n+1} + EZ_{n_1} + EZ_{n_2}] + 2KC.$$

If $R_1 = R_2$ (i.e. Property (2) holds), then, using (a),

$$E|Z_{n_1} - Z'_{n_1}| \leq 2KC,$$

and the tightness follows by standard arguments, since $E|Z_{n-1} - Z'_{n-1}| \geq E|Z_{n-1} - EZ'_{n-1}| = E|Z_{n-1} - EZ_{n-1}|$ by the independence of Z_{n-1} and Z'_{n-1}, and Jensen's inequality. Otherwise, assume Property (3) with $n_2 > n_1$. By (a'), we can construct a version of Z'_{n_1}, independent of Z_{n_1}, so that $|Z'_{n_2} - Z'_{n_1}| \leq K(n_2 - n_1)$. Therefore,

$$E|Z_{n_1} - Z'_{n_1}| \leq [-2EZ_{n+1} + EZ_{n_1} + EZ_{n_2}] + 2KC + K(R_1 - R_2).$$

Applying again (a) we get, for some constant C',

$$E|Z_{n_1} - Z'_{n_1}| \leq 2KC + K(R_1 - R_2) \leq C'K,$$

and as before it is standard that this implies tightness.   □

4 Questions

Question 1: Extend the theorem to the lamplighter group over Γ, for any finitely generated group Γ; start with $\mathbb{Z}$.

Question 2: Show that tightness of fluctuations is a quasi-isometric invariant. In particular, show this in the class of Cayley graphs.

Question 3: The lamplighter over $\mathbb{Z}$ is a rather small group among the finitely generated groups with exponential growth. It is solvable, amenable and Liouville. This suggests that all Cayley graphs of exponential growth are tight. We ask then which Cayley graphs admit tightness; is there an infinite Cayley graph, which is not quasi-isometric to $\mathbb{Z}$ or $\mathbb{Z}^2$, for which tightness does not hold? Start with a sub exponential example with tightness or even only variance smaller than on $\mathbb{Z}^2$.

Question 4: (Gabor Pete) Note that requiring (1) only quasi-isometrically (plus the root condition of (2)) does not imply exponential growth, because when one iterates, one may collect a factor (from quasiness) each time, killing the exponential growth. E.g., branch groups like Grigorchuk's group [7], where $G \times G$ is a subgroup of G, may have intermediate growth, see e.g. [11]. This condition is somewhat in the spirit of Property (1). Bound the variance for FPP on the Grigorchuk's group.

Maybe ideas related to the one above will be useful in proving at least a sublinear variance?

The last two questions are regarding point to point FPP.

Question 5: We conjecture that in any hyperbolic lattice the point to point FPP fluctuations admit a central limit theorem with variance proportional to the distance. This is motivated by the fact that, due to the Morse lemma, the minimal path will be in a bounded neighborhood of the hyperbolic geodesic, and for cylinders a CLT is known to hold [5].

A related question is the following. Assume that for any pair of vertices in a Cayley graph the variance of point to point FPP is proportional to the distance, is the Cayley graph hyperbolic? Alternatively, what point to point variances can be achieved for Cayley graphs? As pointed out above, the only behavior known is linear in the distance (for $\mathbb{Z}$), the conjectured (and proved in some cases) behavior for $\mathbb{Z}^2$, which is the distance to the power $2/3$. Can the bound or proof of Theorem 2.1 be adapted to give point to point order 1 variance for $T \times \mathbb{Z}^d$ or $T \times T$ or some other graphs? Are other behaviors possible?

Question 6: In [2] tightness was proved for point to point FPP between random vertices in the configuration model of random d-regular graph. Does tightness hold for point to point FPP between random vertices on expanders?

All the questions above are regarding the second order issue of bounding fluctuations. The fundamental fact regarding FPP on $\mathbb{Z}^d$ is the shape theorem, see e.g. [9]. That is, rescale the random FPP metric then the limiting metric space a.s. exists and is $\mathbb{R}^d$ with some deterministic norm. The subadditive ergodic theorem is a key in the proof. We conjecture that FPP on Cayley graph of groups of polynomial growth also admits a shape theorem. What can replace the subadditive ergodic theorem in the proof? Start with

Question 7: Prove a shape theorem for FPP on the Cayley graph of the discrete Heisenberg group.

Acknowledgements Thanks to Pierre Pansu and Gabor Pete for very useful discussions.

References

1. I. Benjamini, G. Kalai, O. Schramm, First passage percolation has sublinear distance variance. Ann. Prob. **31**, 1970–1978 (2003)
2. S. Bhamidi, R. van der Hofstad, G. Hooghiemstra, First passage percolation on the Erdos-Rényi random graph. Combin. Probab. Comput. **20**, 683–707 (2011)

3. E. Bolthausen, J.-D. Deuschel, O. Zeitouni, Recursions and tightness for the maximum of the discrete, two dimensional Gaussian Free Field. Elect. Comm. Prob. **16**, 114–119 (2011)
4. D. Burago, Y. Burago, S. Ivanov, in *A Course in Metric Geometry*. Graduate Studies in Mathematics, vol. 33 (American Mathematical Society, Providence, 2001)
5. S. Chatterjee, P. Dey, Central limit theorem for first-passage percolation time across thin cylinders. Preprint (2010). http://arxiv.org/abs/0911.5702. To appear, Prob. Th. Rel. Fields (2012)
6. M. Dekking, B. Host, Limit distributions for minimal displacement of branching random walks. Probab. Theor. Relat. Fields **90**, 403–426 (1991)
7. R. Grigorchuk, I. Pak, Groups of intermediate growth: An introduction for beginners. Enseign. Math. (2) **54**, 251–272 (2008). http://arxiv.org/abs/math/0607384
8. K. Johansson, On some special directed last-passage percolation models. Contemp. Math. **458**, 333–346 (2008)
9. H. Kesten, in *Aspects of First Passage Percolation*. Lecture Notes in Math., vol. 1180 (Springer, Berlin, 1986), pp. 125–264
10. R. Lyons, Y. Peres, *Probability on Trees and Networks*. In preparation, to be published by Cambridge University Press. Current version available at http://mypage.iu.edu/~rdlyons/.
11. V. Nekrashevych, *Self-Similar Groups*. A.M.S. Mathematical Surveys and Monographs, vol. 117 (2005), Providence, RI
12. C. Newman, M. Piza, Divergence of shape fluctuations in two dimensions. Ann. Probab. **23**, 977–1005 (1995)
13. R. Pemantle, Y. Peres, Planar first-passage percolation times are not tight, in *Probability and Phase Transition (Cambridge, 1993)*. Nato Adv. Sci. Inst. Ser. C., Math Phys. Sci., Volume 420, Kluwer Academic Pub., Dordrecht, 261–264

Finitely Supported Measures on $SL_2(\mathbb{R})$ Which are Absolutely Continuous at Infinity

Jean Bourgain

Abstract We construct finitely supported symmetric probability measures on $SL_2(\mathbb{R})$ for which the Furstenberg measure on $\mathbb{P}_1(\mathbb{R})$ has a smooth density.

1 Introduction

In this note, we give explicit examples of finitely supported symmetric probability measures v on $SL_2(\mathbb{R})$ for which the corresponding Furstenberg measure μ on $\mathbb{P}_1(\mathbb{R})$ is absolutely continuous wrt to Haar measure $d\theta$, and moreover $\frac{d\mu}{d\theta}$ is of class C^r, with r any given positive integer. Probabilistic constructions of finitely supported (non-symmetric measures v on $SL_2(\mathbb{R})$ with absolutely continuous Furstenberg measure appear in the paper [1], setting (in the negative) a conjecture from [4]. The construction in [1] may be viewed as a non-commutative analogue of the theory of random Bernoulli convolutions and uses methods from [5,6].

It is not clear if this technique may produce Furstenberg measures with say C^1-density. Our method also addresses the issue of obtaining a symmetric v (raised in [4]), which seems problematic with the [1] technique.

Our starting point is a construction from [2] of certain Hecke operators on $SL_2(\mathbb{R})$ whose projective action exhibits a spectral gap. The mathematics underlying [2] is closely related to the paper [3] and makes essential use of results and techniques from arithmetic combinatorics. In particular, it should be pointed out that

J. Bourgain (✉)
Institute for Advanced Study, Princeton, NJ 08540, USA
e-mail: bourgain@ias.edu

B. Klartag et al. (eds.), *Geometric Aspects of Functional Analysis*, Lecture Notes in Mathematics 2050, DOI 10.1007/978-3-642-29849-3_7,
© Springer-Verlag Berlin Heidelberg 2012

the spectral gap is not achieved by exploiting hyperbolicity, at least not in the usual way. Our measure ν has in fact a Lyapounov exponent that can be made arbitrary small, while the spectral gap (in an appropriate restricted sense) remains uniformly controlled (the size of supp ν becomes larger of course).

We believe that similar constructions are possible also in the $SL_d(\mathbb{R})$-setting, for $d > 2$ (cf. [4]). In fact, such Hecke operators can be produced using the construction from Lemmas 1 and 2 below in $SL_2(\mathbb{R})$ and considering a suitable family of $SL_2(\mathbb{R})$-embeddings in SL_d. We do not present the details here.

Acknowledgements The author is grateful to C. McMullen and P. Varju for several related discussions. Research was partially supported by NSF grants DMS-0808042 and DMS-0835373

2 Preliminaries

We recall Lemmas 2.1 and 2.2 from [2].

Lemma 1. *Given $\varepsilon > 0$, there is $Q \in \mathbb{Z}_+$ and $\mathcal{G} \subset SL_2(\mathbb{R}) \cap \left(\frac{1}{Q} Mat_2(\mathbb{Z})\right)$ with the following properties*

$$\frac{1}{\varepsilon} < Q < \left(\frac{1}{\varepsilon}\right)^{c_1} \tag{1}$$

$$|\mathcal{G}| > Q^{c_2} \tag{2}$$

The elements of $\mathcal{G}$ are free generators of a free group $\tag{3}$

$$\|g - 1\| < \varepsilon \text{ for } g \in \mathcal{G} \tag{4}$$

Here c_1, c_2 are constants independent of ε.
Define the probability measure ν on $SL_2(\mathbb{R})$ as

$$\nu = \frac{1}{2|\mathcal{G}|} \sum_{g \in \mathcal{G}} (\delta_g + \delta_{g^{-1}}). \tag{5}$$

Denote also $P_\delta, \delta > 0$, an approximate identity on $SL_2(\mathbb{R})$. For instance, one may take $P_\delta = \frac{1_{B_\delta(1)}}{|B_\delta(1)|}$ where $B_\delta(1)$ is the ball of radius δ around 1 in $SL_2(\mathbb{R})$.

Lemma 2. *Fix $\tau > 0$. Then we have*

$$\|\nu^{(\ell)} * P_\delta\|_\infty < \delta^{-\tau} \tag{6}$$

provided

$$\ell > c_3(\tau) \frac{\log 1/\delta}{\log 1/\varepsilon} \tag{7}$$

and assuming δ small enough (depending on Q and τ).

3 Furstenberg Measure

Denote for $g \in SL_2(\mathbb{R})$ by τ_g the action on $P_1(\mathbb{R})$ that we identify with the circle $\mathbb{R}/\mathbb{Z} = \mathbb{T}$. Thus if $g = \begin{pmatrix} a & b \\ c & d \end{pmatrix}, ad - bc = 1$, then

$$e^{i\tau_g(\theta)} = \frac{(a\cos\theta + b\sin\theta) + i(c\cos\theta + d\sin\theta)}{[(a\cos\theta + b\sin\theta)^2 + (c\cos\theta + d\sin\theta)^2]^{\frac{1}{2}}}. \tag{8}$$

Assume μ on $P_1(\mathbb{R})$ is ν-stationary, i.e.

$$\mu = \sum \nu(g)g_*[\mu]. \tag{9}$$

4 A Restricted Spectral Gap

Take $\mathcal{G}$ as in Lemma 1 and $\nu = \frac{1}{2r} \sum_{g \in \mathcal{G}} (\delta_g + \delta_{g^{-1}})$ with $r = |\mathcal{G}|$.

Lemma 3. *There is some constant $K > 0$ (depending on ν), such that if $f \in L^2(\mathbb{T})$ satisfies*

$$\|f\|_2 \leq 1 \text{ and } \hat{f}(n) = 0 \text{ for } |n| < K \tag{10}$$

then

$$\left\| \int (f \circ \tau_g) d\nu \right\|_2 < \frac{1}{2}. \tag{11}$$

Proof. Define $\rho_g f = (\tau_g')^{1/2}(f \circ \tau_g)$, hence ρ is the projective representation. Since $\|1 - g\| < \varepsilon$, $|\tau_g' - 1| \lesssim \varepsilon$ and (11) will follow from

$$\left\| \int (\rho_g f)\nu(dg) \right\|_2 < \frac{1}{3}. \tag{12}$$

Assume (12) fails. By almost orthogonality, there is $f \in L^2(\mathbb{T})$ such that

$$\text{supp } \hat{f} \subset [2^k, 2^{k+1}] \tag{13}$$

$$\|f\|_2 = 1 \tag{14}$$

$$\left\| \int (\rho_g f)\nu(dg) \right\|_2 > c \text{(for some } c > 0). \tag{15}$$

Let $\ell < k$ to be specified. From (15), since ν is symmetric,

$$\left\| \int (\rho_g f)\nu^{(\ell)}(dg) \right\|_2 > c^\ell \tag{16}$$

and hence

$$\int |\langle \rho_g f, f \rangle| \nu^{(2\ell)}(dg) = \iint |\langle \rho_g f, \rho_h f \rangle| \nu^{(\ell)}(dg) \nu^{(\ell)}(dh) > c^{2\ell}. \qquad (17)$$

Take $\delta = 10^{-k}$. Recalling (13), straightforward approximation permits us to replace in (16) the discrete measure $\nu^{(\ell)}$ by $\nu^{(\ell)} * P_\delta$, where $P_\delta (\delta > 0)$ denotes the approximate identity on $SL_2(\mathbb{R})$. Hence (17) becomes

$$\int_{SL_2(\mathbb{R})} |\langle \rho_g f, f \rangle| (\nu^{2\ell} * P_\delta)(g) dg + 2^{-k} > c^{2\ell}. \qquad (18)$$

Fix a small constant $\tau > 0$ and apply Lemma 2. This gives

$$\ell \sim C(\tau) \frac{\log \frac{1}{\delta}}{\log \frac{1}{\varepsilon}} \qquad (19)$$

such that

$$\| \nu^{(\ell)} * P_\delta \|_\infty < \delta^{-\tau}. \qquad (20)$$

Note that supp $\nu^{(\ell)}$ is contained in a ball of radius at most $(1 + \varepsilon)^\ell$, by (4).

Introduce a smooth function $0 \le \omega \le 1$ on $\mathbb{R}, \omega = 1$ on $[-(1 + \varepsilon)^{4\ell}, (1 + \varepsilon)^{4\ell}]$ and $\omega = 0$ outside $[-2(1 + \varepsilon)^{4\ell}, 2(1 + \varepsilon)^{4\ell}]$.

Let $\omega_1(g) = \omega(a^2 + b^2 + c^2 + d^2)$ for $g = \begin{pmatrix} a & b \\ c & d \end{pmatrix}$.

From (20), the first term of (18) is bounded by

$$\delta^{-\tau} \int_{SL_2(\mathbb{R})} |\langle \rho_g f, f \rangle| \omega_1(g) dg. \qquad (21)$$

Note also that by assuming ε a sufficiently small constant, we can ensure that $\ell \ll k$ and $2^{-k} < c^{2\ell}$. Thus

$$\int_{SL_2(\mathbb{R})} |\langle \rho_g f, f \rangle| \omega_1(g) dg > \frac{1}{2} \delta^\tau c^{2\ell} \qquad (22)$$

and applying Cauchy-Schwarz

$$c^{4\ell} \delta^{2\tau} (1 + \varepsilon)^{-6\ell} \le \int_{SL_2(\mathbb{R})} |\langle \rho_g f, f \rangle|^2 \omega_1(g) dg$$

$$= \left| \int_{SL_2(\mathbb{R})} \int_{\mathbb{T}} \int_{\mathbb{T}} f(x) \overline{f(y)} \, \overline{f(\tau_g x)} f(\tau_g y) (\tau_g'(x))^{1/2} (\tau_g'(y))^{1/2} \omega_1(g) dg dx dy \right|$$

$$\le \int_{\mathbb{T}} \int_{\mathbb{T}} |f(x)| \, |f(y)| \left| \int_{SL_2(\mathbb{R})} f(\tau_g x) \overline{f(\tau_g y)} (\tau_g'(x))^{1/2} (\tau_g'(y))^{\frac{1}{2}} \omega_1(g) dg \right| dx dy.$$

$$(23)$$

Fix $x \neq y$ and consider the inner integral. If we restrict $g \in SL_2(\mathbb{R})$ s.t. $\tau_g x = \theta$ (fixed), there is still an averaging in $\psi = \tau_g y$ that can be exploited together with (13). By rotations, we may assume $x = \theta = 0$. Write $g = \begin{pmatrix} a & b \\ c & d \end{pmatrix} \in SL_2(\mathbb{R})$, $dg = \frac{da\,db\,dc}{a}$ on the chart $a \neq 0$. Since

$$e^{i\tau_g x} = \frac{(a\cos x + b\sin x) + i(c\cos x + d\sin x)}{[(a\cos x + b\sin x)^2 + (c\cos x + d\sin x)^2]^{1/2}}$$

the condition $\tau_g 0 = 0$ means $c = 0$ and thus

$$e^{i\psi} = e^{i\tau_g y} = \frac{(a\cos y + b\sin y) + \frac{i}{a}\sin y}{[(a\cos y + b\sin y)^2 + \frac{1}{a^2}\sin^2 y]^{\frac{1}{2}}}.$$

Hence, fixing a

$$\frac{\partial \psi}{\partial b} = -a\sin^2 \psi. \tag{24}$$

Also

$$\tau_g'(z) = \frac{\cos^2 \tau_g(z)}{(a\cos z + b\sin z)^2} = a^2 \frac{\sin^2 \tau_g(z)}{\sin^2 z} \tag{25}$$

implying

$$\tau_g'(0) = \frac{1}{a^2} \text{ and } \tau_g'(y) = \frac{a^2 \sin^2 \psi}{\sin^2 y}. \tag{26}$$

Substituting (24), (26) in (23) gives for the inner integral the bound

$$\frac{1}{|\sin(x-y)|} \iint d\theta \frac{da}{a^2} |f(\theta)|$$

$$\cdot \left| \int f(\psi) \frac{1}{|\sin(\theta - \psi)|} \omega\left(a^2 + \frac{1}{a^2} + \left(\frac{1}{a}\cotg(\psi - \theta) - a\cotg(y-x)\right)^2\right) d\psi \right|. \tag{27}$$

The weight function restricts a to $(1+\varepsilon)^{-2\ell} \lesssim |a| \lesssim (1+\varepsilon)^{2\ell}$ and clearly

$$|\sin(\theta - \psi)| \gtrsim (1+\varepsilon)^{-4\ell} |\sin(x-y)|. \tag{28}$$

If we restrict $|\sin(x - y)| > 2^{-\frac{k}{10}}$, Assumption (13) gives a bound at most $2^{-k}\|f\|_1$ for the ψ-integral in (27). Indeed, if β is a smooth function vanishing on a neighborhood of 0 and $|n| \sim 2^k$, partial integration implies that for any given $A > 0$

$$\int e^{-in\psi} \frac{1}{\sin(\theta - \psi)} \beta\left(2^{\frac{k}{10}}(\theta - \psi)\right) d\psi \lesssim 2^{-Ak}.$$

Thus
$$(27) < 2^{k/10}(1 + \varepsilon)^{2\ell} 2^{-k} \| f \|_1^2. \tag{29}$$

The contribution to (23) is at most

$$2^{-k/2}(1 + \varepsilon)^{2\ell} \| f \|_1^4. \tag{30}$$

Next we consider, the contribution of $| \sin(x - y) | \le 2^{-\frac{k}{10}}$ to (23).
    First, from (25), we have that

$$| \tau_g' | \lesssim a^2 + \frac{1}{a^2} + b^2 \lesssim \| g \|^2 < (1 + \varepsilon)^{4\ell}.$$

By Cauchy-Schwarz, the inner integral in (23) is at most

$$(1 + \varepsilon)^{4\ell} \left(\int | f(\tau_g x) |^2 \omega_1(g) dg \right) < (1 + \varepsilon)^{10\ell} \| f \|_2^2.$$

Hence, we obtain

$$\left[\iint_{|x-y|<2^{-k/10}} | f(x) | \, | f(y) | dx dy \right] (1 + \varepsilon)^{10\ell} \| f \|_2^2$$
$$< 2^{-k/20}(1 + \varepsilon)^{10\ell} \| f \|_2^4. \tag{31}$$

From (30), (31),
$$(23) \le 2^{-k/20}(1 + \varepsilon)^{10\ell}$$

and hence, by (19)
$$2^{k/10} < 100^{k\tau} . C^{C(\tau)(\log \frac{1}{\varepsilon})^{-1} k}. \tag{32}$$

Taking (in order) τ and ε small enough, a contradiction follows.
    This proves Lemma 3.

5 Absolute Continuity of the Furstenberg Measure and Smoothness of the Density

Our aim is to establish the following.

Theorem. *Let μ be the stationary measure introduced in (9). Given $r \in \mathbb{Z}_+$ and taking ε in Lemma 1 small enough will ensure that $\frac{d\mu}{d\theta} \in C^r$.*

    This will be an immediate consequence of

Lemma 4. *Let $k > k(\varepsilon)$ be sufficiently large and $f \in L^\infty(\mathbb{T}), | f | \le 1$ such that $\operatorname{supp} \hat{f} \subset [2^{k-1}, 2^k]$. Then*

$$|\langle f, \mu \rangle| < C_\varepsilon^{-k} \tag{33}$$

where $C_\varepsilon \xrightarrow{\varepsilon \to 0} \infty$.

Proof. Clearly, for any $\ell \in \mathbb{Z}_+$

$$|\langle f, \mu \rangle| \leq \left\| \sum_g v^{(\ell)}(g)(f \circ \tau_g) \right\|_\infty. \tag{34}$$

We will iterate Lemma 3 and let $K = K(\varepsilon)$ satisfy (10), (11).
We assume $2^k > 10K^{10}$. For $m < \ell$ and $|n| < K$, we evaluate $|\widehat{F_m}(n)|$, denoting

$$F_m = \sum_g v^{(m)}(g)(f \circ \tau_g). \tag{35}$$

Clearly $|\widehat{F_m}(n)| \leq \max_{g \in \mathrm{supp}\, v^{(m)}} |(f \circ \tau_g)^\wedge(n)|$ and by assumption on supp $\hat{f}$

$$|(f \circ \tau_g)^\wedge(n)| = \left| \int f(\tau_g(x)) e^{-2\pi i n x} dx \right|$$

$$\leq 2^{k/2} \|f\|_2 \max_{n' \in [2^{k-1}, 2^k]} \left| \int e^{2\pi i (n' \tau_g(x) - nx)} dx \right|.$$

Performing a change of variables gives

$$\left| \int e^{2\pi i (n' \tau_g(x) - nx)} dx \right| = \left| \int e^{2\pi i (n' y - n \tau_{g^{-1}}(y))} \tau'_{g^{-1}}(y) dy \right|$$

$$\ll_r \|e^{-2\pi i n \tau_{g^{-1}}} \tau'_{g^{-1}}\|_{C^r} |n'|^{-r}$$

$$\ll_r \frac{K^r}{|n'|^r}(1 + \varepsilon)^{2m(r+1)} \ll_r 2^{-\frac{3}{4}kr}(1 + \varepsilon)^{2\ell(r+1)} \tag{36}$$

by partial integration and our assumptions. It follows from (36) that if ℓ satisfies

$$\ell < \frac{k}{100\varepsilon} \tag{37}$$

then for $m < \ell$ and $k > k(r)$

$$\max_{|n| < K} |\widehat{F_m}(n)| < 2^{-\frac{kr}{2}} \tag{38}$$

(with r a fixed large integer).
Next, decompose

$$F_m = F_m^{(1)} + F_m^{(2)} \text{ where } F_m^{(1)}(x) = \sum_{|n| < K} \widehat{F_m}(n) e^{2\pi i n x}.$$

Hence, by (38)

$$\|F_m^{(1)}\|_\infty < 2K2^{-\frac{kr}{2}}. \tag{39}$$

Estimate using (39) and Lemma 3

$$\|F_{m+1}\|_2 \leq \left\|\int (F_m^{(1)} \circ \tau_g)dv\right\|_\infty + \left\|\int (F_m^{(2)} \circ \tau_g)dv\right\|_2$$

$$\leq \|F_m^{(1)}\|_\infty + \frac{1}{2}\|F_m^{(2)}\|_2$$

$$\leq 3K2^{-\frac{kr}{2}} + \frac{1}{2}\|F_m\|_2. \tag{40}$$

Iteration of (40) implies by (37)

$$\|F_\ell\|_2 \leq 4K2^{-\frac{kr}{2}} + 2^{-\ell} \lesssim 2^{-\frac{kr}{2}} + 2^{-\frac{k}{100\varepsilon}}. \tag{41}$$

Also

$$|F_\ell'| \leq \max_{g \in \text{supp } v^{(\ell)}} \|(f \circ \tau_g)'\|_\infty \leq \|f'\|_\infty (1+\varepsilon)^{2\ell} \lesssim 5^k \tag{42}$$

and interpolation between (41), (42) implies for r (resp. ε) large (resp. small) enough

$$\|F_\ell\|_\infty \lesssim (41)^{1/2}.(42)^{1/2} < 2^{-\frac{kr}{5}} + 2^{-\frac{k}{300\varepsilon}} \tag{43}$$

provided $k > k(\varepsilon, r)$.

In view of (34), this proves (33).

Remark. For v finitely supported (with positive Lyapounov exponent), one cannot obtain a Furstenberg measure μ that equals Haar measure on $\mathbb{P}_1(\mathbb{R}) \simeq \mathbb{T}$. Indeed, otherwise for any f on $\mathbb{T}$, we would have

$$\hat{f}(0) = \int_\mathbb{T} fd\mu = \int v(dg)\left[\int (f \circ \tau_g)d\mu\right]$$

$$= \int v(dg)\left[\int f(x)(\tau_{g^{-1}})'(x)dx\right]. \tag{44}$$

For $g \in SL_2(\mathbb{R})$,

$$\int f(x)(\tau_{g^{-1}})'(x)dx = \int f(\theta)P_z(2\theta)d\theta$$

$$= \hat{f}(0) + \sum_{n\neq 0}|z|^{|n|}e^{2\pi in(Arg z)}\hat{f}(-2n) \tag{45}$$

for some $z \in D = \{z \in \mathbb{C}; |z| < 1\}$, with $P_z(\theta) = \frac{1-|z|^2}{|1-\bar{z}e^{i\theta}|^2}$ the Poisson kernel.

From (44), (45), taking $\nu = \sum_{j=1}^{r} c_j \delta_{g_j}, c_j > 0$ and $\sum c_j = 1$ and $\{z_j\}$ the corresponding points in D, we get

$$\sum_{1}^{r} c_j |z_j|^n e^{2\pi i n (Arg z_j)} = 0 \text{ for all } n \neq 0. \tag{46}$$

This easily implies that $z_1 = \cdots = z_r = 0$. But then each g_j has unimodular spectrum and ν vanishing Lyapounov exponent.

References

1. B. Barany, M. Pollicott, K. Simon, Stationary measures for projective transformations: The Blackwell and Furstenberg measures. Preprint (2010)
2. J. Bourgain, Expanders and dimensional expansion. C.R. Math. Acad. Sci. Paris **347**(7–8), 356–362 (2000)
3. J. Bourgain, A. Gamburd, On the spectral gap for finitely generated subgroups of $SU(2)$. Invent. Math. **171**(1), 83–121 (2008)
4. V. Kaimanovich, V. Le Prince, Matrix random products with singular harmonic measure. Geom. Ded. **150**, 257–279 (2011)
5. K. Simon, B. Solomyak, M. Urbanski, Hausdorff dimension of limit sets for parabolic IFS with overlap. Pac. J. Math. **201**(2), 441–478 (2001)
6. K. Simon, B. Solomyak, M. Urbanski, Invariant measurés for parabolic IFS with overlaps and random continued fractions. Trans. Am. Math. Soc. **353**(12), 5145–5164 (2001)

Moebius Schrödinger

Jean Bourgain

Abstract Consider the one-dimensional lattice Schrödinger operator with potential given by the Moebius function. It is shown that the Lyapounov exponent is strictly positive for almost all energies, answering a question posed by P. Sarnak.

1 Statement of the Result

Let $\mu(n)$ be the Moebius function and consider the Schrödinger operator on $\mathbb{Z}_+$

$$H = \Delta + \lambda\mu \qquad (\lambda \neq 0 \text{ arbitrary}). \tag{1}$$

We prove the following

Theorem 1. *For $E \in \mathbb{R}$ outside a set of 0-measure, any solution $\psi = (\psi_n)_{n\geq 0}$, $\psi_0 = 0, \psi \neq 0$ of*

$$H\psi = E\psi$$

satisfies

$$\overline{\lim} \frac{\log^+ |\psi_n|}{n} > 0. \tag{2}$$

Recalling the spectral theory of 1D Schrödinger operators with a random potential, Theorem 1 fits the general heuristic, known as the 'Moebius randomness law' (cf. [4]). The question whether (1) satisfies Anderson localization remains open and is probably difficult.

The fact that H has no ac-spectrum is actually immediate from the following result of Remling.

J. Bourgain (✉)
Institute for Advanced Study, Princeton, NJ 08540, USA
e-mail: bourgain@math.ias.edu

B. Klartag et al. (eds.), *Geometric Aspects of Functional Analysis*, Lecture Notes in Mathematics 2050, DOI 10.1007/978-3-642-29849-3_8,
© Springer-Verlag Berlin Heidelberg 2012

Proposition 1 ([3, Theorem 1.1]). *Suppose that the (half line) potential $V(n)$ takes only finitely many values and $\sigma_{ac} \neq \phi$. Then V is eventually periodic.*

We will use again Proposition 1 later on, in the proof of the Theorem.

2 Proof of the Theorem (I)

Let $X \subset \{0, 1, -1\}^{\mathbb{Z}}$ be the point-wise closure of the set $\{T^j \overline{\omega}; j \in \mathbb{Z}\}$, where T is the left shift and $\overline{\omega}$ defined by

$$\overline{\omega}_n = \begin{cases} \mu(n) \text{ for } n \in \mathbb{Z}_+ \\ 0 \text{ for } n \in \mathbb{Z}_-. \end{cases} \tag{3}$$

Let

$$\nu_N = \frac{1}{N} \sum_{j=0}^{N-1} \delta_{T^j \overline{\omega}} \qquad (\delta_x = \text{Dirac measure at } x)$$

and $\nu \in \mathcal{P}(X)$ a weak*-limit point of $\{\nu_N\}$.

Then ν is a T-invariant probability measure on X.

The only property of the Moebius function exploited in the proof of Theorem 1 is the following fact.

Lemma 1. *For no element $\omega \in X, (\omega_n)_{n \geq 0}$ is eventually periodic, unless $\omega_n = 0$ for n large enough. Similarly for $(\omega_n)_{n \leq 0}$.*

Proof. Suppose ω eventually periodic. Hence there is $n_0 \in \mathbb{Z}_+$ and $d \in \mathbb{Z}_+$ such that

$$\omega(n + d) = \omega(n) \text{ for } n \geq n_0. \tag{4}$$

Take $N = 10^3(n_1 + d^3)$ and choose $n_1 \geq n_0$ and $k \in \mathbb{Z}_+$ such that

$$\omega(n) = \mu(k + n) \text{ for } n \in [n_1, n_1 + N]. \tag{5}$$

Let $d < p < 10d$ be a prime. Taking $n \in [n_1, n_1 + \frac{N}{2}]$, there is $0 \leq j < p^2$ such that $k + n + jd \equiv 0 \pmod{p^2}$ and thus $\mu(k + n + jd) = 0$. Since $n + jd \in [n_1, n_1 + N]$, (5), (4) imply that $\mu(k + n + jd) = \omega(n + jd) = \omega(n)$ and therefore $\omega = 0$ on $[n_1, n_1 + \frac{N}{2}]$, hence on $[n_1, \infty[$. $\qquad \square$

Denote for $\omega \in X$

$$H_\omega = \Delta + \lambda\omega. \tag{6}$$

Combined with Proposition 1, Lemma 1 implies

Lemma 2.

$$\sigma_{ac}(H_\omega) = \phi \qquad (\nu - a.e.)$$

Proof. Denoting $H_\omega^\pm$ the corresponding halfline SO's, we have

$$\sigma_{ac}(H_\omega) = \sigma_{ac}(H_\omega^+) \cup \sigma_{ac}(H_\omega^-)$$

and these sets are empty, unless

$$\omega \in \bigcup_{k=1}^{\infty} \{\omega \in X; \omega_n = 0 \text{ for all } n \geq k \text{ or all } n \leq -k\}. \tag{7}$$

Clearly $\nu(7) = 0$.   □

The measure ν need not be T-ergodic, so we consider its ergodic decomposition

$$\nu = \int \nu_\alpha d\alpha. \tag{8}$$

For each α, let $\gamma_\alpha(E)$ be the Lyapounov exponent of H_ω, i.e.

$$\gamma_\alpha(E) = \lim_{N \to \infty} \frac{1}{N} \log \left\| \prod_N^0 \begin{pmatrix} E - \lambda\omega_n & -1 \\ 1 & 0 \end{pmatrix} \right\| \quad (\nu_\alpha \ a.e.). \tag{9}$$

Next, we apply Kotani's theorem (for stochastic Jacobi matrices, as proven in [5, Theorem 2]).

Proposition 2 (Assuming (Ω, μ, T) ergodic). *If $\gamma(E) = 0$ on a subset A of $\mathbb{R}$ with positive Lebesgue measure, then $E_\omega^{ac}(A) \neq 0$ for a.e. ω.*

(E^{ac} denote the projection on the ac-spectrum).

Apply Proposition 2 to H_ω on (X, ν_α). By Lemma 2, $E_\omega^{ac} = 0$, ν_α a.e., hence $\{E \in \mathbb{R}; \gamma_\alpha(E) = 0\}$ is a set of zero Lebesgue measure. For E outside a subset $\mathcal{E}_* \subset \mathbb{R}$ of zero Lebesgue measure, we have that $\gamma_\alpha(E) > 0$ for almost all α in (8), therefore

$$\liminf_{N \to \infty} \int \frac{1}{N} \log \left\| \prod_N^0 \begin{pmatrix} E - \lambda\omega_n & -1 \\ 1 & 0 \end{pmatrix} \right\| \nu(d\omega)$$

$$\geq \int \left\{ \liminf_{N \to \infty} \int \left[\frac{1}{N} \log \left\| \prod_N^0 \begin{pmatrix} E - \lambda\omega_n & -1 \\ 1 & 0 \end{pmatrix} \right\| \right] \nu_\alpha(d\omega) \right\} d\alpha$$

$$\geq \int \gamma_\alpha(E) d\alpha > 0. \tag{10}$$

Denoting R_N the restriction operator to $[1, N]$, let

$$H_\omega^{(N)} = R_N H_\omega R_N$$

$$G_\omega^{(N)}(E) = (H_\omega^{(N)} - E + i0)^{-1} \qquad (= \text{restricted Green's function}).$$

Recall that by Cramer's rule, for $1 \leq k_1 \leq k_2 \leq N$

$$|G_\omega^{(N)}(E)(k_1, k_2)| = \frac{\det[H_\omega^{(k_1-1)} - E].|\det[H_{T^{k_2}\omega}^{(N-k_2)} - E]|}{|\det[H_\omega^{(N)} - E]|} \qquad (11)$$

and also the formula

$$M_N(E, \omega) = \prod_N^1 \begin{pmatrix} E - \lambda\omega_n & -1 \\ 1 & 0 \end{pmatrix}$$

$$= \begin{bmatrix} \det[E - H_\omega^{(N)}] & -\det[E - H_{T_\omega}^{(N-1)}] \\ \det[E - H_\omega^{(N-1)}] & -\det[E - H_{T_\omega}^{(N-2)}] \end{bmatrix}. \qquad (12)$$

Using the above formalism, it is well-known how to derive from positivity of the Lyapounov exponent, bounds and decay estimates on the restricted Green's functions. Since ergodicity of the measure is used, application to the preceding requires to start from the ν_α.

For $E \in \mathbb{R}, \delta, c > 0, M \in \mathbb{Z}_+$, define

$$\Omega_{E,\delta,c,M} = \{\omega \in X; \|G_\omega^{(M)}(E)\| < e^{\delta M} \text{ and } |G_\omega^{(M)}(E)(k, k')| < e^{-c|k-k'|}$$

$$\text{if } 1 \leq k, k' \leq M \text{ and } |k - k'| > \delta M\}. \qquad (13)$$

Fix α and $\delta > 0$. Then E a.e

$$\lim_{M \to \infty} \nu_\alpha(\Omega_{E,\delta,\frac{1}{2}\gamma_\alpha(E),M}) = 1. \qquad (14)$$

Using Fubini arguments and (8), we derive the following

Lemma 3. *Given $\varepsilon > 0$, there is $b > 0$, such that for all $\delta > 0$, there is a subset $\mathcal{E}_\varepsilon \subset \mathbb{R}$, mes $\mathcal{E}_\varepsilon < \varepsilon$ and some scale M satisfying*

$$\nu(\Omega_{E,\delta,b,N}) > 1 - \varepsilon \text{ for } E \notin \mathcal{E}_\varepsilon \text{ and } N > M. \qquad (15)$$

3   Proof of the Theorem (II)

Using the definition of v, we re-express (15) in terms of the Moebius function.
Let H be as in (1). For $I \subset \mathbb{Z}_+$ an interval, denote

$$H_I = R_I H R_I \tag{16}$$

and

$$G_I(E) = (H_I - E + io)^{-1}. \tag{17}$$

Let $S = S_{E,\delta,N}$ be defined by

$$S = \{k \in \mathbb{Z}; \|G_{[k,k+N[}(E)\| < e^{\delta N} \text{ and}$$

$$|G_{[k,k+N[}(E)(k',k'')| < e^{-b|k'-k''|} \text{ if } k \le k', k'' \le k + N, |k' - k''| > \delta N. \tag{18}$$

Property (15) then translates as follows

$$\lim_{\substack{\ell \to \infty \\ \ell \gg N}} \frac{1}{\ell} |S \cap [1, \ell]| > \frac{1}{2} \tag{19}$$

for $E \notin \mathcal{E}_\varepsilon$ and $N > M$. Here "lim" refers to the Banach limit in the definition of v.

Fix $\varepsilon > 0$ a small number, take $0 < b < \frac{1}{10}$ as in Lemma 3 and let $\delta = b^{10}$. Let $\mathcal{E}_\varepsilon \subset \mathbb{R}$, $M > \delta^{-2} + \frac{1}{\varepsilon}$, satisfy the lemma. Hence, from (19)

$$\lim_{\substack{\ell \to \infty \\ \ell \gg M}} \frac{1}{\ell} |S_{E,\delta,M} \cap [1, \ell]| > \frac{1}{2} \text{ for } E \notin \mathcal{E}_\varepsilon. \tag{20}$$

Choose $\ell \gg M$ such that

$$\frac{1}{\ell} |S_{E,\delta,M} \cap [1, \ell]| > \frac{1}{2} \text{ for } E \notin \mathcal{E}'_\varepsilon \tag{21}$$

where $\mathcal{E}_\varepsilon \subset \mathcal{E}'_\varepsilon \subset \mathbb{R}$ satisfies

$$\text{mes } \mathcal{E}'_\varepsilon < 2\varepsilon.$$

Next we rely on a construction from [1, Lemma 6.1 and Corollary 6.54]. We recall the statement

Lemma 4. Let $0 < c_0 < 1$, $0 < c_1 < \frac{1}{10}$ be constants, $0 < \delta < c_1^{10}$ and $\ell \gg M > \delta^{-2}$.

Let

$$A = v_n \delta_{nn'} + \Delta \quad (1 \le n, n' \le \ell) \tag{22}$$

(hence A is an $\ell \times \ell$ matrix) with diagonal v_n arbitrary, bounded, $|v_n| = 0(1)$.

Let $\mathcal{U} \subset \mathbb{R}$ be a set of energies E such that for each $E \in \mathcal{U}$, the following holds: There is a collection $\{I_\alpha\}$ of disjoint intervals in $[1, \ell], |I_\alpha| = M$ such that for each α

$$\|(R_{I_\alpha}(A - E)R_{I_\alpha})^{-1}\| < e^{\delta M} \tag{23}$$

and

$$|(R_{I_\alpha}(A - E)R_{I_\alpha})^{-1}(k, k')| < e^{-c_1|k-k'|} \text{ for } k, k' \in I_\alpha, |k - k'| > \delta M \tag{24}$$

holds, and

$$\sum_\alpha |I_\alpha| > c_0 \ell. \tag{25}$$

Then there is a set $\mathcal{E}'' \subset \mathbb{R}$ so that

$$mes\,(\mathcal{E}'') < \frac{1}{M} \tag{26}$$

and for $E \in \mathcal{U} \backslash \mathcal{E}''$,

$$\max_{\substack{1 \leq x \leq \frac{c_0}{10}\ell \\ \ell \geq y \geq \ell - \frac{c_0}{10}\ell}} |(A - E)^{-1}(x, y)| < e^{-\frac{1}{8}c_0 c_1 \ell}. \tag{27}$$

The proof of Lemma 4 is a bit technical, but uses nothing more than the resolvent identity and energy perturbation.

Let $v_n = \lambda \mu(n)$.

Take $c_0 = \frac{1}{2}, c_1 = b, \mathcal{U} = \mathbb{R} \backslash \mathcal{E}'_\varepsilon$ with $\mathcal{E}'_\varepsilon$ as above:

Let $\ell_0 \gg M$ satisfy (21). From the definition (18) of $S_{E,\delta,M}$ and (21), we clearly obtain a collection $\{I_\alpha\}$ of M-intervals in $[1, \ell]$ such that (23)–(25) hold.

It follows that for E outside of the set $\mathcal{E}''_\varepsilon = \mathcal{E}'_\varepsilon \cup \mathcal{E}''$ of measure at most $2\varepsilon + \frac{1}{M} < 3\varepsilon$, one has for $b' \sim b$ that

$$\max_{\substack{1 \leq x \leq \frac{c_0}{10}\ell \\ \ell \geq y \geq \ell - \frac{c_0}{10}\ell}} |G_{[1,\ell]}(E)(x, y)| < e^{-b'\ell}. \tag{28}$$

Note that $b' > 0$ depends on ε and v and $\mathcal{E}''_\varepsilon$ depends on ℓ, which can be taken arbitrarily large in the subsequence of $\mathbb{Z}_+$ used to define v. Since this subsequence is arbitrary, it follows that there is some $b' = b_\varepsilon$ and $\ell_\varepsilon \in \mathbb{Z}_+$ such that for $\ell > \ell_\varepsilon$

$$mes\,[E \in \mathbb{R}; \max_{\substack{1 \leq x \leq \frac{c_0}{10}\ell \\ \ell \geq y \geq \ell - \frac{c_0}{10}\ell}} |G_{[1,\ell]}(E)(x, y)| > e^{-b'\ell}] = mes\,\tilde{\mathcal{E}}_\ell < \varepsilon. \tag{29}$$

Assume $\psi = (\psi_n)_{n \geq 0}, \psi_0 = 0$ a solution of

$$H\psi = E\psi.$$

Taking ℓ large, one has by projection

$$H_{[1,\ell]}\psi^{(\ell)} + \psi_{\ell+1}e_\ell = E\psi^{(\ell)} \tag{30}$$

where $\psi^{(\ell)} = \sum_{1 \le x \le \ell} \psi_x e_x$, $\{e_x\}$ the unit vector basis.
Hence

$$\psi^{(\ell)} = -\psi_{\ell+1}G_{[1,\ell]}(E)e_\ell$$

and fixing some coordinate $x \ge 1$, for ℓ large enough

$$|\psi_x| \le |\psi_{\ell+1}| \, |G_{[1,\ell]}(E)(x,\ell)|. \tag{31}$$

Take x with $\psi_x \ne 0$. Assuming

$$\varlimsup_n \frac{\log^+ |\psi_n|}{n} = 0$$

it follows from (31) that

$$\varlimsup_\ell \frac{1}{\ell} \log^+ |G_{[1,\ell]}(E)(x,\ell)|^{-1} = 0. \tag{32}$$

From the definition of $\tilde{\mathcal{E}}_\ell$ in (29), this means that

$$E \in \bigcup_{\ell_0} \bigcap_{\ell \ge \ell_0} \tilde{\mathcal{E}}_\ell \tag{33}$$

which is a set of measure $\le \varepsilon$.
Letting $\varepsilon \to 0$, Theorem 1 follows.

4 Further Comments

Taking into account the comment made prior to Lemma 1, our argument gives the
following more general result, that can be viewed as a refinement of [3].

Theorem 2. *Suppose that the (half line) potential $(V_n)_{n \ge 0}$ takes only finitely many
values and satisfies the following property*

$$\varlimsup_{r \to \infty} \varlimsup_{N \to \infty} \frac{1}{N} |\{1 \le k \le N; V_k = \omega_0, V_{k+1} = \omega_1, \ldots, V_{k+r} = \omega_r\}| = 0 \quad (34)$$

*whenever $\overline{\omega} = (\omega_r)_{r \ge 0}$ is a periodic sequence in the pointwise closure of the
sequences $(V_{n+j})_{n \in \mathbb{Z}_+}$ $(j \in \mathbb{Z}_+)$.*
   *Then the Schrödinger operator $H = \Delta + V$ satisfies the conclusion of
Theorem 1.*

Acknowledgements The author is grateful to P. Sarnak for bringing the problem to his attention and several discussions. He also thanks H. Krüger for comments on how Theorem 1 in this note may be derived directly from Lemma 2 and the results in [2]. The author was partially supported by NSF Grants DMS-0808042 and DMS 0835373.

References

1. J. Bourgain, *Positive Lyapounov Exponents for Most Energies*. Geometric Aspects of Functional Analysis, Lecture Notes in Math., vol. 1745 (Springer, Berlin, 2000), pp. 37–66
2. H. Krüger, Probabilistic averages of Jacobi operators. CMP **295**, 853–875 (2010)
3. C. Remling, The absolutely continuous spectrum of Jacobi matrices. Ann. Math. (1) **174**(1), 125–171 (2011)
4. P. Sarnak, Moebius randomness law. Notes
5. B. Simon, Kotani theory for one-dimensional stochastic Jacobi matrices. Comm. Math. Phys. **89**(2), 227–234 (1983)

Interpolations, Convexity and Geometric Inequalities

Dario Cordero-Erausquin and Bo'az Klartag

Abstract We survey some interplays between spectral estimates of Hörmander-type, degenerate Monge-Ampère equations and geometric inequalities related to log-concavity such as Brunn-Minkowski, Santaló or Busemann inequalities.

1 Introduction

The Brunn-Minkowski inequality has an L^2 interpretation, an observation that can be traced back to the proof provided by Hilbert. More recently, it has been noted that the Brunn-Minkowski inequality for convex bodies is related, in its local form, to spectral inequalities. In fact, the Prékopa theorem, which is the function form of the Brunn-Minkowski inequality for convex sets, is *equivalent* to spectral inequalities of Brascam-Lieb type. The local derivation of Prékopa's theorem from spectral L^2 inequalities was described in the more general complex setting in [13] and then extended further in [6, 7].

Let $K_0, K_1 \subset \mathbb{R}^n$ be two convex bodies (i.e., compact convex sets with non-empty interior) and denote, for $t \in [0, 1]$,

$$K(t) := (1-t)K_0 + tK_1 = \{z \in \mathbb{R}^n ; \exists (a,b) \in K_0 \times K_1, z = (1-t)a + tb\}. \quad (1)$$

D. Cordero-Erausquin (✉)
Institut de Mathématiques de Jussieu, Université Pierre et Marie Curie (Paris 6), 4 place Jussieu, 75252 Paris, France

Institut Universitaire de France
e-mail: cordero@math.jussieu.fr

B. Klartag
School of Mathematical Sciences, Sackler Faculty of Exact Sciences, Tel-Aviv University, Tel Aviv 69978, Israel
e-mail: klartagb@post.tau.ac.il

B. Klartag et al. (eds.), *Geometric Aspects of Functional Analysis*, Lecture Notes in Mathematics 2050, DOI 10.1007/978-3-642-29849-3_9,
© Springer-Verlag Berlin Heidelberg 2012

The Brunn-Minkowski inequality is central in the theory of convex bodies. Denoting the Lebesgue measure by $|\cdot|$, it states that

$$|K(t)| \geq |K_0|^{1-t} |K_1|^t,$$

with equality if and only if $K_0 = K_1 + x_0$ for $x_0 \in \mathbb{R}^n$. Introducing the convex body

$$K := \bigcup_{t \in [0,1]} \{t\} \times K(t) \subset \mathbb{R}^{n+1},$$

then $K(t)$ is the section over t, and the Brunn-Minkowski inequality expresses the log-concavity of the marginal measure. Namely, it shows that the function

$$\alpha(t) := -\log |K(t)|$$

is convex. The Brunn-Minkowski inequality for convex bodies admits the following useful functional form, which states that marginals of log-concave functions are log-concave.

Theorem 1 (Prékopa). *Let $F : \mathbb{R}^{n+1} \to \mathbb{R} \cup \{+\infty\}$ be convex with $\int \exp(-F) < \infty$ and define $\alpha : \mathbb{R} \longrightarrow \mathbb{R} \cup \{+\infty\}$ by*

$$e^{-\alpha(t)} = \int_{\mathbb{R}^n} e^{-F(t,x)} \, dx.$$

Then α is convex.

The Brunn-Minkowski inequality then follows by considering, for a given convex set $K \subset \mathbb{R}^{n+1} = \mathbb{R} \times \mathbb{R}^n$, the convex function F defined by

$$e^{-F(t,x)} = \mathbf{1}_K(t,x) = \mathbf{1}_{K(t)}(x). \tag{2}$$

The standard proofs of the Brunn-Minkowski inequality rely on parameterization or mass transport techniques between K_0 and K_1, with the parameter $t \in [0,1]$ being fixed. A natural question is whether one can provide a direct local approach by proving $\alpha''(t) \geq 0$? The answer is affirmative and this was shown recently by Ball, Barthe and Naor [4]. As mentioned earlier, this local approach was put forward in an L^2 framework, for analogous complex versions, in Cordero-Erausquin [13] and in subsequent far-reaching works by Berndtsson [6, 7]. We can also point out that this local approach was implicitly initiated in the paper by Brascamp and Lieb [9] from which the knowledgeable reader can extract the equivalence between Prekopa's inequality and Brascamp-Lieb's inequality (17).

Another essential concept in the theory of convex bodies is duality. This requires us to fix a center and a scalar product. Let $x \cdot y$ stand for the standard scalar product of $x, y \in \mathbb{R}^n$. We write $|x|^2 = x \cdot x$ and $B_2^n = \{x \in \mathbb{R}^n; \ x \cdot x \leq 1\}$, the associated unit ball. Recall that $K \subset \mathbb{R}^n$ is a centrally-symmetric convex body if and only if K

is the unit ball of some norm $\| \cdot \|$ on $\mathbb{R}^n$, a relation denoted by $K = B_{\|\cdot\|} := \{x \in \mathbb{R}^n ; \|x\| \leq 1\}$. The polar of K is defined as the unit ball of the dual norm $\| \cdot \|_*$,

$$K^\circ = B_{\|\cdot\|_*} = \{y \in \mathbb{R}^n ; x \cdot y \leq 1, \forall x \in K\}.$$

We have the following beautiful result:

Theorem 2 (Blaschke-Santaló inequality). *For every centrally-symmetric convex body $K \subset \mathbb{R}^n$, we have*

$$|K| \, |K^\circ| \leq |B_2^n|^2 \tag{3}$$

with equality holding true if and only if K is an ellipsoid (i.e. a linear image of B_2^n).

The corresponding functional form reads as follows (see [1, 2]): for an even function $f : \mathbb{R}^n \to \mathbb{R}$ with $0 < \int e^{-f} < \infty$, if $\mathcal{L}f$ denotes its Legendre transform, then

$$\int e^{-f} \int e^{-\mathcal{L}f} \leq \left(\int e^{-|x|^2/2} \, dx \right)^2 = (2\pi)^n. \tag{4}$$

Note that the Brunn-Minkowski inequality entails

$$\sqrt{|K| \, |K^\circ|} \leq \left| \frac{K + K^\circ}{2} \right|. \tag{5}$$

In general we have $\frac{K+K^\circ}{2} \supset B_2^n$, since $(\|x\| + \|x\|_*)/2 \geq \sqrt{\|x\| \|x\|_*} \geq 1$ for any vector $x \in \mathbb{R}^n$ and a norm $\| \cdot \|$. However, typically, $(K + K^\circ)/2$ is much larger than B_2^n, and (5) is weaker than (3). For instance, take $K = T(B_2^n)$, where $T \neq \mathrm{Id}_{\mathbb{R}^n}$ is a positive-definite symmetric operator. Then $K^\circ = T^{-1}(B_2^n)$. Observe that $\frac{K+K^\circ}{2} \supset \frac{T+T^{-1}}{2}(B_2^n)$ and

$$\frac{T + T^{-1}}{2} > \sqrt{T \, T^{-1}} = \mathrm{Id}_{\mathbb{R}^n}$$

in the sense of symmetric matrices. This suggest that instead of taking convex combinations, as in the Brunn-Minkowski theory, we would like to consider geometric means of convex bodies. It turns out that this is exactly what complex interpolation does, and it is a challenging question to understand real analogues of this procedure.

In this note we will consider several ways of going from K_0 to K_1, or equivalently from a norm $\| \cdot \|_0$ to another norm $\| \cdot \|_1$. There are many ways to recover the volume of K from the associated norm $\| \cdot \|$. Let $p > 0$ and $n \geq 1$. There exists an explicit constant $c_{n,p} > 0$ such that for every centrally-symmetric convex body $K \subset \mathbb{R}^n$, with associated norm $\| \cdot \|_K$, we have

$$\int_{\mathbb{R}^n} e^{-\|x\|_K^p/p} \, dx = c_{n,p} \, |K|. \tag{6}$$

Note that the procedure (2) corresponds to the case $p \to +\infty$.

We aim to find ways of interpolating between norms in order to recover, among other things, the Brunn-Minkowski and the Santaló inequalities.

Let us next put forward some notation as well as a formula that we shall use throughout the paper.

Notation 3. For a function $F : \mathbb{R}^n \to \mathbb{R}$ such that $\int e^{-F(x)} dx < +\infty$, we denote by μ_F the *probability* measure on $\mathbb{R}^n$ given by

$$d\mu_F(x) := \frac{de^{-F(x)}}{\int e^{-F}} dx.$$

For a function of $n + 1$ variables $F : I \times \mathbb{R}^n \to \mathbb{R}$, where I is an interval of $\mathbb{R}$, we denote, for a fixed $t \in I$, $F_t := F(t, \cdot) : \mathbb{R}^n \to \mathbb{R}$ and then by μ_{F_t} the corresponding probability measure on $\mathbb{R}^n$. We also set

$$\alpha(t) = -\log \int_{\mathbb{R}^n} e^{-F_t(x)} dx.$$

The variance with respect to a *probability* measure μ of a function $u \in L^2(\mu)$ —where, depending on the context, we consider either real-valued or complex-valued functions—is defined as the L^2 norm of the projection of u onto the space of functions orthogonal to constant functions, i.e.

$$\mathrm{Var}_\mu(u) := \int \left| u - \int u \, d\mu \right|^2 d\mu = \int |u|^2 d\mu - \left| \int u \, d\mu \right|^2.$$

A straightforward computation yields:

Fact 4. *With Notation 3, we have for every $t \in I$,*

$$\alpha''(t) = \int_{\mathbb{R}^n} \partial_{tt}^2 F \, d\mu_{F_t}(x) - \left[\int_{\mathbb{R}^n} \left(\partial_t F(t, x) \right)^2 d\mu_{F_t}(x) \right.$$
$$\left. - \left(\int_{\mathbb{R}^n} \partial_t F(t, x) \, d\mu_{F_t}(x) \right)^2 \right]$$
$$= \int_{\mathbb{R}^n} \partial_{tt}^2 F \, d\mu_{F_t} - \mathrm{Var}_{\mu_{F_t}}(\partial_t F), \tag{7}$$

assuming that F is sufficiently regular to allow for the differentiations under the integral sign.

Our goal is to understand for which families of functions F the function α is convex, by looking at α''. Actually, we will first discuss the complex case, where convexity is replaced by plurisubharmonicity. We will recover the fact that families given by complex interpolation, or equivalently by degenerate Monge-Ampére equations, lead to subharmonic functions α. Then we will try to see, at a very heuristic level, what can be said in the real case. A final section proposes a local L^2 approach, to the Busemann inequality similar to that used in the preceding sections.

Acknowledgements We thank Yanir Rubinstein and Bo Berndtsson for interesting, related discussions. Bo'az Klartag was supported in part by the Israel Science Foundation and by a Marie Curie Reintegration Grant from the Commission of the European Communities.

2   The Complex Case

Let K_0 and K_1 be two unit balls of $\mathbb{C}^n$ associated with the (complex vector space) norms $\|\cdot\|_0$ and $\|\cdot\|_1$. Note that here we are working with the class of convex bodies K of $\mathbb{R}^{2n}$ that are *circled*, meaning that $e^{i\theta}K = K$ for every $\theta \in \mathbb{R}$. We think of a normed space as a triplet consisting of a vector space, a norm and its unit ball. Consider the complex normed spaces $X_0 = (\mathbb{C}^n, \|\cdot\|_0, K_0)$ and $X_1 = (\mathbb{C}^n, \|\cdot\|_1, K_1)$ and write

$$X_z = (\mathbb{C}^n, \|\cdot\|_z, K_z)$$

for the complex Calderón interpolated space at

$$z \in C := \{w \in \mathbb{C} \,;\, \Re(w) \in [0, 1]\}$$

where $\Re(w)$ is the real part of $w \in \mathbb{C}$. Recall that $X_z = X_{\Re(z)}$ and therefore $K_z = K_t$ with $t = \Re(z) \in [0, 1]$. We have:

Theorem 5 ([12]). *The function* $t \to |K_t|$ *is log-concave on* $[0, 1]$ *and so*

$$|K_0|^{1-t} \, |K_1|^t \le |K_t|. \tag{8}$$

In the case of complex unit balls, this result improves upon the Brunn-Minkowski inequality since it can be verified, by using the Poisson kernel on $[0, 1] \times \mathbb{C}^n$ and the definition of the interpolated norm, that

$$K_t \subset (1 - t)K_0 + tK_1 = K(t).$$

In this setting, it also gives the Santaló inequality. Indeed, for a given complex unit ball $K \subset \mathbb{C}^n$, let X_0 be the associated complex normed space, and let X_1 be the dual conjugate space which has $K^\circ \subset \mathbb{C}^n$ as its unit ball. Then it is well known that

$$X_{1/2} = \ell_2^n(\mathbb{C}) = \ell_2^{2n}(\mathbb{R}) \tag{9}$$

and therefore we obtain

$$\sqrt{|K||K^\circ|} \le |B_2^{2n}|.$$

(Let us mention here that the conjugation bar in the statements of [12] is superfluous according to standard definitions).

In order to have a better grasp on complex interpolation, let us write an explicit formula in the specific case of Reinhardt domains. A subset $K \subset \mathbb{C}^n$ is *Reinhardt* if for any $z = (z_1, \ldots, z_n) \in \mathbb{C}^n$,

$$(z_1, \ldots, z_n) \in K \quad \Longleftrightarrow \quad (|z_1|, \ldots, |z_n|) \in K.$$

Note that a Reinhardt convex set is necessarily circled. In the case where $X_0 = (\mathbb{C}^n, \| \cdot \|_0, K_0)$ and $X_1 = (\mathbb{C}^n, \| \cdot \|_1, K_1)$ are such that K_0 and K_1 are Reinhardt, the interpolated space $X_z = (\mathbb{C}^n, \| \cdot \|_z, K_z)$ satisfies

$$K_z = \left\{ z \in \mathbb{C}^n ; \exists (a, b) \in K_0 \times K_1, |z_j| = |a_j|^{1-t} |b_j|^t \text{ for } j = 1, \ldots, n \right\}$$

with $t = \Re(z)$. The case of Reinhardt unit balls is particularly simple and easy to analyze, but it has its limitations. Still, the idea is that in general, K_t should be understood as a "geometric mean" of the bodies K_0 and K_1, whereas the Minkowski sum (1) reminds us of an arithmetic mean.

Theorem 5 was proved using the complex version of the Prékopa theorem obtained by Berndtsson [5], which was derived in [13] using a local computation and L^2 spectral inequalities of Hördmander type. Here, we would like to provide a different direct proof, by combining the results of Rochberg and Hörmander's *a priori* L^2-estimates. Let $\| \cdot \|_z$ be a family of interpolated norms on $\mathbb{C}^n$ and $K_z = B_{\|\cdot\|_z}$. We assume for simplicity that these norms are smooth and strictly convex, so that we will not have to worry about justification of the differentiations under the integral signs. In fact, by approximation we can assume that $1/R \leq \text{Hess} \| \cdot \|_k^2 \leq R$ (for some large constant $R > 1$) for $k = 1, 2$, and these bounds remain valid for the interpolated norms. Introduce the function $F : C \times \mathbb{C}^n \to \mathbb{R}$,

$$F(z, w) := \frac{1}{2} \|w\|_z^2.$$

Denote the Lebesgue measure on $\mathbb{C}^n \simeq \mathbb{R}^{2n}$ by λ, and introduce, in view of (6),

$$\alpha(z) = -\log \int_{\mathbb{C}^n} e^{-F(z,w)} d\lambda(w) = -\log |K_z| - \log(c_{2n,2})$$

for $z \in C$. Our goal is to prove that $t \to \alpha(t)$ is convex on $[0, 1]$. Since $\alpha(z) = \alpha(\Re(z))$, this is equivalent to proving that α is subharmonic on the strip C. The following analogue of (7) is also straightforward:

$$\frac{1}{4} \Delta \alpha(z) = \partial_{z\bar{z}}^2 \alpha(z) = \int_{\mathbb{C}^n} \partial_{z\bar{z}}^2 F \, d\mu_{F_z} - \int_{\mathbb{C}^n} \left| \partial_z F(w) - \int \partial_z F \, d\mu_{F_z} \right|^2 d\mu_{F_z}(w),$$

where μ_{F_z} is the probability measure on $\mathbb{C}^n$ given by

$$\mathrm{d}d\mu_{F_z}(w) = \frac{e^{-F(z,w)}}{\int e^{-F(z,\zeta)}d\lambda(\zeta)}d\lambda(w).$$

It was explained by Rochberg [17] that complex interpolation is characterized by the following differential equation:

$$\partial_{z\bar{z}}^2 F = \sum_{j,k=1}^{n} F^{j\bar{k}}(z,w)\partial_{\overline{w_j}}(\partial_z F)\overline{\partial_{\overline{w_k}}(\partial_z F)} \tag{10}$$

where $(F^{j\bar{k}})_{j,k\le n}$ is the inverse of the complex Hessian in the w-variables of $F(z,w)$, that is

$$\left(F^{j\bar{k}}\right)_{j,k\le n} = \left(\mathrm{Hess}_w^{\mathbb{C}} F\right)^{-1} := \left[\left(\partial_{w_j \overline{w_k}}^2 F\right)_{j,k\le n}\right]^{-1}.$$

Actually, the function F is plurisubharmonic on $C \times \mathbb{C}^n \subset \mathbb{C}^{n+1}$ and (10) expresses the fact that it is a solution of the degenerate Monge-Ampère equation

$$\det\left(\mathrm{Hess}_{z,w}^{\mathbb{C}} F\right) = 0$$

where $\mathrm{Hess}_{z,w}^{\mathbb{C}} F$ is the full complex Hessian of $F(z,w)$, an $(n+1) \times (n+1)$ matrix.

As a consequence of the previous discussion, we have that, for a fixed $z \in C$ and setting $u := \partial_z F(z,\cdot) : \mathbb{C}^n \to \mathbb{C}$,

$$\Delta\alpha(z)/4 = \int_{\mathbb{C}^n} \sum_{j,k=1}^{n} F^{j\bar{k}}\partial_{\overline{w_j}}u\,\overline{\partial_{\overline{w_k}}u}\,d\mu_{F_z} - \int \left|u - \int u\,d\mu_{F_z}\right|^2 d\mu_{F_z}. \tag{11}$$

Of course, it is now irresistible to appeal to Hörmander's *a priori* estimate (see e.g. [15]). It states that if $F : \mathbb{C}^n \to \mathbb{R}$ is a (strictly) plurisubharmonic function and if u is a (smooth enough) function, then

$$\int_{\mathbb{C}^n} |u - P_H u|^2 \, d\mu_F \le \int_{\mathbb{C}^n} \sum_{j,k=1}^{n} F^{j\bar{k}}\partial_{\overline{w_j}}u\,\overline{\partial_{\overline{w_k}}u}\,d\mu_F \tag{12}$$

where $\mathrm{d}d\mu_F(w) = \frac{e^{-F(w)}}{\int e^{-F} d\lambda}d\lambda(w)$ and $P_H : L^2(\mu_F) \to L^2(\mu_F)$ is the orthogonal projection onto the closed space $H = \{h \in L^2(\mu_F) \ ; \ \bar{\partial}h = 0\}$ of holomorphic functions. Actually, this *a priori* estimate on $\mathbb{C}^n$ is rather easy to prove by duality and integration by parts. We now apply this result to $F = F(z,\cdot)$, $\mu_F = \mu_{F_z}$ and $u = \partial_z F$. Note that F (and thus μ_F) and u are invariant under the action of S^1: $F(z,e^{i\theta}w) = F(z,w)$ and the same is true for $\partial_z F$. This implies that the function $P_H u$ has the same invariance, but since it is a holomorphic function on $\mathbb{C}^n$, it has to be constant. Therefore $P_H u = \int u d\mu_{F_z}$ and we indeed obtain that $\Delta\alpha(z) \geq 0$ by combining (11) and (12), as desired.

Here, we reproved (8) without using explicitly [5], but rather by combining the local computations of [13] and the degenerate Monge-Ampère equation satisfied by the complex interpolation. In fact, this computation also appears, in a much more general and deep form, in recent works by Berndtsson [6, 7]. The reason is that complex interpolation corresponds to a geodesic in the space of metrics, and therefore enters Berndtsson's abstract theorems. Also, it can be noticed that complex interpolation corresponds to an extremal construction (for given boundary data), in the sense that it can be viewed as a plurisubharmonic hull. Equivalently, plurisubharmonic functions may be viewed as sub-solutions of degenerate Monge-Ampère equations.

Following our presentation, it is very tempting to develop an analogous presentation for convex bodies in $\mathbb{R}^n$. However, the real case is more complex, as we shall now see.

3   Real Interpolations

The concept of interpolation and the basic properties we present here are due to Semmes [18], building on previous work by Rochberg [17]. Semmes indeed raised the question of whether such interpolations (which are not interpolations in the operator sense) could be used to prove inequalities, by showing that certain functionals are convex along the interpolation. Our main contribution here is to explain that this is indeed the case, by connecting this interpolation with some well-known spectral inequalities. However, some discussions will remain at a heuristic level, as it is not the purpose of this note to discuss existence, unicity and regularity of solutions to the partial differential equations we refer to.

Definition 1 (Rochberg–Semmes interpolation [18]). Let I be an interval of $\mathbb{R}$ and $p \in [1, +\infty]$. We say that a smooth function $F : I \times \mathbb{R}^n \to \mathbb{R}$ is a family of p-interpolation if for any $t \in I$, the function $F(t, \cdot)$ is (strongly) convex on $\mathbb{R}^n$ and for $(t, x) \in I \times \mathbb{R}^n$

$$\partial_{tt}^2 F = \frac{1}{p} \left(\text{Hess}_x F \right)^{-1} \nabla \partial_t F \cdot \nabla \partial_t F. \tag{13}$$

Accordingly, when $\partial_{tt}^2 F \geq \frac{1}{p} \left(\text{Hess}_x F \right)^{-1} \nabla \partial_t F \cdot \nabla \partial_t F$, we say that F is a sub-family of p-interpolation.

In Definition 1, we denote by ∇F the gradient of $F(t, x)$ in the x variables, and a function is strongly convex when $\text{Hess}_x F > 0$. By standard linear algebra we have the following equivalent formulation in terms of the degenerate Monge-Ampère equation:

Proposition 6 (Interpolation and degenerate Monge-Ampère equation). *Let* $F : I \times \mathbb{R}^n \to \mathbb{R}$ *be a smooth function such that* $F(t, \cdot)$ *is (strongly) convex on* $\mathbb{R}^n$ *and introduce, for* $(t, x) \in I \times \mathbb{R}^n$, *the* $(n + 1) \times (n + 1)$ *matrix*

$$H = H_p F(t, x) := \begin{pmatrix} \partial_{tt}^2 F & (\nabla_x \partial_t F)^* \\ \nabla_x \partial_t F & \mathrm{d} p \mathrm{Hess}_x F \end{pmatrix}. \tag{14}$$

Then, F *is a family (resp. a sub-family) of* p-*interpolation if and only if* $\mathrm{d} \det H = 0$ *(resp.* $\det H \geq 0$) *on* $I \times \mathbb{R}^n$.

In particular, 1-interpolation corresponds exactly to the degenerate Monge-Ampère equation on $I \times \mathbb{R}^n$. In fact, we see p-interpolation as a (Dirichlet) boundary value problem.

Definition 2. Let F_0 and F_1 be two smooth convex functions on $\mathbb{R}^n$. We say that $\{F_t : \mathbb{R}^n \to \mathbb{R}\}_{t \in [0,1]}$ is a p-interpolated family associated with $\{F_0, F_1\}$ if $F(t, x) = F_t(x)$ is a family of p-interpolation on $[0, 1] \times \mathbb{R}^n$ with boundary value $F(0, \cdot) = F_0$ and $F(1, \cdot) = F_1$.

As we said earlier, we will not discuss in this exposition questions related to existence, uniqueness and regularity of solutions to this Dirichlet problem (except for the easy case $p = 1$, explained below). However, it is reasonable to expect that generalized solutions, which are sufficient for our purposes, can be constructed by using Perron processes, as mentioned by Semmes [18].

Using Notation 3, given a family or a sub-family of p-interpolation F, we aim to understand the convexity of the function on I,

$$\alpha(t) = -\log \int_{\mathbb{R}^n} e^{-F(t,x)} \, dx. \tag{15}$$

In view of (7), we see that for every fixed $t \in I$ we have the implication

$$\mathrm{Var}_{\mu_{F_t}} (\partial_t F) \leq \frac{1}{p} \int_{\mathbb{R}^n} \left(\mathrm{Hess}_x F \right)^{-1} \nabla \partial_t F \cdot \nabla \partial_t F \, d\mu_{F_t} \quad \Longrightarrow \quad \alpha''(t) \geq 0, \tag{16}$$

under some mild regularity assumptions. The left-hand side is of course reminiscent of the real version of Hörmander's estimate (12), which is known as the Brascamp-Lieb inequality from [9]. Recall that this inequality states that if $F : \mathbb{R}^n \to \mathbb{R}$ is a (strongly) convex function and if $u \in L^2(\mu_F)$ is a locally Lipschitz function, then

$$\mathrm{Var}_{\mu_F} (u) \leq \int_{\mathbb{R}^n} \left(\mathrm{Hess}_x F \right)^{-1} \nabla u \cdot \nabla u \, d\mu_F, \tag{17}$$

with our notation $\mathrm{d} d\mu_F(x) = \frac{e^{-F(x)}}{\int e^{-F}} \, dx$. Again, this inequality can easily be proven along the lines of Hörmander's approach (see below).

Applying the Brascamp-Lieb inequality (17) to $F = F(t, \cdot)$ and $u = \partial_t F$ when F is a 1-interpolation sub-family, we obtain, in view of (16), the following statement:

Proposition 7. *If F is a sub-family of 1-interpolation, then α is convex.*

The first comment is that we have not proved anything new! Indeed, it is directly verified below that for any C^2-smooth function F,

$$F \text{ is a sub-family of 1-interpolation} \iff F \text{ is convex on } I \times \mathbb{R}^n. \tag{18}$$

Therefore, we have reproduced Prékopa's Theorem 1. In order to demonstrate (18), observe that the positive semi-definiteness of the matrix $H_1 F(t, x)$ amounts to the inequality

$$(\text{Hess}_x F)y \cdot y + 2\nabla_x(\partial_t F) \cdot y + \partial_{tt}^2 F \geq 0 \qquad \text{for all } y \in \mathbb{R}^n,$$

or equivalently,

$$\partial_{tt}^2 F \geq \sup_{y \in \mathbb{R}^n} [2\nabla_x(\partial_t F) \cdot y - (\text{Hess}_x)Fy \cdot y] = (\text{Hess}_x F)^{-1}\nabla_x \partial_t F \cdot \nabla_x \partial_t F,$$

as $\text{Hess}_x F$ is positive definite. Let us note that if F_0 and F_1 are given, then the associated family of 1-interpolation—equivalently, the unique solution to the degenerate Monge-Ampère equation on $[0, 1] \times \mathbb{R}^n$ with $F(t, x)$ convex in x—is

$$F(t, w) = \inf_{w=(1-t)x+ty} \left\{ (1 - t)F_0(x) + tF_1(y) \right\}. \tag{19}$$

Every sub-family of 1-interpolation is *above* this F, and thus the statement of Prékopa's Theorem reduces to 1-interpolation families (an argument that is standard in the study of functional Brunn-Minkowski inequalities). One way to recover the Brunn-Minkowski inequality directly from this family F of 1-interpolation, is to take, as in the derivation from Prékopa's theorem, something like $F_0(x) = \|x\|_{K_0}^q/q$, $F_1(y) := \|y\|_{K_1}^q/q$ and let $q \to +\infty$.

We have just shown that Prékopa's theorem reduces, locally, to the Brascamp-Lieb inequality, an observation that is already implicitly present in [9] . This is parallel to the complex setting, i.e to the local L^2-proof of the complex Prékopa theorem of Berndtsson given in [13] and extended in [6,7]. The converse procedure was known, starting from the work of Brascamp and Lieb; more explicitly, Bobkov and Ledoux [8] noted that the Prékopa-Leindler inequality (an extension of Prékopa's result to the case that the fibers are not convex) indeed implies the Brascamp-Lieb inequality. We also emphasize Colesanti's work [11], where, starting from the Brunn-Minkowski inequality, spectral inequalities of Brascamp-Lieb type on the boundary ∂K of a convex body $K \subset \mathbb{R}^n$ are obtained. This can also be recovered by applying the Brascamp-Lieb inequality to homogeneous functions.

The conclusion is that all of these results are the global/local versions of the same phenomenon. At the local level, we have reduced the problem to the inequality (17) which expresses a spectral bound in $L^2(\mu_F)$ for the elliptic operator associated with the Dirichlet form on the right-hand side of (17).

For completeness, we would like to briefly recall here Hörmander's original approach to (17). Consider the Laplace-type operator on $L^2(\mu_F)$,

$$L := \Delta - \nabla F \cdot \nabla,$$

that we define, say, on the space of C^2-smooth compactly supported functions. First, recall the integration by parts formulae, $\int uL\varphi \, d\mu_F = -\int \nabla u \cdot \nabla \varphi \, d\mu_F$ and

$$\int_{\mathbb{R}^n} (L\varphi)^2 \, d\mu_F = \int_{\mathbb{R}^n} (\text{Hess}_x F)\nabla \varphi \cdot \nabla \varphi \, d\mu_F + \int_{\mathbb{R}^n} \| \text{Hess}_x \varphi \|_2^2 \, d\mu_F, \qquad (20)$$

where $\| \text{Hess} \, \varphi \|_2^2 = \sum_{i,j \leq n}(\partial^2_{i,j}\varphi)^2$. Let u be a locally-Lipschitz function on $\mathbb{R}^n$. We use the (rather weak) standard observation that the image by L of the C^2-smooth compactly supported functions is dense in the space of $L^2(\mu_F)$ functions orthogonal to constants (see e.g. [14]). For $\varepsilon > 0$ let φ be a C^2-smooth, compactly-supported function such that $L\varphi - (u - \int ud\mu_F)$ has $L^2(\mu_F)$-norm smaller than ε. Then, by integration by parts and using (20) we get

$$\text{Var}_{\mu_F}(u) = 2\int \left(u - \int u \, d\mu_F\right)L\varphi \, d\mu_F - \int (L\varphi)^2 d\mu_F + \int \left(L\varphi - \left(u - \int u \, d\mu_F\right)\right)^2 d\mu_F$$

$$\leq -2\int \nabla u \cdot \nabla \varphi \, d\mu_F - \int (\text{Hess}_x F)\nabla \varphi \cdot \nabla \varphi \, d\mu_F - \int \| \text{Hess}_x \varphi \|_2^2 d\mu_F + \varepsilon^2$$

$$\leq -2\int \nabla u \cdot \nabla \varphi - \int (\text{Hess}_x F)\nabla \varphi \cdot \nabla \varphi \, d\mu_F + \varepsilon^2$$

$$\leq \int (\text{Hess}_x F)^{-1}\nabla u \cdot \nabla u \, d\mu_F + \varepsilon^2,$$

and (17) follows by letting ε tend to zero.

Let us go back to interpolation families. As we said, 1-sub-interpolation corresponds to a function F that is convex on $I \times \mathbb{R}^n$. More generally, we have the following characterization, proved by Semmes:

Proposition 8. *For a smooth function* $F : I \times \mathbb{R}^n \to \mathbb{R}$, *the following are equivalent:*

- *F is a sub-family of p-interpolation.*
- *With the notation (14), we have,* $\forall (t, x) \in I \times \mathbb{R}^n$, $dH_p F(t, x) \geq 0$.

- *For all $x_0, y_0 \in \mathbb{R}^n$, the function*

$$(s, t) \longrightarrow F\left(t, x_0 + (t + \sqrt{p-1}\,s)y_0\right)$$

is subharmonic on the subset of $\mathbb{R}^2$ where it is defined.

Note that the third condition in Proposition 8 needs only a minimal level of smoothness. We may thus speak of a sub-family F of p-interpolation even when F is not very smooth.

We turn now to duality, which was part of the motivation of Semmes. We shall denote by $\mathcal{L}$ the Legendre transform in space, i.e. on $\mathbb{R}^n$. In particular, for $F : I \times \mathbb{R}^n$, we shall write

$$\mathcal{L}F(t, x) = \mathcal{L}(F_t)(x) = \sup_{y \in \mathbb{R}^n} \{x \cdot y - F(t, y)\}.$$

It is classical that if F is the family of 1-interpolation given by (19), then $\mathcal{L}F$ is a family of ∞-interpolation, meaning that $\mathcal{L}F$ is affine in t:

$$\mathcal{L}F_t(x) = (1 - t)\mathcal{L}F_0(x) + t\mathcal{L}F_1(x).$$

So in this case, when we move to the dual setting, Brunn-Minkowski or Prékopa's inequality is replaced by the trivial fact that $\alpha(t) = -\log \int e^{-\mathcal{L}_t F(x)} dx$ is *concave* by Hölder's inequality.

More general duality relations hold for p-interpolations. Suppose $F(t, x) = F_t(x)$ is convex in x, and denote $G(t, y) = \mathcal{L}F_t(y)$. We have the identity (proved below):

$$\partial_{tt}^2 F + \partial_{tt}^2 G = (\mathrm{Hess}_x F)^{-1} \nabla \partial_t F \cdot \nabla \partial_t F = (\mathrm{Hess}_y G)^{-1} \nabla \partial_t G \cdot \nabla \partial_t G, \quad (21)$$

where F and its derivatives are evaluated at (t, x), while G and its derivatives are evaluated at $(t, y) = (t, \nabla F(x))$. From this identity, we immediately conclude

Proposition 9. *If F is a family of p-interpolation, then $\mathcal{L}F$ is a family of p'-interpolation, where $\frac{1}{p'} + \frac{1}{p} = 1$.*

We now present the details of the straightforward proof of (21). From the definition,

$$G(t, \nabla F(t, x)) = \langle x, \nabla F(t, x) \rangle - F(t, x), \quad (22)$$

$$\nabla G_t(\nabla F_t(x)) = (\nabla G)(t, \nabla F(x)) = x \quad (23)$$

$$\mathrm{Hess}_y G(t, \nabla F(x, t)) = (\mathrm{Hess}_x F(t, x))^{-1}. \quad (24)$$

where the gradients and the hessians refer only to the space variables x, y. By differentiating (23) with respect to t, we see that

$$\nabla \partial_t G = -(\text{Hess}_y\, G)(\nabla \partial_t F) \tag{25}$$

where G and its derivatives are evaluated at $(t, y) = (t, \nabla F(x))$, while F and its derivatives are evaluated at (t, x). From (24) and (25),

$$-\nabla \partial_t G \cdot \nabla \partial_t F = (\text{Hess}_x\, F)^{-1} \nabla \partial_t F \cdot \nabla \partial_t F = (\text{Hess}_y\, G)^{-1} \nabla \partial_t G \cdot \nabla \partial_t G. \tag{26}$$

Differentiating (22) with respect to t and using (23) we get that $\partial_t G(t, \nabla F(x))$ $= -\partial_t F(t, x)$. If we differentiate this last equality one more time with respect to t, we find

$$\partial_{tt}^2 G + \nabla \partial_t G \cdot \nabla \partial_t F = -\partial_{tt}^2 F,$$

which combined with (26) yields the desired formula (21).

As a consequence of Proposition 8, we see that 2-interpolation families satisfy an interpolation duality theorem. Let f be a convex function on $\mathbb{R}^n$, and suppose that $F_t(x) = F(t, x)$ is the 2-interpolation family F with $F_0 = f$ and $F_1 = \mathcal{L}f$. Then,

$$F(t, x) = \mathcal{L}F(1 - t, x)$$

provided we have unicity for the 2-interpolation problem, and therefore we have

$$F\left(\frac{1}{2}, x\right) = \frac{|x|^2}{2}.$$

If we take $f(x) = \|x\|_K^2 / 2$, then $\mathcal{L}f(x) = \|x\|_{K^\circ}^2 / 2$. Thus, if we could prove that for a 2-interpolation family F, the associated function α from (15) is convex, as it is for 1-interpolations, then we would recover Santaló's inequality. This would be the case if we had a Brascamp-Lieb inequality with a factor $1/2$ on the right-hand side of (17) for every convex function $F : \mathbb{R}^n \to \mathbb{R}$. However, this is of course false in general. Recall that even for the Santaló inequality, some "center" must be fixed or some symmetry must be assumed. Therefore, a more reasonable question to ask, is whether α is convex when the initial data f is even. This guarantees that F_t is even for all $t \in [0, 1]$. However, it is again false in general that the Brascamp-Lieb inequality holds with factor $1/2$ in the right-hand side of (17) when F and u are even, as can be shown by taking a perturbation of the Gaussian measure. This suggests that the answer to the question could be negative in general. A reasonable conjecture, perhaps, is:

Conjecture 1. Assume F_0 and F_1 are even, convex and 2-homogeneous (i.e. $F_i(x) = \lambda_i \|x\|_{K_i}^2$ for some centrally-symmetric convex bodies $K_i \subset \mathbb{R}^n$), properties that propagate along the interpolation. Then, the function α associated with the 2-interpolation family is convex.

Here is a much more modest result:

Fact 11. *Assume that f is convex and even, and let F be a 2-interpolation family with $F_0 = f$ and $F_1 = \mathcal{L}f$, with the associated function α as in* (15). *Then, one has*

$$\alpha''(1/2) \geq 0.$$

Proof. Since $F(\frac{1}{2}, x) = |x|^2/2$, the probability measure $\mu_{F_{1/2}}$ is exactly the Gaussian measure on $\mathbb{R}^n$, which we denote by γ. Note also that $\mathrm{Hess}_x \, F_{1/2} = \mathrm{Id}_{\mathbb{R}^n}$. Therefore, if we denote $u = \partial_t F(\frac{1}{2}, \cdot)$, we need to check that

$$\mathrm{Var}_\gamma(u) \leq \frac{1}{2} \int_{\mathbb{R}^n} |\nabla u|^2 \, d\gamma.$$

The function $v := u - \int u \, d\gamma$ is by construction orthogonal to constant functions in $L^2(\gamma)$. But since u is even (because F_t is even for all t, and so is $\partial_t F$), this function v is also orthogonal to linear functions. Recall that the Hermite (or Ornstein-Uhlenbeck) operator $L = \Delta - x \cdot \nabla$ has non-positive integers as eigenvalues, and that the eigenspaces (generated by Hermite polynomials) associated with the eigenvalues 0 and -1 are formed by the constant and linear functions. Therefore, v belongs to the subspace where $-L \geq 2 \, \mathrm{Id}$ and so

$$\mathrm{Var}_\gamma(u) = \int |v|^2 \, d\gamma \leq -\frac{1}{2} \int v L v \, d\gamma = \frac{1}{2} \int |\nabla u|^2 \, d\gamma.$$

We conclude this section by mentioning that we have analogous formulas in the case where we work with some fixed measure ν on $\mathbb{R}^n$, in place of the Lebesgue measure. Then, for a function $F : \mathbb{R}^n \to \mathbb{R}$ such that $\int e^{-F} \, d\nu < +\infty$, we denote by $\mu_{\nu, F}$ the probability measure on $\mathbb{R}^n$ given by

$$d\mu_{\nu, F}(x) := \frac{de^{-F(x)}}{\int e^{-F} \, d\nu} \, d\nu(x).$$

For a function of $n + 1$ variables $F : I \times \mathbb{R}^n \to \mathbb{R}$, we denote as before $F_t := F(t, \cdot) : \mathbb{R}^n \to \mathbb{R}$ and then μ_{ν, F_t} is the corresponding probability measure on $\mathbb{R}^n$. We are then interested in the convexity of the function

$$\alpha_\nu(t) := -\log \int_{\mathbb{R}^n} e^{-F(t,x)} \, d\nu(x) = -\log \int_{\mathbb{R}^n} e^{-F_t} \, d\nu.$$

The computation is identical:

$$\alpha_\nu''(t) = \int_{\mathbb{R}^n} \partial_{tt}^2 F \, d\mu_{\nu, F_t} - \mathrm{Var}_{\mu_{\nu, F_t}}(\partial_t F).$$

Here is an illustration. Let ν be a symmetric log-concave measure on $\mathbb{R}^n$: $d\nu(x) = e^{-W(w)}\, dx$ with W being convex and even on $\mathbb{R}^n$, and consider the family

$$F(t, x) = e^t\, |x|^2/2.$$

This is a typical example of a 2-interpolation family. Then, the fact that the corresponding α_ν is convex is equivalent to the B-conjecture proved in [14]. The argument there begins with the computation above. It turns out that for this particular family F, the required Brascamp-Lieb inequality reduces to a Poincaré inequality for the measure μ_{ν, F_t}, which holds precisely with a constant $1/2$ when restricted to even functions.

Let us also mention in this direction that the Santaló inequality in its functional form (4) also holds if the Lebesgue measure is, in the three integrals, replaced by an even log-concave measure of $\mathbb{R}^n$, as noted in Klartag [16]. Several examples of this type suggest that the Lebesgue measure can often be replaced by a more general log-concave measure.

4 The Busemann Inequality

We conclude this survey with a proof of the Busemann inequality via L^2 inequalities. The Busemann inequality [10] is concerned with non-parallel hyperplane sections of a convex body $K \subset \mathbb{R}^n$. In the particular case where K is centrally-symmetric, the Busemann inequality states that

$$g(x) = \frac{|x|}{|K \cap x^\perp|} \qquad (x \in \mathbb{R}^n)$$

is a norm on $\mathbb{R}^n$. Here $|K \cap x^\perp|$ is the $(n-1)$-dimensional volume of the hyperplane section $K \cap x^\perp = \{y \in K; y \cdot x = 0\}$, and $g(0) = 0$ as interpreted by continuity. The convexity of the function g is a non-trivial fact. Using the Brunn-Minkowski inequality, the convexity of g reduces to a statement about log-concave functions in the plane, as observed by Busemann. Indeed, the convexity of g has to be checked along affine lines, and therefore on 2-dimensional vector subspaces. Specifically, let $E \subset \mathbb{R}^n$ be a two-dimensional plane, which we conveniently identify with $\mathbb{R}^2$. For $y \in \mathbb{R}^2 = E$ set

$$e^{-w(y)} = |K \cap (y + E^\perp)|,$$

the $(n-2)$-dimensional volume of the section of K. Then $w : \mathbb{R}^2 \to \mathbb{R} \cup \{+\infty\}$ is a convex function, according to the Brunn-Minkowski inequality. For $p > 0$ and $t \in \mathbb{R}$ define

$$\alpha_p(t) = \int_0^\infty e^{-w(ts, s)} s^{p-1} ds. \tag{27}$$

Note that when K is centrally-symmetric, $2\sqrt{1+t^2}\alpha_1(t) = |K \cap (1,-t)^\perp|$. We therefore see that Busemann's inequality amounts to the convexity of the function $1/\alpha_1(t)$ on $\mathbb{R}$. Next we will prove the following more general statement, which is due to Ball [3] when $p \geq 1$:

Theorem 12. *Let X be an n-dimensional real linear space and let $w : X \to \mathbb{R}$ be a convex function with $\int e^{-w} < \infty$. For $p > 0$ and $0 \neq x \in X$ denote*

$$h(x) = \left(\int_0^\infty e^{-w(sx)} s^{p-1} ds \right)^{-1/p}$$

with $h(0) = 0$. Then h is a convex function on X.

Busemann's proof of the case $p = 1$ of Theorem 12, and the generalization to $p \geq 1$ by Ball, rely on transportation of measure in one dimension. The proof we present below may be viewed as an infinitesimal version of Busemann's transportation argument. This is reminiscent of the proof given in Ball, Barthe and Naor [4] of the Prékopa inequality, which may be viewed as an infinitesimal version of the transportation proof of the latter inequality.

Proof of Theorem 12: By a standard approximation argument, we may assume that w is smooth and $1/R \leq \mathrm{Hess}(w) \leq R$ at all points of $\mathbb{R}^n$, for some large constant $R > 1$. Therefore h is a continuous function, smooth outside the origin, and homogeneous of degree one. Since convexity of a function involves three collinear points contained in a two-dimensional subspace, we may assume that $n = 2$. Thus, selecting a point $0 \neq z \in X$ and a direction $\theta \in X$, our goal is to show that $\partial^2_{\theta\theta} h(z) \geq 0$ (since h is homogeneous of degree one, it suffices to consider the case $z \neq 0$). If θ is proportional to z, then the second derivative vanishes as h is homogeneous of degree one. We may therefore select coordinates $(t, x) \in \mathbb{R}^2 = X$, and identify $z = (0, 1)$ and $\theta = (1, 0)$. With this identification, in order to prove the theorem we need to show that

$$\left(\alpha_p^{-1/p} \right)'' (0) \geq 0,$$

where α_p is defined in (27). Equivalently, we need to prove that at the origin,

$$\partial^2_{tt} \alpha_p \leq \left(1 + \frac{1}{p} \right) \left(\partial_t \alpha_p \right)^2 / \alpha_p. \tag{28}$$

We denote by μ the probability measure on $[0, \infty)$ whose density is proportional to the integrable function $\exp(-w(0, x))x^{p-1}$. Similarly to Fact (7) above with $F(t, x) = w(tx, x)$, the desired inequality (28) is equivalent to

$$\mathrm{Var}_\mu(x \partial_t w) \leq \int_0^\infty x^2 (\partial^2_{tt} w) d\mu(x) + \frac{1}{p} \left(\int_0^\infty x (\partial_t w) d\mu(x) \right)^2. \tag{29}$$

We will use the convexity of $w(t, x)$ via the inequality $\partial_{tt}^2 w \geq \left(\partial_{tx}^2 w\right)^2 / \partial_{xx}^2 w$, which expresses the fact that $w_t(x) = w(t, x)$ is a sub-family of 1-interpolation. Denote $u(x) = x \partial_t w(0, x)$ and compute that $x \partial_{tx}^2 w = u' - u(x)/x$ for $x > 0$. Hence, in order to prove (29), it suffices to show that

$$\text{Var}_\mu(u) \leq \int_0^\infty \frac{1}{\partial_{xx}^2 w} \left(u'(x) - \frac{u(x)}{x}\right)^2 d\mu(x) + \frac{1}{p} \left(\int_0^\infty u \, d\mu(x)\right)^2. \tag{30}$$

We will prove (30) for any smooth function $u \in L^2(\mu)$ (it is clear that the function $x \partial_t w(0, x)$ grows at most polynomially at infinity, and hence belongs to $L^2(\mu)$). By approximation (e.g., multiply u by an appropriate cutoff function), it suffices to restrict our attention to smooth functions such that $u - \int u \, d\mu$ is compactly-supported in $[0, \infty)$. Consider the Laplace-type operator

$$L\varphi = \varphi'' - \left(\partial_x w(0, x) - \frac{p-1}{x}\right)\varphi' = \varphi'' - \partial_x\left(w(0, x) - (p-1)\log(x)\right)\varphi'.$$

Integrating the ordinary differential equation, we find a smooth function φ, with $\varphi'(0) = 0$ and φ' compactly-supported in $[0, \infty)$, such that $L\varphi = u - \int u \, d\mu$. As before, we have the integration by parts $\int (L\varphi) u \, d\mu = -\int \varphi' u' \, d\mu$ and

$$\int_0^\infty (L\varphi)^2 d\mu = -\int_0^\infty \varphi'(x) u'(x) d\mu$$

$$= \int_0^\infty (\varphi''(x))^2 d\mu + \int_0^\infty \left(\partial_{xx}^2 w + \frac{p-1}{x^2}\right) (\varphi'(x))^2 d\mu.$$

Let us abbreviate $w'' = \partial_{xx}^2 w(0, x)$, $E = \int u \, d\mu$ and also $\langle f \rangle = \int_0^\infty f(x) d\mu(x)$. Then, by using the above identities and by completing three squares (marked by wavy underline),

$$\text{Var}_\mu(u) = -2\langle u'\varphi'\rangle - \langle (L\varphi)^2\rangle$$

$$= \left\langle -2\varphi'\left(u' - \frac{u}{x}\right)\right\rangle - \left\langle \frac{2\varphi'u}{x}\right\rangle - \left\langle (\varphi'')^2 + w''(\varphi')^2 + \frac{p-1}{x^2}(\varphi')^2\right\rangle$$

$$\leq \left\langle \frac{1}{w''}\left(u' - \frac{u}{x}\right)^2\right\rangle - 2\left\langle \frac{\varphi'(L\varphi + E)}{x}\right\rangle - \left\langle (\varphi'')^2 + \frac{p-1}{x^2}(\varphi')^2\right\rangle$$

$$= \left\langle \frac{1}{w''}\left(u' - \frac{u}{x}\right)^2\right\rangle + \langle 2\varphi''\varphi'/x\rangle - \left\langle \frac{2\varphi'E}{x} + (\varphi'')^2 + (p+1)\frac{(\varphi')^2}{x^2}\right\rangle$$

$$\leq \left\langle \frac{1}{w''}\left(u' - \frac{u}{x}\right)^2\right\rangle - \left\langle \frac{2\varphi'E}{x} + p\frac{(\varphi')^2}{x^2}\right\rangle \leq \left\langle \frac{1}{w''}\left(u' - \frac{u}{x}\right)^2\right\rangle + \frac{E^2}{p},$$

and (30) is proven. $\qquad\square$

References

1. S. Artstein-Avidan, B. Klartag, V.D. Milman, The santaló point of a function, and a functional form of the santaló inequality. Mathematika **51**(1–2), 33–48 (2004)
2. K. Ball, Isometric problems in ℓ_p and sections of convex sets. Ph.D. dissertation, Cambridge (1986)
3. K. Ball, Logarithmically concave functions and sections of convex sets in $\mathbb{R}^n$. Studia Math. **88**(1), 69–84 (1988)
4. K. Ball, F. Barthe, A. Naor, Entropy jumps in the presence of a spectral gap. Duke Math. J. **119**(1), 41–63 (2003)
5. B. Berndtsson, Prekopa's theorem and kiselman's minimum principle for plurisubharmonic functions. Math. Ann. **312**(4), 785–792 (1998)
6. B. Berndtsson, Subharmonicity properties of the Bergman kernel and some other functions associated to pseudoconvex domains. Ann. Inst. Fourier (Grenoble) **56**, 1633–1662 (2006)
7. B. Berndtsson, Curvature of vector bundles associated to holomorphic fibrations. Ann. Math. (2) **169**(2), 531–560 (2009)
8. S.G. Bobkov, M. Ledoux, From Brunn-Minkowski to Brascamp-Lieb and to logarithmic sobolev inequalities. Geom. Funct. Anal. **10**(5), 1028–1052 (2000)
9. H.J. Brascamp, E.H. Lieb, On extensions of the Brunn-Minkowski and Prékopa-Leindler theorems, including inequalities for log concave functions, and with an application to the diffusion equation. J. Funct. Anal. **22**(4), 366–389 (1976)
10. H. Busemann, A theorem on convex bodies of the Brunn-Minkowski type. Proc. Nat. Acad. Sci. U.S.A. **35**, 27–31 (1949)
11. A. Colesanti, From the Brunn-Minkowski inequality to a class of Poincaré-type inequalities. Comm. Contemp. Math. **10**(5), 765–772 (2008)
12. D. Cordero-Erausquin, Santaló's inequality on $\mathbb{C}^n$ by complex interpolation. C. R. Math. Acad. Sci. Paris **334**, 767–772 (2002)
13. D. Cordero-Erausquin, On Berndtsson's generalization of Prékopa's theorem. Math. Z. **249**(2), 401–410 (2005)
14. D. Cordero-Erausquin, M. Fradelizi, B. Maurey, The (B) conjecture for the Gaussian measure of dilates of symmetric convex sets and related problems. J. Funct. Anal. **214**(2), 410–427 (2004)
15. L. Hörmander, in *Notions of Convexity*. Progress in Mathematics, vol. 127 (Birkhäuser, Boston, 1994)
16. B. Klartag, in *Marginals of Geometric Inequalities*. Geometric Aspects of Functional Analysis, Lecture Notes in Math., vol. 1910 (Springer, Berlin, 2007), pp. 133–166
17. R. Rochberg, Interpolation of Banach spaces and negatively curved vector bundles. Pac. J. Math. **110**(2), 355–376 (1984)
18. S. Semmes, Interpolation of Banach spaces, differential geometry and differential equations, Rev. Mat. Iberoamericana **4**, 155–176 (1988)

Hypercontractive Measures, Talagrand's Inequality, and Influences

Dario Cordero-Erausquin and Michel Ledoux

Abstract We survey several Talagrand type inequalities and their application to influences with the tool of hypercontractivity for both discrete and continuous, and product and non-product models. The approach covers similarly by a simple interpolation the framework of geometric influences recently developed by N. Keller, E. Mossel and A. Sen. Geometric Brascamp-Lieb decompositions are also considered in this context.

1 Introduction

In the famous paper [24], Talagrand showed that for every function f on the discrete cube $X = \{-1, +1\}^N$ equipped with the uniform probability measure μ,

$$\mathrm{Var}_\mu(f) = \int_X f^2 d\mu - \left(\int_X f d\mu \right)^2 \leq C \sum_{i=1}^{N} \frac{\|D_i f\|_2^2}{1 + \log\left(\|D_i f\|_2 / \|D_i f\|_1\right)} \quad (1)$$

for some numerical constant $C \geq 1$, where $\|\cdot\|_p$ denote the norms in $L^p(\mu)$, $1 \leq p \leq \infty$, and for every $i = 1, \ldots, n$ and every $x = (x_1, \ldots, x_N) \in \{-1, +1\}^N$,

D. Cordero-Erausquin (✉)
Institut de Mathématiques, Université Pierre et Marie Curie (Paris 6 – Jussieu), 4 place Jussieu, 75005 Paris, France

Institut Universitaire de France
e-mail: cordero@math.jussieu.fr

M. Ledoux
Institut de Mathématiques de Toulouse, Université de Toulouse, 31062 Toulouse, France

Institut Universitaire de France
e-mail: ledoux@math.univ-toulouse.fr

B. Klartag et al. (eds.), *Geometric Aspects of Functional Analysis*, Lecture Notes in Mathematics 2050, DOI 10.1007/978-3-642-29849-3_10,
© Springer-Verlag Berlin Heidelberg 2012

$$D_i f(x) = f(\tau_i x) - f(x) \tag{2}$$

with $\tau_i x = (x_1, \ldots, x_{i-1}, -x_i, x_{i+1}, \ldots, x_N)$. Up to the numerical constant, this inequality improves upon the classical spectral gap inequality (see below)

$$\mathrm{Var}_\mu(f) \le \frac{1}{4} \sum_{i=1}^{N} \|D_i f\|_2^2. \tag{3}$$

The proof of (1) is based on an hypercontractivity estimate known as the Bonami-Beckner inequality [7,9] (see below). Inequality (1) was actually deviced to recover (and extend) a famous result of Kahn et al. [12] about influences on the cube. Namely, applying (1) to the Boolean function $f = \mathbf{1}_A$ for some set $A \subset \{-1, +1\}^N$, it follows that

$$\mu(A)(1 - \mu(A)) \le C \sum_{i=1}^{N} \frac{2I_i(A)}{1 + \log\left(1/\sqrt{2I_i(A)}\right)} \tag{4}$$

where, for each $i = 1, \ldots, N$,

$$I_i(A) = \mu(\{x \in A, \tau_i x \notin A\})$$

is the so-called influence of the i-th coordinate on the set A (noticing that $\|D_i \mathbf{1}_A\|_p^p = 2I_i(A)$ for every $p \ge 1$). In particular, for a set A with $\mu(A) = a$, there is a coordinate i, $1 \le i \le N$, such that

$$I_i(A) \ge \frac{a(1-a)}{8CN} \log\left(\frac{N}{a(1-a)}\right) \ge \frac{a(1-a) \log N}{8CN} \tag{5}$$

which is the main result of [12]. (To deduce (5) from (4), assume for example that $I_i(A) \le \left(\frac{a(1-a)}{N}\right)^{1/2}$ for every $i = 1, \ldots, N$, since if not the result holds. Then, from (4), there exists i, $1 \le i \le N$, such that

$$\frac{a(1-a)}{CN} \le \frac{2I_i(A)}{1 + \log\left(1/\sqrt{2I_i(A)}\right)} \le \frac{8I_i(A)}{4 + \log(N/4a(1-a))}$$

which yields (5)). Note that (5) remarkably improves by a (optimal) factor $\log N$ what would follow from the spectral gap inequality (3) applied to $f = \mathbf{1}_A$. The numerical constants like C throughout this text are not sharp.

The aim of this note is to amplify the hypercontractive proof of Talagrand's original inequality (1) to various settings, including non-product spaces and continuous variables, and in particular to address versions suitable to geometric influences. It is part of the folklore indeed (cf. e.g. [8]) that an inequality similar to (1), with the same hypercontractive proof, holds for the standard Gaussian measure μ on $\mathbb{R}^N$

(viewed as a product measure of one-dimensional factors), that is, for every smooth enough function f on $\mathbb{R}^N$ and some constant $C > 0$,

$$\text{Var}_\mu(f) \leq C \sum_{i=1}^{N} \frac{\|\partial_i f\|_2^2}{1 + \log(\|\partial_i f\|_2 / \|\partial_i f\|_1)}. \tag{6}$$

(A proof will be given in Sect. 2 below.) However, the significance of the latter for influences is not clear, since its application to characteristic functions is not immediate (and requires notions of capacities). Recently, Keller et al. [13] introduced a notion of geometric influence of a Borel set A in $\mathbb{R}^N$ with respect to a measure μ (such as the Gaussian measure) simply as $\|\partial_i f\|_1$ for some smooth approximation f of $\mathbf{1}_A$, and proved for it the analogue of (5) (with $\sqrt{\log N}$ instead of $\log N$) for the standard Gaussian measure on $\mathbb{R}^N$. It is therefore of interest to seek for suitable versions of Talagrand's inequality involving only L^1-norms $\|\partial_i f\|_1$ of the partial derivatives. While the authors of [13] use isoperimetric properties, we show here how the common hypercontractive tool together with a simple interpolation argument may be developed similarly to reach the same conclusion. In particular, for the standard Gaussian measure μ on $\mathbb{R}^N$, we will see that for every smooth enough function f on $\mathbb{R}^N$ such that $|f| \leq 1$,

$$\text{Var}_\mu(f) \leq C \sum_{i=1}^{N} \frac{\|\partial_i f\|_1 (1 + \|\partial_i f\|_1)}{[1 + \log^+ (1/\|\partial_i f\|_1)]^{1/2}}. \tag{7}$$

Applied to $f = \mathbf{1}_A$, this inequality indeed ensures the existence of a coordinate i, $1 \leq i \leq N$, such that the geometric influence of A along i is at least of the order of $\frac{\sqrt{\log N}}{N}$, that is one of the main conclusions of [13] (where it is shown moreover that the bound is sharp). In this continuous setting, the hypercontractive approach yields more general examples of measures with such an influence property in the range between exponential and Gaussian for which only a logarithmic Sobolev type inequality is needed while [13] required an isoperimetric inequality for the individual measures μ_i.

This note is divided into two main parts. In the first one, we present Talagrand type inequalities for various models, from the discrete cube to Gaussian and more general product measures, by the general principle of hypercontractivity of Markov semigroups. The method of proof, originating in Talagrand's work, has been used recently by O'Donnell and Wimmer [20, 21] to investigate non-product models such as random walks on some graphs which enter the general presentation below. Actually, most of the Talagrand inequalities we present in the discrete setting are already contained in the work by O'Donnell and Wimmer. It is worth mentioning that an approach to the Talagrand inequality (1) rather based on the logarithmic Sobolev inequality was deviced in [22] and [11] a few years ago. The abstract semigroup approach applies in the same way on the sphere along the decomposition of the Laplacian. Geometric Brascamp-Lieb decompositions within this setting are also discussed. In the second part, we address our new version (7) of Talagrand's inequality towards geometric influences and the recent results of [13] by a further interpolation step on the hypercontractive proof.

In the last part of this introduction, we describe a convenient framework in order to develop hypercontractive proofs of Talagrand type inequalities. While of some abstract flavor, the setting easily covers two main concrete instances, probability measures on finite state spaces (as invariant measures of some Markov kernels) and continuous probability measures of the form $d\mu(x) = e^{-V(x)}dx$ on the Borel sets of $\mathbb{R}^n$ where V is some (smooth) potential (as invariant measures of the associated diffusion operators $\Delta - \nabla V \cdot \nabla$). We refer for the material below to the general references $[1, 2, 4, 10, 23]$...

Let μ be a probability measure on a measurable space $(X, \mathcal{A})$. For a function $f : X \to \mathbb{R}$ in $L^2(\mu)$, define its variance with respect to μ by

$$\text{Var}_\mu(f) = \int_X f^2 d\mu - \left(\int_X f d\mu \right)^2.$$

Similarly, whenever $f > 0$, define its entropy by

$$\text{Ent}_\mu(f) = \int_X f \log f d\mu - \int_X f d\mu \log \left(\int_X f d\mu \right)$$

provided it is well-defined. The $L^p(\mu)$-norms, $1 \leq p \leq \infty$, will be denoted by $\| \cdot \|_p$.

Let then $(P_t)_{t \geq 0}$ be a Markov semigroup with generator L acting on a suitable class of functions on $(X, \mathcal{A})$. Assume that $(P_t)_{t \geq 0}$ and L have an invariant, reversible and ergodic probability measure μ. This ensures that the operators P_t are contractions in all $L^p(\mu)$-spaces, $1 \leq p \leq \infty$. The Dirichlet form associated to the couple (L, μ) is then defined, on functions f, g of the Dirichlet domain, as

$$\mathcal{E}(f, g) = \int_X f(-Lg) d\mu.$$

Within this framework, the first example of interest is the case of a Markov kernel K on a finite state space X with invariant $(\sum_{x \in X} K(x, y)\mu(x) = \mu(y), y \in X)$ and reversible $(K(x, y)\mu(x) = K(y, x)\mu(y), x, y \in X)$ probability measure μ. The Markov operator $L = K - \text{Id}$ generates the semigroup of operators $P_t = e^{tL}$, $t \geq 0$, and defines the Dirichlet form

$$\mathcal{E}(f, g) = \int_X f(-Lg) d\mu = \frac{1}{2} \sum_{x, y \in X} [f(x) - f(y)][g(x) - g(y)]K(x, y)\mu(x)$$

on functions $f, g : X \to \mathbb{R}$. The second class of examples is the case of $X = \mathbb{R}^n$ equipped with its Borel σ-field. Letting $V : \mathbb{R}^n \to \mathbb{R}$ be such that $\int_{\mathbb{R}^n} e^{-V(x)}dx = 1$, under mild smoothness and growth conditions on the potential V, the second order operator $L = \Delta - \nabla V \cdot \nabla$ admits $d\mu(x) = e^{-V(x)}dx$ as symmetric and invariant probability measure. The operator L generates the Markov semigroup of operators $(P_t)_{t \geq 0}$ and defines by integration by parts the Dirichlet form

$$\mathcal{E}(f, g) = \int_{\mathbb{R}^n} f(-Lg)d\mu = \int_{\mathbb{R}^n} \nabla f \cdot \nabla g \, d\mu$$

for smooth functions f, g on $\mathbb{R}^n$.

Given such a couple (L, μ), it is said to satisfy a spectral gap, or Poincaré, inequality if there is a constant $\lambda > 0$ such that for all functions f of the Dirichlet domain,

$$\lambda \, \mathrm{Var}_\mu(f) \leq \mathcal{E}(f, f). \tag{8}$$

Similarly, it satisfies a logarithmic Sobolev inequality if there is a constant $\rho > 0$ such that for all functions f of the Dirichlet domain,

$$\rho \, \mathrm{Ent}_\mu(f^2) \leq 2\,\mathcal{E}(f, f). \tag{9}$$

One speaks of the spectral gap constant (of (L, μ)) as the best $\lambda > 0$ for which (8) holds, and of the logarithmic Sobolev constant (of (L, μ)) as the best $\rho > 0$ for which (9) holds. We still use λ and ρ for these constants. It is classical that $\rho \leq \lambda$.

Both the spectral gap and logarithmic Sobolev inequalities translate equivalently on the associated semigroup $(P_t)_{t \geq 0}$. Namely, the spectral gap inequality (8) is equivalent to saying that

$$\|P_t f\|_2 \leq e^{-\lambda t} \|f\|_2$$

for every $t \geq 0$ and every mean zero function f in $L^2(\mu)$. Equivalently for the further purposes, for every $f \in L^2(\mu)$ and every $t > 0$,

$$\mathrm{Var}_\mu(f) \leq \frac{1}{1 - e^{-\lambda t}} \left[\|f\|_2^2 - \|P_t f\|_2^2 \right]. \tag{10}$$

On the other hand, the logarithmic Sobolev inequality gives rise to hypercontractivity which is a smoothing property of the semigroup. Precisely, the logarithmic Sobolev inequality (9) is equivalent to saying that, whenever $p \geq 1 + e^{-2\rho t}$, for all functions f in $L^p(\mu)$,

$$\|P_t f\|_2 \leq \|f\|_p. \tag{11}$$

For simplicity, we say below that a probability measure μ in this context is hypercontractive with constant ρ.

A standard operation on Markov operators is the product operation. Let (L_1, μ_1) and (L_2, μ_2) be Markov operators on respective spaces X_1 and X_2. Then

$$L = L_1 \otimes \mathrm{Id} + \mathrm{Id} \otimes L_2$$

is a Markov operator on the product space $X_1 \times X_2$ equipped with the product probability measure $\mu_1 \otimes \mu_2$. The product semigroup $(P_t)_{t \geq 0}$ is similarly obtained as the tensor product $P_t = P_t^1 \otimes P_t^2$ of the semigroups on each factor. For the product Dirichlet form, the spectral gap and logarithmic Sobolev constants are stable in the sense that, with the obvious notation, $\lambda = \min(\lambda_1, \lambda_2)$ and $\rho = \min(\rho_1, \rho_2)$. This basic stability by products will allow for constants independent of the dimension in the Talagrand type inequalities under investigation. For the clarity of the exposition, we will not mix below products of continuous and discrete spaces, although this may easily be considered.

Let us illustrate the preceding definitions and properties on two basic examples. Consider first the two-point space $X = \{-1, +1\}$ with the measure $\mu = p\delta_{+1} + q\delta_{-1}$, $p \in [0, 1]$, $p + q = 1$, and the Markov kernel $K(x, y) = \mu(y)$, $x, y \in X$. Then, for every function $f : X \to \mathbb{R}$,

$$\mathcal{E}(f, f) = \int_X f(-Lf)d\mu = \mathrm{Var}_\mu(f)$$

so that the spectral gap $\lambda = 1$. The logarithmic Sobolev constant is known to be

$$\rho = \frac{2(p-q)}{\log p - \log q} \quad (= 1 \text{ if } p = q). \tag{12}$$

The product chain on the discrete cube $X = \{-1, +1\}^N$ with the product probability measure $\mu = (p\delta_{+1} + q\delta_{-1})^{\otimes N}$ and generator $L = \sum_{i=1}^N L_i$ is associated to the Dirichlet form

$$\mathcal{E}(f, f) = \int_X \sum_{i=1}^N f(-L_i f)d\mu = pq \int_X \sum_{i=1}^N |D_i f|^2 d\mu$$

where $D_i f$ is defined in (2). By the previous product property, it admits 1 as spectral gap and ρ given by (12) as logarithmic Sobolev constant. In its hypercontractive formulation, the case $p = q$ is the content of the Bonami-Beckner inequality [7,9].

As mentioned before, M. Talagrand [24] used this hypercontractivity on the discrete cube $\{-1, +1\}^N$ equipped with the product measure $\mu = (p\delta_{+1} + q\delta_{-1})^{\otimes N}$ to prove that for any function $f : \{-1, +1\}^N \to \mathbb{R}$,

$$\mathrm{Var}_\mu(f) \le \frac{Cpq(\log p - \log q)}{p - q} \sum_{i=1}^N \frac{\|D_i f\|_2^2}{1 + \log(\|D_i f\|_2/2\sqrt{pq}\,\|D_i f\|_1)} \tag{13}$$

for some numerical constant $C > 0$ (this statement will be covered in Sect. 2 below). This in turn yields a version of the influence result of [12] on the biased cube.

In the continuous setting $X = \mathbb{R}^n$, the case of a quadratic potential V amounts to the Hermite or Ornstein-Uhlenbeck operator $L = \Delta - x \cdot \nabla$ with invariant measure the standard Gaussian measure $d\mu(x) = (2\pi)^{-n/2} e^{-|x|^2/2} dx$. It is known here that $\lambda = \rho = 1$ independently of the dimension. (More generally, if $V(x) - c\frac{|x|^2}{2}$

is convex for some $c > 0$, then $\lambda \geq \rho \geq c$.) Actually, L may also be viewed as the sum $\sum_{i=1}^{n} L_i$ of one-dimensional Ornstein-Uhlenbeck operators along each coordinate, and μ as the product measure of standard normal distributions. Within this product structure, the analogue (6) of (13) has been known for some time, and will be recalled below.

2 Hypercontractivity and Talagrand's Inequality

This section presents the general hypercontractive approach to Talagrand type inequalities including the discrete cube, the Gaussian product measure and more general non-product models. The method of proof, directly inspired from [24], has been developed recently by O'Donnell and Wimmer [20, 21] towards non-product extensions on suitable graphs. Besides hypercontractivity, a key feature necessary to develop the argument is a suitable decomposition of the Dirichlet form along "directions" commuting with the Markov operator or its semigroup. These directions are immediate in a product space, but do require additional structure in more general contexts.

In the previous abstract setting of a Markov semigroup $(P_t)_{t \geq 0}$ with generator L, assume thus that the associated Dirichlet form $\mathcal{E}$ may be decomposed along directions Γ_i acting on functions on X as

$$\mathcal{E}(f, f) = \sum_{i=1}^{N} \int_X \Gamma_i(f)^2 d\mu \tag{14}$$

in such a way that, for each $i = 1, \ldots, N$, Γ_i commutes to $(P_t)_{t \geq 0}$ in the sense that, for some constant $\kappa \in \mathbb{R}$, every $t \geq 0$ and every f in a suitable family of functions,

$$\Gamma_i(P_t f) \leq e^{\kappa t} P_t(\Gamma_i(f)). \tag{15}$$

These properties will be clearly illustrated on the main examples of interest below, with in particular explicit descriptions of the classes of functions for which (14) and (15) may hold.

We first present the Talagrand inequality in this context. The proof is the prototype of the hypercontractive argument used throughout this note and applied to various examples.

Theorem 1. *In the preceding setting, assume that* (L, μ) *is hypercontractive with constant* $\rho > 0$ *and that* (14) *and* (15) *hold. Then, for any function* f *in* $L^2(\mu)$,

$$\mathrm{Var}_\mu(f) \leq C(\rho, \kappa) \sum_{i=1}^{N} \frac{\|\Gamma_i f\|_2^2}{1 + \log(\|\Gamma_i f\|_2 / \|\Gamma_i f\|_1)}$$

where $C(\rho, \kappa) = 4 e^{(1 + (\kappa/\rho))^+} / \rho$.

Proof. The starting point is the variance representation along the semigroup $(P_t)_{t\geq 0}$ of a function f in the $L^2(\mu)$-domain of the semigroup as

$$\mathrm{Var}_\mu(f) = -\int_0^\infty \left(\frac{d}{dt}\int_X (P_t f)^2 d\mu\right)dt = -2\int_0^\infty \left(\int_X P_t f\, LP_t f d\mu\right)dt.$$

The time integral has to be handled both for the large and small values. For the large values of t, we make use of the exponential decay provided by the spectral gap in the form of (10) to get that, with $T = 1/2\rho$ for example since $\rho \leq \lambda$,

$$\mathrm{Var}_\mu(f) \leq 2\left[\|f\|_2^2 - \|P_T f\|_2^2\right].$$

We are thus left with the variance representation of

$$\|f\|_2^2 - \|P_T f\|_2^2 = -2\int_0^T \left(\int_X P_t f\, LP_t f d\mu\right)dt = 2\int_0^T \mathcal{E}(P_t f, P_t f)dt.$$

Now by the decomposition (14),

$$\|f\|_2^2 - \|P_T f\|_2^2 = 2\sum_{i=1}^N \int_0^T \left(\int_X (\Gamma_i(P_t f))^2 d\mu\right)dt.$$

Under the commutation assumption (15),

$$\int_X (\Gamma_i(P_t f))^2 d\mu \leq e^{2\kappa t}\int_X (P_t(\Gamma_i(f)))^2 d\mu.$$

Since $(P_t)_{t\geq 0}$ is hypercontractive with constant $\rho > 0$, for every $i = 1,\dots,N$ and $t \geq 0$,

$$\|P_t(\Gamma_i(f))\|_2 \leq \|\Gamma_i(f)\|_p$$

where $p = p(t) = 1 + e^{-2\rho t} \leq 2$. After the change of variables $p(t) = v$, we thus reached at this point the inequality

$$\mathrm{Var}_\mu(f) \leq \frac{2\, e^{(1+(\kappa/\rho))^+}}{\rho}\sum_{i=1}^N \int_1^2 \|\Gamma_i(f)\|_v^2\, dv. \tag{16}$$

This inequality actually basically amounts to Theorem 1. Indeed, by Hölder's inequality,

$$\|\Gamma_i(f)\|_v \leq \|\Gamma_i(f)\|_1^\theta \|\Gamma_i(f)\|_2^{1-\theta}$$

where $\theta = \theta(v) \in [0, 1]$ is defined by $\frac{1}{v} = \frac{\theta}{1} + \frac{1-\theta}{2}$. Hence

$$\int_1^2 \|\Gamma_i(f)\|_v^2 \, dv \leq \|\Gamma_i(f)\|_2^2 \int_1^2 b^{2\theta(v)} dv$$

where $b = \|\Gamma_i(f)\|_1 / \|\Gamma_i(f)\|_2 \leq 1$. It remains to evaluate the latter integral with $2\theta(v) = s$,

$$\int_1^2 b^{2\theta(v)} dv \leq \int_0^2 b^s \, ds \leq \frac{2}{1 + \log(1/b)}$$

from which the conclusion follows.   □

Inequality (16) of the preceding proof may also be used towards a version of Theorem 1 with Orlicz norms as emphasized in [24]. As in [24], let $\varphi : \mathbb{R}_+ \to \mathbb{R}_+$ be convex such that $\varphi(x) = x^2 / \log(e + x)$ for $x \geq 1$, and $\varphi(0) = 0$, and denote by

$$\|g\|_\varphi = \inf \left\{ c > 0; \int_X \varphi(|g|/c) d\mu \leq 1 \right\}$$

the associated Orlicz norm of a measurable function $g : X \to \mathbb{R}$. Then, for some numerical constant $C > 0$,

$$\int_1^2 \|g\|_v^2 \, dv \leq C \|g\|_\varphi^2 \tag{17}$$

so that (16) yields

$$\mathrm{Var}_\mu(f) \leq \frac{2C \, e^{(1+(\kappa/\rho))^+}}{\rho} \sum_{i=1}^N \|\Gamma_i(f)\|_\varphi^2. \tag{18}$$

Since as pointed out in Lemma 2.5 of [24],

$$\|g\|_\varphi^2 \leq \frac{C \|g\|_2^2}{1 + \log(\|g\|_2/\|g\|_1)},$$

we see that (18) improves upon Theorem 1. To briefly check (17), assume by homogeneity that $\int_X g^2 / \log(e + g) d\mu \leq 1$ for some non-negative function g. Then, setting $g_k = g \, 1_{\{2^{k-1} < g \leq 2^k\}}$, $k \geq 1$, and $g_0 = g \, 1_{\{g \leq 1\}}$,

$$\sum_{k \in \mathbb{N}} \frac{1}{k+1} \int_X g_k^2 d\mu \leq C_1 \tag{19}$$

for some numerical constant $C_1 > 0$. Hence, since $g_k \leq 2^k$ for every k,

$$\int_1^2 \|g\|_v^2 \, dv = \int_1^2 \left(\sum_{k \in \mathbb{N}} \int_X g_k^v \, d\mu \right)^{2/v} dv$$

$$\leq 4 \int_1^2 \left(\sum_{k \in \mathbb{N}} 2^{-(2-v)k} \int_X g_k^2 \, d\mu \right)^{2/v} dv$$

$$\leq C_2 \sum_{k \in \mathbb{N}} \left(\int_1^2 (k+1)^{2/v} 2^{-2(2-v)k/v} \, dv \right) \frac{1}{k+1} \int_X g_k^2 \, d\mu$$

where we used (19) as convexity weights in the last step. Now, it is easy to check that

$$\int_1^2 (k+1)^{2/v} 2^{-2(2-v)k/v} \, dv \leq C_3$$

uniformly in k so that $\int_1^2 \|g\|_v^2 \, dv \leq C_1 C_2 C_3$ concluding thus the claim.

We next illustrate the general Theorem 1 on various examples of interest.

On a probability space $(X, \mathcal{A}, \mu)$, consider first the Markov operator $Lf = \int_X f \, d\mu - f$ acting on integrable functions (in other words $Kf = \int_X f \, d\mu$). This operator is symmetric with respect to μ with Dirichlet form

$$\mathcal{E}(f, f) = \int_X f(-Lf) \, d\mu = \mathrm{Var}_\mu(f).$$

In particular, it has spectral gap 1. Let now $X = X_1 \times \cdots \times X_N$ be a product space with product probability measure $\mu = \mu_1 \otimes \cdots \otimes \mu_N$. Consider the product operator $L = \sum_{i=1}^N L_i$ where L_i is acting on the i-th coordinate of a function f as $L_i f = \int_{X_i} f \, d\mu_i - f$. The product operator L has still spectral gap 1. Its Dirichlet form is given by

$$\mathcal{E}(f, f) = \sum_{i=1}^N \int_X f(-L_i f) \, d\mu = \sum_{i=1}^N \int_X (L_i f)^2 \, d\mu.$$

We are therefore in the setting of a decomposition of the type (14). Moreover, it is immediately checked that $L_i L = L L_i$ for every $i = 1, \ldots, N$, and thus the commutation property (15) also holds (with $\kappa = 0$). Hence Theorem 1 applies for this model with hypercontractive constant $\rho = \min_{1 \leq i \leq N} \rho_i > 0$. In particular, Theorem 1 includes Talagrand's inequality (13) for the hypercube $X = \{-1, +1\}^N$ with the product measure $\mu = (p\delta_{+1} + q\delta_{-1})^{\otimes N}$ with hypercontractive constant given by (12), for which it is immediately checked that, for every $r \geq 1$ and every $i = 1, \ldots, N$,

$$\int_X |L_i f|^r \, d\mu = (pq^r + p^r q) \int_X |D_i f|^r \, d\mu.$$

More generally, as pointed out to us by J. van den Berg and D. Kiss (private communication), we may consider similarly products of the complete graph $X_1 = \cdots = X_N = \{0, \ldots, k\}$, each factor being equipped with the probability measure $\mu_1 = \sum_{j=0}^{k} p_j \delta_j$. Talagrand's approach is known to extend to this case, as noted for instance in [14]. The hypercontractive constant of X_1 has been computed in [10] and is given by

$$\rho = \frac{2(1 - 2p^*)}{\log(1/p^* - 1)}$$

with $p^* = \min_{0 \leq j \leq k} p_j$, so that Theorem 1.3 from [14] follows from Theorem 1 above.

Non-product examples may be considered similarly as has been thus emphasized recently in [20,21] with similar arguments. Let for example G be a finite group, and let S be a symmetric set of generators of G. The Cayley graph associated to S is the graph with vertices the element of G and edges the couples (x, xs) where $x \in G$ and $s \in S$. The transition kernel associated to this graph is

$$K(x, y) = \frac{1}{|S|} \mathbf{1}_S(yx^{-1}), \quad x, y \in G,$$

where $|S|$ is the cardinal of S. The uniform probability measure μ on G is an invariant and reversible measure for K. This framework includes the example of $G = S_n$ the symmetric group on n elements with the set of transpositions as generating set and the uniform measure as invariant and symmetric measure.

Given such a finite Cayley graph G with generator set S, kernel K and uniform measure μ as invariant measure, the associated Dirichlet form may be expressed on functions $f : G \to \mathbb{R}$ in the form (14)

$$\mathcal{E}(f, f) = \frac{1}{2|S|} \sum_{s \in S} \sum_{x \in G} [f(sx) - f(x)]^2 \mu(x) = \frac{1}{2|S|} \sum_{s \in S} \|D_s f\|_2^2$$

where for $s \in S$, $D_s f(x) = f(sx) - f(x)$, $x \in G$. In order that the operators D_s commute to K in the sense of (15) (with again $\kappa = 0$), it is necessary to assume that S is stable by conjugacy in the sense that

$$\text{for all } u \in S, \quad u S u^{-1} = S$$

as it is the case for the set of transpositions on the symmetric group S^n. The following statement from [20] is thus an immediate consequence of the general Theorem 1.

Corollary 2. *Under the preceding notation and assumptions, denote by ρ the logarithmic Sobolev constant of the chain (K, μ). Then for every function f on G,*

$$\text{Var}_\mu(f) \leq \frac{2e}{\rho |S|} \sum_{s \in S} \frac{\|D_s f\|_2^2}{1 + \log(\|D_s f\|_2 / \|D_s f\|_1)}.$$

One may wonder for the significance of this Talagrand type inequality for influences. For $A \subset G$ and $s \in S$, define the influence $I_s(A)$ of the direction s on the set A by

$$I_s(A) = \mu(\{x \in G; x \in A, sx \notin A\}).$$

As on the discrete cube, given $A \subset G$ with $\mu(A) = a$, Corollary 2 yields the existence of $s \in S$ such that

$$I_s(A) \geq \frac{1}{C} a(1-a)\rho \log\left(1 + \frac{1}{C\rho a(1-a)}\right) \geq \frac{1}{C} a(1-a)\rho \log\left(1 + \frac{1}{C\rho}\right) \quad (20)$$

(where $C \geq 1$ is numerical). However, with respect to the spectral gap inequality of the chain (K, μ)

$$\lambda \operatorname{Var}_\mu(f) \leq \frac{1}{2|S|} \sum_{s \in S} \|D_s f\|_2^2,$$

we see that (20) is only of interest provided that $\rho \log(1 + (1/\rho)) \gg \lambda$. This is the case on the symmetric discrete cube $\{-1, +1\}^N$ for which, in the Cayley graph normalization of Dirichlet forms, $\lambda = \rho = 1/N$. On the symmetric group, it is known that the spectral gap λ is $\frac{2}{n-1}$ whereas its logarithmic Sobolev constant ρ is of the order of $1/n \log n$ ([10, 17]) so that $\rho \log(1 + (1/\rho))$ and λ are actually of the same order for large n, and hence yield the existence of a transposition τ with influence at least only of the order of $1/n$. It is pointed out in [21] that this result is however optimal. The paper [20] presents examples in the more general context of Schreier graphs for which (20) yields influences strictly better than the ones from the spectral gap inequality.

Theorem 1 may also be illustrated on continuous models such as Gaussian measures. While the next corollary is stated in some generality, it is already of interest for products of one-dimensional factors and covers in particular the example (6) of the standard Gaussian product measure.

Corollary 3. *Let* $d\mu_i(x) = e^{-V_i(x)}dx$, $i = 1, \ldots, N$, *on* $X_i = \mathbb{R}^{n_i}$ *be hypercontractive with constant* $\rho_i > 0$. *Let* $\mu = \mu_1 \otimes \cdots \otimes \mu_N$ *on* $X = X_1 \times \cdots \times X_N$. *Assume in addition that* $V_i'' \geq -\kappa$, $\kappa \in \mathbb{R}$, $i = 1, \ldots, N$. *Then, for any smooth function* f *on* X,

$$\operatorname{Var}_\mu(f) \leq C(\rho, \kappa) \sum_{i=1}^N \frac{\|\nabla_i f\|_2^2}{1 + \log\left(\|\nabla_i f\|_2 / \|\nabla_i f\|_1\right)}$$

where $\rho = \min_{1 \leq i \leq N} \rho_i$, *and where* $\nabla_i f$ *denotes the gradient of* f *in the direction* X_i, $i = 1, \ldots, N$.

Corollary 3 again follows from Theorem 1. Indeed, the product structure immediately allows for the decomposition (14) of the Dirichlet form

$$\mathcal{E}(f, f) = \int_X |\nabla f|^2 d\mu = \sum_{i=1}^N \int_X |\nabla_i f|^2 d\mu$$

along smooth functions with thus $\Gamma_i(f) = |\nabla_i f|$. On the other hand, the basic commutation (15) between the semigroup and the gradients ∇_i is described here as a curvature condition. Namely, whenever the Hessian V'' of a smooth potential V on $\mathbb{R}^n$ is (uniformly) bounded below by $-\kappa$, $\kappa \in \mathbb{R}$, the semigroup $(P_t)_{t \geq 0}$ generated by the operator $L = \Delta - \nabla V \cdot \nabla$ commutes to the gradient in the sense that, for every smooth function f and every $t \geq 0$,

$$|\nabla P_t f| \leq e^{\kappa t} P_t(|\nabla f|). \tag{21}$$

In the product setting of Corollary 3, the semigroup $(P_t)_{t \geq 0}$ is the tensor product of the semigroups along every coordinate so that (21) ensures that

$$|\nabla_i P_t f| \leq e^{\kappa t} P_t(|\nabla_i f|) \tag{22}$$

along the partial gradients ∇_i, $i = 1, \ldots, N$, and hence (15) holds on smooth functions. This commutation property (with $\kappa = -1$) is for example explicit on the integral representation

$$P_t f(x) = \int_{\mathbb{R}^n} f\left(e^{-t}x + (1 - e^{-2t})^{1/2}y\right) d\mu(y), \quad x \in \mathbb{R}^n, \ t \geq 0, \tag{23}$$

of the Ornstein-Uhlenbeck semigroup with generator $L = \Delta - x \cdot \nabla$ and invariant and symmetric measure the standard Gaussian distribution. The assumption $V'' \geq -\kappa$ describes a curvature property of the generator L and is linked to Ricci curvature on Riemannian manifolds. Since only $\kappa \in \mathbb{R}$ is required here, it appears as a mild property, shared by numerous potentials such as for example double-well potentials on the line of the form $V(x) = ax^4 - bx^2$, $a, b > 0$. Recall that the assumption $V'' \geq c > 0$ (for example the quadratic potential with the Gaussian measure as invariant measure) actually implies that μ satisfies a logarithmic Sobolev inequality, and thus hypercontractivity (with constant c). We refer for example to [2,4,15]... for an account on (21) and the preceding discussion.

Corollary 3 admits generalizations in broader settings. Weighted measures on Riemannian manifolds with a lower bound on the Ricci curvature may be considered similarly with the same conclusions. In another direction, the hypercontractive approach may be developed in presence of suitable geometric decompositions. The next statements deal with the example of the sphere and with geometric decompositions of the identity in Euclidean space which are familiar in the context of Brascamp-Lieb inequalities (see [6] for further illustrations in a Markovian framework).

A non-product example in the continuous setting is the one of the standard sphere $\mathbb{S}^{n-1} \subset \mathbb{R}^n$ ($n \geq 2$) equipped with its uniform normalized measure μ. Consider, for every $i, j = 1, \ldots, n$, $D_{ij} = x_i \partial_j - x_j \partial_i$. These will be the directions along which the Talagrand inequality may be considered since

$$\mathcal{E}(f, f) = \int_{\mathbb{S}^{n-1}} f(-\Delta f) d\mu = \frac{1}{2} \sum_{i,j=1}^{n} \int_{\mathbb{S}^{n-1}} (D_{ij} f)^2 d\mu.$$

The operators D_{ij} namely commute in an essential way to the spherical Laplacian $\Delta = \frac{1}{2} \sum_{i,j=1}^{n} D_{ij}^2$ so that (15) holds with $\kappa = 0$. Finally, the logarithmic Sobolev constant is known to be $n - 1$ [2, 4, 15].... Corollary 4 thus again follows from the general Theorem 1.

Corollary 4. *For every smooth enough function* $f : \mathbb{S}^{n-1} \to \mathbb{R}$,

$$\mathrm{Var}_\mu(f) \leq \frac{4e}{n} \sum_{i,j=1}^{n} \frac{\|D_{ij} f\|_2^2}{1 + \log\left(\|D_{ij} f\|_2 / \|D_{ij} f\|_1\right)}.$$

Up to the numerical constant, this inequality improves upon the Poincaré inequality for μ (with constant $\lambda = n - 1$).

We turn to geometric Brascamp-Lieb decompositions. Consider thus E_i, $i = 1, \ldots, m$, subspaces in $\mathbb{R}^n$, and $c_i > 0$, $i = 1, \ldots, m$, such that

$$\mathrm{Id}_{\mathbb{R}^n} = \sum_{i=1}^{m} c_i \, Q_{E_i} \tag{24}$$

where Q_{E_i} is the projection onto E_i. In particular, for every $x \in \mathbb{R}^n$, $|x|^2 = \sum_{i=1}^{m} c_i |Q_{E_i}(x)|^2$ and thus, for every smooth function f on $\mathbb{R}^n$,

$$\mathcal{E}(f, f) = \int_{\mathbb{R}^n} |\nabla f|^2 d\mu = \sum_{i=1}^{m} c_i \left(\int_{\mathbb{R}^n} |Q_{E_i}(\nabla P_t f)|^2 d\mu \right).$$

Furthermore, $Q_{E_i}(\nabla P_t f) = e^{-t} P_t(Q_{E_i}(\nabla f))$ which may be examplified on the representation (23) of the Ornstein-Uhlenbeck semigroup with hypercontractive constant 1. Theorem 1 thus yields the following conclusion.

Corollary 5. *Under the decomposition* (24), *for* μ *the standard Gaussian measure on* $\mathbb{R}^n$, *and for every smooth function* f *on* $\mathbb{R}^n$,

$$\mathrm{Var}_\mu(f) \leq 4 \sum_{i=1}^{m} c_i \frac{\left\|Q_{E_i}(\nabla f)\right\|_2^2}{1 + \log\left(\|Q_{E_i}(\nabla f)\|_2 / \|Q_{E_i}(\nabla f)\|_1\right)}.$$

3 Hypercontractivity and Geometric Influences

In the continuous context of the preceding section, and as discussed in the introduction, the L^2-norms of gradients in Corollary 3 are not well-suited to the (geometric) influences of [13] which require L^1-norms. In order to reach L^1-norms through the hypercontractive argument, a further simple interpolation trick will be necessary.

To this task, we use an additional feature of the curvature condition $V'' \geq -\kappa$, $\kappa \geq 0$, namely that the action of the semigroup $(P_t)_{t \geq 0}$ with generator $L = \Delta - \nabla V \cdot V$ on bounded functions yields functions with bounded gradients. More precisely (cf. [4, 15]...), for every smooth function f with $|f| \leq 1$, and every $0 < t \leq 1/2\kappa$,

$$|\nabla P_t f| \leq \frac{1}{\sqrt{t}}. \tag{25}$$

This property may again be illustrated in case of the Ornstein-Uhlenbeck semigroup (22) for which, by integration by parts,

$$\nabla P_t f(x) = \frac{e^{-t}}{(1 - e^{-2t})^{1/2}} \int_{\mathbb{R}^n} y \, f\big(e^{-t}x + (1 - e^{-2t})^{1/2}y\big) d\mu(y).$$

With this additional tool, the following statement then presents the expected result. The setting is similar to the one of Corollary 3. Dependence on ρ and κ for the constant $C'(\rho, \kappa)$ below may be drawn from the proof. It will of course be independent of N.

Theorem 6. *Let $d\mu_i(x) = e^{-V_i(x)}dx$, $i = 1, \ldots, N$, on $X_i = \mathbb{R}^{n_i}$ be hypercontractive with constant $\rho_i > 0$. Let $\mu = \mu_1 \otimes \cdots \otimes \mu_N$ on $X = X_1 \times \cdots \times X_N$, and set as before $\rho = \min_{1 \leq i \leq N} \rho_i$. Assume in addition that $V_i'' \geq -\kappa$, $\kappa \geq 0$, $i = 1, \ldots, N$. Then, for some constant $C'(\rho, \kappa) \geq 1$ and for any smooth function f on X such that $|f| \leq 1$,*

$$\mathrm{Var}_\mu(f) \leq C'(\rho, \kappa) \sum_{i=1}^{N} \frac{\|\nabla_i f\|_1\big(1 + \|\nabla_i f\|_1\big)}{\big[1 + \log^+\big(1/\|\nabla_i f\|_1\big)\big]^{1/2}}.$$

Proof. We follow the same line of reasoning as in the proof of Theorem 1, starting on the basis of (10) from

$$\|f\|_2^2 - \|P_T f\|_2^2 = 2 \sum_{i=1}^{N} \int_0^T \left(\int_X |\nabla_i P_t f|^2 d\mu \right) dt$$

$$\leq 4 \sum_{i=1}^{N} \int_0^T \left(\int_X |\nabla_i P_{2t} f|^2 d\mu \right) dt$$

for some $T > 0$. By (22) along each coordinate, for each $t \geq 0$,

$$|\nabla_i P_{2t} f| \leq e^{\kappa t} \, P_t\big(|\nabla_i P_t f|\big).$$

Hence, by the hypercontractivity property as in Theorem 1,

$$\|\nabla_i P_{2t} f\|_2 \le e^{\kappa t} \|\nabla_i P_t f\|_p$$

where $p = p(t) = 1 + e^{-2\rho t} \le 2$. We then proceed to the interpolation trick. Namely, by (25) and the tensor product form of the semigroup, $|\nabla_i P_t f| \le t^{-1/2}$ for $0 < t \le 1/2\kappa$, so that in this range,

$$\|\nabla_i P_{2t} f\|_2 \le e^{\kappa(1+1/p)t} t^{-(1-1/p)/2} \|\nabla_i f\|_1^{1/p}$$

(where we used again (22)). As a consequence, provided $T \le 1/2\kappa$,

$$\|f\|_2^2 - \|P_T f\|_2^2 \le 4 e^{4\kappa T} \sum_{i=1}^{N} \|\nabla_i f\|_1 \int_0^T t^{-(1-1/p(t))} \|\nabla_i f\|_1^{(2/p(t))-1} dt.$$

We are then left with the estimate of the latter integral that only requires elementary calculus. Set $b = \|\nabla_i f\|_1$ and $\theta(t) = \frac{2}{p(t)} - 1 \le 1$. Assuming $T \le 1$,

$$\int_0^T t^{-(1-1/p(t))} b^{\theta(t)} dt \le \int_0^T t^{-1/2} b^{\theta(t)} dt.$$

Distinguish between two cases. When $b \ge 1$,

$$\int_0^T t^{-1/2} b^{\theta(t)} dt \le b \int_0^T t^{-1/2} dt \le 2b\sqrt{T}.$$

When $b \le 1$, use that $\theta(t) \ge \rho t/2$ for every $0 \le t \le 1/2\rho$. Hence, provided $T \le 1/2\rho$,

$$\int_0^T t^{-1/2} b^{\theta(t)} dt \le \int_0^T t^{-1/2} b^{\rho t/2} dt \le \frac{C}{\sqrt{\rho}} \cdot \frac{1}{[1 + \log(1/b)]^{1/2}}$$

where $C \ge 1$ is numerical. Summarizing, in all cases, provided T is chosen smaller than $\min\left(1, \frac{1}{2\rho}\right)$, we have

$$\int_0^T t^{-(1-1/p(t))} b^{\theta(t)} dt \le \frac{2C}{\sqrt{\rho}} \cdot \frac{1+b}{[1 + \log^+(1/b)]^{1/2}}.$$

Choosing for example $T = \min\left(1, \frac{1}{2\rho}, \frac{1}{2\kappa}\right)$ and using (10), Theorem 6 follows with $C'(\rho, \kappa) = C'/\rho^{3/2} T$ for some further numerical constant C'. If $\kappa \le c\rho$, then this constant is of order $\rho^{-1/2}$. $\square$

The preceding proof may actually be adapted to interpolate between Corollary 3 and Theorem 6 as

$$\mathrm{Var}_\mu(f) \le C \sum_{i=1}^{N} \frac{\|\nabla_i f\|_q^q \left(1 + \|\nabla_i f\|_1^2/\|\nabla_i f\|_q^q\right)}{\left[1 + \log^+\left(\|\nabla_i f\|_q^q/\|\nabla_i f\|_1^2\right)\right]^{q/2}}$$

for any smooth function f on X such that $|f| \le 1$, and any $1 \le q \le 2$ (where C depends on ρ, κ and q).

As announced in the introduction, the conclusion of Theorem 6 may be interpreted in terms of influences. Namely, for $f = \mathbf{1}_A$ (or some smooth approximation), define $\|\nabla_i f\|_1$ as the geometric influence $I_i(A)$ of the ith coordinate on the set A. In other words, $I_i(A)$ is the surface measure of the section of A along the fiber of $x \in X = X_1 \times \cdots \times X_N$ in the ith direction, $1 \le i \le N$, averaged over the remaining coordinates (see [13]). Then Theorem 6 yields that

$$\mu(A)\big(1 - \mu(A)\big) \le C(\rho, \kappa) \sum_{i=1}^{N} \frac{I_i(A)\big(1 + I_i(A)\big)}{\left[1 + \log^+\left(1/I_i(A)\right)\right]^{1/2}}.$$

Proceeding as in the introduction for influences on the cube, the following consequence holds.

Corollary 7. *In the setting of Theorem 6, for any Borel set A in X with $\mu(A) = a$, there is a coordinate i, $1 \le i \le N$, such that*

$$I_i(A) \ge \frac{a(1-a)}{CN}\left(\log\frac{N}{a(1-a)}\right)^{1/2} \ge \frac{a(1-a)(\log N)^{1/2}}{CN}$$

where C only depends on ρ and κ.

It is worthwhile mentioning that when $N = 1$, $I_1(A)$ corresponds to the surface measure (Minkowski content)

$$\mu^+(A) = \liminf_{\varepsilon \to 0} \frac{1}{\varepsilon}\left[\mu(A_\varepsilon) - \mu(A)\right]$$

of $A \subset \mathbb{R}^{n_1}$, so that Corollary 7 contains the quantitative form of the isoperimetric inequality for Gaussian measures

$$\mu^+(A) \ge \frac{1}{C} a(1-a)\left(\log\frac{1}{a(1-a)}\right)^{1/2}.$$

Recall indeed (cf. e.g. [15, 16]) that the Gaussian isoperimetric inequality indicates that $\mu^+(A) \ge \varphi \circ \Phi^{-1}(a)$ $(a = \mu(A))$ where $\varphi(x) = (2\pi)^{-1/2}\,e^{-x^2/2}$, $x \in \mathbb{R}$, $\Phi(t) = \int_{-\infty}^{t} \varphi(x)dx$, $t \in \mathbb{R}$, and that $\varphi \circ \Phi^{-1}(u) \sim u(2\log\frac{1}{u})^{1/2}$ as $u \to 0$.

This conclusion, for hypercontractive log-concave measures, was established previously in [3]. See [18, 19] for recent improvements in this regard.

Theorem 6 admits also generalizations in broader settings such as weighted measures on Riemannian manifolds with a lower bound on the Ricci curvature (this ensures that both (21) and (25) hold).

Besides the Gaussian measure, Keller et al. [13] also investigate with isoperimetric tools products of one-dimensional distributions of the type $c_\alpha e^{-|x|^\alpha} dx$, $1 < \alpha < \infty$, for which they produce influences at least of the order of $\frac{(\log N)^{\beta/2}}{N}$ where $\beta = 2(1 - \frac{1}{\alpha})$ ($\alpha = 2$ corresponding to the Gaussian case). The proof of Theorem 6 may be adapted to cover this result but only seemingly for $1 < \alpha < 2$. Convexity of the potentials $|x|^\alpha$ ensures (21) and (25). When $1 < \alpha < 2$, measures $c_\alpha e^{-|x|^\alpha} dx$ are not hypercontractive. Nevertheless, the hypercontractive theorems in Orlicz norms of [5] still indicate that the semigroup $(P_t)_{t \geq 0}$ generated by the potential $|x|^\alpha$ is such that, for every bounded function g with $\|g\|_\infty = 1$ and every $0 \leq t \leq 1$,

$$\|P_t g\|_2^2 \leq C \|g\|_1 \exp\left(-ct \log^\beta (1 + (1/\|g\|_1))\right) \tag{26}$$

for $\beta > 0$ and some constants $C, c > 0$, and similarly for the product semigroup with constants independent of N. The hypercontractive step in the proof of Theorem 6 is then modified into

$$\left\|\, |\nabla_i P_{2t} f|\, \right\|_2^2 \leq C \|\nabla_i f\|_1 \int_0^1 t^{-1/2} \exp\left(-ct \log^\beta (1 + (1/\|\nabla_i f\|_1))\right) dt.$$

As a consequence, for any smooth f with $|f| \leq 1$,

$$\mathrm{Var}_\mu(f) \leq C \sum_{i=1}^N \frac{\|\nabla_i f\|_1 \left(1 + \|\nabla_i f\|_1\right)}{\left[1 + \log^+ \left(1/\|\nabla_i f\|_1\right)\right]^{\beta/2}}. \tag{27}$$

We thus conclude to the influence result of [13] in this range. When $\alpha > 2$ ($\beta \in (1, 2)$), the potentials are hypercontractive in the usual sense so that the preceding proofs yield (27) but only for $\beta = 1$. We do not know how to reach the exponent $\beta/2$ in this case by the hypercontractive argument.

We conclude this note by the L^1 versions of Corollaries 4 and 5. In the case of the sphere, the proof is identical to the one of Theorem 6 provided one uses that $|D_{ij} f| \leq |\nabla f|$ which ensures that $|D_{ij} P_t f| \leq 1/\sqrt{t}$. The behavior of the constant is drawn from the proof of Theorem 6.

Theorem 8. *For every smooth enough function* $f : \mathbb{S}^{n-1} \to \mathbb{R}$ *such that* $|f| \leq 1$,

$$\mathrm{Var}_\mu(f) \leq \frac{C}{\sqrt{n}} \sum_{i,j=1}^n \frac{\|D_{ij} f\|_1 \left(1 + \|D_{ij} f\|_1\right)}{\left[1 + \log^+ \left(1/\|D_{ij} f\|_1\right)\right]^{1/2}}.$$

Application to geometric influences $I_{ij}(A)$ as the limit of $\|D_{ij} f\|_1$ as f approaches the characteristic function of the set A may be drawn as in the previous corresponding statements. From a geometric perspective, $I_{ij}(A)$ can be viewed as the average over x of the boundary of the section of A in the 2-plane $x + \mathrm{span}(e_i, e_j)$. We do not know if the order $n^{-1/2}$ of the constant in Theorem 8 is optimal.

As announced, the last statement is the L^1-version of the geometric decompositions of Corollary 5 which seems again of interest for influences. Under the corresponding commutation properties, the proof is developed similarly.

Proposition 9. *Under the decomposition* (24), *for* μ *the standard Gaussian measure on* $\mathbb{R}^n$ *and for every smooth function* f *on* $\mathbb{R}^n$ *such that* $|f| \leq 1$,

$$\mathrm{Var}_\mu(f) \leq C \sum_{i=1}^{m} c_i \frac{\| Q_{E_i}(\nabla f)\|_1 (1 + \| Q_{E_i}(\nabla f)\|_1)}{\left[1 + \log^+ \left(1/\| Q_{E_i}(\nabla f)\|_1 \right) \right]^{1/2}}$$

where $C > 0$ *is numerical.*

Let us illustrate the last statement on a simple decomposition. As in the Loomis-Whitney inequality, consider the decomposition

$$\mathrm{Id}_{\mathbb{R}^n} = \sum_{i=1}^{n} \frac{1}{n-1} Q_{E_i}$$

with $E_i = e_i^\perp$, $i = 1, \ldots, n$, $(e_1, \ldots, e_n)$ orthonormal basis. Proposition 9 applied to $f = \mathbf{1}_A$ for a Borel set A in $\mathbb{R}^n$ with $\mu(A) = a$ then shows that there is a coordinate i, $1 \leq i \leq n$, such that

$$\left\| Q_{E_i}(\nabla f) \right\|_1 \geq \frac{1}{C} a(1-a) \left(\log \frac{1}{a(1-a)} \right)^{1/2}$$

for some constant $C > 0$. Now, $\| Q_{E_i}(\nabla f)\|_1$ may be interpreted as the boundary measure of the hyperplane section

$$A^{x \cdot e_i} = \{ (x \cdot e_1, \ldots, x \cdot e_{i-1}, x \cdot e_{i+1}, \ldots, x \cdot e_n); \ (x \cdot e_1, \ldots, x \cdot e_i, \ldots, x \cdot e_n) \in A \}$$

along the coordinate $x \cdot e_i \in \mathbb{R}$ averaged over the standard Gaussian measure. By Fubini's theorem, there is $x \cdot e_i \in \mathbb{R}$ (or even a set with measure as close to 1 as possible) such that

$$\mu^+(A^{x \cdot e_i}) \geq \frac{1}{C} a(1-a) \left(\log \frac{1}{a(1-a)} \right)^{1/2}. \tag{28}$$

The interesting point here is that a is the full measure of A. Indeed, recall that the isoperimetric inequality for μ indicates that $\mu^+(A) \geq \varphi \circ \Phi^{-1}(a)$, hence a quantitative lower bound for $\mu^+(A)$ of the same form as (28). When A is a half-space in $\mathbb{R}^n$, thus extremal set for the isoperimetric problem and satisfying $\mu^+(A) = \varphi \circ \Phi^{-1}(a)$, it is easy to see that there is indeed a coordinate $x \cdot e_i$ such that $A^{x \cdot e_i}$

is again a half-space in the lower-dimensional space. The preceding (28) therefore extends this property to all sets.

Acknowledgements We thank F. Barthe and P. Cattiaux for their help with the bound (26), and R. Rossignol for pointing out to us the references [20, 21]. We also thank J. van den Berg and D. Kiss for pointing out that the techniques developed here cover the example of the complete graph and for letting us know about [14].

References

1. C. Ané, S. Blachère, D. Chafaï, P. Fougères, I. Gentil, F. Malrieu, C. Roberto, G. Scheffer, *Sur les inégalités de Sobolev logarithmiques.* (French. Frech summary) [*Logarithmic Sobolev inequalities*] Panoramas et Synthèses [Panoramas and Syntheses] vol. 10 (Société Mathématique de France, Paris, 2000)
2. D. Bakry, in *L'hypercontractivité et son utilisation en théorie des semigroupes.* Ecole d'Eté de Probabilités de Saint-Flour, Lecture Notes in Math., vol. 1581 (Springer, Berlin, 1994), pp. 1–114
3. D. Bakry, M. Ledoux, Lévy-Gromov's isoperimetric inequality for an infinite dimensional diffusion generator. Invent. Math. **123**, 259–281 (1996)
4. D. Bakry, I. Gentil, M. Ledoux, Forthcoming monograph (2012)
5. F. Barthe, P. Cattiaux, C. Roberto, Interpolated inequalities between exponential and gaussian Orlicz hypercontractivity and isoperimetry. Revista Mat. Iberoamericana **22**, 993–1067 (2006)
6. F. Barthe, D. Cordero-Erausquin, M. Ledoux, B. Maurey, Correlation and Brascamp-Lieb inequalities for Markov semigroups. Int. Math. Res. Not. IMRN **10**, 2177–2216 (2011)
7. W. Beckner, Inequalities in Fourier analysis. Ann. Math. **102**, 159–182 (1975)
8. S. Bobkov, C. Houdré, A converse Gaussian Poincaré-type inequality for convex functions. Statist. Probab. Lett. **44**, 281–290 (1999)
9. A. Bonami, Étude des coefficients de Fourier des fonctions de Lp(G). Ann. Inst. Fourier **20**, 335–402 (1971)
10. P. Diaconis, L. Saloff-Coste, Logarithmic Sobolev inequalities for finite Markov chains. Ann. Appl. Prob. **6**, 695–750 (1996)
11. D. Falik, A. Samorodnitsky, Edge-isoperimetric inequalities and influences. Comb. Probab. Comp. **16**, 693–712 (2007)
12. J. Kahn, G. Kalai, N. Linial, The Influence of Variables on Boolean Functions. *Foundations of Computer Science*, IEEE Annual Symposium, 29th Annual Symposium on Foundations of Computer Science (FOCS, White Planes, 1988), pp. 68–80
13. N. Keller, E. Mossel, A. Sen, Geometric influences. Ann. Probab. **40**(3), 1135–1166 (2012)
14. D. Kiss, A note on Talagrand's variance bound in terms of influences. Preprint (2011)
15. M. Ledoux, The geometry of Markov diffusion generators. Ann. Fac. Sci. Toulouse **IX**, 305–366 (2000)
16. M. Ledoux, in *The Concentration of Measure Phenomenon*. Math. Surveys and Monographs, vol. 89 (Amer. Math. Soc., Providence, 2001)
17. T.Y. Lee, H.-T. Yau, Logarithmic Sobolev inequality for some models of random walks. Ann. Probab. **26**, 1855–1873 (1998)
18. E. Milman, On the role of convexity in isoperimetry, spectral gap and concentration. Invent. Math. **177**, 1–43 (2009)
19. E. Milman, Isoperimetric and concentration inequalities – Equivalence under curvature lower bound. Duke Math. J. **154**, 207–239 (2010)
20. R. O'Donnell, K. Wimmer, *KKL, Kruskal-Katona, and Monotone Nets.* 2009 50th Annual IEEE Symposium on Foundations of Computer Science (FOCS 2009) (IEEE Computer Soc., Los Alamitos, 2009), pp. 725–734

21. R. O'Donnell, K. Wimmer, Sharpness of KKL on Schreier graphs. Preprint (2011)
22. R. Rossignol, Threshold for monotone symmetric properties through a logarithmic Sobolev inequality. Ann. Probab. **34**, 1707–1725 (2006)
23. G. Royer, An initiation to logarithmic Sobolev inequalities. Translated from the 1999 French original by Donald Babbitt. *SMF/AMS Texts and Monographs*, vol. 14 (American Mathematical Society, Providence, RI (Société Mathématique de France, Paris, 2007)
24. M. Talagrand, On Russo's approximate zero-one law. Ann. Probab. **22**, 1576–1587 (1994)

A Family of Unitary Operators Satisfying a Poisson-Type Summation Formula

Dmitry Faifman

Abstract We consider a weighted form of the Poisson summation formula. We prove that under certain decay rate conditions on the weights, there exists a unique unitary Fourier–Poisson operator which satisfies this formula. We next find the diagonal form of this operator, and prove that under weaker conditions on the weights, a unique unitary operator still exists which satisfies a Poisson summation formula in operator form. We also generalize the interplay between the Fourier transform and derivative to those Fourier–Poisson operators.

1 Introduction

The classical summation formula of Poisson states that, for a well-behaved function $f : \mathbf{R} \to \mathbf{C}$ and its (suitably scaled) Fourier Transform $\hat{f}$ we have the relation

$$\sum_{n=-\infty}^{\infty} f(n) = \sum_{n=-\infty}^{\infty} \hat{f}(n)$$

Fix $x > 0$, and replace $f(t)$ with $\frac{1}{x} f(t/x)$. The Poisson formula is linear and trivial for odd f, so we assume f is even. Also, assume $f(0) = 0$. Then

$$\sum_{n=1}^{\infty} \hat{f}(nx) = \frac{1}{x} \sum_{n=1}^{\infty} f(n/x) \tag{1}$$

D. Faifman (✉)
Sackler Faculty of Exact Sciences, Department of Mathematics, Tel Aviv University, Ramat Aviv, 69978 Tel Aviv, Israel
e-mail: dfaifman@gmail.com

B. Klartag et al. (eds.), *Geometric Aspects of Functional Analysis*, Lecture Notes in Mathematics 2050, DOI 10.1007/978-3-642-29849-3_11,
© Springer-Verlag Berlin Heidelberg 2012

We discussed in [6] the extent to which this summation formula, which involves sums over lattices in $\mathbf{R}$, determines the Fourier transform of a function. Taking a weighted form of the Poisson summation formula as our starting point, we define a generalized Fourier–Poisson transform, and show that under certain conditions it is a unitary operator on $L^2[0, \infty)$. As a sidenote, we show a peculiar family of unitary operators on $L^2[0, \infty)$ defined by series of the type $f(x) \mapsto \sum a_n f(nx)$.

2 Some Notation, and a Summary of Results

The Fourier transform maps odd functions to odd functions, rendering The Poisson summation formula trivial. Thus we only consider square-integrable even functions, or equivalently, all functions belong to $L^2[0, \infty)$.

Denote by δ_n, $n \geq 1$ the sequence given by $\delta(1) = 1$ and $\delta(n) = 0$ for $n > 1$, and the convolution of sequences as $(a * b)_k = \sum_{mn=k} a_m b_n$.

Define the (possibly unbounded) operator

$$T(a_n)f(x) = \sum_{n=1}^{\infty} a_n f(nx) \qquad (2)$$

It holds that $T_{b_n} T_{a_n} f = T_{a_n * b_n} f$ whenever the series in both sides are well defined and absolutely convergent.

Let $a_n, b_n, n \geq 1$ be two sequences, which satisfy $a * b = \delta$.

This is equivalent to saying that $L(s; a_n)L(s; b_n) = 1$ where $L(s; c_n) = \sum_{n=1}^{\infty} \frac{c_n}{n^s}$. For a given a_n with $a_1 \neq 0$, its convolutional inverse is uniquely defined via those formulas.

Then, the formal inverse transform to T_{a_n} is given simply by T_{b_n}.

Note that the convolutional inverse of the sequence $a_n = 1$ is the Möbius function $\mu(n)$, defined as

$$\mu(n) = \begin{cases} (-1)^{\#\{p|n \text{ prime}\}}, & n \text{ square-free} \\ 0, & d^2|n \end{cases}$$

Also, define the operator

$$Sf(x) = \frac{1}{x} f\left(\frac{1}{x}\right)$$

—a unitary involution on $L^2[0, \infty)$, which is straightforward to check.

In terms of S and T, the Poisson summation formula for the Fourier Transform can be written as following:

$$T(e_n)\hat{f}(x) = ST(e_n)f$$

where $e_n = 1$ for all n. This suggests a formula for the Fourier transform:

$$\hat{f}(x) = T(\mu_n)ST(e_n)f(x) \tag{3}$$

with $\mu_n = \mu(n)$ the Möbius function. We would like to mention that Davenport in [4] established certain identities, such as

$$\sum_{n=1}^{\infty} \frac{\mu(n)}{n}\{nx\} = -\frac{1}{\pi}\sin(2\pi x)$$

which could be used to show that formula (3) actually produces the Fourier transform of a (zero-integral) step function.

We define the (possibly unbounded) Fourier–Poisson Transform associated with (a_n) as

$$\mathcal{F}(a_n)f(x) = T(\overline{a_n})^{-1}ST(a_n)f(x) \tag{4}$$

This is clearly an involution, and produces the operator $Sf(x) = \frac{1}{x}f\left(\frac{1}{x}\right)$ for $a_n = \delta_n$, and (non-formally) the Fourier Transform for $a_n = 1$. Note that both are unitary operators on $L_2[0, \infty)$. In the following, we will see how this definition can be carried out rigorously. First we give conditions on (a_n) that produce a unitary operator satisfying a pointwise Poisson summation formula, as was the case with Fourier transform (Theorem 3.3). Then we relax the conditions, which produces a unitary operator satisfying a weaker operator-form Poisson summation formula (Theorem 5.2).

Remark. A similar approach appears in [1,2,5], where it is used to study the Fourier Transform and certain variants of it.

A somewhat different approach is taken in [3], there, arithmetic sequences were characterized through the Fourier transform.

3   The Fourier–Poisson Operator is Unitary

We prove that under certain rate-of-growth assumptions on the coefficients a_n and its convolution-inverse b_n, it holds that $\mathcal{F}(a_n) = T(\overline{a_n})^{-1}ST(a_n)$ is unitary.

In the following, $f(x) = O(g(x))$ will be understood to mean at $x \to \infty$ unless otherwise indicated.

Lemma 3.1. *Assume*

$$\sum \frac{|a_n|}{\sqrt{n}} < \infty \tag{5}$$

holds, and let $f \in C(0, \infty)$ satisfy $f = O(x^{-1-\epsilon})$ for some $\epsilon > 0$. Then $T(a_n)f$ as defined in (2) is a continuous function satisfying $T(a_n)f(x) = O(x^{-1-\epsilon})$. Moreover, $T(a_n)$ extends to a bounded operator on $L^2[0, \infty)$, and $\|T\| \le \sum \frac{|a_n|}{\sqrt{n}}$.

Proof. Consider a continuous function $f = O(x^{-1-\epsilon})$. It is straightforward to verify that $T(a_n)f$ is well-defined, continuous and $T(a_n)f(x) = O(x^{-1-\epsilon})$. Now apply Cauchy-Schwartz:

$$|\langle f(mx), f(nx)\rangle| \le \frac{1}{\sqrt{mn}}\|f\|^2$$

implying

$$\|T(a_n)f\|^2 \le \sum_{m,n} |a_m||a_n||\langle f(mx), f(nx)\rangle| \le \left(\sum_n \frac{|a_n|}{\sqrt{n}}\right)^2 \|f\|^2$$

So $T(a_n)$ can be extended as a bounded operator to all L^2, and $\|T\| \le \sum \frac{|a_n|}{\sqrt{n}}$. $\qquad\square$

Now, consider a sequence a_n together with its convolution-inverse b_n. In all the following, we assume that a_n, b_n both satisfy (5) (as an example, consider $a_n = n^{-\lambda}$ and $b_n = \mu(n)n^{-\lambda}$ with $\lambda > 0.5$).

Then $T(a_n), T(\overline{b_n})$ are both bounded linear operators, and we define the Fourier–Poisson operator

$$\mathcal{F}(a_n) = T(\overline{b_n})ST(a_n)$$

Note that $T(a_n)^{-1} = T(b_n)$ (and likewise $T(\overline{a_n})^{-1} = T(\overline{b_n})$), which is easy to verify on the dense subset of continuous functions with compact support.

Corollary 3.2. *Assume that $\sum |a_n|n^\epsilon < \infty$ for some $\epsilon > 0$ and (b_n) satisfies (5). Take a continuous f satisfying $f(x) = O(x^{-1-\epsilon})$ as $x \to \infty$ and $f(x) = O(x^\epsilon)$ as $x \to 0$ for some $\epsilon > 0$. Then*

(a) $\mathcal{F}(a_n)f$ *is continuous and* $\mathcal{F}(a_n)f(x) = O(x^{-1-\epsilon})$.

(b) *The formula* $\sum \overline{a_n}\mathcal{F}(a_n)f(nx) = (1/x)\sum a_n f(n/x)$ *holds pointwise.*

Proof. (a) It is easy to see that all the properties of the function are preserved by applying $T(a_n)$ (using $\sum |a_n|n^\epsilon < \infty$) and then by S. Then by Lemma 3.1 application of $T(\overline{b_n})$ to $ST(a_n)$ completes the proof.

(b) By Lemma 3.1, we get an equality a.e. of two continuous functions:

$$T(\overline{a_n})\mathcal{F}(a_n)f = ST(a_n)f$$

$\qquad\square$

Theorem 3.3. *Assume that $\sum |a_n|n^\epsilon < \infty$ for some $\epsilon > 0$ and (b_n) satisfies (5). Then $\mathcal{F}(a_n)$ is a unitary operator.*

Proof. Consider $G = ST(\overline{b_n})ST(a_n)$. Take a continuous function f which is compactly supported. Define $g(x) = ST(a_n)f(x) = \frac{1}{x}\sum a_n f(\frac{n}{x})$, and note that g

vanishes for small values of x, and $|g(x)| = O\left(\sum |a_n| n^\epsilon x^{-1-\epsilon}\right)$. Then $T(\overline{b_n})g$ is given by the series (2), and we obtain the absolutely convergent formula

$$Gf(x) = \sum_{m,n} \frac{a_n \overline{b_m}}{m} f\left(\frac{n}{m} x\right)$$

Take two such f_1, f_2 and compute

$$\langle Gf_1, Gf_2 \rangle = \sum_{k,l,m,n} \frac{a_n \overline{b_m} \overline{a_k} b_l}{ml} \left\langle f_1\left(\frac{n}{m} x\right), f_2\left(\frac{k}{l} x\right) \right\rangle$$

the series are absolutely convergent when both a_n and b_n satisfy (5). Now we sum over all co-prime (p,q), such that $\frac{n}{m} = \frac{p}{q} \frac{k}{l}$. So, $\frac{nl}{mk} = \frac{p}{q}$, i.e. $nl = up$ and $mk = uq$ for some integer u. Then

$$\left\langle f_1\left(\frac{n}{m} x\right), f_2\left(\frac{k}{l} x\right) \right\rangle = \frac{l}{k} \left\langle f_1\left(\frac{p}{q} x\right), f_2(x) \right\rangle$$

$$\langle Gf_1, Gf_2 \rangle = \sum_{(p,q)=1} \left\langle f_1\left(\frac{p}{q} x\right), f_2(x) \right\rangle \sum_u \sum_{mk=uq} \sum_{nl=up} \frac{a_n \overline{b_m} \overline{a_k} b_l}{mk}$$

$$= \sum_{(p,q)=1} \frac{1}{q} \left\langle f_1\left(\frac{p}{q} x\right), f_2(x) \right\rangle \sum_u \frac{1}{u} \sum_{mk=uq} \overline{a_k} b_m \sum_{nl=up} a_n b_l$$

and so the only non-zero term corresponds to $p = q = 1$, $u = 1$, $m = n = k = l = 1$, i.e.

$$\langle Gf_1, Gf_2 \rangle = \langle f_1, f_2 \rangle$$

Since $\mathcal{F}(a_n)$ is invertible, we conclude that $\mathcal{F}(a_n) = SG$ is unitary.   □

4   An Example of a Unitary Operator Defined by Series

Let $a_n \in \mathbb{C}$ be a sequence satisfying (5). We denote by $C_0(0, \infty)$ the space of compactly supported continuous functions.

Let $T(a_n) : C_0(0, \infty) \to C_0(0, \infty)$ be given by (2). We will describe conditions on a_n that would imply $\langle Tf, Tg \rangle_{L^2} = \langle f, g \rangle_{L^2}$ for all $f, g \in C_0(0, \infty)$. Then we can conclude that T is an isometric operator on a dense subspace of $L^2[0, \infty)$, and thus can be extended as an isometry of all $L^2[0, \infty)$.

A C-isometric (correspondingly, unitary) operator will mean an isometric (unitary) operator, scaled by a constant factor C.

Proposition 4.1. $\langle Tf, Tg \rangle_{L^2} = C^2 \langle f, g \rangle_{L^2}$ *for all* $f, g \in C_0(0, \infty)$ *if and only if for all co-prime pairs* (m_0, n_0)

$$\sum_{k=1}^{\infty} \frac{a_{m_0 k} \overline{a_{n_0 k}}}{k} = \begin{cases} C^2, & m_0 = n_0 = 1 \\ 0, & m_0 \neq n_0 \end{cases} \tag{6}$$

Proof. Take $\epsilon > 0$. Denote $M = \sup\{x \mid f(x) \neq 0 \vee g(x) \neq 0\}$. Write

$$\int_{\epsilon}^{\infty} Tf(x) \overline{Tg(x)} dx = \int_{\epsilon}^{\infty} \sum_{m,n=1}^{\infty} a_m \overline{a_n} f(mx) \overline{g(nx)} dx$$

It is only necessary to consider $m, n < M/\epsilon$. Thus the sum is finite, and we may write

$$\int_{\epsilon}^{\infty} Tf(x) \overline{Tg(x)} dx = \sum_{m,n=1}^{\infty} a_m \overline{a_n} \int_{\epsilon}^{\infty} f(mx) \overline{g(nx)} dx$$

Note that

$$\int_{0}^{\infty} f(mx) \overline{g(nx)} dx \leq \|f(mx)\| \|g(nx)\| = \frac{1}{\sqrt{mn}} \|f\| \|g\|$$

and therefore

$$\sum_{m,n=1}^{\infty} a_m \overline{a_n} \int_{0}^{\infty} f(mx) \overline{g(nx)} dx$$

is absolutely convergent:

$$\sum_{m,n=1}^{\infty} \left| a_m \overline{a_n} \int_{0}^{\infty} f(mx) \overline{g(nx)} dx \right| \leq \sum_{m,n=1}^{\infty} \frac{|a_m| |a_n|}{\sqrt{mn}} \|f\| \|g\|$$

Therefore, the sum

$$S(\epsilon) = \sum_{m,n=1}^{\infty} a_m \overline{a_n} \int_{0}^{\epsilon} f(mx) \overline{g(nx)} dx$$

is absolutely convergent. We will show that $S(\epsilon) \to 0$ as $\epsilon \to \infty$. Assume $|f| \leq A_f$, $|g| \leq A_g$. Then

$$|S(\epsilon)| \leq \sum_{m,n} a_m \overline{a_n} \int_{0}^{\epsilon} f(mx) \overline{g(nx)} dx \leq \sum_{m,n} \frac{|a_m a_n|}{\sqrt{mn}} \sqrt{\int_{0}^{m\epsilon} |f|^2} \sqrt{\int_{0}^{n\epsilon} |g|^2}$$

And

$$\sum_{m=1}^{\infty} \frac{|a_m|}{\sqrt{m}} \sqrt{\int_0^{m\epsilon} |f|^2} = \sum_{m=1}^{\sqrt{1/\epsilon}} + \sum_{m=\sqrt{1/\epsilon}}^{\infty}$$

$$\leq \sum_{m=1}^{\infty} \frac{|a_m|}{\sqrt{m}} \sqrt{\int_0^{\sqrt{\epsilon}} |f|^2} + \|f\| \sum_{m=\sqrt{1/\epsilon}}^{\infty} \frac{|a_m|}{\sqrt{m}} \longrightarrow 0$$

We conclude that

$$\langle Tf, Tg \rangle = \sum_{m,n=1}^{\infty} \frac{a_m \overline{a_n}}{m} \int_0^{\infty} f(x) \overline{g\left(\frac{n}{m}x\right)} dx$$

$$= \left\langle f(x), \sum_{(m_0,n_0)=1}^{\infty} \frac{1}{m_0} \sum_{k=1}^{\infty} \frac{a_{m_0 k} \overline{a_{n_0 k}}}{k} g\left(\frac{n_0}{m_0}x\right) \right\rangle$$

Therefore, $T(a_n)$ is a C-isometry on $C_0(0, \infty)$ if and only if (a_n) satisfy (6).

□

Example 1. Take $a_n^{(2)} = 0$ for $n \neq 2^k$ and

$$a_{2^k}^{(2)} = \begin{cases} 1, & k = 0 \\ (-1)^{k+1}, & k \geq 1 \end{cases}$$

Then

$$T_2 f(x) = \sum_n a_n^{(2)} f(nx) = f(x) + f(2x) - f(4x) + f(8x) - f(16x) + \dots$$

is a $\sqrt{2}$-isometry.

Example 2. Generalizing Example 1 (and using the already defined $a_n^{(2)}$), we fix a natural number m, and take $a_{m^k}^{(m)} = \left(\frac{m}{2}\right)^{k/2} a_{2^k}^{(2)}$ and $a_n^{(m)} = 0$ for $n \neq m^k$. Then

$$T_m f(x) = \sum_n a_n^{(m)} f(nx)$$

is again a $\sqrt{2}$-isometry.

Example 3. Similarly, we could take $a_n = 0$ for $n \neq 2^k$ and

$$a_{2^k} = \begin{cases} 1, & k = 0 \\ -1, & k \geq 1 \end{cases}$$

Then

$$Tf(x) = f(x) - f(2x) - f(4x) - f(8x) - f(16x) - \cdots$$

is a $\sqrt{2}$-isometry.

Remarks.

- If a_n and b_n satisfy (5), then so does their convolution $c_n = (a * b)_n = \sum_{kl=n} a_k b_l$:

$$\sum_n \frac{|c_n|}{\sqrt{n}} \leq \sum_n \sum_{kl=n} \frac{|a_k||b_l|}{\sqrt{kl}} = \sum_k \frac{|a_k|}{\sqrt{k}} \sum_l \frac{|b_l|}{\sqrt{l}} < \infty$$

- Also, any two scaled isometries of the form $T(a_n)$ commute: If a_n and b_n satisfy (6), $T(a_n)$ and $T(b_n)$ are isometries from $C_0(0, \infty)$ to itself, and thus so is their composition which is easily computed to be $T(a_n * b_n)$.

Proposition 4.2. *When (a_n) satisfies (6), $T(a_n)$ is C-unitary.*

Proof. It is easy to verify that for any $g \in C[0, \infty)$ and a_n satisfying (6),

$$T(a_n)^* g = \sum \frac{\overline{a_n}}{n} g\left(\frac{x}{n}\right)$$

Moreover, $T(a_n)^*$ is a scaled isometry on $\{f \in C_0[0, \infty) : supp(f) \subset [a, b], a > 0\}$ (proof identical to that of $T(a_n)$), and so a scaled isometry on L^2. Thus $T^*T = TT^* = \|T\|^2 I$, and so $T(a_n)$ is C-unitary. $\square$

Remark. We recall the operator $Sf(x) = \frac{1}{x} f\left(\frac{1}{x}\right)$—a unitary operator of $L^2(0, \infty)$. Then for a continuous function f with compact support which is bounded away from 0, we have $Sf \in C_0(0, \infty)$ and so we can use (2) to obtain $ST(\overline{a_n})Sf = T(a_n)^* f$ and therefore $ST(\overline{a_n})S = T(a_n)^*$ on all L^2. In particular, for real sequences (a_n), ST^m and $T^m S$ are unitary involutions (up to scaling) for any integer m.

5   Diagonalizing the Fourier–Poisson Operator

We further generalize the Poisson summation formula: by removing some of the conditions on the sequence (a_n), we are still able to construct a unitary operator satisfying the summation formula, but only in the weaker operator sense. This is done through a natural isometry between $L^2[0, \infty)$ and $L^2(-\infty, \infty)$ which was suggested to us by Bo'az Klartag (see also [7]).

We will denote by dm the Lebesgue measure on $\mathbb{R}$, and $\hat{g}$ will stand for the Fourier transform defined as $\hat{g}(\omega) = \int_{-\infty}^{\infty} g(y)e^{-iy\omega} dy$.

First, define two isometries of spaces:

1. $u : L^2([0,\infty), dm(x)) \to L^2(\mathbb{R}, e^y dm(y))$ given by $f(x) \mapsto g(y) = f(e^y)$.
2. $v : L^2(\mathbb{R}, e^y dm(y)) \to L^2(\mathbb{R}, dm)$ given by $g(y) \mapsto h(x) = (2\pi)^{-\frac{1}{2}} \hat{g}(x + i/2)$.

u is isometric by a simple change of variables.

To see that v is isometric, note that $\widehat{f}(x + i/2) = \widehat{e^{t/2} f(t)}(x)$, and so by Plancherel's formula

$$\int |\hat{f}(x+i/2)|^2 dx = 2\pi \int |f(t)|^2 e^t dt$$

(alternatively, one could decompose v into the composition of two isometries: $f(y) \mapsto e^{y/2} f(y)$, identifying $L^2(\mathbb{R}, e^y dm(y))$ with $L^2(\mathbb{R}, dm)$, and then Fourier transform).

We will denote the composition $v \circ u = w$.

For $A : L^2[0,\infty) \to L^2[0,\infty)$, we write $\widetilde{A} = wAw^{-1} : L^2(\mathbb{R}) \to L^2(\mathbb{R})$—the conjugate operator to A. The conjugate to S is $\tilde{S}(h)(x) = h(-x)$.

Let a_n satisfy (5), implying $|L(1/2 + ix; a_n)|$ is bounded and continuous. Then for $g = u(f)$,

$$(uT(a_n)u^{-1}g)(y) = \sum a_n g(y + \log n) = g * \nu(y)$$

where $\nu(y) = \sum a_n \delta_{-\log n}(y)$ and $\hat{\nu}(z) = \sum a_n e^{iz \log n} = \sum a_n n^{iz} = L(-iz; a_n)$, which converges for $Imz \geq 1/2$ by (5). And so letting $h = vg$,

$$\widetilde{T(a_n)}h(x) = (2\pi)^{-\frac{1}{2}} \widehat{g * \nu}(x + i/2) = L(1/2 - ix; a_n)h(x)$$

thus we proved

Corollary 5.1. *Assume a_n satisfies (5). Then the following are equivalent:*

(a) $|L(1/2 + ix; a_n)| = C$
(b) $T(a_n)$ is C-unitary on $L^2[0,\infty)$
(c) (a_n) satisfies (6).

The equivalence of (a) and (c) can easily be established directly.

For example, the $\sqrt{2}$-unitary $Tf(x) = f(x) + f(2x) - f(4x) + f(8x) - \dots$ discussed previously is associated with $L(s; a_n) = \frac{2+2^s}{1+2^s}$ which has absolute value of $\sqrt{2}$ on $Re(s) = 1/2$.

This suggests that the Fourier–Poisson transform associated with a_n, which was defined in Sect. 3 for some special sequences (a_n), could be generalized as follows: $\mathcal{F}(a_n)f = T(\overline{a_n})^{-1}ST(a_n)f$ should be defined through

$$\widehat{\mathcal{F}(a_n)}h(x) = h(-x)\frac{L(1/2+ix;a_n)}{L(1/2-ix;\overline{a_n})} = h(-x)\frac{L(1/2+ix;a_n)}{\overline{L(1/2+ix;a_n)}} \qquad (7)$$

we arrive at the following:

Theorem 5.2. *Assume* $\sum |a_n|n^{-1/2} < \infty$. *Then*

(a) *There exists a bounded operator* $\mathcal{F}(a_n) : L^2[0,\infty) \to L^2[0,\infty)$ *satisfying the Poisson summation formula (in its operator form)* $T(\overline{a_n})\mathcal{F}(a_n) = ST(a_n)$. *Moreover,* $\mathcal{F}(a_n)$ *is unitary.*

(b) *If for some* $\epsilon > 0$, $\sum |a_n|n^{-1/2+\epsilon} < \infty$, *then a bounded* $\mathcal{F}(a_n)$ *satisfying* $T(\overline{a_n})\mathcal{F}(a_n) = ST(a_n)$ *is unique.*

Proof. (a) We have $L(1/2+ix;a_n)/\overline{L(1/2+ix;a_n)} = e^{2i(\arg L(1/2+ix;a_n))}$ whenever $L(1/2+ix;a_n) \neq 0$. In accordance with (7), define

$$\widehat{\mathcal{F}(a_n)}h(x) = e^{2i(\arg L(1/2+ix;a_n))}h(-x)$$

taking $\arg L(1/2+ix;a_n) = 0$ whenever $L(1/2+ix;a_n) = 0$. We then have

$$L(1/2-ix;\overline{a_n})\widehat{\mathcal{F}(a_n)}h(x) = L(1/2+ix;a_n)h(-x)$$

for all $h \in L^2(\mathbb{R})$, implying $T(\overline{a_n})\mathcal{F}(a_n) = ST(a_n)$ in $L^2[0,\infty)$. Also, $\mathcal{F}(a_n)$ is isometric and invertible, thus unitary.

(b) For uniqueness, observe that $L(s;a_n)$ is analytic in a neighborhood of $Re(s) = 1/2$, and so its set of zeros Z is discrete, and the ratio $L(1/2 + ix;a_n)/\overline{L(1/2+ix;a_n)}$ is continuous and of absolute value 1 outside of Z. Thus for continuous h with $supp(h) \cap Z = \emptyset$, the equation

$$L(1/2-ix;\overline{a_n})\widehat{\mathcal{F}(a_n)}h(x) = L(1/2+ix)h(-x)$$

determines $\widehat{\mathcal{F}(a_n)}h$ uniquely, and all such h are dense in $L^2(\mathbb{R})$. $\square$

By part (b) we conclude that under the conditions of Theorem 3.3, the operator $\mathcal{F}(a_n)$ defined in Sect. 3 coincides with the operator defined here.

Remark. It was pointed out to us by Fedor Nazarov that under the conditions of Theorem 5.2 the Poisson summation formula cannot hold pointwise for all sequences (a_n).

6   A Formula Involving Differentiation

Denote by $B : L^2[0, \infty) \to L^2[0, \infty)$ the unbounded operator

$$Bf(x) = i(xf' + f/2)$$

with $Dom(B) = \{f \in C^\infty : xf' + f/2 \in L^2\}$. It is straightforward to check that B is a symmetric operator.

It is easy to verify that the ordinary Fourier transform $\mathcal{F}$ satisfies, for a well behaved (i.e. Schwartz) function f, the identity $B\mathcal{F}f + \mathcal{F}Bf = 0$. It turns out to be also a consequence of Poisson's formula, and so holds for a large family of operators. We will need the following standard lemma (see [8]).

Lemma 6.1. *Take a function* $g \in L^2(\mathbb{R})$. *The following are equivalent:*

(a) $g \in C^\infty(\mathbb{R})$ *and*

$$\sup_{|y|<b} \sup_{t} e^{yt} |g^{(k)}(t)| < \infty$$

for all $b < B$ *and* $k \geq 0$.

(b) $h = \hat{g}$ *is a Schwartz function, which has an analytic extension to the strip* $|y| < B$ *such that*

$$\sup_{|y|<b} \sup_{x} |x|^k |h(x + iy)| < \infty$$

for all $b < B$ *and* $k \geq 0$.

Proof. (a)$\Rightarrow$(b).   Observe that $\hat{g}(x + iy) = \widehat{e^{yt}g(t)}(x)$. Thus the existence of analytic extension is clear, and we can write

$$|x|^k |h(x + iy)| = |x|^k |\widehat{e^{yt}g(t)}(x)| = |(\widehat{e^{yt}g(t)})^{(k)}(x)|$$

Note that

$$\left(e^{yt}g(t)\right)^{(k)} = e^{yt} \sum_{j=0}^{k} P_{j,k}(y)g^{(j)}(t)$$

where $P_{j,k}$ denotes some universal polynomial of degree $\leq k$. Therefore

$$\sup_{x} |x|^k |h(x + iy)| \leq \int_{-\infty}^{\infty} \left| e^{yt} \sum_{j=0}^{k} P_{j,k}(y)g^{(j)}(t) \right| dy$$

The sum is finite, so we can bound every term separately. Choose $b < Y < B$, $\epsilon = Y - b$. Then

$$\sup_{|y|<b} \int_{-\infty}^{\infty} |e^{yt} g^{(j)}(t)| \leq C(j,Y) \int_{-\infty}^{\infty} e^{-\epsilon|t|} dt < \infty$$

$(b)\Rightarrow(a)$.   Note that g is a Schwartz function since h is. It suffices to show (by induction) that supremums of $|(e^{yt}g)^{(k)}|$ are finite for every k and $b < B$. Notice that $\widehat{g^{(k)}}(x) = (ix)^k \hat{g}(x)$ has an analytic extension to the strip $|y| < B$ (namely: $(iz)^k h(z)$), satisfying the same conditions as h itself. Now take a C^∞ compactly supported function ϕ on $\mathbb{R}$. We will show that

$$\int_{-\infty}^{\infty} e^{yt} g(t)\overline{\phi(t)}dt = \int_{-\infty}^{\infty} h(x+iy)\overline{\hat{\phi}(x)}dx \qquad (8)$$

implying

$$\widehat{e^{yt}g(t)}(x) = h(x+iy)$$

and therefore for any k

$$\widehat{e^{yt} g^{(k)}(t)}(x) = i^k (x+iy)^k h(x+iy)$$

which is equivalent to having

$$\widehat{(e^{yt}g(t))^{(k)}}(x) = i^k x^k h(x+iy)$$

Indeed, $\psi = \hat{\phi}$ is an analytic function satisfying the supremum condition by the "$(a) \Rightarrow (b)$" implication. Then

$$\int_{-\infty}^{\infty} e^{yt} g(t)\overline{\phi(t)} = \int_{-\infty}^{\infty} \overline{\hat{g}e^{yt}\phi(t)}dt = \int_{-\infty}^{\infty} h(x)\overline{\psi(x+iy)}dt$$

Observe that $\lambda(z) = h(z)\overline{\psi(iy+\bar{z})}$ is an analytic function, and the integrals over the intervals $Re(z) = \pm R, -b < Im(z) < b$ of $\lambda(z)$ converge to 0 as $R \to \infty$ by the uniform bounds on h and ψ. Considering the line integral of λ over a rectangle with these vertical sides and horizontal lines at $Im(z) = 0$ and $Im(z) = y$, we get

$$\int_{-\infty}^{\infty} h(x)\overline{\psi(iy+x)} = \int_{-\infty}^{\infty} h(x+iy)\overline{\psi(x)}$$

which proves (8). Finally,

$$\sup_{|y|<b} \sup_{t} |(e^{yt}g(t))^{(k)}| \le \sup_{|y|<b} \int_{-\infty}^{\infty} |x|^k h(x+iy)dx$$

which is finite by the assumptions.

$\square$

Let $\mathcal{S}_0$ be the following class of "Schwartz" functions in $L^2[0,\infty)$

$$S_0 = \{f \in C^\infty : \sup |x|^n |f^{(k)}(x)| < \infty \ \forall k \geq 0, \forall n \in \mathbb{Z}\}$$

Note that $n \in \mathbb{Z}$ can be negative. Observe that $S_0 \subset Dom(B)$.

Proposition 6.2. *Assume (a_n) satisfies $\sum |a_n| n^\epsilon < \infty$ for some $\epsilon > 0$, and the convolution inverse (b_n) satisfies $\sum |b_n|/\sqrt{n} < \infty$. Next, assume that*

$$L(1/2 + iz; a_n)/\overline{L(1/2 + i\overline{z}; a_n)}$$

(which is meromorphic by assumption in the strip $|y| < 1/2 + \epsilon$) satisfies the following polynomial growth condition: there exist constants N and C such that

$$\left| \frac{L(1/2 - y + ix; a_n)}{L(1/2 + y + ix; a_n)} \right| \leq C_0 + C_1 |x|^N$$

for all $x, y \in \mathbb{R}$, $|y| \leq 1/2 + \epsilon/2$. Let $f \in S_0$. Then $\mathcal{F}(a_n)Bf + B\mathcal{F}(a_n)f = 0$.

Proof. Denote $g = \mathcal{F}(a_n)f$. Denote $F(t) = e^{t/2} f(e^t)$ and $G(t) = e^{t/2}g(e^t)$, $h_f = \hat{F}$ and $h_g = \hat{G}$. The condition $f \in S_0$ implies immediately that $F \in C^\infty$ and $\sup_{t \in \mathbb{R}} e^{yt} |F^{(k)}(t)| < \infty$ for all $y \in \mathbb{R}$, since $F^{(k)}(t) = P_k(e^{t/2} f(e^t), \ldots, f^{(k)}(e^t))$ for some fixed polynomial P_k. By Lemma (6.1), h_f is a Schwartz function (on the real line), with an analytic extension to the strip $|y| < 1$ such that

$$\sup_{|y| \leq 1} \sup_x |x|^k |h_f(x + iy)| < \infty$$

for all $k \geq 0$. Next,

$$h_g(x + iy) = \frac{L(1/2 - y + ix; a_n)}{L(1/2 + y + ix; a_n)} h_f(x + iy)$$

is an analytic function in the strip $|y| < 1/2 + \epsilon$. By the assumed bound on the L-function ratio, it is again a Schwartz function when restricted to the real line; and

$$\sup_{|y| < b} \sup_x |x|^k |h_g(x + iy)| < \infty$$

for all $b < 1/2 + \epsilon/2$. Denote $\delta = \epsilon/4$. Again by Lemma (6.1), $G \in C^\infty$ and satisfies $e^{(1/2+\delta)|t|}|G(t)| \leq C \iff |G(t)| \leq Ce^{-(1/2+\delta)|t|}$ and likewise $|G'(t)| \leq Ce^{-(1/2+\delta)t}$ for some constant C. Then, as $t \to -\infty$,

$$|g(e^t)| = e^{-t/2}|G(t)| \leq Ce^{-t/2}e^{(1/2+\delta)t} = O(e^{-\delta t})$$

and as $t \to +\infty$,

$$|g(e^t)| = e^{-t/2}|G(t)| \leq Ce^{-t/2}e^{-(1/2+\delta)t} = O(e^{-(1+\delta)t})$$

Also, as $t \to +\infty$,

$$|g'(e^t)| = \left| G'(t) - \frac{1}{2}G(t) \right| e^{-3t/2} = O(e^{-(2+\delta)t})$$

Thus $g \in C^\infty(0, \infty)$ and $g = O(x^\delta)$ as $x \to 0$, $g = O(x^{-1-\delta})$ as $x \to \infty$ while $g' = O(x^{-2-\delta})$ as $x \to \infty$. By Corollary 3.2 (b) we can write $\sum \overline{a_n} g(nx) = (1/x) \sum a_n f(n/x)$, and then the functions on both sides are C^1, and can be differentiated term-by-term. Carrying the differentiation out, we get

$$\sum \overline{a_n}(nx)g'(nx) = -(1/x) \sum a_n f(n/x) - (1/x^2) \sum a_n(n/x)f'(n/x)$$

Invoke Lemma 3.1 to write

$$T(\overline{a_n})(xg') = -T(\overline{a_n})g - ST(a_n)(xf')$$

and then use Corollary 3.2 applied to xf' to conclude

$$T(\overline{a_n})(xg') = -T(\overline{a_n})g - T(\overline{a_n})\mathcal{F}(a_n)(xf')$$

Finally, apply $T(\overline{b_n})$ to obtain the announced result. $\square$

Remark. As an example of such a sequence, take $a_n = n^\lambda$, $\lambda < -1$.

Acknowledgements I am indebted to Bo'az Klartag for the idea behind Sect. 5, and also for the motivating conversations and reading the drafts. I am grateful to Nir Lev, Fedor Nazarov, Mikhail Sodin and Sasha Sodin for the illuminating conversations and numerous suggestions. Also, I'd like to thank my advisor, Vitali Milman, for the constant encouragement and stimulating talks. Finally, I would like to thank the Fields Institute for the hospitality during the final stages of this work.

References

1. L. Bàez-Duarte, A class of invariant unitary operators. Adv. Math. **144**, 1–12 (1999)
2. J.-F. Burnol, On Fourier and Zeta(s). Habilitationsschrift, 2001–2002, Forum Mathematicum **16**, 789–840 (2004)
3. A. Cordoba, La formule sommatoire de Poisson. C.R. Acad Sci. Paris, Serie I **306**, 373–376 (1988)
4. H. Davenport, On some infinite series involving arithmetical functions. Q. J. Math. **8**(8–13), 313–320 (1937)
5. R.J. Duffin, H.F. Weinberger, Dualizing the Poisson summation formula, Proc. Natl. Acad. Sci. USA **88**, 7348–7350 (1991)
6. D. Faifman, A characterization of Fourier transform by Poisson summation formula. Comptes rendus – Mathematique **348**, 407–410 (2010)
7. A. Korànyi, The Bergman kernel function for tubes over convex cones. Pac. J. Math. **12**(4), 1355–1359 (1962)
8. M. Reed, B. Simon, *Methods of Modern Mathematical Physics II: Fourier Analysis, Self-Adjointness* (Academic, CA, 1975)

Stability of Order Preserving Transforms

Dan Florentin and Alexander Segal

Abstract The purpose of this paper is to show stability of order preserving/reversing transforms on the class of non-negative convex functions in $\mathbb{R}^n$, and its subclass, the class of non-negative convex functions attaining 0 at the origin (these are called "geometric convex functions"). We show that transforms that satisfy conditions which are weaker than order preserving transforms, are essentially close to the order preserving transforms on the mentioned structures.

1 Introduction

The concept of duality was studied by Artstein-Avidan and Milman in recent papers [1, 3, 4] on different classes which arise from geometric problems. Examples of such classes are the class of convex bodies containing zero, the class of all lower semi-continuous convex functions on $\mathbb{R}^n$, which we denote by $Cvx(\mathbb{R}^n)$, and its subclass—the class of all lower semi-continuous *geometric* convex functions denoted by $Cvx_0(\mathbb{R}^n)$. A convex function f is said to be geometric if it is non-negative and $f(0) = 0$.

It turned out that duality on such classes is uniquely defined by simple properties like *order reversion* and *involution* (actually involution is not required and can be replaced by bijectivity). The Legendre transform is an example of such a duality transform that acts on the class of convex functions $Cvx(\mathbb{R}^n)$. When dealing with $Cvx(\mathbb{R}^n)$, it was shown by Artstein-Avidan and Milman [3] that the Legendre transform is essentially the only order reversing transform acting on this class, where "essentially" means up to the choice of scalar product and addition of linear terms.

D. Florentin (✉) · A. Segal
School of Mathematical Science, Sackler Faculty of Exact Science, Tel Aviv University,
Ramat Aviv, 69978 Tel Aviv, Israel
e-mail: danflorentin@gmail.com; segalale@gmail.com

B. Klartag et al. (eds.), *Geometric Aspects of Functional Analysis*, Lecture Notes
in Mathematics 2050, DOI 10.1007/978-3-642-29849-3_12,
© Springer-Verlag Berlin Heidelberg 2012

Note: The properties of order preservation and involution actually imply preservation of supremum and infimum on the classes. It is also known that the mentioned classes can be generated with supremum (or infimum) of an extremal family. This concept is not new, and was used by Kutateladze and Rubinov [7] to discuss Minkowski duality on complete lattices.

Studying the structure of $Cvx_0(\mathbb{R}^n)$ shows that it differs from $Cvx(\mathbb{R}^n)$. As was shown by Artstein-Avidan and Milman in [4], there exist essentially two duality (order reversing) bijective transforms—The Legendre transform, and a "geometric duality" transform called $\mathcal{A}$, on the class of geometric convex functions.

Actually, the authors of [4] showed first that there exist essentially two order *preserving* bijections—identity transform $\mathcal{I}$ and the Gauge transform $\mathcal{J}$ which greatly differs from $\mathcal{I}$. After showing this, using the fact that $\mathcal{L}$ is an involution and the fact that $\mathcal{J} = \mathcal{L}\mathcal{A} = \mathcal{A}\mathcal{L}$, it is easy to see that the order reversing transforms are also uniquely defined. Notice that the results about order reversing transforms are "dual" to the results about order preserving transforms. For details of the mentioned transforms we refer the reader to [4], and provide the basic definitions for completeness.

Definition 1.1. The geometric transform $\mathcal{A} : Cvx_0(\mathbb{R}^n) \to Cvx_0(\mathbb{R}^n)$ is defined as follows:

$$(\mathcal{A}f)(x) = \begin{cases} \sup_{\{y\in\mathbb{R}^n : f(y)>0\}} \frac{<x,y>-1}{f(y)} & \text{if } x \in \{y : f(y) = 0\}^\circ \\ +\infty & \text{if } x \notin \{y : f(y) = 0\}^\circ \end{cases}$$

assuming $\sup \emptyset = 0$.

Definition 1.2. The Legendre transform $\mathcal{L}$ of a function f is defined as follows:

$$(\mathcal{L}f)(x) = \sup_y(< x, y > -f(y)),$$

and the Gauge transform $\mathcal{J}$ is defined as $\mathcal{J}f = \mathcal{A}\mathcal{L}f = \mathcal{L}\mathcal{A}f$, for $f \in Cvx_0(\mathbb{R}^n)$. Notice that the commutativity of $\mathcal{A}$ and $\mathcal{L}$ requires a proof, and is actually a nontrivial fact. The Gauge transform $\mathcal{J}$ can be calculated, and written explicitly:

$$(\mathcal{J}f)(y) = \inf\{1/f(x) : y = tx/f(x), 0 \le t \le 1\},$$

where $\inf \emptyset = +\infty$, and $0/f(0)$ is understood in the sense of limits.

In this paper we discuss the stability of the mentioned transforms on the class $Cvx_0(\mathbb{R}^n)$ and $Cvx_+(\mathbb{R}^n)$ (non-negative convex functions). We do not deal with classes of convex bodies, and refer the reader to [5] for results on such classes. We start with the following definitions:

Definition 1.3. Let $\tilde{C} > 1$, and $\tilde{c} = \tilde{C}^{-1}$. A bijective transform T, on the class $Cvx_0(\mathbb{R}^n)$ that satisfies the following conditions:

$$f \leq g \text{ implies } Tf \leq \tilde{C}Tg, \tag{1a}$$

$$f \leq \tilde{c}g \text{ implies } Tf \leq Tg, \tag{1b}$$

will be called a $\tilde{C}$-*almost order preserving* transformation, or just almost order preserving, in case there exists some unspecified constant that satisfies conditions (1a) and (1b).

Similarly, we can define a $\tilde{C}$-*almost order reversing* transform:

Definition 1.4. Let $\tilde{C} > 1$, and $\tilde{c} = \tilde{C}^{-1}$. A bijective transform T, on the class $Cvx_0(\mathbb{R}^n)$ that satisfies the following conditions:

$$f \leq g \text{ implies } Tf > \tilde{c}Tg, \tag{2a}$$

$$f \leq \tilde{c}g \text{ implies } Tf > Tg \tag{2b}$$

will be called a $\tilde{C}$-*almost order reversing* transformation, or just almost order reversing, in case there exists some constant that satisfies conditions (a) and (b).

Remark 1.5. If T and T^{-1} are almost order preserving transforms, the following hold:

$$Tf \leq Tg \text{ implies } f \leq \tilde{C}g, \tag{3a}$$

$$Tf \leq \tilde{c}Tg \text{ implies } f \leq g \tag{3b}$$

Indeed, since T is bijective, we can write $f = Tf'$ and $g = Tg'$. If Property (1a) holds for f, g and T^{-1}, after substituting Tf' and Tg', we come to $Tf' \leq Tg' \Rightarrow f' \leq \tilde{C}g'$. So condition (1a) on T^{-1} is equivalent to condition (3a) on T. The same applies for (3b) and (1b).

Notice that when $\tilde{C} = 1$, we have order preserving transform. We would like to show that order preserving transforms are stable, i.e. almost order preserving transforms are, in some sense, close to the order preserving transforms discussed above. Our main theorems are the following:

Theorem 1.6. *Let $n \geq 2$. Any $1 - 1$ and onto transform $T : Cvx_0(\mathbb{R}^n) \to Cvx_0(\mathbb{R}^n)$ such that both, T and T^{-1} are $\tilde{C}$-almost order preserving, satisfies one of the following conditions:*
Either

$$\text{For all } f \in Cvx_0(\mathbb{R}^n), \quad cf \circ B \leq Tf \leq Cf \circ B, \tag{4a}$$

or

$$\text{For all } f \in Cvx_0(\mathbb{R}^n), c(\mathcal{J}f) \circ B \leq Tf \leq C(\mathcal{J}f) \circ B, \tag{4b}$$

where $B \in GL(n)$ and c, C are positive constants depending only on $\tilde{C}$.

Remark 1.7. Actually the proof gives $C \leq \lambda\tilde{C}^7$, but it is entirely possible that the dependence on $\tilde{C}$ is linear.

The "dual" of the above statement follows:

Theorem 1.8. *Let $n \geq 2$. Any bijective transform $T : Cvx_0(\mathbb{R}^n) \to Cvx_0(\mathbb{R}^n)$ such that both, T and T^{-1} are almost order reversing, satisfies one of the following conditions:*
Either
$$\text{for all } f \in Cvx_0(\mathbb{R}^n), \, c(\mathcal{A}f) \circ B \leq Tf \leq C(\mathcal{A}f) \circ B, \qquad (5a)$$
or
$$\text{for all } f \in Cvx_0(\mathbb{R}^n), \, c(\mathcal{L}f) \circ B \leq Tf \leq C(\mathcal{L}f) \circ B, \qquad (5b)$$
where $B \in GL(n)$ and c, C are positive constants as above.

In the case of general positive convex functions ($Cvx_+(\mathbb{R}^n)$), we have a similar theorem:

Theorem 1.9. *Let $n \geq 2$. Any bijective transform $T : Cvx_+(\mathbb{R}^n) \to Cvx_+(\mathbb{R}^n)$ such that both, T and T^{-1} are almost order preserving, must be close to the identity transform:*
$$cf(Bx + b_0) \leq (Tf)(x) \leq Cf(Bx + b_0)$$
where $B \in GL(n), b_0 \in \mathbb{R}^n$ and c, C are positive constants.

Notice that there there is no dual statement for the class of general convex nonnegative functions, since there exist no order reversing transformations on this class, as was noted by Artstein-Avidan and Milman in [4].

2 Preliminaries and Notation

Let us state that throughout the article, all the constants c, C, c', C' etc, mostly depend on $\tilde{C}$ which appears in the definition of order almost preserving transforms. The dependence is some power of $\tilde{C}$ which can be seen during the proofs. These constants are not universal and might have a different meaning in different context.

We will use the notation of *convex indicator* functions, 1_K^∞ where K is some convex domain. As our discussion is limited to convex functions, we define it in the following way:
$$1_K^\infty(x) = \begin{cases} 0, & x \in K \\ +\infty, & x \notin K \end{cases}$$

Likewise, we will use *modified Delta* functions denoted by $D_\theta + c$, which equals c when $x = \theta$ and $+\infty$ otherwise.

Next we state a known stability result by Hyers and Ulam [6, 9], which we will use in some of the proofs:

Theorem 2.1. *Let E_1 be a normed vector space, E_2 a Banach space and suppose that the mapping $f : E_1 \to E_2$ satisfies the inequality*

$$\|f(x+y) - f(x) - f(y)\| \le \epsilon$$

for all $x, y \in E_1$, where $\epsilon > 0$ is a constant. Then the limit

$$g(x) = \lim_{n \to \infty} 2^{-n} f(2^n x)$$

exists for each $x \in E_1$, and g is the unique additive mapping satisfying

$$\|f(x) - g(x)\| \le \epsilon$$

for all $x \in E_1$. If f is continuous at a single point of E_1, then g is continuous everywhere.

Lemma 2.2. *Assume we have function* $f : \mathbb{R}^+ \to \mathbb{R}^+$, *which satisfies the following condition of* C-*monotonicity:*

$$x \le y \text{ implies } f(x) \le C f(y) \tag{6}$$

for a constant $C > 1$ *independent of* x *and* y. *Then there exist a monotonic function* $g(x)$ *such that* $C^{-1} g(x) \le f(x) \le g(x)$.

Proof. Define $g(x)$ to be the infimum over all monotone functions which are greater or equal $f(x)$:

$$g(x) = \sup_{0 \le y \le x} f(y).$$

Obviously $g(x) \ge f(x)$ and $g(x)$ is monotone. For any x_0, we know that if $y \le x_0$ then $f(y) \le C f(x_0)$. Therefore this is true after applying sup, which brings us to $g(x_0) \le C f(x_0)$ as desired. $\qquad\square$

Lemma 2.3. *Assume we have a function* $f : \mathbb{R}^n \times \mathbb{R}^+ \to \mathbb{R}^+$ *which satisfies the following inequalities for all* (x, a) *and* (y, b):

$$\frac{1}{C}(\lambda f(x,a) + (1-\lambda)f(y,b)) \le f(\lambda(x,a) + (1-\lambda)(y,b))$$

$$\le C(\lambda f(x,a) + (1-\lambda)f(y,b)), \tag{7}$$

and $f(x, 0) = 0$, *for all* $x \in \mathbb{R}^n$. *Then there exists a constant* C' *such that*

$$\frac{1}{C'} a \le f(x,a) \le C' a$$

Proof. First we check the case where $n = 0$. Substitute $y = 0$ and use the fact that $f(0) = 0$ to conclude:

$$c\lambda f(a) \le f(\lambda a) \le C\lambda f(a). \tag{8}$$

This is true for every $0 \leq \lambda < 1$ and $a \in \mathbb{R}$, so choose $a = 1$ to get almost-linearity of f:

$$c\lambda \leq f(\lambda) \leq C\lambda. \tag{9}$$

Note that this is true for $\lambda \leq 1$, but we can easily conclude it for all λ by taking $a' = \lambda a$ and applying (8) again. First, rewrite (8) in the form

$$\frac{1}{C\lambda} \leq \frac{f(a)}{f(\lambda a)} \leq \frac{1}{c\lambda}.$$

After substituting $a' = \lambda a$, it becomes:

$$\frac{1}{C\lambda} \leq \frac{f(a'/\lambda)}{f(a')} \leq \frac{1}{c\lambda}.$$

Now, if $a' = 1$, we get what we required:

$$\frac{c}{\lambda} \leq f(1/\lambda) \leq \frac{C}{\lambda}.$$

Replace $1/\lambda$ with $t > 1$ and conclude the proof. To prove the general case, notice that if $x = (x_1, x_2, \ldots x_n)$, then using the previous case

$$f(x, a) = f(\frac{1}{2}(2x_1, 2x_2, \ldots 2x_n - 1, 0) + \frac{1}{2}(0, 0, \ldots, 0, 1, 2a))$$

$$\leq \frac{1}{2}Cf(0, 0, \ldots, 1, 2a) \leq C'a$$

In the same way, we see that $f(x, a) \geq \frac{1}{C'}a$. This completes the proof.   □

3   Stability On the Class of Geometric Convex Functions

3.1   *Preservation of* sup *and* $\hat{\inf}$

Since we work with convex functions, taking supremum results in a convex function in our class. However, infimum of convex functions is not necessarily convex, thus we use a modified infimum denoted by $\hat{\inf}$, defined as follows:

$$\hat{\inf_{\alpha}}(f_\alpha) = \sup_{g}(g \in Cvx_0(\mathbb{R}^n) : g \leq f_\alpha \text{ for each } \alpha).$$

Now we can see how almost order preserving transforms act on sup and $\hat{\inf}$.

Lemma 3.1. *If T is almost order preserving transformation, then:*

$$\tilde{c}^2 T(\max f_\alpha) \leq \max Tf_\alpha \leq \tilde{C} T(\max f_\alpha) \tag{10}$$

$$\tilde{c} T(\hat{\inf} f_\alpha) \leq \hat{\inf} Tf_\alpha \leq \tilde{C}^2 T(\hat{\inf} f_\alpha) \tag{11}$$

Proof. Since T is bijective, we may assume that there exists such a function h such that: $\max Tf_\alpha = \tilde{c} T h$. Hence, for each α, $Tf_\alpha \leq \tilde{c} T h$. Condition (3b) implies that $f_\alpha \leq h$ for all α. If we define $h' = \max f_\alpha$, we may write $h' \leq h$. Applying condition (1a) we conclude that $T h' \leq \tilde{C} T h$, which means (by definition of Th) that $\tilde{c}^2 T(\max f_\alpha) \leq \max Tf_\alpha$. To get the right hand side, we write that $f_\alpha \leq h'$, so by condition (1a) we get that $Tf_\alpha \leq \tilde{C} T h'$. This is true for all α, so $\max Tf_\alpha \leq \tilde{C} T h'$. The proof of inequality (11) is similar. $\qquad\square$

3.2 Preservation of Zero and Infinity

Lemma 3.2. *If T is almost order preserving transformation, then $T1^\infty_{\{0\}} = 1^\infty_{\{0\}}$ and $T0 = 0$.*

Proof. Since $1^\infty_{\{0\}}$ is the maximal function on the set $Cvx_0(\mathbb{R}^n)$, we may write: $f \leq \tilde{c} 1^\infty_{\{0\}}$. Using condition (1b), we get $Tf \leq T1^\infty_{\{0\}}$, for every f. Since T is bijective, $T1^\infty_{\{0\}}$ must be the maximal function. In the same way $T0 = 0$. $\qquad\square$

3.3 Ray-wise-ness

Lemma 3.3. *If $T : Cvx_0(\mathbb{R}^n) \to Cvx_0(\mathbb{R}^n)$ is an almost order preserving transformation then there exists some bijection $\Phi : S^{n-1} \to S^{n-1}$ such that a function supported on the ray $\mathbb{R}^+ y$ is mapped to a function supported on the ray $\mathbb{R}^+ \Phi(y)$.*

Proof. Let us check that if $\max(f, h) = 1^\infty_{\{0\}}$, then $\max(Tf, Tg) = 1^\infty_{\{0\}}$. This follows immediately from the fact that $\max(Tf, Tg) \geq \tilde{c}^2 T(\max(f, g)) = 1^\infty_{\{0\}}$ according to the previous lemma. The proof of the lemma follows exactly in the same way as in [4]. $\qquad\square$

Lemma 3.4. *Let $f \in Cvx_0(\mathbb{R}^n)$ and $y \in S^{n-1}$. Then, $(suppTf) \cap \mathbb{R}^+\Phi(y) = suppT(\max(f, 1^\infty_{\mathbb{R}^+ y}))$.*

Proof. Notice that $T(1^\infty_{\mathbb{R}^+ y}) = 1_{\mathbb{R}^+\Phi(y)}$. Indeed, the function $1^\infty_{\mathbb{R}^+ y}$ is supported on a ray, and thus must be mapped to a function supported on a ray. In addition, it is the smallest function on $\mathbb{R}^+ y$ which, by the reasoning of Lemma 3.2 must be mapped to the smallest function on $\mathbb{R}^+ \Phi(y)$.

By Lemma 3.1 we have

$$\tilde{c}^2 \max(Tf, 1^{\infty}_{\mathbb{R}^+ \Phi(y)}) \le T(\max(f, 1^{\infty}_{\mathbb{R}^+ y})) \le \tilde{C} \max(Tf, 1^{\infty}_{\mathbb{R}^+ \Phi(y)}).$$

Hence, we see that on the ray $R^+ \Phi(y)$, the function Tf is finite if and only if the function $T(\max(f, 1^{\infty}_{\mathbb{R}^+ y}))$. This completes the proof.   □

3.4 Convex Functions on $\mathbb{R}^+$

We have seen that due to the ray-wise-ness, the case of $\mathbb{R}^+$ will give us an idea about the general case. We will state and proof this special case, which is actually not required for the general one, but is of independent interest.

Theorem 3.5. *if T and T^{-1} are both $\tilde{C}$-almost order preserving transforms on the class of convex geometric functions $Cvx_0(\mathbb{R}^+)$ and T is bijective, then there exist positive constants α_1, α_2, β_1, β_2 (dependent of $\tilde{C}$), such that either*

$$\text{for all } f \in Cvx_0(\mathbb{R}^+), \ \beta_1 f(x/\alpha_2) \le Tf(x) \le \beta_2 f(x/\alpha_1), \tag{12a}$$

or

$$\text{for all } f \in Cvx_0(\mathbb{R}^+), \ \alpha_1 \mathcal{J} f(x/\beta_2) \le Tf(x) \le \alpha_2 \mathcal{J} f(x/\beta_1). \tag{12b}$$

The proof of this theorem uses a few preliminary constructions and facts which we introduce and study in Sects. 3.5–3.7.

3.5 Property $\tilde{P}$

To study general functions, we need a family of extremal functions which are easy to deal with and can be used to describe the general case. In the case of geometric convex functions the family of indicators and linear functions is most convenient. A property which uniquely defines such a family was introduced in [4] (property P). Due to the modified nature of our problem, we introduce a slightly modified property:

Definition 3.6. We say that a function f satisfies property $\tilde{P}$, if there exist no two functions $g, h \in Cvx_0(\mathbb{R}^n)$, such that $g \not\ge \tilde{c}^3 f, h \not\ge \tilde{c}^3 f$ but $\max(g, h) \ge f$.

Note that Definition 3.6 depends on the constant $\tilde{C}$.

Obviously, if a function satisfies property P, then it satisfies property $\tilde{P}$. This means that the family of functions which satisfy $\tilde{P}$ contains all the indicator functions and linear functions through 0. We will now show that the non-linear functions that have property $\tilde{P}$ cannot differ greatly from the linear ones, unless they are indicators.

Lemma 3.7. *If f has property $\tilde{P}$ and f is not an indicator, then $f'(0) > 0$.*

Proof. Assume otherwise, that is, $f'(0) = 0$. Since f is not an indicator, there exists a $x_0 > 0$ such that $f(x_0) > 0$. Define

$$L(x) = \frac{f(x_0)}{x_0}x,$$

and $L_2(x) := \tilde{c}^3 L_1(x)$. Since $f'(0) = 0$, there exists a point x_1 such that $L_2(x_1) = f(x_1)$. Hence, it holds that $\max(L_1, 1^\infty_{[0,x_1]}) \geq f$. Since f has property $\tilde{P}$ we have that either $1^\infty_{[0,x_1]} \geq \tilde{c}^3 f$ or $L_1 \geq \tilde{c}^3 f$. Clearly, neither of the inequalities holds and this is a contradiction to the fact that f has property $\tilde{P}$. □

Lemma 3.8. *Every function with property $\tilde{P}$ that is not an indicator can be bounded by*

$$f'(0)z \leq f(z) \leq \tilde{C}^3 f'(0)z.$$

Proof. First note that by Lemma 3.7 $f'(0) > 0$. It is clear that $f(z)$ is bounded by $L_1(z) := f'(0)z$ from below. Assume that $L_2(z) := \tilde{C}^3 L_1(z)$ intersects $f(z)$ at some point x_0. This means that $L_1(z)$ intersects $\tilde{c}^3 f(z)$ at x_0. Hence, the derivative of $\tilde{c}^3 f(x)$ at x_0 is bigger than $f'(0)$ (otherwise, they wouldn't intersect). So there exists a constant $a > f'(0)$ such that $L_a(z) = az$ intersects both, $f(z)$ and $\tilde{c}^3 f(z)$. This is a contradiction to property $\tilde{P}$ (take L_a and an indicator function to see this). □

Lemma 3.9. *If a function f with property $\tilde{P}$, equals ∞ for $x \geq x_0$ for some x_0, it must be an indicator.*

Proof. Assume there exists $x_1 < x_0$ such that $f(x_1) = c > 0$. Then, define two functions: $g(x) = \frac{c}{x_1}x$ and $h(x) = 1^\infty_{[0,x_1]}$. Obviously, $g(x) \not\geq \tilde{c}^3 f(x)$ and $h(x) \not\geq \tilde{c}^3 f(x)$, but $\max(g, h) \geq f$, which contradicts f having property $\tilde{P}$. □

From now on, a function f with property $\tilde{P}$ that is not an indicator, will be called *almost linear* function.

3.6 Properties of $\tilde{P}$

Lemma 3.10. *If T is almost order preserving, and f has property P, then Tf has property $\tilde{P}$.*

Proof. Assume that f has property P, but Tf does not have $\tilde{P}$. So there exist g, h such that $g \not\geq \tilde{c}^3 Tf$, $h \not\geq \tilde{c}^3 Tf$, but $\max(g, h) \geq Tf$. Since T is bijective, there exist $\phi, \psi \in Cvx_0(\mathbb{R}^n)$ such that

$$g = \tilde{c}^2 T\phi, \qquad h = \tilde{c}^2 T\psi.$$

Using Property (10) we write:

$$\tilde{c}T(\max(\phi,\psi)) \geq \max(g,h) \geq Tf.$$

Now, applying condition (3a), we conclude that $f \leq \max(\phi, \psi)$. Since f has property P, then either $f \leq \phi$ or $f \leq \psi$. After applying condition (1a), we get that either $Tf \leq \tilde{C}T\phi = \tilde{C}^3 g$, or $Tf \leq \tilde{C}^3 h$, which is a contradiction. The same proof applies for T^{-1}, so T^{-1} also maps functions with property P, to functions with property $\tilde{P}$. $\square$

Lemma 3.11. *T either maps all indicators to indicators or it maps all indicators to almost linear functions.*

Proof. Assume that the claim is false and there exist two indicators $1_{[0,x]}^{\infty}$, $1_{[0,y]}^{\infty}$ which are mapped to an indicator $1_{[0,x']}^{\infty}$ and an almost linear function f, respectively. Assume, without loss of generality, that $x < y$. Then we know that $1_{[0,y]}^{\infty} \leq \tilde{c}1_{[0,x]}^{\infty}$, which by properties of T implies that f and $1_{[0,x']}^{\infty}$ are comparable. But we know that an almost linear function cannot be comparable to an indicator, and the proof is complete. $\square$

Lemma 3.12. *T either maps all linear functions to almost linear functions, or it maps them to the indicators.*

Proof. Assume that there exist linear functions l_a, l_b such that $T(l_a) = 1_{[0,x]}^{\infty}$ and $T(l_b) = f$ where f is almost linear. In addition, assume, without loss of generality, that $l_a < l_b$. Then, by properties of T we have $T(l_a) < \tilde{C}T(l_b)$, which is equivalent to $1_{[0,x]}^{\infty} \leq \tilde{C}f$. But this is a contradiction since indicators and almost linear functions are not comparable. $\square$

Lemma 3.13. *T cannot map all functions with property P to indicators, and it cannot map all functions with property P to almost linear functions.*

Proof. Assume, all functions are mapped to indicators. Then we may write that $T(1_{[0,x]}^{\infty}) = 1_{[0,y]}^{\infty}$ and $T(l_a) = 1_{[0,z]}^{\infty}$. If, without loss of generality, $y < z$, then $T(l_a) \leq \tilde{c}T(1_{[0,x]}^{\infty})$. This implies that l_a and $1_{[0,z]}^{\infty}$ are comparable, which cannot be.

The other option, is that all functions with property P, are mapped to almost linear functions: $T1_{[0,x]}^{\infty} = f$, and $T(l_a) = g$. Since both f and g are almost linear, there exists a linear function l_b such that $f \leq \tilde{c}l_b$ and $g \leq \tilde{c}l_b$. By Lemma 3.10, we know that the function $T^{-1}(l_b)$ is either almost linear or an indicator. If it is an indicator then the inequality $g \leq \tilde{c}l_b$ implies that $l_a \leq T^{-1}(l_b)$ (Property (1.5b)). This is a contradiction as an indicator cannot be comparable to linear function. If it is almost linear, then the inequality $f \leq \tilde{c}l_b$ implies that $1_{[0,x]}^{\infty} \leq T^{-1}(l_b)$. This is again a contradiction since almost linear functions cannot be comparable to indicators. This completes the proof. $\square$

Lemma 3.14. *If T maps linear functions to indicators, T preserves property P.*

Proof. Assume that we are in the case where indicators are mapped to almost-linear functions, and linear functions are mapped to indicators. First we will show that T maps the linear functions onto the indicators. If we have a g which is not linear, but $Tg = 1_{[0,z]}^{\infty}$, then we can find a linear function ax that intersects g. l_a is mapped to some indicator $1_{[0,y]}$. If $y > z$ then $1_{[0,y]}^{\infty} < \tilde{c}1_{[0,z]}^{\infty}$, hence g and l_a are comparable. This is a contradiction, so g must be linear.

Now we know that the indicators are mapped to almost-linear functions under T. But T^{-1} maps property P to $\tilde{P}$, so if we restrict it to linear functions, we get only indicators. So the preimage of every linear function is an indicator. But this also means that there are no non-linear functions with property $\tilde{P}$ other than indicators. The reason is similar: If we have a non-linear function $g = T1_{[0,y]}^{\infty}$, then intersect it with some linear function l_b. They are not comparable, but the sources are, which is a contradiction. This ends the proof. □

Lemma 3.15. *If T maps linear functions to indicators, then it is order preserving on functions that have property P.*

Proof. Assume that $T(1_{[0,z]}^{\infty}) = l_{\phi(z)}$ and $T(l_a) = 1_{[0,c(a)]}^{\infty}$. If $1_{[0,z]}^{\infty} < 1_{[0,z']}^{\infty}$ ($z' < z$), then $1_{[0,z]}^{\infty} < \tilde{c}1_{[0,z']}^{\infty}$ and using Property (1.3a) we write $\phi(z) < \phi(z')$. So we see that $\phi(z)$ is monotone decreasing and continuous. The same applies for $c(a)$. □

3.7 Triangles

Define triangle: $\vartriangleleft_{z,c} = \max(l_c, 1_{[0,z]}^{\infty})$. Using facts shown above, we conclude that:

$$\tilde{c} \vartriangleleft_{z',c'} \leq T(\vartriangleleft_{z,c}) \leq \tilde{C}^2 \vartriangleleft_{z',c'}$$

where $z' = z'(z,c)$ and $c' = c'(z,c)$. Next, we show that triangles determine everything, in the general setting of $Cvx_0(\mathbb{R}^n)$. To this end, we will need a definition of triangle in $\mathbb{R}^n$. Given a vector z and a gradient c, define:

$$\vartriangleleft_{z,c}(x) = \max\{c\frac{x}{|z|}, 1_{[0,z]}^{\infty}\}.$$

Lemma 3.16. *Assume that T is an almost order preserving map defined on $Cvx_0(\mathbb{R}^n)$, and that there exists $B \in GL(n)$ and a constant β such that*

$$\tilde{c}\beta \vartriangleleft_{B^{-1}z,d} \leq T(\vartriangleleft_{z,d}) \leq \tilde{C}\beta \vartriangleleft_{B^{-1}z,d} . \tag{13}$$

Then, T is almost a variation of identity, in the following sense:

$$\frac{1}{C}\varphi(Bx) \leq (T\varphi)(x) \leq C\varphi(Bx)$$

Proof. Any $\varphi(x) \in Cvx_0(\mathbb{R}^n)$ can be written as $\varphi(x) = (\hat{\inf}_y \lhd_{y,\varphi(y)})(x)$. Hence, using (3.1) we have $(T\varphi)(x) \leq \tilde{C}^2(\hat{\inf}_y T(\lhd_{y,\varphi(y)}))(x)$. Using assumption (13) we conclude that

$$(T\varphi)(x) \leq \tilde{C}^3 \beta(\hat{\inf}_y(\lhd_{B^{-1}y,\varphi(y)}))(x) = \tilde{C}^3 \beta(\hat{\inf}_z \lhd_{z,\varphi(Bz)})(x) = \tilde{C}^3 \beta\varphi(Bx).$$

The other part is concluded in a similar way.   □

Lemma 3.17. *Assume that T is an almost order preserving map defined on $Cvx_0(\mathbb{R}^n)$, and that there exist $B \in GL(n)$ and a constant β such that*

$$\tilde{c}\beta \lhd_{\frac{B^{-1}z}{d|z|},\frac{1}{|z|}} \leq T(\lhd_{z,d}) \leq \tilde{C}\beta \lhd_{\frac{B^{-1}z}{d|z|},\frac{1}{|z|}}. \tag{14}$$

Then, T is almost a variation of $\mathcal{J}$, in the following sense:

$$\frac{1}{C}(\mathcal{J}\varphi)(Bx) \leq (T\varphi)(x) \leq C(\mathcal{J}\varphi)(Bx)$$

The proof is similar to Lemma 3.16.

3.8   Proof of Theorem 3.5

3.8.1   The Case of "J"

We stay with the notation of Lemma 3.15. First we deal with the case we identify with $\mathcal{J}$, i.e linear functions are mapped to indicators and vice versa. Define $g = \inf\{1^\infty_{[0,tz]}, l_{a/(1-t)}\}$, for some $0 < t < 1$. Now, $g \leq \lhd_{z,a}$, but it does not hold for any $a' < a$ or $z' > z$. Applying T to the last inequality we get that $T(g) \leq \tilde{C}T(\lhd_{z,a})$. Using Property (11):

$$\tilde{c}^2 \hat{\inf}(1^\infty_{[0,c(a/(1-t))]}, l_{\phi(tz)}) \leq T(g) \leq \tilde{C}\hat{\inf}(1^\infty_{[0,c(a/(1-t))]}, l_{\phi(tz)}), \tag{15}$$

and using (10) for the triangle:

$$\tilde{c} \lhd_{c(a),\phi(z)} \leq T(\lhd_{z,a}) \leq \tilde{C}^2 \lhd_{c(a),\phi(z)}. \tag{16}$$

Plugging (15) and (16) into our inequality, we conclude that

$$\tilde{c}^2 \hat{\inf}(1^\infty_{[0,c(a/(1-t))]}, l_{\phi(tz)}) \leq \tilde{C}^2 \lhd_{c(a),\phi(z)}, \tag{17}$$

which means that

$$(c(a) - c(a/(1-t)))\phi(tz) \le \tilde{C}^4 c(a)\phi(z). \tag{18}$$

We know that $g \not\lessdot_{z,a'}$ for $a' < a$. This means that $T(g) \not\lessdot \tilde{c}T(\lessdot_{z,a'})$. Using the right-hand side of (15) and the left hand side of (16) we conclude

$$\tilde{c}^3 c(a')\phi(z) \le (c(a) - c(a/(1-t)))\phi(tz). \tag{19}$$

This is true for every $a' < a$, so using continuity, we may right the above with a instead. Inequalities (18) and (19) can be rewritten together:

$$\tilde{c}^3 \frac{c(a)}{c(a) - c(a/(1-t))} \le \frac{\phi(tz)}{\phi(z)} \le \tilde{C}^4 \frac{c(a)}{c(a) - c(a/(1-t))}. \tag{20}$$

Choose $z = 1$ to get:

$$\tilde{c}^3 \frac{c(a)}{c(a) - c(a/(1-t))} \le \frac{\phi(t)}{\phi(1)} \le \tilde{C}^4 \frac{c(a)}{c(a) - c(a/(1-t))}. \tag{21}$$

Combining (20) and (21), we come to the inequality

$$c\phi(t)\phi(z) \le \phi(tz) \le C\phi(t)\phi(z), \tag{22}$$

for some constant C and $c = C^{-1}$. To solve this, substitue $t = \exp\alpha_1$ and $z = \exp\alpha_2$, and define $h(s) = \log(\phi(e^s))$. Then after applying log, (22) becomes:

$$-C' + h(\alpha_1) + h(\alpha_2) \le h(\alpha_1 + \alpha_2) \le C' + h(\alpha_1) + h(\alpha_2), \tag{23}$$

or equivalently:

$$|h(\alpha_1 + \alpha_2) - h(\alpha_1) - h(\alpha_2)| \le C'. \tag{24}$$

The Hyers–Ulam theorem (2.1), implies that there exists a linear $g(\alpha) = \gamma\alpha$, such that $|h(\alpha) - g(\alpha)| < C'$. This means that $|\log(\phi(e^s)) - \log(e^{\gamma s})|$ and it is easy to check that this implies the following on ϕ:

$$c''z^\gamma \le \phi(z) \le C''z^\gamma, \qquad c'' = \frac{1}{C''} \tag{25}$$

Notice that $\gamma < 0$, since we know that ϕ is decreasing. Let use now proceed to estimate $c(a)$. Notice that all the arguments applied so far can be reused for T^{-1}, hence there exists $\gamma' < 0$ and $c'z^{\gamma'} \le \psi(z) \le C'z^{\gamma'}$ such that $T^{-1}1_{[0,z]}^\infty = l_{\psi(z)}$. We know that $Tl_a = 1_{[0,c(a)]}^\infty$, or $T^{-1}1_{[0,c(a)]}^\infty = l_a$, which is equivalent. But we have just shown that $T^{-1}1_{[0,c(a)]}^\infty = l_{\psi(c(a))}$. So,

$$c'(c(a))^{\gamma'} \le a = \psi(c(a)) \le C'(c(a))^{\gamma'}. \tag{26}$$

Rewriting (26) gives

$$c'a^{1/\gamma'} \le c(a) \le C'a^{1/\gamma'}. \tag{27}$$

Using (18), (19) and the estimate we have for $\phi(z)$, we write:

$$c''^2\tilde{c}^3 t^{-\gamma} \le \tilde{c}^3 \frac{\phi(z)}{\phi(tz)} \le \frac{c(a) - c(a/(1-t))}{c(a)} \le \tilde{C}^4 \frac{\phi(z)}{\phi(tz)} \le C''^2\tilde{C}^4 t^{-\gamma}, \tag{28}$$

which is equivalent to

$$1 - C'''t^{-\gamma} \le \frac{c(a/(1-t))}{c(a)} \le 1 - c'''t^{-\gamma}. \tag{29}$$

Choose $a = 1$ to get the following:

$$1 - C'''t^{-\gamma} \le \frac{c(1/(1-t))}{c(1)} \le 1 - c'''t^{-\gamma}. \tag{30}$$

Using the bounds we got in (27), we see that $\gamma = \gamma' = -1$. To summarize, there exist positive constants α_1, α_2 and β_1, β_2, such that

$$\frac{\alpha_1}{z} \le \phi(z) \le \frac{\alpha_2}{z}, \qquad \frac{\beta_1}{a} \le c(a) \le \frac{\beta_2}{a} \tag{31}$$

To conclude the proof, notice that we have:

$$T(\lhd_{z,a}) \le \tilde{C}^2 \lhd_{\beta_1/a,\alpha_2/z} = \tilde{C}^2 \mathcal{J}(\lhd_{z/\alpha_2,a/\beta_1}). \tag{32}$$

Also, notice that $f(x) = (\hat{\inf}_y \lhd_{y,f(y)})(x)$, so

$$(Tf)(x) \le \tilde{C}\hat{\inf}T(\lhd_{y,f(y)}) \le \tilde{C}^3\hat{\inf}(\lhd_{\beta_1/f(y),\alpha_2/y}) = \alpha_2\tilde{C}^3\hat{\inf}\mathcal{J}(\lhd_{y,f(y)/\beta_1})$$

$$= \alpha_2\tilde{C}^3 \mathcal{J}(\hat{\inf} \lhd_{y,f(y)/\beta_1}) = \tilde{C}^3\alpha_2\mathcal{J}(\frac{1}{\beta_1}f(x)) = C'\mathcal{J}(f(x/\beta_1)).$$

The same applies for the lower bound.

3.8.2 The Case of "I"

In the case T maps indicators to themselves, we do not know that it preserves property P. Assume now that $T1^\infty_{[0,z]} = 1^\infty_{[0,\hat{\phi}(z)]}$. We also know, due to Lemma 3.8, that if $T(l_a) = f$, then $f'(0)x \le f(x) \le \tilde{C}^3 f'(0)x$. If we define $c(a) = f'(0)$, then, $l_{c(a)} \le T(l_a) \le \tilde{C}^3 l_{c(a)}$. Now, we can estimate how triangles are mapped:

$$T(\lhd_{z,a}) \le \tilde{C}^2 \max(1^\infty_{[0,\phi(z)]}, T(l_a)) \le \tilde{C}^5 \lhd_{\phi(z),c(a)}, \tag{33}$$

and

$$T(\lhd_{z,a}) \ge \tilde{c} \max(1^\infty_{[0,\phi(z)]}, T(l_a)) \ge \tilde{c} \lhd_{\phi(z),c(a)}. \tag{34}$$

Now, let us rewrite the bounds for g, from the previous case:

$$T(g) \le \tilde{C}\hat{\inf}(1^\infty_{[0,\phi(z)]}, T(l_{a/(1-t)})) \le \tilde{C}^4\hat{\inf}(1^\infty_{[0,\phi(z)]}, l_{c(a/(1-t))}), \tag{35}$$

and

$$T(g) \ge \tilde{c}^2\hat{\inf}(1^\infty_{[0,\phi(z)]}, l_{c(a/(1-t))}). \tag{36}$$

Using the fact that $T(g) \le \tilde{C}T(\lhd_{z,a})$, we get a similar inequality:

$$(\phi(z) - \phi(tz))c(a/(1-t)) \le \tilde{C}^7\phi(z)c(a), \tag{37}$$

and using the same methods as before (this time taking $z' > z$, since we only know that ϕ is continuous), we get the lower bound:

$$(\phi(z) - \phi(tz))c(a/(1-t)) \ge \tilde{c}^6\phi(z)c(a). \tag{38}$$

We can see that we have the same inequalities we had in the previous case, but with ϕ and c interchanged, and we come to the inequality

$$\frac{1}{C_1}c(1/(1-t)) \le \frac{c(a/(1-t))}{c(a)} \le C_1c(1/(1-t)). \tag{39}$$

After substituting $s = 1/(1-t)$, we come to an inequality we already know how to solve:

$$c_1c(s)c(a) \le c(as) \le C_1c(s)c(a). \tag{40}$$

Using Hyers–Ulam thorem again (2.1), we conclude again that $ct^\gamma \le c(t) \le Ct^\gamma$, for some γ. To find bounds for ϕ, substitute this in the original inequality, and conclude that $\alpha_1 z \le \phi(z) \le \alpha_2 z$. After we have this estimate it is easy to conclude that $\gamma = 1$.

To conclude the proof, notice that we have:

$$T(\lhd_{z,a}) \le \tilde{C}^5 \lhd_{\alpha_1 z, \beta_2 a} \tag{41}$$

Also, notice that $f(x) = (\hat{\inf}_y \lhd_{y,f(y)})(x)$, so

$$(Tf)(x) \le \tilde{C}\hat{\inf}T(\lhd_{y,f(y)}) \le \tilde{C}^6\hat{\inf} \lhd_{\alpha_1 y, \beta_2 f(y)} = \tilde{C}^6\beta_2 f(x/\alpha_1).$$

The same applies for the lower bound.

3.9 Completing the Proof in $\mathbb{R}^n$

Recall, that we know there exists a function $\Phi : S^{n-1} \to S^{n-1}$ (1-1 and onto), such that any function supported on $\mathbb{R}^+ y$ is mapped to a function supported on $\mathbb{R}^+ \Phi(y)$. Let us define for each $y \in S^{n-1}$ a number $j(y)$ that equals 0 if T, restricted to $\mathbb{R}^+ y$ behaves like the identity (indicators are mapped to indicators) and 1 if T, restricted to $\mathbb{R}^+ y$ behaves like $\mathcal{J}$ (indicators are mapped to linear functions and vice versa).

Lemma 3.18. *Denote* $S_0 = \{y \in S^{n-1} : j(y) = 0\}$, $S_1 = \{y \in S^{n-1} : j(y) = 1\}$. *Then, either* $S_0 = S^{n-1}$, *or* $S_1 = S^{n-1}$.

Proof. Let us see that $S_i, \Phi(S_i)$ are convex. To see this, consider the function $f = 1_B^\infty$ where B is the n-dimensional unit ball. Let $x \in S_1$. By Lemma 3.4 we know that the support of Tf on $\mathbb{R}^+ \Phi(x)$ is the same as the support of $T(\max(f, 1_{\mathbb{R}^+ x}^\infty))$, and the latter is $\mathbb{R}^+ \Phi(x)$. Since Tf is a convex function with a convex support, we get that for every $x, y \in S_1$, the support of Tf must contain every ray $\mathbb{R}^+ \Phi(z)$ such that $\Phi(z)$ is contained in $\Phi(x) \vee \Phi(y)$. Hence, $\Phi(S_1)$ is convex. Choosing $g(x) = |x|$, by the same argument, we get that $\Phi(S_0)$ is also convex. Since T and T^{-1} have the same properties we may conclude that S_i are also convex.

Notice that $S_0 \cup S_1 = S^{n-1}$, so either one of the sets is empty and we are done, or S_0 and S_1 are both half-spheres, and likewise $\Phi(S_i)$. In this case let us check how T acts on the function f. Denote by H_1 the half-space $\bigcup_{y \in \Phi(S_1)} \mathbb{R}^+ y$, and by H_0 the half-space $\bigcup_{y \in \Phi(S_0)} \mathbb{R}^+ y$. We know that for $y \in S_1, \mathbb{R}^+ \Phi(y) \subset supp(Tf)$. Hence, $supp(Tf)$ contains H_1. This means that the support of $Tf|_{H_0}$ cannot be bounded. But then, we could choose a convex, bounded set $K \subset H_0$ (containing zero in the interior of the boundary) and consider the pre-image of 1_K^∞. Since T^{-1} preserves order on indicators we know that the support M of $T^{-1}(1_K^\infty)$ will be contained in the unit ball B. Consider the function $h = 1_{M \vee (-M)}^\infty$. By the preceding argument we know that $supp(Th)$ contains H_1 and $supp(Th) \cap H_0$ is bounded, which is a contradiction to the fact that $supp(Th)$ is convex. $\square$

Now we proceed to analyze the behavior of T in each case.

The case of $j \equiv 0$. Define $\varphi : \mathbb{R}^n \to \mathbb{R}^n$ by $T1_{[0,x]}^\infty = 1_{[0,\varphi(x)]}^\infty$. We know that T preserves sup and $\hat{\text{inf}}$ on indicators with equality (compare to Lemma 3.1), thus for any convex body K with $0 \in K$, $T1_K^\infty = 1_{\varphi(K)}^\infty$, and $\varphi(K)$ is also convex. The point map φ therefore induces an order preserving isomorphism on $\mathcal{K}_0^n$, the class of convex bodies containing the origin, and by known results (see [2]), this implies φ is a linear.

Take two triangles $\lhd_1, \lhd_2$ with bases x, x' and heights a, a' accordingly. The largest triangle $\lhd_\lambda$ which is smaller than $\hat{\text{inf}}(\lhd_1, \lhd_2)$ with the base $\lambda x + (1 - \lambda)x'$ has the height $\lambda a + (1 - \lambda)a'$. Denote by $h(x, a)$ the height of the maximal triangle which bounds $T(\lhd_1)$ from below. We have shown before that $\tilde{C}^3 h(x, a)$ will be the height of the triangle that bounds $T(\lhd_1)$ from above. Since $T(\lhd_\lambda) \le \tilde{C}^2 \hat{\text{inf}}(T \lhd_1, T \lhd_2)$, we may write (using Lemma 3.1):

$$h(\lambda(x, a) + (1 - \lambda)(x', a')) \le \tilde{C}^7 (\lambda h(x, a) + (1 - \lambda)h(x', a')).$$

Notice that for a given x, h satisfies conditions of Lemma 2.2. This is verified by choosing $a' = 1$. Applying this lemma we know that for every x there exists a monotone function $\omega_x(a)$ such that $\tilde{c}\omega_x(a) \leq h(x, a) \leq \omega_x(a)$. We know that if we increase the height of the triangle $\lhd_\lambda$ by some $\epsilon > 0$ (denote this triangle by $\lhd_\epsilon$), then $T(\lhd_\epsilon) \not\leq \tilde{c}\hat{\inf}(T \lhd_1, T \lhd_2)$. Hence, by Lemma 3.1 combined with properties of T, we have

$$h(\lambda(x, a) + (1 - \lambda)(x', a') + (0, \epsilon)) \geq \tilde{c}^5(\lambda h(x, a) + (1 - \lambda)h(x', a')). \quad (42)$$

We would like now to say that the inequality holds when $\epsilon \to 0$, but we don't know that h is continuous. We do know however, that in the worst case, the right hand side of (42) is multiplied by $\tilde{C}$ after taking the limit (due to existence of ω_x which is monotone and continuous). Hence

$$h(\lambda(x, a) + (1 - \lambda)(x', a')) \geq \tilde{c}^6(\lambda h(x, a) + (1 - \lambda)h(x', a')).$$

Applying Lemma 2.3 on $h(x, a)$, we conclude that there exists a constant β such that

$$\tilde{c}\beta a \leq h(x, a) \leq \tilde{C}\beta a.$$

To sum it up, we know that for a triangle f, $\tilde{c}\beta f \circ A \leq T(f) \leq \tilde{C}\beta f \circ A$. Using Lemma 3.16 we conclude the same inequality for every $f \in Cvx_0(\mathbb{R}^n)$.

The case of $j \equiv 1$. In this case we know that lines are mapped to indicators and vice-versa. Notice, that we cannot compose T with $\mathcal{J}$ and apply the previous case, since $\mathcal{J} \circ T$ would not necessarily satisfy the conditions of almost order preserving transform. However, we do know that in this case $\mathcal{J} \circ T$ is order preserving on the extremal family of indicators and rays. Thus, as explained above (the case of $j \equiv 0$), $(\mathcal{J} \circ T)1_{[0,z]}^\infty = 1_{[0,Bz]}^\infty$ for some $B \in GL(n)$. Composing both sides with $\mathcal{J}$ (recall that $\mathcal{J}$ is an involution), we see that T sends the indicator $1_{[0,z]}^\infty$ to a ray in direction Bz. Due to the ray-wise-ness of the problem we may conclude that any function supported on a ray in direction z is mapped to a function supported by the ray $\mathbb{R}^+ Bz(\Phi(z) = Bz)$.

Take two indicators $I_1 = 1_{[0,z]}^\infty$ and $I_2 = 1_{[0,z']}^\infty$ and define the function $g = \hat{\inf}(I_1, I_2)$. The indicator $I_\lambda = 1_{[0,\lambda z + (1-\lambda)z']}^\infty$ is bigger then g, but for every $\epsilon > 0$ the indicator $I_\epsilon = 1_{[0,(1+\epsilon)(\lambda z + (1-\lambda)z')]}^\infty$ is not comparable to g. Since T is order preserving on indicators and rays, it preserves the inf, so $\hat{\inf}(TI_1, TI_2) = Tg \leq TI_\lambda$, but the same is not true for any $T_I\epsilon$. Define $\psi(z)$ by the way T maps indicators to lines: $T1_{[0,z]}^\infty = l_{Bz/|z|,\psi(z)}$. Hence the ray TI_λ is comparable to the sector Tg that is spanned by the rays TI_1, TI_2. Since the same is not true for T_ϵ, and ψ is monotone in every direction, we conclude that Tg is a linear combination of TI_1, TI_2. Using this fact we come to the following property of ψ:

$$\psi(\lambda z + (1 - \lambda)z') = \frac{\lambda|z|}{|\lambda z + (1 - \lambda)z'|}\psi(z) + \frac{(1 - \lambda)|z'|}{|\lambda z + (1 - \lambda)z'|}\psi(z'). \quad (43)$$

Define the function $h(z) := |z|\psi(z)$. It follows that $h(z)$ satisfies: $h(\lambda z + (1-\lambda)z') = \lambda h(z) + (1-\lambda)h(z')$, from which it follows that h is linear: $h(z) = <u_0, z> + \beta$ for some vector $u_0 \in \mathbb{R}^n$ and a constant β. Since $\psi(z)$ cannot be zero, $u_0 = 0$, it means that $\psi(z) = \beta/|z|$.

Now, define $\theta(z, a)$ by $Tl_{z,a} = 1^\infty_{[0, Bz\theta(z,a)/|z|]}$. Finding $\theta(z, a)$ can be accomplished directly as with ψ, but it is simpler to notice that T and T^{-1} have the same properties. On one hand we know that $T^{-1}1^\infty_{[0, Bz/|z|\theta(z,a)]} = l_{z,a}$. On the other hand, applying the same arguments used for ψ, we have $T^{-1}1^\infty_{[0, Bz/|z|\theta(z,a)]} = l_{z/|z|,\gamma/\theta(z,a)}$, for some $\gamma > 0$. Thus,

$$a = \frac{\gamma}{\theta(z, a)},$$

or equivalently, $\theta(z, a) = \gamma/a$. Using Lemma 3.17 we conclude the theorem.

To show the dual statement (1.8), apply $\mathcal{A}$ to T, and use the homogeneity of $\mathcal{A}$ to conclude that T is almost order preserving. Now we know that $\mathcal{A}T$ is either almost-$\mathcal{J}$ or almost identity. Applying $\mathcal{A}$ again, and using the fact that it is an involution and that $\mathcal{A}\mathcal{J} = \mathcal{L}$ we finish the proof.

Remark 3.19. In case $n \geq 3$, we could use a shorter proof to see that Φ is linear. Notice that Φ sends cones to cones, and preserves intersections and convex-hulls of unions of cones. This is shown easily by using properties of sup and înf from Lemma 3.1: Define functions which are zero on the cone and ∞ everywhere else. The intersection is given by sup of the functions, and the convex hull is given by înf. Observe that the functions have values of 0 and ∞ only, the inequalities in Lemma 3.1 become equalities, and the property holds. Using Schneider's theorem [8], we conclude that Φ is linear. This means that $\Phi(x) = Bx$ for some $B \in GL(n)$.

4   Stability on the Class of Non-negative Convex Functions

We now proceed to the proof of theorem (1.9). Again, like in the previous case, we will need a family of extremal functions and some properties of their behaviour under our transform. The extremal family of function we will use in the case are what we call here "delta" functions $D_\theta + c$, mentioned before.

4.1   *Preservation of* sûp *and* înf

Clearly, properties (10) and (11) hold in this case too, and the proof of Lemma 3.1 can be applied verbatim.

4.2   Behaviour of "Delta" Functions

4.2.1   Delta Functions are Mapped to Delta Functions

We will show that T maps the class of "delta" functions $\{D_\theta + c\}$ to itself and does so bijectively . Assume $T(D_\theta + c) = f$. We want to show that the support of f has exactly one point. Assume there exist two functions g and h such that $g \geq f$ and $h \geq f$. Due to surjectivity we may write: $g = \tilde{c}T\varphi$ and $h = \tilde{c}T\psi$. Hence,

$$T(D_\theta + c) \leq \tilde{c}T\varphi \tag{44a}$$

$$T(D_\theta + c) \leq \tilde{c}T\psi. \tag{44b}$$

Condition (3b) now implies that $D_\theta + c \leq \varphi$ and $D_\theta + c \leq \psi$. This means that both φ and ψ, are of the form $D_\theta + \alpha_i$. Thus, they are comparable, and without loss of generality we may assume that $\varphi > \psi$. Applying condition (1a), we get that $h \leq \tilde{C}g$. But, if the support of f has two or more points, we can easily find two functions greater than f, but not comparable up to $\tilde{C}$. So we conclude that f is supported at one point only, and has the form $D_\theta + c'$.

4.2.2   Only Delta Functions are Mapped to Delta Functions

Now assume that $Tf = D_\theta + c$ and that the support of f has at least two points x_0 and x_1, with values c_0 and c_1. Then $D_{x_0} + c_0 \geq f$ and $D_{x_1} + c_1 \geq f$. Applying condition (1a), we get $\tilde{C}T(D_{x_i} + c_i) \geq D_\theta + c$. According to the previous lemma $T(D_{x_i} + c_i) = D_{y_i} + a_i$, but they must be comparable (since they are greater than $D_\theta + c$), so $y_1 = y_2 = \theta$. But this also implies that the sources are comparable, up to a constant $\tilde{C}$, hence $x_1 = x_2$.

4.2.3   Delta Functions are Mapped in Fibres

Since $D_\theta + c > D_\theta$, we get that $T(D_\theta) \leq \tilde{C}T(D_\theta + c) = D_{\theta'} + c'$. We know that $T(D_\theta) = D_\varphi + \alpha$. So $D_\varphi + \alpha \leq D_{\theta'} + c'$, which means that $\varphi = \alpha$ and $T(D_\theta + c) = T(D_\theta) + c''$. We see that all the delta functions on the fibre $x = \theta$ are mapped to delta functions on the fiber $T(D_\theta)$.

4.3   The Mapping Rule for $D_\theta + c$

Assume that the delta functions are mapped by the rule $T(D_\theta + c) = D_{\phi(\theta)} + \psi(\theta, c)$. Notice that due to the property of mapping in fibres, ϕ does not depend on c. Now we analyze the behavior of $\phi(\theta)$. Take $D_{\theta_0} + c_0$ and $D_{\theta_1} + c_1$, and define

$g = \hat{\inf}(D_{\theta_0} + c_0, D_{\theta_1} + c_1)$. Consider $\theta = \lambda\theta_0 + (1-\lambda)\theta_1$, and $c = \lambda c_0 + (1-\lambda)c_1$. Obviously $g \leq D_\theta + c$, and this is not true for any $D_\theta + c'$ where $c' < c$. Apply Property (1a) and use (11) to write:

$$\tilde{c}^2\hat{\inf}(D_{\phi(\theta_0)} + \psi(\theta_0, c_0), D_{\phi(\theta_1)} + \psi(\theta_1, c_1)) \leq T(g) \leq \tilde{C}T(D_\theta + c)$$

$$= D_{\phi(\theta)} + \tilde{C}\psi(\theta, c).$$

This inequality implies that $\phi(\theta)$ is on the interval $[\phi(\theta_0), \phi(\theta_1)]$, since otherwise $T(g)$ and $\tilde{C}T(D_\theta + c)$ would not be comparable. So $\phi : \mathbb{R}^n \to \mathbb{R}^n$ sends intervals to intervals, hence it must be affine (see [2]). Thus, there exists $A \in GL(n)$ and $b \in R^n$ such that $\phi(x) = Ax + b$.

We know that according to properties (1a) and (1b), for a given θ, ψ satisfies (20) and (21). Applying Lemma 2.2 we find a monotone function ω_θ that satisfies the following:

$$\tilde{c}\omega_\theta(t) \leq \psi(\theta, t) \leq \omega_\theta(t).$$

If $c' < c$, then $g \not\leq D_\theta + c$, and $Tg \not\leq \tilde{c}(D_{\phi(\theta)} + \psi(c'))$. Hence by Property (11) $\tilde{C}\hat{\inf}(D_{\phi(\theta_0)} + \psi(\theta_0, c_0), D_{\phi(\theta_1)} + \psi(\theta_1, 2c_1)) \not\leq \tilde{c}^2(D_{\phi(\theta)} + \omega_\theta(c'))$. Since ω_θ is monotone and the last statement applies for all $c' < c$, we can conclude that

$$\psi(\theta, c) \leq \omega_\theta(c) \leq \tilde{C}^3(\lambda(\psi(\theta_0, c_0) + (1-\lambda)\psi(\theta_1, c_1)).$$

But, on the other hand, we have

$$\psi(\theta, c) \geq \tilde{c}^3(\lambda(\psi(\theta_0, c_0) + (1-\lambda)\psi(\theta_1, c_1)).$$

So, we know that for every (x, c) and (y, d) in $\mathbb{R}^n \times \mathbb{R}^+$, ψ satisfies the following:

$$c(\lambda\psi(x, c) + (1-\lambda)\psi(y, d)) \leq \psi(\lambda(x, c) + (1-\lambda)(y, d))$$

$$\leq C(\lambda\psi(x, c) + (1-\lambda)\psi(y, d)), \quad (45)$$

and $\psi(x, 0) = 0$.

4.4 Proving Stability

According to Lemma 2.3, we know that there exists a constant such that $\tilde{c}\beta d \leq \psi(\theta, d) \leq \tilde{C}\beta d$. Recall also that there exists $A \in GL(n)$ and a vector b, such that $T(D_\theta) = D_{A\theta+b}$. This means that $D_{A\theta+b} + \tilde{c}\beta d \leq T(D_\theta + d) \leq D_{A\theta+b} + \tilde{C}\beta d$.

We know that any function $f(x) \in Cvx^+(\mathbb{R}^n)$ can be described by "Delta" functions: $f(x) = (\hat{\inf}(D_y + f(y)))(x)$. So,

$$(Tf)(x) = T(\hat{\inf_y}(D_y + f(y)))(x) \leq \tilde{C}(\hat{\inf_y}T(D_y + f(y)))(x)$$

$$\leq \tilde{C}(\hat{\inf_y}D_{Ay+b} + C\beta f(y))(x) = \tilde{C}(\beta f(A^{-1}(x-b))$$

The lower bound is obtained in the same way, so we come to:

$$\tilde{c}\beta f(A^{-1}(x-b)) \leq (Tf)(x) \leq \tilde{C}\beta f(A^{-1}(x-b)),$$

as required.

Acknowledgements The authors would like to express their sincere appreciation to Prof. Vitali Milman and Prof. Shiri Artstein-Avidan for their support, advice and discussions. Dan Florentin was Partially supported by the Israel Science Foundation Grant 865/07. Alexander Segal was Partially supported by the Israel Science Foundation Grant 387/09.

References

1. S. Artstein-Avidan, V. Milman, A new duality transform, C. R. Acad. Sci. Paris **346**, 1143–1148 (2008)
2. S. Artstein-Avidan, V. Milman, The concept of duality for measure projections of convex bodies. J. Funct. Anal. **254**, 2648–2666 (2008)
3. S. Artstein-Avidan, V. Milman, The concept of duality in asymptotic geometric analysis, and the characterization of the Legendre transform. Ann. Math. **169**(2), 661–674 (2009)
4. S. Artstein-Avidan, V. Milman, Hidden structures in the class of convex functions and a new duality transform. J. Eur. Math. Soc. **13**, 975–1004
5. S. Artstein-Avidan, V. Milman, Stability results for some classical convexity operations. To appear in Advances in Geometry
6. D.H. Hyers, On the stability of the linear functional equation. Proc. Nat. Acad. Sci. U.S.A. **27**, 222–224 (1941)
7. S.S. Kutateladze, A.M. Rubinov, *The Minkowski Duality and Its Applications* (Russian) (Nauka, Novosibirsk, 1976)
8. R. Schneider, The endomorphisms of the lattice of closed convex cones. Beitr. Algebra Geom. **49**, 541–547 (2008)
9. S.M. Ulam, *A Collection of Mathematical Problems* (Interscience Publ., New York, 1960)

On the Distribution of the ψ_2-Norm of Linear Functionals on Isotropic Convex Bodies

Apostolos Giannopoulos, Grigoris Paouris, and Petros Valettas

Abstract It is known that every isotropic convex body K in $\mathbb{R}^n$ has a "subgaussian" direction with constant $r = O(\sqrt{\log n})$. This follows from the upper bound $|\Psi_2(K)|^{1/n} \leqslant \frac{c\sqrt{\log n}}{\sqrt{n}} L_K$ for the volume of the body $\Psi_2(K)$ with support function $h_{\Psi_2(K)}(\theta) := \sup_{2 \leqslant q \leqslant n} \frac{\|\langle \cdot, \theta \rangle\|_q}{\sqrt{q}}$. The approach in all the related works does not provide estimates on the measure of directions satisfying a ψ_2-estimate with a given constant r. We introduce the function $\psi_K(t) := \sigma(\{\theta \in S^{n-1} : h_{\Psi_2(K)}(\theta) \leqslant ct\sqrt{\log n} L_K\})$ and we discuss lower bounds for $\psi_K(t)$, $t \geqslant 1$. Information on the distribution of the ψ_2-norm of linear functionals is closely related to the problem of bounding from above the mean width of isotropic convex bodies.

1 Introduction

A convex body K in $\mathbb{R}^n$ is called isotropic if it has volume 1, it is centered (i.e. it has its center of mass at the origin), and there exists a constant $L_K > 0$ such that

$$\int_K \langle x, \theta \rangle^2 dx = L_K^2 \tag{1}$$

for every $\theta \in S^{n-1}$. It is known (see [19]) that for every convex body K in $\mathbb{R}^n$ there exists an invertible affine transformation T such that $T(K)$ is isotropic.

A. Giannopoulos (✉) · P. Valettas
Department of Mathematics, University of Athens, Panepistimioupolis 157 84, Athens, Greece
e-mail: apgiannop@math.uoa.gr; petvalet@math.uoa.gr

G. Paouris
Department of Mathematics, Texas A & M University, College Station, TX 77843, USA
e-mail: grigoris_paouris@yahoo.co.uk

B. Klartag et al. (eds.), *Geometric Aspects of Functional Analysis*, Lecture Notes in Mathematics 2050, DOI 10.1007/978-3-642-29849-3_13,
© Springer-Verlag Berlin Heidelberg 2012

Moreover, this isotropic position of K is uniquely determined up to orthogonal transformations; therefore, if we define $L_K = L_{\tilde{K}}$ where $\tilde{K}$ is an isotropic affine image of K, then L_K is well defined for the affine class of K.

A central question in asymptotic convex geometry asks if there exists an absolute constant $C > 0$ such that $L_K \leqslant C$ for every convex body K. Bourgain [4] proved that $L_K \leqslant c\sqrt[4]{n}\log n$ for every symmetric convex body K in $\mathbb{R}^n$. The best known general estimate is currently $L_K \leqslant c\sqrt[4]{n}$; this was proved by Klartag in [11]—see also [13].

Let K be a centered convex body of volume 1 in $\mathbb{R}^n$. We say that $\theta \in S^{n-1}$ is a subgaussian direction for K with constant $r > 0$ if $\|\langle \cdot, \theta \rangle\|_{\psi_2} \leqslant r\|\langle \cdot, \theta \rangle\|_2$, where

$$\|f\|_{\psi_\alpha} = \inf\left\{t > 0 : \int_K \exp\left((|f(x)|/t)^\alpha\right) dx \leqslant 2\right\}, \qquad \alpha \in [1,2]. \quad (2)$$

V. Milman asked if every centered convex body K has at least one "subgaussian" direction (with constant $r = O(1)$). By the formulation of the problem, it is clear that one can work within the class of isotropic convex bodies. Affirmative answers have been given in some special cases. Bobkov and Nazarov (see [2,3]) proved that if K is an isotropic 1-unconditional convex body, then $\|\langle \cdot, \theta \rangle\|_{\psi_2} \leqslant c\sqrt{n}\|\theta\|_\infty$ for every $\theta \in S^{n-1}$; a direct consequence is that the diagonal direction is a subgaussian direction with constant $O(1)$. In [23] it is proved that every zonoid has a subgaussian direction with a uniformly bounded constant. Another partial result was obtained in [24]: if K is isotropic and $K \subseteq (\gamma\sqrt{n}L_K)B_2^n$ for some $\gamma > 0$, then

$$\sigma\left(\theta \in S^{n-1} : \|\langle \cdot, \theta \rangle\|_{\psi_2} \geqslant c_1\gamma t L_K\right) \leqslant \exp(-c_2\sqrt{n}t^2/\gamma) \quad (3)$$

for every $t \geqslant 1$, where σ is the rotationally invariant probability measure on S^{n-1} and $c_1, c_2 > 0$ are absolute constants.

The first general answer to the question was given by Klartag who proved in [12] that every isotropic convex body K in $\mathbb{R}^n$ has a "subgaussian" direction with a constant which is logarithmic in the dimension. An alternative proof with a slightly better estimate was given in [6]. The best known estimate, which appears in [7], follows from an upper bound for the volume of the body $\Psi_2(K)$ with support function

$$h_{\Psi_2(K)}(\theta) := \sup_{2 \leqslant q \leqslant n} \frac{\|\langle \cdot, \theta \rangle\|_q}{\sqrt{q}}. \quad (4)$$

It is known that $\|\langle \cdot, \theta \rangle\|_{\psi_2} \simeq \sup_{2 \leqslant q \leqslant n} \frac{\|\langle \cdot, \theta \rangle\|_q}{\sqrt{q}}$, and hence, $h_{\Psi_2(K)}(\theta) \simeq \|\langle \cdot, \theta \rangle\|_{\psi_2}$. The main result in [7] states that

$$\frac{c_1}{\sqrt{n}}L_K \leqslant |\Psi_2(K)|^{1/n} \leqslant \frac{c_2\sqrt{\log n}}{\sqrt{n}}L_K, \quad (5)$$

where $c_1, c_2 > 0$ are absolute constants. A direct consequence of the right hand side inequality in (5) is the existence of subgaussian directions for K with constant $r = O(\sqrt{\log n})$. With a small amount of extra work, one can also show that if K is a centered convex body of volume 1 in $\mathbb{R}^n$, then there exists $\theta \in S^{n-1}$ such that

$$|\{x \in K : |\langle x, \theta \rangle| \geq ct \| \langle \cdot, \theta \rangle \|_2 \}| \leq e^{-\frac{t^2}{\log(t+1)}} \tag{6}$$

for all $t \geq 1$, where $c > 0$ is an absolute constant.

The approach in [6,7,12] does not provide estimates on the measure of directions for which an isotropic convex body satisfies a ψ_2-estimate with a given constant r. Klartag obtains some information on this question, but for a different position of K. More precisely, in [12] he proves that if K is a centered convex body of volume 1 in $\mathbb{R}^n$ then, there exists $T \in SL(n)$ such that the body $K_1 = T(K)$ has the following property: there exists $A \subseteq S^{n-1}$ with measure $\sigma(A) \geq \frac{4}{5}$ such that, for every $\theta \in A$ and every $t \geq 1$,

$$|\{x \in K_1 : |\langle x, \theta \rangle| \geq ct \| \langle \cdot, \theta \rangle \|_2 \}| \leq e^{-\frac{ct^2}{\log^2 n \log^5 (t+1)}} \tag{7}$$

In this result, K_1 is the ℓ-position of K (this is the position of the body which essentially minimizes its mean width; see [27]). The first aim of this note is to pose the problem of the distribution of the ψ_2-norm of linear functionals on isotropic convex bodies and to provide some first measure estimates. To this end, we introduce the function

$$\psi_K(t) := \sigma \left(\{ \theta \in S^{n-1} : h_{\psi_2(K)}(\theta) \leq ct \sqrt{\log n} L_K \} \right). \tag{8}$$

The problem is to give lower bounds for $\psi_K(t), t \geq 1$. We present a general estimate in Sect. 4:

Theorem 1.1. *Let K be an isotropic convex body in $\mathbb{R}^n$. For every $t \geq 1$ we have*

$$\psi_K(t) \geq \exp(-cn/t^2), \tag{9}$$

where $c > 0$ is an absolute constant.

For the proof of Theorem 1.1 we first obtain, for every $1 \leq k \leq n$, some information on the ψ_2-behavior of directions in an arbitrary k-dimensional subspace of $\mathbb{R}^n$:

Theorem 1.2. *Let K be an isotropic convex body in $\mathbb{R}^n$.*

(i) *For every $\log^2 n \leq k \leq n/\log n$ and every $F \in G_{n,k}$ there exists $\theta \in S_F$ such that*

$$\| \langle \cdot, \theta \rangle \|_{\psi_2} \leq C \sqrt{n/k} \, L_K, \tag{10}$$

(ii) *For every* $1 \leqslant k \leqslant \log^2 n$ *and every* $F \in G_{n,k}$ *there exists* $\theta \in S_F$ *such that*

$$\|\langle \cdot, \theta \rangle\|_{\psi_2} \leqslant C \sqrt{n/k} \sqrt{\log 2k} \, L_K, \tag{11}$$

(iii) *For every* $n/\log n \leqslant k \leqslant n$ *and every* $F \in G_{n,k}$ *there exists* $\theta \in S_F$ *such that*

$$\|\langle \cdot, \theta \rangle\|_{\psi_2} \leqslant C \sqrt{\log n} \, L_K, \tag{12}$$

where $C > 0$ *is an absolute constant.*

It is known (for example, see [10]) that every isotropic convex body K is contained in $[(n + 1)L_K]B_2^n$. This implies that the ψ_2-norm is Lipschitz with constant $O(\sqrt{n}L_K)$. Then, Theorem 1.2 is combined with a simple argument which is based on the fact that the ψ_2-norm is stable on a spherical cap of the appropriate radius.

Note that $\psi_K(t) = 1$ if $t \geqslant c\sqrt{n/\log n}$. Therefore, the bound of Theorem 1.1 is of some interest only when $1 \leqslant t \leqslant c\sqrt{n/\log n}$. Actually, if $t \geq c\sqrt[4]{n}$ then we have much better information. In Sect. 5 we give some estimates on the mean width of the L_q–centroid bodies of K and of $\Psi_2(K)$; as a consequence, we get:

Proposition 1.3. *Let* K *be an isotropic convex body in* $\mathbb{R}^n$. *For every* $t \geqslant c_1\sqrt[4]{n}/\sqrt{\log n}$ *one has*

$$\psi_K(t) \geqslant 1 - e^{-c_2 t^2 \log n}, \tag{13}$$

where $c_1, c_2 > 0$ *are absolute constants.*

Deeper understanding of the function $\psi_K(t)$ would have important applications. The strength of the available information can be measured on the problem of bounding from above the mean width of isotropic convex bodies. From the inclusion $K \subseteq [(n + 1)L_K]B_2^n$, one has the obvious bound $w(K) \leqslant cnL_K$. However, a better estimate is always possible: for every isotropic convex body K in $\mathbb{R}^n$ one has

$$w(K) \leqslant cn^{3/4}L_K, \tag{14}$$

where $c > 0$ is an absolute constant. There are several approaches that lead to the estimate (14). The first one appeared in the PhD Thesis of Hartzoulaki [9] and was based on a result from [5] regarding the mean width of a convex body under assumptions on the regularity of its covering numbers. The second one is more recent and is due to Pivovarov [28]; it relates the question to the geometry of random polytopes with vertices independently and uniformly distributed in K and makes use of the concentration inequality of [25]. A third—very direct—proof of this bound can be based on the "theory of L_q-centroid bodies" which was developed by the second named author (see Sect. 5). In Sect. 6 we propose one more approach, which can exploit our knowledge on $\psi_K(t)$.

2 Background Material

2.1 Notation

We work in $\mathbb{R}^n$, which is equipped with a Euclidean structure $\langle \cdot, \cdot \rangle$. We denote by $\| \cdot \|_2$ the corresponding Euclidean norm, and write B_2^n for the Euclidean unit ball, and S^{n-1} for the unit sphere. Volume is denoted by $| \cdot |$. We write ω_n for the volume of B_2^n and σ for the rotationally invariant probability measure on S^{n-1}. The Grassmann manifold $G_{n,k}$ of k-dimensional subspaces of $\mathbb{R}^n$ is equipped with the Haar probability measure $\mu_{n,k}$. Let $k \leqslant n$ and $F \in G_{n,k}$. We will denote by P_F the orthogonal projection from $\mathbb{R}^n$ onto F. We also define $B_F := B_2^n \cap F$ and $S_F := S^{n-1} \cap F$.

The letters c, c', c_1, c_2 etc. denote absolute positive constants which may change from line to line. Whenever we write $a \simeq b$, we mean that there exist absolute constants $c_1, c_2 > 0$ such that $c_1 a \leqslant b \leqslant c_2 a$. Also if $K, L \subseteq \mathbb{R}^n$ we will write $K \simeq L$ if there exist absolute constants $c_1, c_2 > 0$ such that $c_1 K \subseteq L \subseteq c_2 K$.

2.2 Convex bodies

A convex body in $\mathbb{R}^n$ is a compact convex subset C of $\mathbb{R}^n$ with non-empty interior. We say that C is symmetric if $x \in C$ implies that $-x \in C$. We say that C is centered if it has center of mass at the origin, i.e. $\int_C \langle x, \theta \rangle \, dx = 0$ for every $\theta \in S^{n-1}$. The support function of a convex body C is defined by

$$h_C(y) = \max\{\langle x, y \rangle : x \in C\}, \tag{15}$$

and the mean width of C is

$$w(C) = \int_{S^{n-1}} h_C(\theta)\sigma(d\theta). \tag{16}$$

For each $-\infty < p < \infty$, $p \neq 0$, we define the p-mean width of C by

$$w_p(C) = \left(\int_{S^{n-1}} h_C^p(\theta)\sigma(d\theta) \right)^{1/p}. \tag{17}$$

The radius of C is the quantity $R(C) = \max\{\|x\|_2 : x \in C\}$ and, if the origin is an interior point of C, the polar body C° of C is

$$C^\circ := \{y \in \mathbb{R}^n : \langle x, y \rangle \leqslant 1 \text{ for all } x \in C\}. \tag{18}$$

A centered convex body C is called *almost isotropic* if C has volume one and $C \simeq T(C)$ where $T(C)$ is an isotropic linear transformation of C. Finally, we write $\overline{C}$ for the homothetic image of volume 1 of a convex body $C \subseteq \mathbb{R}^n$, i.e. $\overline{C} := \frac{C}{|C|^{1/n}}$.

2.3 L_q-Centroid Bodies

Let K be a convex body of volume 1 in $\mathbb{R}^n$. For every $q \geqslant 1$ and $y \in \mathbb{R}^n$ we define

$$h_{Z_q(K)}(y) := \left(\int_K |\langle x, y \rangle|^q dx \right)^{1/q}. \tag{19}$$

We define the L_q-centroid body $Z_q(K)$ of K to be the centrally symmetric convex set with support function $h_{Z_q(K)}$. Note that K is isotropic if and only if $Z_2(K) = L_K B_2^n$. It is clear that $Z_1(K) \subseteq Z_p(K) \subseteq Z_q(K) \subseteq Z_\infty(K)$ for every $1 \leqslant p \leqslant q \leqslant \infty$, where $Z_\infty(K) = \operatorname{conv}\{K, -K\}$. If $T \in SL(n)$ then $Z_p(T(K)) = T(Z_p(K))$. Moreover, as a consequence of Borell's lemma (see [20, Appendix III]), one can check that

$$Z_q(K) \subseteq cq \, Z_2(K) \tag{20}$$

for every $q \geqslant 2$ and, more generally,

$$Z_q(K) \subseteq c\frac{q}{p} Z_p(K) \tag{21}$$

for all $1 \leqslant p < q$, where $c \geqslant 1$ is an absolute constant. Also, if K is centered, then

$$Z_q(K) \supseteq c_1 K \tag{22}$$

for all $q \geqslant n$, where $c_1 > 0$ is an absolute constant.

2.4 The Parameter $k_*(C)$

Let C be a symmetric convex body in $\mathbb{R}^n$. We write $\| \cdot \|_C$ for the norm induced on $\mathbb{R}^n$ by C. We also define $k_*(C)$ as the largest positive integer $k \leq n$ for which the measure of $F \in G_{n,k}$ for which $\frac{1}{2}w(C)B_F \subseteq P_F(C) \subseteq 2w(C)B_F$ is greater than $\frac{n}{n+k}$. The parameter $k_*(C)$ is determined, up to an absolute constant, by the mean width and the radius of C: There exist $c_1, c_2 > 0$ such that

$$c_1 n \frac{w(C)^2}{R(C)^2} \leq k_*(C) \leq c_2 n \frac{w(C)^2}{R(C)^2} \tag{23}$$

for every symmetric convex body C in $\mathbb{R}^n$. The lower bound follows from Milman's proof of Dvoretzky's theorem (see [18]) and the upper bound was proved in [21].

The q-mean width $w_q(C)$ is equivalent to $w(C)$ as long as $q \leq k_*(C)$. As Litvak et al. prove in [16], there exist $c_1, c_2, c_3 > 0$ such that for every symmetric convex body C in $\mathbb{R}^n$ we have:

1. If $1 \leq q \leq k_*(C)$ then $w(C) \leq w_q(C) \leq c_1 w(C)$.
2. If $k_*(C) \leq q \leq n$ then $c_2 \sqrt{q/n}\, R(C) \leq w_q(C) \leq c_3 \sqrt{q/n}\, R(C)$.

2.5 Moments of the Euclidean Norm

For every $q > -n$, $q \neq 0$, we define the quantities $I_q(K)$ by

$$I_q(K) := \left(\int_K \|x\|_2^q\, dx \right)^{1/q}. \tag{24}$$

In [26] and [25] it is proved that for every $1 \leq q \leq n/2$,

$$I_{-q}(K) \simeq \sqrt{n/q}\, w_{-q}(Z_q(K)) \tag{25}$$

and

$$I_q(K) \simeq \sqrt{n/q}\, w_q(Z_q(K)). \tag{26}$$

We define

$$q_*(K) := \max\{k \leq n : k_*(Z_k(K)) \geq k\}. \tag{27}$$

Then, the main result of [26] states that, for every centered convex body K of volume 1 in $\mathbb{R}^n$, one has

$$I_{-q}(K) \simeq I_q(K) \tag{28}$$

for every $1 \leq q \leq q_*(K)$. In particular, for all $q \leq q_*(K)$ one has $I_q(K) \leq CI_2(K)$, where $C > 0$ is an absolute constant.

If K is isotropic, one can check that $q_*(K) \geq c\sqrt{n}$, where $c > 0$ is an absolute constant (for a proof, see [25]). Therefore,

$$I_q(K) \leq C\sqrt{n}L_K \quad \text{for every } q \leq \sqrt{n}. \tag{29}$$

In particular, from (26) and (29) we see that, for all $q \leq \sqrt{n}$,

$$w(Z_q(K)) \simeq w_q(Z_q(K)) \simeq \sqrt{q}L_K. \tag{30}$$

2.6 The Parameter $d_*(C)$

Let C be a symmetric convex body in $\mathbb{R}^n$. For every $\delta \geq 1$ we define

$$d_*(C, \delta) = \max\{q \geq 1 : w(C) \leq \delta w_{-q}(C)\}. \tag{31}$$

It was proved in [14, 15] that

$$k_*(C) \leq cd_*(C, 2) \tag{32}$$

2.7 Keith Ball's Bodies

For every k-dimensional subspace F of $\mathbb{R}^n$ we denote by E the orthogonal subspace of F. For every $\phi \in F \setminus \{0\}$ we define $E^+(\phi) = \{x \in \mathrm{span}\{E, \phi\} : \langle x, \phi \rangle \geq 0\}$. K. Ball (see [1, 19]) proved that, if K is a centered convex body of volume 1 in $\mathbb{R}^n$ then, for every $q \geq 0$, the function

$$\phi \mapsto \|\phi\|_2^{1 + \frac{q}{q+1}} \left(\int_{K \cap E^+(\phi)} \langle x, \phi \rangle^q dx \right)^{-\frac{1}{q+1}} \tag{33}$$

is the gauge function of a convex body $B_q(K, F)$ on F. A basic identity from [25] states that for every $F \in G_{n,k}$ and every $q \geq 1$ we have that

$$P_F(Z_q(K)) = \left(\frac{k+q}{2} \right)^{1/q} |B_{k+q-1}(K, F)|^{1/k+1/q} Z_q(\overline{B}_{k+q-1}(K, F)). \tag{34}$$

It is a simple consequence of Fubini's theorem that if K is isotropic then $\overline{B}_{k+1}(K, F)$ is almost isotropic. Moreover, using (34) one can check that

$$c_1 \frac{k}{k+q} \frac{Z_q(\overline{B}_{k+1}(K, F))}{L_{\overline{B}_{k+1}(K,F)}} \subseteq \frac{P_F(Z_q(K))}{L_K} \subseteq c_2 \frac{k+q}{k} \frac{Z_q(\overline{B}_{k+1}(K, F))}{L_{\overline{B}_{k+1}(K,F)}} \tag{35}$$

for all $1 \leq k, q \leq n$. In particular, for all $q \leq k$ we have

$$\frac{Z_q(\overline{B}_{k+1}(K, F))}{L_{\overline{B}_{k+1}(K,F)}} \simeq \frac{P_F(Z_q(K))}{L_K}. \tag{36}$$

2.8 Covering Numbers

Recall that if A and B are convex bodies in $\mathbb{R}^n$, then the covering number $N(A, B)$ of A by B is the smallest number of translates of B whose union covers A. A simple and useful observation is that, if A and B are both symmetric and if $S_t(A, B)$ is the maximal number of points $z_i \in A$ which satisfy $\|z_i - z_j\|_B \geq t$ for all $i \neq j$, then

$$N(A, tB) \leq S_t(A, B) \leq N(A, (t/2)B). \tag{37}$$

3 Covering Numbers of Projections of L_q-Centroid Bodies

Let K be an isotropic convex body in $\mathbb{R}^n$. We first give an alternative proof of some estimates on the covering numbers $N(Z_q(K), t\sqrt{q}L_K B_2^n)$ that were recently obtained in [7]; they improve upon previous estimates from [6].

Proposition 3.1. *Let K be an isotropic convex body in $\mathbb{R}^n$, let $1 \leq q \leq n$ and $t \geq 1$. Then,*

$$\log N\left(Z_q(K), c_1 t\sqrt{q}L_K B_2^n\right) \leq c_2 \frac{n}{t^2} + c_3 \frac{\sqrt{qn}}{t}, \tag{38}$$

where $c_1, c_2, c_3 > 0$ are absolute constants.

Note that the upper bound in (38) is of the order n/t^2 if $t \leq \sqrt{n/q}$ and of the order $\sqrt{qn}/t$ if $t \geq \sqrt{n/q}$. Our starting point is a "small ball probability" type estimate which appears in [22, Fact 3.2(c)]:

Lemma 3.2. *Let $\theta \in S^{n-1}$, $1 \leq k \leq n-1$ and $r \geq \sqrt{e}$. Then,*

$$\mu_{n,k}\left(\left\{F \in G_{n,k} : \|P_F(\theta)\|_2 \leq \frac{1}{r}\sqrt{\frac{k}{n}}\right\}\right) \leq \left(\frac{\sqrt{e}}{r}\right)^k. \tag{39}$$

Under the restriction $\log N(C, tB_2^n) \leq k$, Lemma 3.2 allows us to compare the covering numbers $N(C, tB_2^n)$ of a convex body C with the covering numbers of its random k-dimensional projections.

Lemma 3.3. *Let C be a convex body in $\mathbb{R}^n$, let $r \geq \sqrt{e}$, $s > 0$ and $1 \leq k \leq n-1$. If $N_s := N(C, sB_2^n)$, then there exists $\mathcal{F} \subseteq G_{n,k}$ such that $\mu_{n,k}(\mathcal{F}) \geq 1 - N_s^2 e^{k/2} r^{-k}$ and*

$$N\left(P_F(C), \frac{s}{2r}\sqrt{\frac{k}{n}} B_F\right) \geq N_s \tag{40}$$

for all $F \in \mathcal{F}$.

Proof. Let $N_s = N(C, sB_2^n)$. From (37) we see that there exist $z_1, \ldots, z_{N_s} \in C$ such that $\|z_i - z_j\|_2 \geq s$ for all $1 \leq i, j \leq N_s, i \neq j$. Consider the set $\{w_m : 1 \leq m \leq \frac{N_s(N_s-1)}{2}\}$ of all differences $z_i - z_j$ ($i \neq j$). Note that $\|w_m\|_2 \geq s$ for all m. Lemma 3.2 shows that

$$\mu_{n,k}\left(\left\{F \in G_{n,k} : \|P_F(w_m)\|_2 \leq \frac{1}{r}\sqrt{\frac{k}{n}}\|w_m\|_2\right\}\right) \leq \left(\frac{\sqrt{e}}{r}\right)^k, \quad (41)$$

and hence,

$$\mu_{n,k}\left(\left\{F : \|P_F(w_m)\|_2 \geq \frac{1}{r}\sqrt{\frac{k}{n}}\|w_m\|_2 \text{ for all } m\right\}\right) \geq 1 - N_s^2 e^{k/2} r^{-k}. \quad (42)$$

Let $\mathcal{F}$ be the subset of $G_{n,k}$ described in (42). Then, for every $F \in \mathcal{F}$ and all $i \neq j$,

$$\|P_F(z_i) - P_F(z_j)\|_2 \geq \frac{1}{r}\sqrt{\frac{k}{n}}\|z_i - z_j\|_2 \geq \frac{s}{r}\sqrt{\frac{k}{n}}. \quad (43)$$

Since $P_F(z_i) \in P_F(C)$, the right hand side inequality of (37) implies that

$$N\left(P_F(C), \frac{s}{2r}\sqrt{\frac{k}{n}}B_F\right) \geq N_s, \quad (44)$$

as claimed. □

Finally, we will use the following regularity estimate for the covering numbers of L_q-centroid bodies (see [6, Proposition 3.1] for a proof of the first inequality and [9] for a proof of the second one): For all $t > 0$ and $1 \leq q \leq n$,

$$\log N\left(Z_q(K), ct\sqrt{q}L_K B_2^n\right) \leq \frac{\sqrt{qn}}{\sqrt{t}} + \frac{n}{t} \text{ and } \log N\left(K - K, t\sqrt{n}L_K B_2^n\right) \leq \frac{n}{t}, \quad (45)$$

where $c > 0$ is an absolute constant. Note that the upper bound in (45) is of the order n/t if $t \leq n/q$ and of the order $\sqrt{qn}/\sqrt{t}$ if $t \geq n/q$.

Proof of Proposition 3.1. We set $s = ct\sqrt{q}L_K$ and $N_s := N(Z_q(K), sB_2^n)$. Because of (45) we may assume that $3 \leq N_s \leq e^{cn}$, and then, we choose $1 \leq k \leq n$ so that $\log N_s \leq k \leq 2\log N_s$. We distinguish two cases:

(a) Assume that $1 \leq t \leq \sqrt{n/q}$. Applying Lemma 3.3 with $r = e^3$ we have that, with probability greater than $1 - N_s^2 e^{-5k/2} \geq 1 - e^{-k/2}$, a random subspace $F \in G_{n,k}$ satisfies

$$\frac{k}{2} \leqslant \log N_s \leqslant \log N\left(P_F(Z_q(K)), c_1 s \sqrt{\frac{k}{n}} B_F\right), \tag{46}$$

where $c_1 > 0$ is an absolute constant.

If $\log N_s \leqslant q$ then we trivially get $\log N_s \leqslant n/t^2$ because $q \leqslant n/t^2$. So, we may assume that $\log N_s \geqslant q$; in particular, $q \leqslant k$. Then, using (35) we get

$$\frac{k}{2} \leqslant \log N\left(Z_q(\overline{B}_{k+1}(K, F)), c\frac{L_{\overline{B}_{k+1}(K,F)}}{L_K}\sqrt{\frac{k}{n}}s B_F\right). \tag{47}$$

Observe that $\frac{s\sqrt{k/n}}{\sqrt{q}L_K} = ct\sqrt{k/n} \leqslant ct \leqslant cn/q$. Therefore, applying the estimate (45) for the k-dimensional isotropic convex body $\overline{B}_{k+1}(K, F)$, we get

$$\frac{k}{2} \leqslant c_2\frac{k}{t\sqrt{k/n}} = c_2\frac{\sqrt{kn}}{t}, \tag{48}$$

which shows that

$$\log N(Z_q(K), t\sqrt{q}L_K B_2^n) = \log N_s \leqslant k \leqslant c_3\frac{n}{t^2}, \tag{49}$$

where $c_3 = 4c_2^2$.

(b) Assume that $t \geqslant \sqrt{n/q}$. We set $p := \frac{\sqrt{qn}}{t} \leqslant q$. Then, using (22), we have that

$$N\left(Z_q(K), t\sqrt{q}L_K B_2^n\right) \leqslant N\left(\frac{q}{p}Z_p(K), c_4 t\sqrt{q}L_K B_2^n\right)$$

$$\leqslant N\left(Z_p(K), c_4 t\sqrt{\frac{p}{q}}\sqrt{p}L_K B_2^n\right)$$

$$= N\left(Z_p(K), c_4\sqrt{\frac{n}{p}}\sqrt{p}L_K B_2^n\right).$$

Applying the result of case (a) for $Z_p(K)$ with $t = \sqrt{n/p}$, we see that

$$N\left(Z_q(K), t\sqrt{q}L_K B_2^n\right) \leqslant N\left(Z_p(K), c_4\sqrt{\frac{n}{p}}\sqrt{p}L_K B_2^n\right)$$

$$\leqslant e^{c_5 p} = \exp\left(c_5\frac{\sqrt{qn}}{t}\right),$$

and the proof is complete. $\square$

Using Proposition 3.1 we can obtain analogous upper bounds for the covering numbers of $P_F(Z_q(K))$, where $F \in G_{n,k}$.

Proposition 3.4. *Let K be an isotropic convex body in $\mathbb{R}^n$. For every $1 \leqslant q < k \leqslant n$, for every $F \in G_{n,k}$ and every $t \geqslant 1$, we have*

$$\log N\left(P_F(Z_q(K)), t\sqrt{q}L_K B_F\right) \leqslant \frac{c_1 k}{t^2} + \frac{c_2\sqrt{qk}}{t}, \tag{50}$$

where $c_1, c_2 > 0$ are absolute constants. Also, for every $k \leqslant q \leqslant n$, $F \in G_{n,k}$ and $t \geqslant 1$,

$$\log N\left(P_F(Z_q(K)), t\sqrt{q}L_K B_F\right) \leqslant \frac{c_3\sqrt{qk}}{t}, \tag{51}$$

where $c_3 > 0$ is an absolute constant.

Proof. (i) Let $1 \leqslant q \leqslant k$, $F \in G_{n,k}$ and $t \geqslant 1$. From (36) we see that

$$\log N\left(P_F(Z_q(K)), t\sqrt{q}L_K B_F\right)$$
$$\leqslant \log N\left(Z_q(\overline{B}_{k+1}(K, F)), ct\sqrt{q}L_{\overline{B}_{k+1}(K,F)}B_F\right), \tag{52}$$

where $c > 0$ is an absolute constant. Since $\overline{B}_{k+1}(K, F)$ is almost isotropic, we may apply Proposition 3.1 for $\overline{B}_{k+1}(K, F)$ in F: we have

$$\log N\left(Z_q(\overline{B}_{k+1}(K, F)), ct\sqrt{q}L_{\overline{B}_{k+1}(K,F)}B_F\right) \leqslant \frac{c_1 k}{t^2} + \frac{c_2\sqrt{qk}}{t}, \tag{53}$$

and hence,

$$\log N\left(P_F(Z_q(K)), t\sqrt{q}L_K B_F\right) \leqslant \frac{c_1 k}{t^2} + \frac{c_2\sqrt{qk}}{t}. \tag{54}$$

(ii) Assume that $k \leqslant q \leqslant n$ and $F \in G_{n,k}$. Then, using (35) and the fact that $Z_q(C) \subseteq \text{conv}\{C, -C\}$, for every $t \geqslant 1$ we write

$$\log N\left(P_F(Z_q(K)), t\sqrt{q}L_K B_F\right)$$
$$\leqslant \log N\left(\frac{cq}{k}D_{k+1}(K, F), t\sqrt{q}L_{\overline{B}_{k+1}(K,F)}B_F\right)$$
$$\leqslant \log N\left(D_{k+1}(K, F), t\sqrt{\frac{k}{q}}\sqrt{k}L_{\overline{B}_{k+1}(K,F)}B_F\right)$$
$$\leqslant c_3\frac{\sqrt{qk}}{t},$$

where $D_{k+1}(K, F) = \overline{B}_{k+1}(K, F) - \underline{B}_{k+1}(K, F)$, using in the end the second estimate of (45) for the isotropic convex body $\overline{B}_{k+1}(K, F)$. This completes the proof. □

Using these bounds we can prove the existence of directions with relatively small ψ_2-norm on any subspace of $\mathbb{R}^n$. The dependence is better as the dimension increases.

Theorem 3.5. *Let K be an isotropic convex body in $\mathbb{R}^n$.*

(i) *For every $\log^2 n \leqslant k \leqslant n/\log n$ and every $F \in G_{n,k}$ there exists $\theta \in S_F$ such that*

$$\|\langle \cdot, \theta \rangle\|_{\psi_2} \leqslant C \sqrt{n/k}\, L_K, \tag{55}$$

(ii) *For every $n/\log n \leqslant k \leqslant n$ and every $F \in G_{n,k}$ there exists $\theta \in S_F$ such that*

$$\|\langle \cdot, \theta \rangle\|_{\psi_2} \leqslant C \sqrt{\log n}\, L_K, \tag{56}$$

where $C > 0$ is an absolute constant.

Proof. For every integer $q \geq 1$ we define the *normalized L_q-centroid body K_q of K* by

$$K_q = \frac{1}{\sqrt{q} L_K} Z_q(K), \tag{57}$$

and we consider the convex body

$$T = \text{conv}\left(\bigcup_{i=1}^{\lfloor \log_2 n \rfloor} K_{2^i} \right). \tag{58}$$

Then, for every $F \in G_{n,k}$ we have

$$P_F(T) = \text{conv}\left(\bigcup_{i=1}^{\lfloor \log_2 n \rfloor} P_F(K_{2^i}) \right). \tag{59}$$

We will use the following standard fact (see [6] for a proof): If $A_1, \ldots, A_s$ are subsets of $R B_2^k$, then for every $t > 0$ we have

$$N(\text{conv}(A_1 \cup \cdots \cup A_s), 2t B_2^k) \leqslant \left(\frac{cR}{t} \right)^s \prod_{i=1}^{s} N(A_i, t B_2^k). \tag{60}$$

We apply this to the sets $A_i = P_F(K_{2^i})$. Observe that $K_{2^i} \subseteq c_1 2^{i/2} B_2^n$, and hence, $N(A_i, t B_F) = 1$ if $c_1 2^{i/2} \leqslant t$. Also, $A_i \subseteq c_2 \sqrt{n} B_F$ for all i.

Using Proposition 3.4, for every $t \geq 1$ we can write

$$N(P_F(T), 2tB_F) \leq (c_2\sqrt{n})^{\lfloor \log_2 n \rfloor} \left[\prod_{i=1}^{\lfloor \log_2 n \rfloor} N(P_F(K_{2^i}), tB_F) \right]$$

$$\leq e^{c_3 \log^2 n} \exp\left(C \sum_{i=1}^{\lfloor \log_2 n \rfloor} \frac{2^{i/2}\sqrt{k}}{t} + C \sum_{t^2 \leq 2^i \leq k} \frac{k}{t^2} \right)$$

$$\leq e^{c_3 \log^2 n} \exp\left(C \frac{\sqrt{nk}}{t} + C \frac{k}{t^2} \log(k/t^2) \right),$$

where the second term appears only if $k \geq ct^2$.

Now, we distinguish two cases:

(i) If $\log^2 n \leq k \leq n/\log n$ we choose $t_0 = \sqrt{n/k}$. Observe that $\frac{\sqrt{nk}}{t_0} = k$ and

$$\frac{k}{t_0^2} \log\left(\frac{k}{t_0^2} \right) = \frac{k^2}{n} \log\left(\frac{k^2}{n} \right) \leq \frac{k}{\log n} \log\left(\frac{k^2}{n} \right) \leq k. \tag{61}$$

This implies that $N(P_F(T), \sqrt{n/k}B_F) \leq e^{ck}$. It follows that

$$|P_F(T)| \leq |C\sqrt{n/k} B_F|. \tag{62}$$

Therefore, there exists $\theta \in S_F$ such that

$$h_T(\theta) = h_{P_F(T)}(\theta) \leq C\sqrt{n/k}, \tag{63}$$

which implies

$$\|\langle \cdot, \theta \rangle\|_{2^i} \leq C \, 2^{i/2} \sqrt{n/k} \, L_K \tag{64}$$

for every $i = 1, 2, \ldots, \lfloor \log_2 n \rfloor$. This easily implies (55).

(ii) If $n/\log n \leq k \leq n$ we choose $t_0 = \sqrt{\log n} \simeq \sqrt{\log k}$. Observe that $\frac{\sqrt{nk}}{t_0} = k\sqrt{\frac{n}{k \log n}} \leq k$ and

$$\frac{k}{t_0^2} \log\left(\frac{k}{t_0^2} \right) = \frac{k}{\log n} \log\left(\frac{k}{\log n} \right) \leq \frac{k}{\log n} \log\left(\frac{n}{\log n} \right) \leq k. \tag{65}$$

This implies that $N(P_F(T), \sqrt{\log n}B_F) \leq e^{ck}$ and, as in case (i), we see that

$$\|\langle \cdot, \theta \rangle\|_{2^i} \leq C \, 2^{i/2} \sqrt{\log n} \, L_K \tag{66}$$

for every $i = 1, 2, \ldots, \lfloor \log_2 n \rfloor$. The result follows. $\square$

We close this section with a sketch of the proof of an analogue of the estimate of Proposition 3.1 for $N(Z_q(K), t\sqrt{q}L_K B_2^n)$ for $t \in (0, 1)$.

Proposition 3.6. *Let K be an isotropic convex body in $\mathbb{R}^n$. If $1 \leqslant q \leqslant n$ and $t \in (0, 1)$, then*

$$N\left(Z_q(K), c_1 t\sqrt{q}L_K B_2^n\right) \leqslant \left(\frac{c_2}{t}\right)^n \tag{67}$$

and

$$N\left(Z_q(K), c_3 t\sqrt{q}B_2^n\right) \geqslant \left(\frac{c_4}{t}\right)^n, \tag{68}$$

where $c_i > 0$ are absolute constants.

Proof. The lower bound is a consequence of the estimate $|Z_q(K)|^{1/n} \geqslant c\sqrt{q}|B_2^n|^{1/n}$ (see [17]). Then, we write

$$N\left(Z_q(K), c_1 t\sqrt{q}B_2^n\right) \geqslant \frac{|Z_q(K)|}{|c_1 t\sqrt{q}B_2^n|} \geqslant \left(\frac{c_2}{t}\right)^n. \tag{69}$$

For the upper bound, we will use the fact (see [7, Sect. 3] for the idea of this construction) that there exists an isotropic convex body K_1 in $\mathbb{R}^n$ with the following properties:

(i) $N\left(Z_q(K), t\sqrt{q}L_K B_2^n\right) \leqslant N\left(Z_q(K_1), c_1 t\sqrt{q}B_2^n\right)$ for every $t > 0$.
(ii) $c_2\sqrt{q}B_2^n \subseteq Z_q(K_1)$ for all $1 \leqslant q \leqslant n$.
(iii) $|Z_q(K_1)|^{1/n} \leqslant c_3\sqrt{q/n}$ for all $1 \leqslant q \leqslant n$.

Therefore, for every $t \in (0, 1)$ we have

$$N\left(Z_q(K), \frac{t}{2}\sqrt{q}L_K B_2^n\right) \leqslant \frac{|Z_q(K_1) + t\sqrt{q}B_2^n|}{|t\sqrt{q}B_2^n|}$$

$$\leqslant \frac{|cZ_q(K_1)|}{|t\sqrt{q}B_2^n|}$$

$$\leqslant \left(\frac{c}{t}\right)^n,$$

and (67) is proved. $\qquad\square$

4   On the Distribution of the ψ_2-Norm

From Theorem 3.5 we can deduce a measure estimate for the set of directions which satisfy a given ψ_2-bound. We start with a simple lemma.

Lemma 4.1. *Let $1 \leqslant k \leqslant n$ and let A be a subset of S^{n-1} which satisfies $A \cap F \neq \emptyset$ for every $F \in G_{n,k}$. Then, for every $\varepsilon > 0$ we have*

$$\sigma(A_\varepsilon) \geq \frac{1}{2}\left(\frac{\varepsilon}{2}\right)^{k-1}, \tag{70}$$

where

$$A_\varepsilon = \{y \in S^{n-1} : \inf\{\|y - \theta\|_2 : \theta \in A\} \leq \varepsilon\}. \tag{71}$$

Proof. We write

$$\sigma(A_\varepsilon) = \int_{S^{n-1}} \chi_{A_\varepsilon}(y)\, d\sigma(y) = \int_{G_{n,k}} \int_{S_F} \chi_{A_\varepsilon}(y)\, d\sigma_F(y)\, d\mu_{n,k}(F), \tag{72}$$

and observe that, since $A \cap S_F \neq \emptyset$, the set $A_\varepsilon \cap S_F$ contains a cap $C_F(\varepsilon) = \{y \in S_F : \|y - \theta_0\|_2 \leq \varepsilon\}$ of Euclidean radius ε in S_F. It follows that

$$\int_{S_F} \chi_{A_\varepsilon}(y)\, d\sigma_F(y) \geq \sigma_F(C_F(\varepsilon)) \geq \frac{1}{2}\left(\frac{\varepsilon}{2}\right)^{k-1}, \tag{73}$$

by a well-known estimate on the area of spherical caps, and the result follows. $\square$

Remark. As the proof of the Lemma shows, the strong assumption that $A \cap F \neq \emptyset$ for every $F \in G_{n,k}$ is not really needed for the estimate on $\sigma(A_\varepsilon)$. One can have practically the same lower bound for $\sigma(A_\varepsilon)$ under the weaker assumption that $A \cap F \neq \emptyset$ for every F in a subset $\mathcal{F}_{n,k}$ of $G_{n,k}$ with measure $\mu_{n,k}(\mathcal{F}_{n,k}) \geq c^{-k}$.

Theorem 4.2. *Let K be an isotropic convex body in $\mathbb{R}^n$. For every $\log^2 n \leq k \leq n$ there exists $A_k \subseteq S^{n-1}$ such that*

$$\sigma(A_k) \geq e^{-c_1 k \log k} \tag{74}$$

where $c_1 > 0$ is an absolute constant, and

$$\|\langle \cdot, y \rangle\|_{\psi_2} \leq C \max\left\{\sqrt{n/k}, \sqrt{\log n}\right\} L_K \tag{75}$$

for all $y \in A_k$.

Proof. We fix $\log^2 n \leq k \leq n/\log n$ and define A to be the set of $\theta \in S^{n-1}$ which satisfy (55). By Theorem 3.5 we have $A \cap S_F \neq \emptyset$ for every $F \in G_{n,k}$. Therefore, we can apply Lemma 4.1 with $\varepsilon = \frac{1}{\sqrt{k}}$. If $y \in A_\varepsilon$ then there exists $\theta \in A$ such that $\|y - \theta\|_2 \leq \varepsilon$, which implies

$$\|\langle \cdot, y - \theta \rangle\|_{\psi_2} \leq \left(\|\langle \cdot, y - \theta \rangle\|_\infty \|\langle \cdot, y - \theta \rangle\|_{\psi_1}\right)^{1/2} \leq c\sqrt{n}\varepsilon L_K, \tag{76}$$

if we take into account the well-known fact that $\|\langle \cdot, \theta \rangle\|_{\psi_1} \leq c\|\langle \cdot, \theta \rangle\|_1 \leq cL_K$ (see [19]) and the fact that $\|\langle \cdot, \theta \rangle\|_\infty \leq (n+1)L_K$. It follows that

$$\|\langle \cdot, y \rangle\|_{\psi_2} \leqslant \|\langle \cdot, \theta \rangle\|_{\psi_2} + \|\langle \cdot, y - \theta \rangle\|_{\psi_2}$$
$$\leqslant \|\langle \cdot, \theta \rangle\|_{\psi_2} + c\sqrt{n/k}\,L_K.$$

Since θ satisfies (55), we get (75)—with a different absolute constant C—for all $y \in A_k := A_{1/\sqrt{k}}$. Finally, Lemma 4.1 shows that

$$\sigma(A_k) \geqslant \frac{1}{2}\left(\frac{1}{2\sqrt{k}}\right)^{k-1} \geqslant e^{-c_1 k \log k}, \tag{77}$$

which completes the proof in this case. A similar argument works for $k \geqslant n/\log n$: in this case, we apply Lemma 4.1 with $\varepsilon = \sqrt{\log n/n}$ and the measure estimate for A_k is the same. $\qquad\square$

Proof of Theorem 1.1. Let $t \geqslant 1$ and consider the largest k for which $\sqrt{n/k} \geqslant t\sqrt{\log n}$. Then,

$$\frac{n}{t^2} \simeq k \log n \geqslant k \log k, \tag{78}$$

and hence, $e^{-c_1 k \log k} \geqslant e^{-c_2 n/t^2}$. Theorem 4.2 shows that

$$\psi_K(t) \geqslant \sigma(A_k) \geqslant e^{-c_2 n/t^2}. \tag{79}$$

This proves our claim. $\qquad\square$

5   On the Mean Width of L_q-Centroid Bodies

5.1   Mean Width of $Z_q(K)$

Let K be an isotropic convex body in $\mathbb{R}^n$. For every $q \leqslant q_*(K)$ we have

$$w(Z_q(K)) \simeq w_q(Z_q(K)) \simeq \sqrt{q/n}\,I_q(K) \leqslant c\sqrt{q}L_K. \tag{80}$$

Since $q_*(K) \geqslant c\sqrt{n}$, (80) holds at least for all $q \leqslant \sqrt{n}$. For $q \geqslant \sqrt{n}$, we may use the fact that $Z_q(K) \subseteq c(q/\sqrt{n})Z_{\sqrt{n}}(K)$ to write

$$w(Z_q(K)) \leqslant c\frac{q}{\sqrt{n}}w(Z_{\sqrt{n}}(K)) \leqslant c_1\frac{q}{\sqrt[4]{n}}L_K. \tag{81}$$

In other words, for all $q \geqslant 1$ we have

$$w(Z_q(K)) \leqslant c\sqrt{q}L_K\left(1 + \frac{\sqrt{q}}{\sqrt[4]{n}}\right). \tag{82}$$

Setting $q = n$ and taking into account (22) we get the general upper bound

$$w(K) \leqslant c_1 w(Z_n(K)) \leqslant c_2 n^{3/4} L_K \tag{83}$$

for the mean width of K.

In the next Proposition we slightly improve these estimates, taking into account the radius of $Z_q(K)$ or K.

Proposition 5.1. *Let K be an isotropic convex body in $\mathbb{R}^n$ and let $1 \leqslant q \leqslant n/2$. Then,*

$$w(Z_q(K)) \leqslant c\sqrt{q}L_K\left(1 + \sqrt{R(Z_q(K))/\sqrt{n}L_K}\right). \tag{84}$$

In particular,

$$w(K) \leqslant c\sqrt{n}L_K\left(1 + \sqrt{R(K)/\sqrt{n}L_K}\right). \tag{85}$$

Proof. Recall that, for all $1 \leqslant q \leqslant n/2$,

$$I_{-q}(K) \simeq \sqrt{n/q}w_{-q}(Z_q(K)). \tag{86}$$

We first observe that, for every $t \geqslant 1$,

$$w_{-q/t^2}(Z_q(K)) \leqslant ct^2 w_{-q/t^2}(Z_{q/t^2}(K)) \simeq t^2\sqrt{\frac{q}{t^2 n}}I_{-q/t^2}(K) \leqslant ct\sqrt{q}L_K. \tag{87}$$

Let $\delta \geqslant 1$. Recall that $d_*(C,\delta) = \max\{q \geqslant 1 : w(C) \leqslant \delta w_{-q}(C)\}$. We distinguish two cases:

(a) If $q \leqslant d_*(Z_q(K),\delta)$ then, by (86), we have that

$$w(Z_q(K)) \leqslant \delta w_{-q}(Z_q(K)) \simeq \delta\sqrt{q}I_{-q}(K)/\sqrt{n} \leqslant c\delta\sqrt{q}L_K. \tag{88}$$

(b) If $q \geqslant d_*(Z_q(K),\delta)$, we set $d := d_*(Z_q(K),\delta)$ and define $t \geqslant 1$ by the equation $q/t^2 = d$. Then, using (87), we have

$$w(Z_q(K)) \leqslant \delta w_{-d}(Z_q(K)) = \delta w_{-q/t^2}(Z_q(K)) \leqslant c\delta t\sqrt{q}L_K. \tag{89}$$

This gives the bound

$$w(Z_q(K)) \leqslant c\delta\frac{q}{\sqrt{d_*(Z_q(K),\delta)}}L_K. \tag{90}$$

Moreover, using the fact that

$$d_*(Z_q(K),c_2) \geqslant k_*(Z_q(K)) \simeq n\frac{w(Z_q(K))2}{R(Z_q(K))^2}, \tag{91}$$

we see that if if $q \geqslant c_1 d_*(Z_q(K), c_2)$ then

$$w(Z_q(K)) \leqslant c \frac{\sqrt{q}\sqrt{R(Z_q(K))}}{\sqrt[4]{n}}\sqrt{L_K}. \tag{92}$$

Choosing $\delta = 2$ and combining the estimates (88) and (92) we get (84). Setting $q = n$ and using (22) we obtain (85). $\qquad\square$

Recall that K is called a ψ_α-body with constant b_α if

$$\|\langle \cdot, \theta \rangle\|_{\psi_\alpha} \leqslant b_\alpha \|\langle \cdot, \theta \rangle\|_1 \tag{93}$$

for all $\theta \in S^{n-1}$. If we assume that K is a ψ_α body for some $\alpha \in [1, 2]$ then $R(Z_q(K)) \leqslant R(b_\alpha q^{1/\alpha} Z_2(K)) = b_\alpha q^{1/\alpha} L_K$, and Proposition 5.1 gives immediately the following.

Proposition 5.2. *Let K be an isotropic convex body in $\mathbb{R}^n$. If K is a ψ_α-body with constant b_α for some $\alpha \in [1, 2]$ then, for all $1 \leqslant q \leqslant n$,*

$$w(Z_q(K)) \leqslant c\sqrt{q}L_K \left(1 + \frac{\sqrt{b_\alpha}q^{\frac{1}{2\alpha}}}{\sqrt[4]{n}}\right) \tag{94}$$

and

$$w(K) \leqslant c\sqrt{b_\alpha}n^{\frac{\alpha+2}{4\alpha}}L_K. \tag{95}$$

5.2   Mean Width of $\Psi_2(K)$

As an application of Theorem 1.1 we can give the following estimate for the q-width of $\Psi_2(K)$ for negative values of q.

Proposition 5.3. *Let K be an isotropic convex body in $\mathbb{R}^n$ and $t \geqslant 1$. Then*

$$w_{-\frac{n}{t^2}}(\Psi_2(K)) \leqslant ct\sqrt{\log n}L_K. \tag{96}$$

Proof. Observe that, by Markov's inequality,

$$\sigma\left(\{\theta \in S^{n-1} : h_{\Psi_2(K)}(\theta) \leqslant \frac{1}{e}w_{-\frac{n}{t^2}}(\Psi_2(K))\}\right) \leqslant e^{-\frac{n}{t^2}}. \tag{97}$$

From Theorem 1.1 we know that

$$e^{-\frac{n}{t^2}} \leqslant \sigma\left(\{\theta \in S^{n-1} : h_{\Psi_2(K)}(\theta) \leqslant ct\sqrt{\log n}L_K\}\right), \tag{98}$$

for some absolute constant $c > 0$. This proves (96). $\qquad\square$

We can also give an upper bound for the mean width of $\Psi_2(K)$:

Proposition 5.4. *Let K be an isotropic convex body in $\mathbb{R}^n$. Then,*

$$w(\Psi_2(K)) \leqslant c \sqrt[4]{n \log n} L_K. \tag{99}$$

Proof. Let $w := w(\Psi_2(K))$. Since $R(\Psi_2(K)) \leqslant c\sqrt{n}L_K$, using (32) we see that

$$d_*(\Psi_2(K)) \geqslant ck_*(\Psi_2(K)) \geqslant c\frac{w^2}{L_K^2}. \tag{100}$$

We choose t so that $\frac{n}{t^2} = c\frac{w^2}{L_K^2}$, i.e.

$$t = \frac{c\sqrt{n}L_K}{w} \geqslant 1. \tag{101}$$

Then, from Proposition 5.3 we see that

$$w \leqslant cw_{-d_*}(\Psi_2(K)) \leqslant w_{-\frac{cw^2}{L_K^2}}(\Psi_2(K)) = w_{-\frac{n}{t^2}}(\Psi_2(K))$$

$$\leqslant c_1\frac{\sqrt{n}}{w}\sqrt{\log n}L_K^2,$$

and (99) follows. $\square$

Actually, we can remove the logarithmic term, starting with the next lemma:

Lemma 5.5. *Let K be an isotropic convex body in $\mathbb{R}^n$ and let $1 \leqslant k \leqslant n-1$. Then for every $F \in G_{n,k}$,*

$$P_F(\Psi_2(K)) \subseteq c\sqrt{n/k}\frac{L_K}{L_{\overline{B}_{k+1}(K,F)}}\Psi_2(\overline{B}_{k+1}(K,F)), \tag{102}$$

where $c > 0$ is an absolute constant.

Proof. Indeed, because of (35) and (36), for every $\theta \in S_F$ we can write

$$\frac{h_{\Psi_2(K)}(\theta)}{L_K} \leqslant \sup_{1 \leqslant q \leqslant k}\frac{h_{Z_q(K)}(\theta)}{\sqrt{q}L_K} + \sup_{k \leqslant q \leqslant n}\frac{h_{Z_q(K)}(\theta)}{\sqrt{q}L_K}$$

$$= \sup_{1 \leqslant q \leqslant k}\frac{h_{P_F(Z_q(K))}(\theta)}{\sqrt{q}L_K} + \sup_{k \leqslant q \leqslant n}\frac{h_{P_F(Z_q(K))}(\theta)}{\sqrt{q}L_K}$$

$$\leqslant c_1\sup_{1 \leqslant q \leqslant k}\frac{h_{Z_q(\overline{B}_{k+1}(K,F))}(\theta)}{\sqrt{q}L_{\overline{B}_{k+1}(K,F)}} + c_2\sup_{k \leqslant q \leqslant n}\frac{q}{k}\frac{h_{P_F(Z_k(K))}(\theta)}{\sqrt{q}L_K}$$

$$= c_1 \sup_{1 \leq q \leq k} \frac{h_{Z_q(\overline{B}_{k+1}(K,F))}(\theta)}{\sqrt{q}L_{\overline{B}_{k+1}(K,F)}} + c_2 \sup_{k \leq q \leq n} \sqrt{\frac{q}{k}} \frac{h_{Z_k(\overline{B}_{k+1}(K,F))}(\theta)}{\sqrt{k}L_{\overline{B}_{k+1}(K,F)}}$$

$$\leq c_3 \frac{h_{\Psi_2(\overline{B}_{k+1}(K,F))}(\theta)}{L_{\overline{B}_{k+1}(K,F)}} + c_4 \sup_{k \leq q \leq n} \sqrt{\frac{q}{k}} \frac{h_{\Psi_2(\overline{B}_{k+1}(K,F))}(\theta)}{L_{\overline{B}_{k+1}(K,F)}}$$

$$\leq c_5 \sqrt{\frac{n}{k}} \frac{h_{\Psi_2(\overline{B}_{k+1}(K,F))}(\theta)}{L_{\overline{B}_{k+1}(K,F)}}.$$

$\square$

Proposition 5.6. *Let K be an isotropic convex body in $\mathbb{R}^n$. Then*

$$w(\Psi_2(K)) \leq c \sqrt[4]{n} L_K. \tag{103}$$

Proof. Let $k = \sqrt{n}$. Using Lemma 5.5 we see that

$$w(\Psi_2(K)) = \int_{G_{n,k}} w(P_F(\Psi_2(K))) d\mu_{n,k}(F)$$

$$\leq c \sqrt{\frac{n}{k}} \int_{G_{n,k}} \frac{L_K}{L_{\overline{B}_{k+1}(K,F)}} w(\Psi_2(\overline{B}_{k+1}(K,F))) d\mu_{n,k}(F).$$

Since $k = \sqrt{n} \leq q_*(K)$, we know that a "random" $\overline{B}_{k+1}(K,F)$ is "ψ_2" (see [8]), and the result follows. $\square$

Applying Lemma 5.5 we can cover the case $1 \leq k \leq \log^2 n$ in Theorem 3.5:

Corollary 5.7. *Let K be an isotropic convex body in $\mathbb{R}^n$. For every $1 \leq k \leq \log^2 n$ and every $F \in G_{n,k}$ there exists $\theta \in S_F$ such that*

$$\|\langle \cdot, \theta \rangle\|_{\psi_2} \leq C \sqrt{n/k} \sqrt{\log 2k} L_K, \tag{104}$$

where $C > 0$ is an absolute constant. In fact, for a random $F \in G_{n,k}$ the term $\sqrt{\log 2k}$ is not needed in (104).

Proof. Let $1 \leq k \leq \log^2 n$ and $F \in G_{n,k}$. Since $\overline{B}_{k+1}(K,F)$ is isotropic, Theorem 3.5(ii) shows that there exists $\theta \in S_F$ such that

$$h_{\psi_2(\overline{B}_{k+1}(K,F))}(\theta) \leq c_1 \sqrt{\log 2k} L_{\overline{B}_{k+1}(K,F)}. \tag{105}$$

Then, Lemma 5.5 shows that

$$\|\langle \cdot, \theta \rangle\|_{\psi_2} \simeq h_{\Psi_2(K)}(\theta) = h_{P_F(\Psi_2(K))}(\theta)$$

$$\leq c \sqrt{n/k} \frac{L_K}{L_{\overline{B}_{k+1}(K,F)}} h_{\psi_2(\overline{B}_{k+1}(K,F))}(\theta) \leq C \sqrt{n/k} \sqrt{\log 2k} L_K.$$

In fact, since $k \leqslant \log^2 n \leqslant q_*(K)$, for a random $F \in G_{n,k}$ we know that $\overline{B}_{k+1}(K, F))$ is a ψ_2-body (see [8]), and hence, $h_{\psi_2(\overline{B}_{k+1}(K,F))}(\theta) \leqslant c_2 L_{\overline{B}_{k+1}(K,F)}$ for all $\theta \in S_F$. Using this estimate instead of (105) we may remove the $\sqrt{\log 2k}$-term in (104) for a random $F \in G_{n,k}$. $\qquad\square$

Proof of Proposition 1.3. Since $h_{\psi_2(K)}$ is $\sqrt{n}L_K$-Lipschitz, we have that

$$\sigma\left(\{\theta \in S^{n-1} : h_{\psi_2(K)}(\theta) - w(\psi_2(K)) \geqslant sw(\psi_2(K))\}\right) \leqslant e^{-cns^2\left(\frac{w(\psi_2(K))}{\sqrt{n}L_K}\right)^2}. \quad (106)$$

Let $u \geqslant 2w(\psi_2(K))$. Then, $u = (1 + s)w(\psi_2(K))$ for some $s \geqslant 1$ and $sw(\psi_2(K)) \geqslant u/2$. From (106) it follows that

$$\sigma\left(\{\theta \in S^{n-1} : h_{\psi_2(K)}(\theta) \geqslant u\}\right) \leqslant \exp\left(-cu^2/L_K^2\right). \quad (107)$$

If $t \geqslant c_1 \sqrt[4]{n}/\sqrt{\log n}$, then Proposition 5.6 shows that $u = t\sqrt{\log n}L_K \geqslant 2w(\psi_2(K))$. Then, we can apply (107) to get the result. $\qquad\square$

The estimate of Proposition 1.3 holds true for all $t \geqslant cw(\psi_2(K))/\sqrt{\log n}L_K$; this is easily checked from the proof. This shows that better lower bounds for $\psi_K(t)$ would follow from a better upper estimate for $w(\psi_2(K))$ and vice versa.

6 On the Mean Width of Isotropic Convex Bodies

Let K be an isotropic convex body in $\mathbb{R}^n$. For every $2 \leqslant q \leqslant n$ we define

$$k_*(q) = n\left(\frac{w(Z_q(K))}{R(Z_q(K))}\right)^2. \quad (108)$$

Since $\|\langle \cdot, \theta \rangle\|_q \leqslant cqL_K$ for all $\theta \in S^{n-1}$, we have $R(Z_q(K)) \leqslant cqL_K$. Therefore,

$$w(Z_q(K)) \leqslant cqL_K \frac{\sqrt{k_*(q)}}{\sqrt{n}} \quad (109)$$

Then, from (22) we see that

$$w(K) \simeq w(Z_n(K)) \leqslant \frac{cn}{q}w(Z_q(K)) \leqslant c\sqrt{n}\sqrt{k_*(q)}\, L_K. \quad (110)$$

Define

$$\rho_* = \rho_*(K) = \min_{2 \leqslant q \leqslant n} k_*(q). \quad (111)$$

Since q was arbitrary in (110), we get the following:

Proposition 6.1. *For every isotropic convex body K in $\mathbb{R}^n$ one has*

$$w(K) \leqslant c\sqrt{n}\sqrt{\rho_*(K)}\,L_K. \tag{112}$$

Our next observation is the following: by the isoperimetric inequality on S^{n-1}, for every $q \geqslant 1$ one has

$$\sigma\left(\mid \|\langle \cdot, \theta \rangle\|_q - w(Z_q) \mid \geqslant \frac{w(Z_q)}{2} \right) \leqslant \exp(-ck_*(q)) \leqslant \exp(-2c\rho_*) \tag{113}$$

where $c > 0$ is an absolute constant. Assume that $\log n \leqslant e^{c\rho_*}$. Then,

$$\|\langle \cdot, \theta \rangle\| \simeq w(Z_q) \tag{114}$$

for all θ on a subset A_q of S^{n-1} of measure $\sigma(A_q) \geqslant 1 - \exp(-c\rho_*)$. Taking $q_i = 2^i$, $i \leqslant \log_2 n$ and setting $A = \bigcap A_{q_i}$, we have the following:

Lemma 6.2. *For every isotropic convex body K in $\mathbb{R}^n$ with $\rho_*(K) \geqslant C\log\log n$ one can find $A \subset S^{n-1}$ with $\sigma(A) \geqslant 1 - e^{-c\rho_*}$ such that*

$$\|\langle \cdot, \theta \rangle\|_q \simeq w(Z_q) \tag{115}$$

for all $\theta \in A$ and all $2 \leqslant q \leqslant n$. In particular,

$$\|\langle \cdot, \theta \rangle\|_{\psi_2} \simeq \max_{2 \leqslant q \leqslant n} \frac{w(Z_q)}{\sqrt{q}} \tag{116}$$

for all $\theta \in A$.

Lemma 6.2 implies that if $\rho_*(K)$ is "large" and $\|\langle \cdot, \theta \rangle\|_{\psi_2}$ is well-bounded on a "relatively large" subset of the sphere, then a similar bound holds true for "almost all" directions. As a consequence, we get a good bound for the mean width of K. The precise statement is the following.

Proposition 6.3. *Let K be an isotropic convex body in $\mathbb{R}^n$ which satisfies the following two conditions:*

(1) $\rho_*(K) \geqslant C\log\log n$.
(2)   For some $b_n > 0$ we have $\|\langle \cdot, \theta \rangle\|_{\psi_2} \leqslant b_n L_K$ for all θ in a set $B \subseteq S^{n-1}$ with $\sigma(B) > e^{-c\rho_*}$.

Then,

$$\|\langle \cdot, \theta \rangle\|_{\psi_2} \leqslant Cb_n L_K \tag{117}$$

for all θ in a set $A \subseteq S^{n-1}$ with $\sigma(A) > 1 - e^{-c\rho_*}$. Also,

$$w(Z_q(K)) \leqslant c\sqrt{q}\,b_n L_K \tag{118}$$

for all $2 \leqslant q \leqslant n$ *and*

$$w(K) \leqslant C \sqrt{n} b_n L_K. \tag{119}$$

Proof. We can find $u \in A \cap B$, where A is the set in Lemma 6.2. Since $u \in B$, we have

$$\|\langle \cdot, u \rangle\|_q \leqslant C_1 \sqrt{q} b_n L_K \tag{120}$$

for all $2 \leqslant q \leqslant n$, and (115) shows that

$$w(Z_q(K)) \leqslant C_2 \sqrt{q} b_n L_K \tag{121}$$

for all $2 \leqslant q \leqslant n$. Going back to (115) we see that if $\theta \in A$ then

$$\|\langle \cdot, \theta \rangle\|_q \leqslant c w(Z_q) \leqslant C_3 \sqrt{q} b_n L_K \tag{122}$$

for all $2 \leqslant q \leqslant n$. For $q = n$ we get (119).

Finally, for every $\theta \in A$ we have

$$\|\langle \cdot, \theta \rangle\|_{\psi_2} \simeq \max_{2 \leqslant q \leqslant n} \frac{\|\langle \cdot, \theta \rangle\|_q}{\sqrt{q}} \leqslant C b_n L_K. \tag{123}$$

This completes the proof.   □

Propositions 6.1 and 6.3 provide a dichotomy. If $\rho_*(K)$ is small then we can use Proposition 6.1 to get an upper bound for $w(K)$. If $\rho_*(K)$ is large then we can use Proposition 6.3 provided that we have some sufficiently good lower bound for $\psi_K(t)$: what we have is

$$\psi_K(t) \geqslant e^{-c_1 n/t^2} \geqslant e^{-c\rho_*}, \tag{124}$$

if $t \simeq \sqrt{n/\rho_*}$. Therefore, we obtain the estimate

$$w(K) \leqslant C \sqrt{n \log n} \sqrt{n/\rho_*} L_K. \tag{125}$$

Combining the previous results, we deduce one more general upper bound for the mean width of K.

Theorem 6.4. *For every isotropic convex body K in $\mathbb{R}^n$ we have*

$$w(K) \leqslant C \sqrt{n} \min \left\{ \sqrt{\rho_*}, \sqrt{n \log n/\rho_*} \right\} L_K, \tag{126}$$

where $c > 0$ is an absolute constant.

The estimate in Theorem 6.4 depends on our knowledge for the behavior of $\psi_K(t)$; as it stands, it only recovers the $O(n^{3/4} L_K)$ bound for the mean width of K. Actually, the logarithmic term in (126) makes it slightly worse. However, we can remove this logarithmic term, starting with the following modification of Proposition 5.1.

Proposition 6.5. *Let K be an isotropic convex body in $\mathbb{R}^n$ and $1 \leqslant q \leqslant n$. Then,*

$$w(Z_q(K)) \leqslant c\sqrt{q}L_K\left(1 + \sqrt{\frac{q}{k_*(Z_q(K))}}\right), \tag{127}$$

where $c > 0$ is an absolute constant.

Proof. If $R(Z_q(K)) \leqslant c\sqrt{n}L_K$ then (127) is a direct consequence of (84). So, we assume that $R(Z_q(K)) \geqslant c\sqrt{n}L_K$. Then, writing (84) in the form

$$w(Z_q(K)) \leqslant c\frac{\sqrt{q}}{\sqrt[4]{n}}\sqrt{R(Z_q(K))}\sqrt{L_K}, \tag{128}$$

and taking into account the definition of $k_*(Z_q(K))$ we see that

$$\frac{R(Z_q(K))}{\sqrt{n}L_K} \leqslant c_1\frac{q}{k_*(Z_q(K))}, \tag{129}$$

and (127) follows from (84) again. $\qquad\square$

Theorem 6.6. *Let K be an isotropic convex body in $\mathbb{R}^n$. Then,*

$$w(K) \leqslant c\sqrt{n}L_K \min\left\{\sqrt{\rho_*}, \sqrt{\frac{n}{\rho_*}}\right\}, \tag{130}$$

where $c > 0$ is an absolute constant.

Proof. From Proposition 6.1 we know that

$$w(K) \leqslant c\sqrt{n}L_K\sqrt{\rho_*}. \tag{131}$$

Let q_0 satisfy $\rho_* = k_*(Z_{q_0}(K))$. From Proposition 6.5 and from (21) and (22) we have that, for all $1 \leqslant q \leqslant n$,

$$w(K) \leqslant c\frac{n}{q}w(Z_q(K)) \leqslant c_1\sqrt{n}L_K\left(\sqrt{\frac{n}{q}} + \sqrt{\frac{n}{k_*(Z_q(K))}}\right). \tag{132}$$

Recall that q_* is the parameter $q_*(K) := \max\{q \in [1, n] : k_*(Z_q(K)) \geqslant q\}$. We distinguish two cases.

(i) Assume that $q_0 \leqslant q_*$. Then we apply (132) for q_*; since $q_* = k_*(Z_{q_*}(K)) \geqslant \rho_*$, we get

$$w(K) \leqslant 2c_1\sqrt{n}L_K\sqrt{\frac{n}{q_*}} \leqslant 2c_1\sqrt{n}L_K\sqrt{\frac{n}{\rho_*}}. \tag{133}$$

(ii) Assume that $q_0 \geq q_*$. Then, $q_0 \geq k_*(Z_{q_0}(K)) = \rho_*$. Applying (132) for q_0, we get

$$w(K) \leq 2c_1 \sqrt{n} L_K \sqrt{\frac{n}{k_*(Z_{q_0}(K))}} = 2c_1 \sqrt{n} L_K \sqrt{\frac{n}{\rho_*}}. \tag{134}$$

In both cases, we have

$$w(K) \leq c \sqrt{n} L_K \sqrt{\frac{n}{\rho_*}}. \tag{135}$$

Combining (135) with (131) we get the result.

$\square$

Acknowledgements We would like to thank the referee for useful comments regarding the presentation of this paper.

References

1. K.M. Ball, Logarithmically concave functions and sections of convex sets in $\mathbb{R}^n$. Studia Math. **88**, 69–84 (1988)
2. S.G. Bobkov, F.L. Nazarov, in *Large Deviations of Typical Linear Functionals on a Convex Body with Unconditional Basis*. Stochastic Inequalities and Applications. Progr. Probab., vol. 56 (Birkhauser, Basel, 2003), pp. 3–13
3. S.G. Bobkov, F.L. Nazarov, in *On Convex Bodies and Log-Concave Probability Measures with Unconditional Basis*, ed. by V. Milman, G. Schechtman. Geometric Aspects of Functional Analysis. Lecture Notes in Math., vol. 1807 (Springer, Berlin, 2003), pp. 53–69
4. J. Bourgain, in *On the Distribution of Polynomials on High Dimensional Convex Sets*. Lecture Notes in Math., vol. 1469 (Springer, Berlin, 1991), pp. 127–137
5. A. Giannopoulos, V.D. Milman, Mean width and diameter of proportional sections of a symmetric convex body. J. Reine Angew. Math. **497**, 113–139 (1998)
6. A. Giannopoulos, A. Pajor, G. Paouris, A note on subgaussian estimates for linear functionals on convex bodies. Proc. Am. Math. Soc. **135**, 2599–2606 (2007)
7. A. Giannopoulos, G. Paouris, P. Valettas, On the existence of subgaussian directions for log-concave measures. Contemp. Math. **545**, 103–122 (2011)
8. A. Giannopoulos, G. Paouris, P. Valettas, Ψ_α-estimates for marginals of log-concave probability measures. Proc. Am. Math. Soc. **140**, 1297–1308 (2012)
9. M. Hartzoulaki, *Probabilistic Methods in the Theory of Convex Bodies*, PhD Thesis, University of Crete (2003)
10. R. Kannan, L. Lovasz, M. Simonovits, Isoperimetric problems for convex bodies and a localization lemma. Discrete Comput. Geom. **13**, 541–559 (1995)
11. B. Klartag, On convex perturbations with a bounded isotropic constant. Geom. Funct. Anal. **16**, 1274–1290 (2006)
12. B. Klartag, Uniform almost sub-gaussian estimates for linear functionals on convex sets. Algebra i Analiz (St. Petersburg Math. J.) **19**, 109–148 (2007)
13. B. Klartag, E. Milman, Centroid bodies and the logarithmic Laplace transform – A unified approach. J. Funct. Anal. **262** 10–34 (2012)
14. B. Klartag, R. Vershynin, Small ball probability and Dvoretzky theorem. Israel J. Math. **157**(1), 193–207 (2007)
15. R. Latala, K. Oleszkiewicz, Small ball probability estimates in terms of width. Studia Math. **169**, 305–314 (2005)

16. A. Litvak, V.D. Milman, G. Schechtman, Averages of norms and quasi-norms. Math. Ann. **312**, 95–124 (1998)
17. E. Lutwak, D. Yang, G. Zhang, L^p affine isoperimetric inequalities. J. Diff. Geom. **56**, 111–132 (2000)
18. V.D. Milman, A new proof of A. Dvoretzky's theorem in cross-sections of convex bodies (Russian). Funkcional. Anal. i Prilozen. **5**(4), 28–37 (1971)
19. V.D. Milman, A. Pajor, in *Isotropic Position and Inertia Ellipsoids and Zonoids of the Unit Ball of a Normed n-Dimensional Space*, ed. by J. Lindenstrauss, V.D. Milman. Geometric Aspects of Functional Analysis. Lecture Notes in Math., vol. 1376 (Springer, Berlin, 1989), pp. 64–104
20. V.D. Milman, G. Schechtman, in *Asymptotic Theory of Finite Dimensional Normed Spaces*. Lecture Notes in Math., vol. 1200 (Springer, Berlin, 1986)
21. V.D. Milman, G. Schechtman, Global versus Local asymptotic theories of finite-dimensional normed spaces. Duke Math. J. **90**, 73–93 (1997)
22. V.D. Milman, S.J. Szarek, in *A Geometric Lemma and Duality of Entropy*. GAFA Seminar Notes. Lecture Notes in Math., vol. 1745 (Springer, Berlin, 2000), pp. 191–222
23. G. Paouris, in *ψ_2-Estimates for Linear Functionals on Zonoids*. Geometric Aspects of Functional Analysis. Lecture Notes in Math., vol. 1807 (Springer, Berlin, 2003), pp. 211–222
24. G. Paouris, On the ψ_2-behavior of linear functionals on isotropic convex bodies. Studia Math. **168**(3), 285–299 (2005)
25. G. Paouris, Concentration of mass on convex bodies. Geom. Funct. Anal. **16**, 1021–1049 (2006)
26. G. Paouris, Small ball probability estimates for log–concave measures. Trans. Am. Math. Soc. **364**, 287–308 (2012)
27. G. Pisier, *The Volume of Convex Bodies and Banach Space Geometry*. Cambridge Tracts in Mathematics, vol. 94 (1989)
28. P. Pivovarov, On the volume of caps and bounding the mean-width of an isotropic convex body. Math. Proc. Camb. Philos. Soc. **149**, 317–331 (2010)

A Remark on Vertex Index of the Convex Bodies

Efim D. Gluskin and Alexander E. Litvak

Abstract The vertex index of a symmetric convex body $\mathbf{K} \subset \mathbb{R}^n$, vein($\mathbf{K}$), was introduced in [Bezdek, Litvak, Adv. Math. **215**, 626–641 (2007)]. Bounds on the vertex index were given in the general case as well as for some basic examples. In this note we improve these bounds and discuss their sharpness. We show that

$$\text{vein}(\mathbf{K}) \leq 24n^{3/2},$$

which is asymptotically sharp. We also show that the estimate

$$\frac{n^{3/2}}{\sqrt{2\pi e}\ \text{ovr}(\mathbf{K})} \leq \text{vein}(\mathbf{K}),$$

obtained in [Bezdek, Litvak, Adv. Math. **215**, 626–641 (2007)] (here ovr($\mathbf{K}$) denotes the outer volume ratio of $\mathbf{K}$), is not always sharp. Namely, we construct an example showing that there exists a symmetric convex body $\mathbf{K}$ which simultaneously has large outer volume ratio and large vertex index. Finally, we improve the constant in the latter bound for the case of the Euclidean ball from $\sqrt{2\pi e}$ to $\sqrt{3}$, providing a completely new approach to the problem.

E.D. Gluskin (✉)
Sackler Faculty of Exact Sciences, Department of Mathematics, Tel Aviv University,
Ramat Aviv, 69978 Tel Aviv, Israel
e-mail: gluskin@post.tau.ac.il

A.E. Litvak
Department of Mathematical and Statistical Sciences, University of Alberta, Edmonton, AB,
Canada T6G 2G1
e-mail: alexandr@math.ualberta.ca

B. Klartag et al. (eds.), *Geometric Aspects of Functional Analysis*, Lecture Notes
in Mathematics 2050, DOI 10.1007/978-3-642-29849-3_14,
© Springer-Verlag Berlin Heidelberg 2012

1 Introduction

Let $\mathbf{K}$ be a convex body symmetric about the origin 0 in $\mathbb{R}^n$ (such bodies below we call 0-symmetric convex bodies). The vertex index of $\mathbf{K}$, vein($\mathbf{K}$), was introduced in [6] as

$$\text{vein}(\mathbf{K}) = \inf\left\{\sum_i \|x_i\|_{\mathbf{K}} \mid \mathbf{K} \subset \text{conv}\{x_i\}\right\},$$

where $\|x\|_{\mathbf{K}} = \inf\{\lambda > 0 \mid x \in \lambda\mathbf{K}\}$ denotes the Minkowski functional of $\mathbf{K}$. In other words, given $\mathbf{K}$ one looks for the convex polytope that contains $\mathbf{K}$ and whose vertex set has the smallest possible closeness to 0 in metric generated by $\mathbf{K}$. Let us note that vein($\mathbf{K}$) is an affine invariant of $\mathbf{K}$, i.e. if $T : \mathbb{R}^n \to \mathbb{R}^n$ is an invertible linear map, then vein($\mathbf{K}$) = vein($T(\mathbf{K})$).

The vertex index is closely connected to some important quantities in analysis and geometry including the illumination parameter of convex bodies, introduced by Bezdek; the Boltyanski-Hadwiger illumination conjecture, which says that every convex body in $\mathbb{R}$ can be illuminated by 2^n sources; the Gohberg-Marcus conjecture, which avers that a convex body can be covered by 2^n smaller positive homothetic copies of itself). We refer to [4–6, 11] for the related discussions, history, and references.

Denote the volume by $|\cdot|$, the canonical Euclidean ball in $\mathbb{R}^n$ by $\mathbf{B}_2^n$, and as usual define the outer volume ratio of $\mathbf{K}$ by ovr($\mathbf{K}$) $= \inf\left(|\mathcal{E}|/|\mathbf{K}|\right)^{1/n}$, where the infimum is taken over all ellipsoids $\mathcal{E} \supset \mathbf{K}$. In [6] the following theorem has been proved.

Theorem 1.1. *There exists a positive absolute constant C such that for every $n \geq 1$ and every 0-symmetric convex body $\mathbf{K}$ in $\mathbb{R}^n$ one has*

$$\frac{n^{3/2}}{\sqrt{2\pi e}\ \text{ovr}(\mathbf{K})} \leq \text{vein}(\mathbf{K}) \tag{1}$$

and

$$\text{vein}(\mathbf{K}) \leq C\ n^{3/2}\ \ln(2n). \tag{2}$$

Moreover, in [9] it was shown that vein($\mathbf{K}$) $\geq 2n$ for every n-dimensional 0-symmetric convex body $\mathbf{K}$.

The purpose of this note is to discuss sharpness of estimates 1 and 2. We start our discussion with the first estimate. Note that it is sharp (especially in view of estimate (3) below) for the class of bodies with finite outer volume ratio, that is bodies such that ovr($\mathbf{K}$) $\leq C$, where C is a positive absolute constant (fixed in advance). This class is very large, it includes in particular the unit balls of ℓ_p-spaces for $p \geq 2$ as well as 0-symmetric convex polytopes having at most $C_1 n$ facets (here C_1 is another absolute constant). In Sect. 3 we show that in fact (1) is not sharp, i.e. that in general vein($\mathbf{K}$) is not equivalent to $n^{3/2}/\text{ovr}(\mathbf{K})$. Namely, we construct a 0-symmetric convex body $\mathbf{K}$ which has simultaneously large outer volume ratio and large vertex index (in fact both are largest possible up to a logarithmic factor): $\mathrm{vein}(\mathbf{K}) \approx n^{3/2}$ and $\mathrm{ovr}(\mathbf{K}) \approx \sqrt{n}/\sqrt{\ln(2n)}$. It shows that for some bodies the gap in (1) can be of the order $\sqrt{n}/\sqrt{\ln(2n)}$. Note that despite of our example, there are bodies with large outer volume ratio for which (1) is sharp, e.g. for the n-dimensional octahedron $\mathbf{B}_1^n$ we have $\mathrm{vein}(\mathbf{B}_1^n) = 2n$ [6] and

$$\mathrm{ovr}(\mathbf{B}_1^n) = \frac{\sqrt{\pi}}{2}\left(\frac{n}{\Gamma(1+n/2)}\right)^{1/n} \approx \frac{\sqrt{\pi}}{\sqrt{2e}}\sqrt{n}.$$

The construction of our example is of the random nature, essentially we take the absolute convex hull of n^2 random points on the sphere and show that it works with high probability.

Next, in Sect. 4, we remove the logarithmic factor in the estimate (2), improving it to the asymptotically best possible one. The main new ingredient in our improvement is a recent result of Batson et al. [3] on the decomposition of a linear operator acting on $\mathbb{R}^n$ (see Theorem 4.1 below). The application of their theorem instead of corresponding Rudelson's Theorem used in [6] allows us to remove the unnecessary logarithm.

In Sect. 5 we turn to the vertex index of the Euclidean ball. In [6] it was conjectured that

$$\mathrm{vein}(\mathbf{B}_2^n) = 2n^{3/2},$$

i.e., the best configuration for the Euclidean ball is provided by the vertices of the n-dimensional octahedron. The conjecture was verified for $n = 2$ and $n = 3$. Note that by (1)

$$\frac{n^{3/2}}{\sqrt{2\pi e}} \leq \mathrm{vein}(\mathbf{B}_2^n).$$

We improve this bound to $n^{3/2}/\sqrt{3}$. Our proof uses completely different approach via operator theory (recall that in [6] the approach via volumes was used). We think that this new approach is interesting by itself and could lead to more results. Thus the results of Sects. 4 and 5 can be summarized in the following theorem.

Theorem 1.2. *For every $n \geq 1$ and every 0-symmetric convex body $\mathbf{K}$ in $\mathbb{R}^n$ one has*

$$\mathrm{vein}(\mathbf{K}) \leq 24\, n^{3/2}. \tag{3}$$

Moreover

$$\mathrm{vein}(\mathbf{B}_2^n) \geq n^{3/2}/\sqrt{3}. \tag{4}$$

Acknowledgements Part of this research was conducted while the second named author participated in the Thematic Program on Asymptotic Geometric Analysis at the Fields Institute in Toronto in Fall 2010. He thanks the Institute for the hospitality. His research partially supported by the E.W.R. Steacie Memorial Fellowship.

2 Preliminaries and Notation

By $|\cdot|$ and $\langle\cdot,\cdot\rangle$ we denote the canonical Euclidean norm and the canonical inner product on $\mathbb{R}^n$. The canonical basis of $\mathbb{R}^n$ we denote by $e_1,\ldots,e_n$. By $\|\cdot\|_p$, $1 \le p \le \infty$, we denote the ℓ_p-norm, i.e.

$$\|x\|_p = \left(\sum_{i\ge 1}|x_i|^p\right)^{1/p} \quad \text{for} \quad p < \infty \quad \text{and} \quad \|x\|_\infty = \sup_{i\ge 1}|x_i|.$$

In particular, $\|\cdot\|_2 = |\cdot|$. As usual, $\ell_p^n = (\mathbb{R}^n, \|\cdot\|_p)$, and the unit ball of ℓ_p^n is denoted by $\mathbf{B}_p^n$.

Given points $x_1,\ldots,x_k$ in $\mathbb{R}^n$ we denote their convex hull by $\operatorname{conv}\{x_i\}_{i\le k}$ and their absolute convex hull by $\operatorname{abs\,conv}\{x_i\}_{i\le k} = \operatorname{conv}\{\pm x_i\}_{i\le k}$. Similarly, the convex hull of a set $A \subset \mathbb{R}^n$ is denoted by $\operatorname{conv} A$ and absolute convex hull of A is denoted by $\operatorname{abs\,conv} A$ $(= \operatorname{conv}\{A \cup -A\})$.

Given convex compact body $\mathbf{K} \subset \mathbb{R}^n$ with 0 in its interior by $|\mathbf{K}|$ we denote its volume and by $\|\cdot\|_{\mathbf{K}}$ its Minkowski functional. $\mathbf{K}^\circ$ denotes the polar of $\mathbf{K}$, i.e.

$$\mathbf{K}^\circ = \{x \mid \langle x,y\rangle \le 1 \text{ for every } y \in \mathbf{K}\}.$$

The outer volume ratio of $\mathbf{K}$ is

$$\operatorname{ovr}(\mathbf{K}) = \inf\left(\frac{|\mathcal{E}|}{|\mathbf{K}|}\right)^{1/n},$$

where infimum is taken over all 0-symmetric ellipsoids in $\mathbb{R}^n$ containing $\mathbf{K}$. It is well-known that

$$\operatorname{ovr}(\mathbf{K}) \le \sqrt{n}$$

for every convex symmetric about the origin body $\mathbf{K}$.

Finally we recall some notations from the Operator Theory. Given $u,v \in \mathbb{R}^n$, $u \otimes v$ denotes the operator from $\mathbb{R}^n$ to $\mathbb{R}^n$ defined by $(u \otimes v)(x) = \langle u,x\rangle v$ for every $x \in \mathbb{R}^n$. The identity operator on $\mathbb{R}^n$ is denoted by Id. Given two operators $T, S : \mathbb{R}^n \to \mathbb{R}^n$ we write $T \le S$ if $S - T$ is positive semidefinite, i.e., $\langle (S-T)x,x\rangle \ge 0$ for every $x \in \mathbb{R}^n$.

3 Example

Theorem 3.1. *There exists an absolute positive constant c such that for every $n \ge 1$ there exists a convex symmetric body $\mathbf{K}$ satisfying*

$$\operatorname{ovr}(\mathbf{K}) \ge c\sqrt{\frac{n}{\ln(2n)}} \quad \text{and} \quad \operatorname{vein}\mathbf{K} \ge cn^{3/2}.$$

Proof. Let $m = n^2$ and $u_1, u_2, \ldots, u_m$ be independent random vectors uniformly distributed on S^{n-1}. Let

$$\mathbf{K} := \operatorname{abs conv} \{u_i, e_j\}_{i \le m, j \le n}.$$

Clearly

$$\frac{1}{\sqrt{n}} \mathbf{B}_2^n \subset \mathbf{K} \subset \mathbf{B}_2^n.$$

Moreover, it is well-known [2, 7, 8] that there exists an absolute positive constant C_0 such that for every linear transformation T satisfying $T\mathbf{K} \subset \mathbf{B}_2^n$ one has

$$|T\mathbf{K}| \le C_0 \frac{\sqrt{\ln(2(m+n)/n)}}{n},$$

which immediately implies that

$$\operatorname{ovr}(\mathbf{K}) \ge c_0 \sqrt{\frac{n}{\ln(2n)}}$$

for an absolute positive constant c_0.

Now we prove the lower bound on $\operatorname{vein}(\mathbf{K})$. First note that if $\mathbf{T}$ is an absolute convex hull of vectors $x_1, x_2, \ldots, x_M$ satisfying

$$a := \sum_{i=1}^{M} |x_i| \le \frac{n^{3/2}}{4\sqrt{2\pi e}}$$

then by Santaló inequality and a result of Ball and Pajor (Theorem 2 in [1]) we have

$$\frac{|\mathbf{T}|}{|\mathbf{B}_2^n|} \le \frac{|\mathbf{B}_2^n|}{|\mathbf{T}^0|} \le \left(\frac{\sqrt{2\pi e}}{\sqrt{n}}\right)^n \left(\frac{a}{n}\right)^n \le 4^{-n}.$$

It implies that the probability

$$\mathbb{P}(\{\mathbf{K} \subset 2\mathbf{T}\}) \le \mathbb{P}(\{\forall i \le m : u_i \in 2\mathbf{T}\}) = (\mathbb{P}(\{\forall i : u_i \in 2\mathbf{T}\}))^m$$

$$= (|2\mathbf{T} \cap S^{n-1}|)^m \le \left(\frac{|2\mathbf{T} \cap \mathbf{B}_2^n|}{|\mathbf{B}_2^n|}\right)^m \le 2^{-n^3}.$$

Now we consider a $\frac{1}{2\sqrt{n}}$-net (in the Euclidean metric) $\mathcal{N}$ in $n^{3/2}\mathbf{B}_2^n$ of cardinality less than $A = (6n^2)^n$ (it is well known that such a net exists). We fix $M = [n^{3/2}/8\sqrt{2\pi e}]$ (assuming without loss of generality $M \ge 3$) and consider

$$C_M = \left\{ \mathbf{T} \mid \mathbf{T} = \operatorname{abs conv}\{x_i\}_{i \le N}, N \le M, x_i \in \mathcal{N}, \sum_{i=1}^{N} |x_i| \le \frac{n^{3/2}}{4\sqrt{2\pi e}} \right\}.$$

Then the cardinality of C_M is

$$|C_M| \leq \sum_{i=1}^{M} \binom{A}{i} \leq \left(\frac{eA}{M}\right)^M \leq (6n^2)^{nM}.$$

It implies that

$$\mathbb{P}\left(\{\exists \mathbf{T} \text{ such that } \mathbf{K} \subset 2\mathbf{T}\}\right) \leq (6n^2)^{nM} 2^{-n^3} < 1.$$

This proves that there exists $\mathbf{K}$ such that

$$\forall \mathbf{T} \in C_M \ : \ \mathbf{K} \not\subset 2\mathbf{T}. \tag{5}$$

Finally fix $\mathbf{K}$ satisfying (5) and assume

$$\text{vein}\, \mathbf{K} < \frac{n^{3/2}}{8\sqrt{2\pi e}},$$

i.e., that there exists $\mathbf{L} = \text{conv}\{x_i\}_{i \leq k}$ with $\mathbf{K} \subset \mathbf{L}$ and

$$k \leq \sum_{i=1}^{k} \|x_i\|_{\mathbf{K}} < \frac{n^{3/2}}{8\sqrt{2\pi e}}.$$

Since $\mathbf{K} \subset \mathbf{B}_2^n$, we observe that

$$\sum_{i=1}^{k} |x_i| < \frac{n^{3/2}}{8\sqrt{2\pi e}},$$

in particular $x_i \in n^{3/2}\mathbf{B}_2^n$, $i \leq k$. Then for every i there exist $y_i \in \mathcal{N}$ such that

$$|x_i - y_i| \leq \frac{1}{2\sqrt{n}}.$$

Therefore

$$\sum_{i=1}^{k} |y_i| \leq \sum_{i=1}^{k} |x_i| + \sum_{i=1}^{k} |x_i - y_i| \leq \frac{n^{3/2}}{8\sqrt{2\pi e}} + \frac{k}{2\sqrt{n}} \leq \frac{n^{3/2}}{4\sqrt{2\pi e}}.$$

Thus $\mathbf{P} = \text{abs conv}\{y_i\}_{i \leq N} \in C_M$, so, by (5) one has $\mathbf{K} \not\subset 2\mathbf{P}$. On the other hand we have for every x

$$\|x\|_{\mathbf{L}^0} = \max_{i \leq k} \langle x, x_i \rangle \leq \max_{i \leq k} \langle x, y_i \rangle + \max_{i \leq k} \langle x, x_i - y_i \rangle$$

$$\leq \|x\|_{\mathbf{P}^0} + \frac{1}{2\sqrt{n}}|x| \leq \|x\|_{\mathbf{P}^0} + \frac{1}{2}\|x\|_{\mathbf{L}^0},$$

where the latter inequality holds because $\frac{1}{\sqrt{n}}\mathbf{B}_2^n \subset \mathbf{K} \subset \mathbf{L}$. The above inequality means that $\mathbf{L} \subset 2\mathbf{P}$, which contradicts the fact that $\mathbf{K} \not\subset 2\mathbf{P}$. Hence

$$\text{vein } \mathbf{K} \geq \frac{n^{3/2}}{8\sqrt{2\pi e}},$$

which proves the theorem.   □

4   An Upper Bound for the Vertex Index

In this section we prove the inequality (3), i.e. we prove the sharp (up to an absolute constant) upper estimate for the vein of a convex symmetric body in the general case, removing the unnecessary logarithmic term from (1). Recall that such bound is attained for any body with a bounded volume ratio as well as for the body from Theorem 3.1.

In [6] the Rudelson theorem on decomposition of identity was essentially used. It contains a logarithmic term which appeared in the upper bound on the vertex index. Here we use a recent result of Batson, Spielman, and Srivastava instead of Rudelson's theorem. In [3], they proved the following theorem.

Theorem 4.1. *Let $m \geq n \geq 1$, $\lambda > 1$, and $u_i \in \mathbb{R}^n$, $i \leq m$ be such that*

$$Id = \sum_{i=1}^{m} u_i \otimes u_i.$$

Then there exist non-negative numbers $c_1, c_2, \ldots, c_m$ such that at most λn of them non-zero and

$$Id \leq \sum_{i=1}^{m} c_i u_i \otimes u_i \leq \left(\frac{\sqrt{\lambda} + 1}{\sqrt{\lambda} - 1}\right)^2 Id.$$

To obtain the upper bound it is enough to apply this theorem combined with the standard John decomposition instead of Rudelson theorem in the proof given in Sect. 5 of [6]. For the sake of completeness we provide the details. The following standard lemma proves (3).

Lemma 4.2. *Let $\lambda > 1$, $n \geq 1$, and $\mathbf{K}$ be a 0-symmetric convex body in $\mathbb{R}^n$ such that its minimal volume ellipsoid is $\mathbf{B}_2^n$. Then there exists a 0-symmetric convex polytope $\mathbf{P}$ in $\mathbb{R}^n$ with at most λn vertices such that*

$$\mathbf{P} \subset \mathbf{K} \subset \mathbf{B}_2^n \subset \frac{\sqrt{\lambda}+1}{\sqrt{\lambda}-1} \sqrt{n}\, \mathbf{P}.$$

In particular

$$\mathrm{vein}(\mathbf{K}) \leq 24 n^{3/2}.$$

Proof. The John decomposition [10] states that there exist points v_i, $i \leq m$, with $\|v_i\|_{\mathbf{K}} = |v_i| = 1$ and scalars $\lambda_i > 0$ such that

$$Id = \sum_{i=1}^{m} \lambda_i v_i \otimes v_i.$$

Then Theorem 4.1 applied to $u_i = \sqrt{\lambda_i} v_i$ implies that there exist non-negative numbers $c_1, c_2, \ldots, c_m$ such that at most λn of them non-zero and

$$Id \leq \sum_{i=1}^{m} c_i \lambda_i v_i \otimes v_i \leq \left(\frac{\sqrt{\lambda}+1}{\sqrt{\lambda}-1} \right)^2 Id. \tag{6}$$

Let I denotes the set of indeces i such that $c_i \neq 0$. Consider $\mathbf{P} = \mathrm{abs}\,\mathrm{conv}\{v_i\}_{i \in I}$. Since $v_i \in \mathbf{K} = -\mathbf{K}$, $i \leq m$, we observe

$$\mathbf{P} \subset \mathbf{K} \subset \mathbf{B}_2^n.$$

By (6) we also have for every $x \in \mathbb{R}^n$

$$|x|^2 = \langle Id\, x, x \rangle \leq \left\langle \sum_{i=1}^{m} c_i \lambda_i \langle v_i, x \rangle v_i, x \right\rangle = \sum_{i=1}^{m} c_i \lambda_i \langle v_i, x \rangle^2$$

$$\leq \max_{i \leq m} \langle v_i, x \rangle^2 \sum_{i=1}^{m} c_i \lambda_i = \|x\|_{\mathbf{P}^\circ}^2 \sum_{i=1}^{m} c_i \lambda_i$$

and

$$\sum_{i=1}^{m} c_i \lambda_i = \sum_{i=1}^{m} c_i \lambda_i \langle v_i, v_i \rangle = \mathrm{trace} \sum_{i=1}^{m} c_i \lambda_i v_i \otimes v_i$$

$$\leq \left(\frac{\sqrt{\lambda}+1}{\sqrt{\lambda}-1} \right)^2 \mathrm{trace}\, Id = \left(\frac{\sqrt{\lambda}+1}{\sqrt{\lambda}-1} \right)^2 n.$$

It implies that $|x| \leq \frac{\sqrt{\lambda}+1}{\sqrt{\lambda}-1} \sqrt{n}\, \|x\|_{\mathbf{P}^\circ}$, which means $\mathbf{B}_2^n \subset \frac{\sqrt{\lambda}+1}{\sqrt{\lambda}-1} \sqrt{n}\, \mathbf{P}$. This proves

$$\mathbf{P} \subset \mathbf{K} \subset \mathbf{B}_2^n \subset \frac{\sqrt{\lambda}+1}{\sqrt{\lambda}-1} \sqrt{n}\, \mathbf{P}$$

and in particular implies

$$\text{vein}(\mathbf{K}) \leq 2\sqrt{n} \frac{\sqrt{\lambda}+1}{\sqrt{\lambda}-1} \sum_{i \in I} \|v_i\|_{\mathbf{K}} \leq 2\lambda \frac{\sqrt{\lambda}+1}{\sqrt{\lambda}-1} n^{3/2}.$$

Choosing $\lambda = 4$ we obtain the result. □

5 A Lower Bound for the Vertex Index of $\mathbf{B}_2^n$

In this section we prove estimate (4), i.e. we improve the constant in the estimate

$$cn^{3/2} \leq \text{vein}(\mathbf{B}_2^n) \leq 2n^{3/2}$$

from $c = 1/\sqrt{2\pi e}$ proved obtained [6] to $c = 1/\sqrt{3}$. Recall that the proof in [6] was based on volume estimates. We use here completely different approach.

Proof. Assume that $\mathbf{B}_2^n \subset \mathbf{L} = \text{conv}\{x_i\}_{i \leq N}$ for some non zero x_i's and denote

$$a = \sum_{i=1}^n |x_i|.$$

Our goal is to show that $a^2 \geq n^3/3$.

Define the operator $T : \mathbb{R}^N \to \mathbb{R}^n$ by $Te_i = x_i, i \leq N$. Then the rank of T is n (since $\mathbf{B}_2^n \subset \mathbf{L}$), $a = \sum_{i=1}^n |Te_i|$ and for every $x \in \mathbb{R}^n$

$$|x| \leq \|x\|_{\mathbf{L}^\circ} = \max_{i \leq N} \langle x, x_i \rangle = \max_{i \leq N} \langle T^*x, e_i \rangle. \qquad (7)$$

For $i \leq N$ denote

$$\lambda_i = \sqrt{|Te_i|/a} \qquad \text{and} \qquad v_i = \frac{Te_i}{a\lambda_i}.$$

Then

$$\sum_{i=1}^n \lambda_i^2 = 1 \qquad \text{and} \qquad \sum_{i=1}^n |v_i|^2 = 1.$$

We also observe that T^* can be presented as $T^* = a\Lambda S$, where Λ is the diagonal matrix with λ_i's on the diagonal and

$$S = \sum_{i=1}^N v_i \otimes e_i.$$

Note that the rank of S equals n. Let $s_1 \geq s_2 \geq \ldots \geq s_n > 0$ be the singular values of S and let $\{w_i\}_{i \leq n}$, $\{z_i\}_{i \leq n}$ be orthonormal systems such that

$$S = \sum_{i=1}^{n} s_n w_i \otimes z_i.$$

Then

$$\sum_{i=1}^{n} s_i^2 = \|S\|_{HS}^2 = \sum_{i=1}^{n} |v_i|^2 = 1,$$

where $\|S\|_{HS}$ is the Hilbert-Schmidt norm of S. Now for $m \leq n$ denote

$$S_m = \sum_{i=m}^{n} s_n w_i \otimes z_i$$

and consider the $(n + 1 - m)$-dimensional subspace

$$E_m = \mathrm{Im}\,(\Lambda S_m) \subset \mathrm{Im}\,T^*.$$

Considering the extreme points of the section of the cube $\mathbf{B}_{\infty}^{N} \cap E_m$ we observe that there exists a vector $y = \{y_i\}_{i \leq N} \in \mathbf{B}_{\infty}^{N} \cap E_m$ such that the set $A = \{i \mid |y_i| = 1\}$ has cardinality at least $n + 1 - m$. Without loss of generality we assume that $|A| = n + 1 - m$ (otherwise we choose an arbitrary subset of A with such cardinality). We observe

$$|(a\Lambda)^{-1}y| = \frac{1}{a}\sqrt{\sum_{i=1}^{N}\frac{y_i^2}{\lambda_i^2}} \geq \frac{1}{a}\sqrt{\sum_{i \in A}\frac{1}{\lambda_i^2}}$$

$$\geq \frac{n+1-m}{a\sqrt{\sum_{i \in A}\lambda_i^2}} \geq \frac{n+1-m}{a\sqrt{\sum_{i=1}^{N}\lambda_i^2}} = \frac{n+1-m}{a}.$$

Note that by construction $y \in E_m \subset \mathrm{Im}\,T^*$, so denoting the inverse of T^* from the image by $(T^*)^{-1}$ we have

$$|(T^*)^{-1}y| = |S^{-1}(a\Lambda)^{-1}y| = |S_m^{-1}(a\Lambda)^{-1}y| \geq \frac{|(a\Lambda)^{-1}y|}{\|S_m\|} \geq \frac{n+1-m}{as_m}.$$

Using (7) we obtain

$$\frac{n+1-m}{as_m} \leq |(T^*)^{-1}y| \leq \max_{i \leq N}\langle T^*(T^*)^{-1}y, e_i\rangle = \|y\|_{\infty} = 1.$$

This shows $s_m \geq (n + 1 - m)/a$ and implies

$$\frac{n^3}{3a^2} \leq \frac{1}{a^2} \sum_{m=1}^{n} (n + 1 - m)^2 \leq \sum_{m=1}^{n} s_m^2 = 1,$$

which proves the desired result.   □

References

1. K. Ball, A. Pajor, Convex bodies with few faces. Proc. Am. Math. Soc. **110**, 225–231 (1990)
2. I. Bárány, Z. Füredy, Approximation of the sphere by polytopes having few vertices. Proc. Am. Math. Soc. **102**(3), 651–659 (1988)
3. J.D. Batson, D.A. Spielman, N. Srivastava, Twice-Ramanujan Sparsifiers. STOC'09 – Proceedings of 2009 ACM International Symposium on Theory of Computing, 255–262, ACM, New York, 2009
4. K. Bezdek, in *Hadwiger-Levi's Covering Problem Revisited*, ed. by J. Pach. New Trends in Discrete and Computational Geometry. Algorithms Comb., vol. 10 (Springer, Berlin, 1993), pp. 199–233
5. K. Bezdek, The illumination conjecture and its extensions. Period. Math. Hungar. **53**, 59–69 (2006)
6. K. Bezdek, A.E. Litvak, On the vertex index of convex bodies. Adv. Math. **215**, 626–641 (2007)
7. B. Carl, A. Pajor, Gelfand numbers of operators with values in a Hilbert space. Invent. Math. **94**, 479–504 (1988)
8. E.D. Gluskin, Extremal properties of orthogonal parallelepipeds and their applications to the geometry of Banach spaces (Russian). Mat. Sb. (N.S.) **136**(178:1), 85–96 (1988); English translation in Math. USSR-Sb. **64**(1), 85–96 (1989)
9. E.D. Gluskin, A.E. Litvak, Asymmetry of convex polytopes and vertex index of symmetric convex bodies. Discrete Comput. Geom. **40**, 528–536 (2008)
10. F. John, in *Extremum Problems with Inequalities as Subsidiary Conditions*. Studies and Essays Presented to R. Courant on his 60th Birthday, January 8, 1948 (Interscience Publishers Inc., New York, 1948), pp. 187–204
11. H. Martini, V. Soltan, Combinatorial problems on the illumination of convex bodies. Aequationes Math. **57**, 121–152 (1999)

Inner Regularization of Log-Concave Measures and Small-Ball Estimates

Bo'az Klartag and Emanuel Milman

Abstract In the study of concentration properties of isotropic log-concave measures, it is often useful to first ensure that the measure has super-Gaussian marginals. To this end, a standard preprocessing step is to convolve with a Gaussian measure, but this has the disadvantage of destroying small-ball information. We propose an alternative preprocessing step for making the measure seem super-Gaussian, at least up to reasonably high moments, which does not suffer from this caveat: namely, convolving the measure with a random orthogonal image of itself. As an application of this "inner-thickening", we recover Paouris' small-ball estimates.

1 Introduction

Fix a Euclidean norm $|\cdot|$ on $\mathbb{R}^n$, and let X denote an isotropic random vector in $\mathbb{R}^n$ with log-concave density g. Recall that a random vector X in $\mathbb{R}^n$ (and its density) is called isotropic if $\mathrm{e}X = 0$ and $\mathrm{e}X \otimes X = Id$, i.e. its barycenter is at the origin and its covariance matrix is equal to the identity one. Taking traces, we observe that $\mathrm{e}|X|^2 = n$. Here and throughout we use e to denote expectation and $\mathbb{P}$ to denote probability. A function $g : \mathbb{R}^n \to \mathbb{R}_+$ is called log-concave if $-\log g : \mathbb{R}^n \to \mathbb{R} \cup \{+\infty\}$ is convex. Throughout this work, C, c, c_2, C', etc. denote universal positive numeric constants, independent of any other parameter and in particular the dimension n, whose value may change from one occurrence to the next.

B. Klartag (✉)
School of Mathematical Sciences, Sackler Faculty of Exact Science, Tel-Aviv University,
Tel Aviv 69978, Israel
e-mail: klartagb@tau.ac.il

E. Milman
Department of Mathematics, Technion – Israel Institute of Technology, Haifa 32000, Israel
e-mail: emilman@tx.technion.ac.il.

B. Klartag et al. (eds.), *Geometric Aspects of Functional Analysis*, Lecture Notes
in Mathematics 2050, DOI 10.1007/978-3-642-29849-3_15,
© Springer-Verlag Berlin Heidelberg 2012

Any high-dimensional probability distribution which is absolutely continuous has at least one super-Gaussian marginal (e.g. [14]). Still, in the study of concentration properties of X as above, it is many times advantageous to know that *all* of the one-dimensional marginals of X are super-Gaussian, at least up to some level (see e.g. [10, 15, 25]). By this we mean that for some $p_0 \geq 2$:

$$\forall 2 \leq p \leq p_0 \quad \forall \theta \in S^{n-1} \quad (E \, |\langle X, \theta \rangle|^p)^{\frac{1}{p}} \geq c(E \, |G_1|^p)^{\frac{1}{p}} , \tag{1}$$

where G_1 denotes a one-dimensional standard Gaussian random variable and S^{n-1} is the Euclidean unit sphere in $\mathbb{R}^n$. It is convenient to reformulate this using the language of L_p-centroid bodies, which were introduced by Lutwak and Zhang in [17] (under a different normalization). Given a random vector X with density g on $\mathbb{R}^n$ and $p \geq 1$, the L_p-centroid body $Z_p(X) = Z_p(g) \subset \mathbb{R}^n$ is the convex set defined via its support functional $h_{Z_p(X)}$ by:

$$h_{Z_p(X)}(y) = \left(\int_{\mathbb{R}^n} |\langle x, y \rangle|^p g(x) dx \right)^{1/p} , \quad y \in \mathbb{R}^n .$$

More generally, the *one-sided L_p-centroid body*, denoted $Z_p^+(X)$, was defined in [10] (cf. [11]) by:

$$h_{Z_p^+(X)}(y) = \left(2 \int_{\mathbb{R}^n} \langle x, y \rangle_+^p g(x) dx \right)^{1/p} , \quad y \in \mathbb{R}^n ,$$

where as usual $a_+ := \max(a, 0)$. Note that when g is even then both definitions above coincide, and that when the barycenter of X is at the origin, $Z_2(X)$ is the Euclidean ball B_2^n if and only X is isotropic. Observing that the right-hand side of (1) is of the order of $\sqrt{p}$, we would like to have:

$$\forall 2 \leq p \leq p_0 \quad Z_p^+(X) \supset c \sqrt{p} B_2^n , \tag{2}$$

where $B_2^n = \{x \in \mathbb{R}^n; |x| \leq 1\}$ is the unit Euclidean ball.

Unfortunately, we cannot in general expect to satisfy (2) for p_0 which grows with the dimension n. This is witnessed by X which is uniformly distributed on the n-dimensional cube $[-\sqrt{3}, \sqrt{3}]^n$ (the normalization ensures that X is isotropic), whose marginals in the directions of the axes are uniform on a constant-sized interval. Consequently, some preprocessing on X is required, which on one hand transforms it into another random variable Y whose density g satisfies (2), and on the other enables deducing back the desired concentration properties of X from those of Y.

A very common such construction is to convolve with a Gaussian, i.e. define $Y := (X + G_n)/\sqrt{2}$, where G_n denotes an independent standard Gaussian random vector in $\mathbb{R}^n$. In [12] (and in subsequent works like [5, 13]), the Gaussian played more of a regularizing role, but in [10], its purpose was to "thicken from inside" the distribution of X, ensuring that (2) is satisfied for all $p \geq 2$ (see [10, Lemma 2.3]). Regarding the transference of concentration properties, it follows from the argument in the proof of [12, Proposition 4.1] that:

$$\mathbb{P}(|X| \geq (1+t)\sqrt{n}) \leq C\mathbb{P}\left(|Y| \geq \sqrt{\frac{(1+t)^2+1}{2}}\sqrt{n}\right) \quad \forall t \geq 0 , \qquad (3)$$

and:

$$\mathbb{P}(|X| \leq (1-t)\sqrt{n}) \leq C\mathbb{P}\left(|Y| \leq \sqrt{\frac{(1-t)^2+1}{2}}\sqrt{n}\right) \quad \forall t \in [0,1] , \qquad (4)$$

for some universal constant $C > 1$. The estimate (3) is perfectly satisfactory for transferring (after an adjustment of constants) deviation estimates above the expectation from $|Y|$ to $|X|$. However, note that the right-hand side of (4) is bounded below by $P(|Y| \leq \sqrt{n/2})$ (and in particular does not decay to 0 when $t \to 1$), and so (4) is meaningless for transferring *small-ball* estimates from $|Y|$ to $|X|$. Consequently, the strategies employed in [5,10,12,13] did not and could not deduce the concentration properties of $|X|$ in the small-ball regime. This seems an inherent problem of adding an independent Gaussian: small-ball information is lost due to the "Gaussian-thickening".

The purpose of this note is to introduce a different inner-thickening step, which does not have the above mentioned drawback. Before formulating it, recall that X (or its density) is said to be "ψ_α with constant $D > 0$" if:

$$Z_p(X) \subset Dp^{1/\alpha} Z_2(X) \quad \forall p \geq 2 . \qquad (5)$$

We will simply say that "X is ψ_α", if it is ψ_α with constant $D \leq C$, and not specify explicitly the dependence of the estimates on the parameter D. By a result of Berwald [1] (or applying Borell's Lemma [3] as in [22, Appendix III]), it is well known that any X with log-concave density satisfies:

$$1 \leq p \leq q \quad \Rightarrow \quad Z_p(X) \subset Z_q(X) \subset C\frac{q}{p}Z_p(X) . \qquad (6)$$

In particular, such an X is always ψ_1 with some universal constant, and so we only gain additional information when $\alpha > 1$.

Theorem 1.1. *Let X denote an isotropic random vector in $\mathbb{R}^n$ with a log-concave density, which is in addition ψ_α ($\alpha \in [1,2]$), and let X' denote an independent copy of X. Given $U \in O(n)$, the group of orthogonal linear maps in $\mathbb{R}^n$, denote:*

$$Y_\pm^U := \frac{X \pm U(X')}{\sqrt{2}} .$$

Then:

1. *For any $U \in O(n)$, the concentration properties of $|Y_{\pm}^{U}|$ are transferred to $|X|$ as follows:*

$$P(|X| \geq (1+t)\sqrt{n}) \leq (2\max(P(|Y_{+}^{U}| \geq (1+t)\sqrt{n}), P(|Y_{-}^{U}| \geq (1+t)\sqrt{n})))^{\frac{1}{2}}$$
$$\forall t \geq 0,$$

and:

$$P(|X| \leq (1-t)\sqrt{n}) \leq (2\max(P(|Y_{+}^{U}| \leq (1-t)\sqrt{n}), P(|Y_{-}^{U}| \leq (1-t)\sqrt{n})))^{\frac{1}{2}}$$
$$\forall t \in [0,1].$$

2. *For any $U \in O(n)$:*

$$Z_{p}^{+}(Y_{\pm}^{U}) \subset C p^{1/\alpha} B_{2}^{n} \quad \forall p \geq 2. \tag{7}$$

3. *There exists a subset $A \subset O(n)$ with:*

$$\mu_{O(n)}(A) \geq 1 - \exp(-cn),$$

where $\mu_{O(n)}$ denotes the Haar measure on $O(n)$ normalized to have total mass 1, so that if $U \in A$ then:

$$Z_{p}^{+}(Y_{\pm}^{U}) \supset c_{1}\sqrt{p}B_{2}^{n} \quad \forall p \in [2, c_{2}n^{\frac{\alpha}{2}}]. \tag{8}$$

Remark 1.2. Note that when the density of X is even, then Y_{+}^{U} and Y_{-}^{U} in Theorem 1.1 are identically distributed, which renders the formulation of the conclusion more natural. However, we do not know how to make the formulation simpler in the non-even case.

Remark 1.3. Also note that $Y_{\pm}^{U}$ are isotropic random vectors, and that by the Prékopa–Leindler Theorem (e.g. [7]), they have log-concave densities.

As our main application, we manage to extend the strategy in the second named author's previous work with Guédon [10] to the small-ball regime, and obtain:

Corollary 1.4. *Let X denote an isotropic random vector in $\mathbb{R}^{n}$ with log-concave density, which is in addition ψ_{α} ($\alpha \in [1,2]$). Then:*

$$\mathbb{P}(|\,|X| - \sqrt{n}| \geq t\sqrt{n}) \leq C \exp(-cn^{\frac{\alpha}{2}}\min(t^{2+\alpha}, t)) \quad \forall t \geq 0, \tag{9}$$

and:

$$\mathbb{P}(|X| \leq \varepsilon\sqrt{n}) \leq (C\varepsilon)^{cn^{\frac{\alpha}{2}}} \quad \forall \varepsilon \in [0, 1/C]. \tag{10}$$

Corollary 1.4 is an immediate consequence of Theorem 1.1 and the following result, which is the content of [10, Theorem 4.1] (our formulation below is slightly more general, but this is what the proof gives):

Theorem (Guédon–Milman). *Let Y denote an isotropic random vector in $\mathbb{R}^n$ with a log-concave density, so that in addition:*

$$c_1 \sqrt{p} B_2^n \subset Z_p^+(Y) \subset c_2 p^{1/\alpha} B_2^n \quad \forall p \in [2, c_3 n^{\frac{\alpha}{2}}] \,, \tag{11}$$

for some $\alpha \in [1, 2]$. Then (9) and (10) hold with $X = Y$ (and perhaps different constants $C, c > 0$).

We thus obtain a preprocessing step which fuses perfectly with the approach in [10], allowing us to treat all deviation regimes *simultaneously* in a single unified framework. We point out that Corollary 1.4 by itself is not new. The *large* positive-deviation estimate:

$$P(|X| \geq (1 + t)\sqrt{n}) \leq \exp(-cn^{\frac{\alpha}{2}}t) \quad \forall t \geq C \,,$$

was first obtained by Paouris in [23]; it is known to be sharp, up to the value of the constants. The more general deviation estimate (9) was obtained in [10], improving when $t \in [0, C]$ all previously known results due to the first named author and to Fleury [5, 12, 13] (we refer to [10] for a more detailed account of these previous estimates). In that work, the convolution with Gaussian preprocessing was used, and so it was not possible to independently deduce the small-ball estimate (10). The latter estimate was first obtained by Paouris in [24], using the reverse Blaschke–Santaló inequality of Bourgain and Milman [4]. In comparison, our main tool in the proof of Theorem 1.1 is a covering argument in the spirit of Milman's M-position [18–20] (see also [26]), together with a recent lower-bound on the volume of Z_p bodies obtained in our previous joint work [15].

Acknowledgements We thank Olivier Guédon and Vitali Milman for discussions. Bo'az Klartag was supported in part by the Israel Science Foundation and by a Marie Curie Reintegration Grant from the Commission of the European Communities. Emanuel Milman was supported by the Israel Science Foundation (grant no. 900/10), the German Israeli Foundation's Young Scientist Program (grant no. I-2228-2040.6/2009), the Binational Science Foundation (grant no. 2010288), and the Taub Foundation (Landau Fellow).

2 Key Proposition

In this section, we prove the following key proposition:

Proposition 2.1. *Let X, X' be as in Theorem 1.1, let U be uniformly distributed on $O(n)$, and set:*

$$Y := \frac{X + U(X')}{\sqrt{2}}.$$

Then there exists a $c > 0$, so that:

$$\forall C_1 > 0 \ \exists c_1 > 0 \ \forall p \in [2, cn^{\alpha/2}] \quad \mathbb{P}(Z_p^+(Y) \supset c_1 \sqrt{p} B_2^n) \geq 1 - \exp(-C_1 n).$$

Here, as elsewhere, "uniformly distributed on $O(n)$" is with respect to the probability measure $\mu_{O(n)}$.

We begin with the following estimate due to Grünbaum [9] (see also [6, Formula (10)] or [2, Lemma 3.3] for simplified proofs):

Lemma 2.2 (Grünbaum). *Let X_1 denote a random variable on $\mathbb{R}$ with log-concave density and barycenter at the origin. Then $\frac{1}{e} \leq \mathbb{P}(X_1 \geq 0) \leq 1 - \frac{1}{e}$.*

Recall that the Minkowski sum $K + L$ of two compact sets $K, L \subset \mathbb{R}^n$ is defined as the compact set given by $\{x + y; x \in K, y \in L\}$. When K, L are convex, the support functional satisfies $h_{K+L} = h_K + h_L$.

Lemma 2.3. *With the same notations as in Proposition 2.1:*

$$Z_p^+(Y) \supset \frac{1}{2\sqrt{2}e^{1/p}}(Z_p^+(X) + U(Z_p^+(X))).$$

Proof. Given $\theta \in S^{n-1}$, denote $Y_1 = \langle Y, \theta \rangle$, $X_1 = \langle X, \theta \rangle$ and $X_1' = \langle U(X'), \theta \rangle$. By the Prékopa–Leindler theorem (e.g. [7]), all these one-dimensional random variables have log-concave densities, and since their barycenter is at the origin, we obtain by Lemma 2.2:

$$h_{Z_p^+(Y)}^p(\theta) = 2\mathbb{E}(Y_1)_+^p = \frac{2}{2^{p/2}}\mathbb{E}(X_1 + X_1')_+^p \geq \frac{2}{2^{p/2}}\mathbb{E}(X_1)_+^p \mathbb{P}(X_1' \geq 0)$$

$$\geq \frac{2}{\mathbb{E}2^{p/2}}\mathbb{E}(X_1)_+^p.$$

Exchanging the roles of X_1 and X_1' above, we obtain:

$$h_{Z_p^+(Y)}^p(\theta) \geq \frac{1}{e2^{p/2}} \max\left(h_{Z_p^+(X)}^p(\theta), h_{Z_p^+(U(X'))}^p(\theta)\right).$$

Consequently:

$$h_{Z_p^+(Y)}(\theta) \geq \frac{1}{\sqrt{2}e^{1/p}} \frac{h_{Z_p^+(X)}(\theta) + h_{Z_p^+(U(X'))}(\theta)}{2},$$

and since $Z_p^+(U(X')) = U(Z_p^+(X')) = U(Z_p^+(X))$, the assertion follows. $\square$

Next, recall that given two compact subsets $K, L \subset \mathbb{R}^n$, the covering number $N(K, L)$ is defined as the minimum number of translates of L required to cover K. The volume-radius of a compact set $K \subset \mathbb{R}^n$ is defined as:

$$\text{V.Rad.}(K) = \left(\frac{\text{Vol}(K)}{\text{Vol}(B_2^n)} \right)^{\frac{1}{n}} ,$$

measuring the radius of the Euclidean ball whose volume equals the volume of K. A convex compact set with non-empty interior is called a convex body, and given a convex body K with the origin in its interior, its polar K° is the convex body given by:

$$K^\circ := \{ y \in \mathbb{R}^n ; \langle x, y \rangle \leq 1 \quad \forall x \in K \} .$$

Finally, the mean-width of a convex body K, denoted $W(K)$, is defined as $W(K) = 2 \int_{S^{n-1}} h_K(\theta) d\mu_{S^{n-1}}(\theta)$, where $\mu_{S^{n-1}}$ denotes the Haar probability measure on S^{n-1}. The following two lemmas are certainly well-known; we provide a proof for completeness.

Lemma 2.4. *Let $K \subset \mathbb{R}^n$ be a convex body with barycenter at the origin, so that:*

$$N(K, B_2^n) \leq \exp(A_1 n) \quad \text{and} \quad \text{V.Rad.}(K) \geq a_1 > 0 .$$

Then:

$$N(K^\circ, B_2^n) \leq \exp(A_2 n) ,$$

where $A_2 \leq A_1 + \log(C/a_1)$, and $C > 0$ is a universal constant.

Proof. Set $K_s = K \cap -K$. By the covering estimate of König and Milman [16], it follows that:

$$N(K^\circ, B_2^n) \leq N(K_s^\circ, B_2^n) \leq C^n N(B_2^n, K_s) .$$

Using standard volumetric covering estimates (e.g. [26, Chap. 7]), we deduce:

$$N(K^\circ, B_2^n) \leq C^n \left(\frac{\text{Vol}(B_2^n + K_s/2)}{\text{Vol}(K_s/2)} \right) \leq C^n N(K_s/2, B_2^n) \frac{\text{Vol}(2B_2^n)}{\text{Vol}(K_s/2)} .$$

By a result of Milman and Pajor [21], it is known that $\text{Vol}(K_s) \geq 2^{-n} \text{Vol}(K)$, and hence:

$$N(K^\circ, B_2^n) \leq (8C)^n N(K, B_2^n) \text{V.Rad.}(K)^{-n} \leq (8C/a_1)^n \exp(A_1 n) ,$$

as required. $\qquad\square$

Lemma 2.5. *Let L denote any compact set in $\mathbb{R}^n$ ($n \geq 2$), so that $N(L, B_2^n) \leq \exp(A_1 n)$. If U is uniformly distributed on $O(n)$, then:*

$$P(L \cap U(L) \subset A_3 B_2^n) \geq 1 - \exp(-A_2 n) ,$$

where $A_2 = A_1 + (\log 2)/2$ and $A_3 = C' \exp(6A_1)$, for some universal constant $C' > 0$.

Proof Sketch. Assume that $L \subset \cup_{i=1}^{\exp(A_1 n)}(x_i + B_2^n)$. Set $R = 4C \exp(6A_1)$, for some large enough constant $C > 0$, and without loss of generality, assume that among all translates $\{x_i\}$, $\{x_i\}_{i=1}^N$ are precisely those points lying outside of RB_2^n. Observe that for each $i = 1, \ldots, N$, the cone $\{t(x_i + B_2^n); t \geq 0\}$ carves a spherical cap of Euclidean radius at most $1/R$ on S^{n-1}. By the invariance of the Haar measures on S^{n-1} and $O(n)$ under the action of $O(n)$, it follows that for every $i, j \in \{1, \ldots, N\}$:

$$P(U(x_i + B_2^n) \cap (x_j + B_2^n) \neq \emptyset) \leq \mu_{S^{n-1}}(B_{2/R}) \,,$$

where B_ε denotes a spherical cap on S^{n-1} of Euclidean radius ε, and recall $\mu_{S^{n-1}}$ denotes the normalized Haar measure on S^{n-1}. When $\varepsilon < 1/(2C)$, it is easy to verify that:

$$\mu_{S^{n-1}}(B_\varepsilon) \leq (C\varepsilon)^{n-1} \,,$$

and so it follows by the union-bound that:

$$P(L \cap U(L) \subset (R+1)B_2^n) \geq P(\forall i, j \in \{1, \ldots, N\}$$
$$U(x_i + B_2^n) \cap (x_j + B_2^n) = \emptyset) \geq 1 - N^2(2C/R)^{n-1} \,.$$

Since $N \leq \exp(2A_1(n-1))$, our choice of R yields the desired assertion with $C' = 5C$.

It is also useful to state:

Lemma 2.6. *For any density g on $\mathbb{R}^n$ and $p \geq 1$:*

$$Z_p^+(g) \subset 2^{1/p} Z_p(g) \subset Z_p^+(g) - Z_p^+(g) \,. \tag{12}$$

Proof. The first inclusion is trivial. The second follows since $a^{1/p} + b^{1/p} \geq (a + b)^{1/p}$ for $a, b \geq 0$, and hence for all $\theta \in S^{n-1}$:

$$h_{Z_p^+(g) - Z_p^+(g)}(\theta) = h_{Z_p^+(g)}(\theta) + h_{Z_p^+(g)}(-\theta) \geq 2^{1/p} h_{Z_p(g)}(\theta) \,.$$

$\square$

The next two theorems play a crucial role in our argument. The first is due to Paouris [23], and the second to the authors [15]:

Theorem (Paouris). *With the same assumptions as in Theorem 1.1:*

$$W(Z_p(X)) \leq C\sqrt{p} \quad \forall p \in [2, cn^{\alpha/2}] \,. \tag{13}$$

Theorem (Klartag–Milman). *With the same assumptions as in Theorem 1.1:*

$$V.Rad.(Z_p(X)) \geq c\sqrt{p} \quad \forall p \in [2, cn^{\alpha/2}]. \tag{14}$$

We are finally ready to provide a proof of Proposition 2.1:

Proof of Proposition 2.1. Let $p \in [2, cn^{\alpha/2}]$, where $c > 0$ is some small enough constant so that (13) and (14) hold. We will ensure that $c \leq 1$, so there is nothing to prove if $n = 1$. By (12), Sudakov's entropy estimate (e.g. [26]) and (13), we have (see also [8, Proposition 5.1]):

$$N(Z_p^+(X)/\sqrt{p}, B_2^n) \leq N(2^{1/p}Z_p(X)/\sqrt{p}, B_2^n)$$

$$\leq \exp(\tilde{C}nW(2^{1/p}Z_p(X)/\sqrt{p})^2) \leq \exp(Cn). \tag{15}$$

Note that by (12) and the Rogers–Shephard inequality [27], we have:

$$2^{n/p}\,\text{Vol}(Z_p(X)) \leq \text{Vol}(Z_p^+(X) - Z_p^+(X)) \leq 4^n\,\text{Vol}(Z_p^+(X)).$$

Consequently, the volume bound in (14) also applies to $Z_p^+(X)$:

$$V.Rad.(Z_p^+(X)) \geq c_1\sqrt{p}. \tag{16}$$

By Lemma 2.4, (15) and (16) imply that:

$$N(\sqrt{p}(Z_p^+(X))^\circ, B_2^n) \leq \exp(C_2 n).$$

Consequently, Lemma 2.5 implies that if U is uniformly distributed on $O(n)$, then for any $C_1 \geq C_2 + (\log 2)/2$, there exists a $C_3 > 0$, so that:

$$\mathbb{P}\left(Z_p^+(X)^\circ \cap U(Z_p^+(X)^\circ) \subset \frac{C_3}{\sqrt{p}}B_2^n\right) \geq 1 - \exp(-C_1 n),$$

or by duality (since $T(K)^\circ = (T^{-1})^*(K^\circ)$ for any linear map T of full rank), that:

$$\mathbb{P}\left(Z_p^+(X) + U(Z_p^+(X)) \supset C_3^{-1}\sqrt{p}B_2^n\right)$$

$$\geq \mathbb{P}\left(\text{conv}(Z_p^+(X) \cup U(Z_p^+(X))) \supset C_3^{-1}\sqrt{p}B_2^n\right) \geq 1 - \exp(-C_1 n).$$

Lemma 2.3 now concludes the proof. $\qquad\qquad\square$

3 Remaining Details

We now complete the remaining (standard) details in the proof of Theorem 1.1.

Proof of Theorem 1.1.

1. For any $U \in O(n)$ and $t \geq 0$, observe that:

$$2 \max \left(\mathbb{P} \left(\left| \frac{X + U(X')}{\sqrt{2}} \right| \leq t \right), \mathbb{P} \left(\left| \frac{X - U(X')}{\sqrt{2}} \right| \leq t \right) \right)$$

$$\geq \mathbb{P} \left(\left| \frac{X + U(X')}{\sqrt{2}} \right| \leq t \right) + \mathbb{P} \left(\left| \frac{X - U(X')}{\sqrt{2}} \right| \leq t \right)$$

$$= \mathbb{P} \left(\frac{|X|^2 + |X'|^2}{2} + \langle X, U(X') \rangle \leq t^2 \right)$$

$$+ \mathbb{P} \left(\frac{|X|^2 + |X'|^2}{2} - \langle X, U(X') \rangle \leq t^2 \right)$$

$$\geq \mathbb{P} \left(|X| \leq t \text{ and } |X'| \leq t \text{ and } \langle X, U(X') \rangle \leq 0 \right)$$

$$+ \mathbb{P} \left(|X| \leq t \text{ and } |X'| \leq t \text{ and } \langle X, U(X') \rangle > 0 \right)$$

$$= \mathbb{P} \left(|X| \leq t \text{ and } |X'| \leq t \right) = \mathbb{P} \left(|X| \leq t \right)^2 .$$

Similarly:

$$2 \max \left(\mathbb{P} \left(\left| \frac{X + U(X')}{\sqrt{2}} \right| \geq t \right), \mathbb{P} \left(\left| \frac{X - U(X')}{\sqrt{2}} \right| \geq t \right) \right) \geq \mathbb{P} \left(|X| \geq t \right)^2 .$$

This is precisely the content of the first assertion of Theorem 1.1.

2. Given $\theta \in S^{n-1}$, denote $Y_1 = P_\theta Y_+^U$, $X_1 = P_\theta X$ and $X_2 = P_\theta U(X')$, where P_θ denotes orthogonal projection onto the one-dimensional subspace spanned by θ. We have:

$$h_{Z_p(Y_+^U)}(\theta) = (\mathrm{e}|Y_1|^p)^{\frac{1}{p}} = \left(\mathrm{e} \left| \frac{X_1 + X_2}{\sqrt{2}} \right|^p \right)^{\frac{1}{p}}$$

$$\leq \frac{1}{\sqrt{2}} \left((\mathrm{e}|X_1|^p)^{\frac{1}{p}} + (\mathrm{e}|X_2|^p)^{\frac{1}{p}} \right) = \frac{1}{\sqrt{2}} \left(h_{Z_p(X)}(\theta) + h_{Z_p(U(X))}(\theta) \right) .$$

Employing in addition (12), it follows that:

$$Z_p^+(Y_+^U) \subset 2^{1/p} Z_p(Y_+^U) \subset \frac{2^{1/p}}{\sqrt{2}} \left(Z_p(X) + U(Z_p(X)) \right) ,$$

and the second assertion for Y_+^U follows since $Z_p(X) \subset C p^{\frac{1}{\alpha}} B_2^n$ by assumption. Similarly for Y_-^U.

3. Given a natural number i, set $p_i = 2^i$. Proposition 2.1 ensures the existence of a constant $c > 0$, so that for any $C_1 > 0$, there exists a constant $c_1 > 0$, so that for any $p_i \in [2, cn^{\frac{\alpha}{2}}]$, there exists a subset $A_i \subset O(n)$ with:

$$\mu_{O(n)}(A_i) \geq 1 - \exp(-C_1 n),$$

so that:

$$\forall U \in A_i \quad Z_{p_i}(Y_+^U) \supset c_1 \sqrt{p_i} B_2^n.$$

Denoting $A_0 := \cap \{A_i \; ; \; p_i \in [2, cn^{\frac{\alpha}{2}}]\}$, and setting $A = A_0 \cap -A_0$, where $-A_0 := \{-U \in O(n); U \in A_0\}$, it follow by the union-bound that:

$$\mu_{O(n)}(A) \geq 1 - 2\log(C_2 + n)\exp(-C_1 n).$$

By choosing the constant $C_1 > 0$ large enough, we conclude that:

$$\mu_{O(n)}(A) \geq 1 - \exp(-C_3 n).$$

By construction, the set A has the property that:

$$\forall U \in A \quad \forall p_i \in [2, cn^{\frac{\alpha}{2}}] \quad Z_{p_i}(Y_\pm^U) \supset c_1 \sqrt{p_i} B_2^n.$$

Using (6), it follows that:

$$\forall U \in A \quad \forall p \in [2, cn^{\frac{\alpha}{2}}] \quad Z_p(Y_\pm^U) \supset \frac{c_1}{\sqrt{2}} \sqrt{p} B_2^n,$$

thereby concluding the proof of the third assertion. $\qquad \Box$

References

1. L. Berwald, Verallgemeinerung eines Mittelwertsatzes von J. Favard für positive konkave Funktionen. Acta Math. **79**, 17–37 (1947)
2. S.G. Bobkov, On concentration of distributions of random weighted sums. Ann. Probab. **31**(1), 195–215 (2003)
3. Ch. Borell, Convex measures on locally convex spaces. Ark. Mat. **12**, 239–252 (1974)
4. J. Bourgain, V.D. Milman, New volume ratio properties for convex symmetric bodies in $\mathbb{R}^n$. Invent. Math. **88**, 319–340 (1987)
5. B. Fleury, Concentration in a thin euclidean shell for log-concave measures. J. Func. Anal. **259**, 832–841 (2010)
6. M. Fradelizi, Contributions à la géométrie des convexes. Méthodes fonctionnelles et probabilistes. Habilitation à Diriger des Recherches de l'Université Paris-Est Marne La Vallée (2008). http://perso-math.univ-mlv.fr/users/fradelizi.matthieu/pdf/HDR.pdf.

7. R.J. Gardner, The Brunn-Minkowski inequality. Bull. Am. Math. Soc. (N.S.) **39**(3), 355–405 (2002)
8. A. Giannopoulos, G. Paouris, P. Valettas, On the existence of subgaussian directions for log-concave measures. Contemp. Math. **545**, 103–122 (2011)
9. B. Grünbaum, Partitions of mass-distributions and of convex bodies by hyperplanes. Pac. J. Math. **10**, 1257–1261 (1960)
10. O. Guédon, E. Milman, Interpolating thin-shell and sharp large-deviation estimates for isotropic log-concave measures. Geom. Funct. Anal. **21**(5), 1043–1068 (2011)
11. C. Haberl, L_p intersection bodies. Adv. Math. **217**(6), 2599–2624 (2008)
12. B. Klartag, A central limit theorem for convex sets. Invent. Math. **168**, 91–131 (2007)
13. B. Klartag, Power-law estimates for the central limit theorem for convex sets. J. Funct. Anal. **245**, 284–310 (2007)
14. B. Klartag, On nearly radial marginals of high-dimensional probability measures. J. Eur. Math. Soc. **12**, 723–754 (2010)
15. B. Klartag, E. Milman, *Centroid bodies and the logarithmic Laplace transform a unified approach*. J. Func. Anal. **262**(1), 10–34 (2012)
16. H. König, V.D. Milman, in *On the Covering Numbers of Convex Bodies*. Geometrical Aspects of Functional Analysis (1985/86). Lecture Notes in Math., vol. 1267 (Springer, Berlin, 1987), pp. 82–95
17. E. Lutwak, G. Zhang, Blaschke-Santaló inequalities. J. Diff. Geom. **47**(1), 1–16 (1997)
18. V.D. Milman, Inégalité de Brunn-Minkowski inverse et applications à la théorie locale des espaces normés. C. R. Acad. Sci. Paris Sér. I Math. **302**(1), 25–28 (1986)
19. V.D. Milman, Entropy point of view on some geometric inequalities. C. R. Acad. Sci. Paris Sér. I Math. **306**(14), 611–615 (1988)
20. V.D. Milman, *Isomorphic Symmetrizations and Geometric Inequalities*. Geometric Aspects of Functional Analysis (1986/87). Lecture Notes in Math., vol. 1317 (Springer, Berlin, 1988), pp. 107–131
21. V.D. Milman, A. Pajor, Entropy and asymptotic geometry of non-symmetric convex bodies. Adv. Math. **152**(2), 314–335 (2000)
22. V.D. Milman, G. Schechtman, *Asymptotic Theory of Finite-Dimensional Normed Spaces*, with an appendix by M. Gromov. Lecture Notes in Mathematics, vol. 1200 (Springer, Berlin, 1986)
23. G. Paouris, Concentration of mass on convex bodies. Geom. Funct. Anal. **16**(5), 1021–1049 (2006)
24. G. Paouris, *Small ball probability estimates for log-concave measures*, Trans. Amer. Math. Soc. **364**, 287–308 (2012)
25. G. Paouris, *On the existence of supergaussian directions on convex bodies*. Mathematika (to appear).
26. G. Pisier, *The Volume of Convex Bodies and Banach Space Geometry*. Cambridge Tracts in Mathematics, vol. 94 (Cambridge University Press, Cambridge, 1989)
27. C.A. Rogers, G.C. Shephard, The difference body of a convex body. Arch. Math. **8**, 220–233 (1957)

An Operator Equation Generalizing the Leibniz Rule for the Second Derivative

Hermann König and Vitali Milman*

Abstract We determine all operators $T : C^2(\mathbb{R}) \to C(\mathbb{R})$ and $A : C^1(\mathbb{R}) \to C(\mathbb{R})$ which satisfy the equation

$$T(f \cdot g) = (Tf) \cdot g + f \cdot (Tg) + (Af) \cdot (Ag) ; \quad f, g \in C^2(\mathbb{R}) . \quad (1)$$

This operator equation models the second order Leibniz rule for $(f \cdot g)''$ with $Af = \sqrt{2} f'$. Under a mild regularity and non-degeneracy assumption on A, we show that the operators T and A have to be of a very restricted type. In addition to the operator solutions S of the Leibniz rule derivation equation corresponding to $A = 0$,

$$S(f \cdot g) = (Sf) \cdot g + f \cdot (Sg) ; \quad f, g \in C^2(\mathbb{R}) \text{ or } C^1(\mathbb{R}) , \quad (2)$$

which are of the form

$$Sf = bf' + af \ln|f|, \quad a, b \in C(\mathbb{R}),$$

*Supported in part by the Alexander von Humboldt Foundation, by ISF grant 387/09 and BSF grant 2006079.

H. König (✉)
Mathematisches Seminar, Christian-Albrechts-Universität zu Kiel, Ludewig-Meyn-Str. 4, 24118 Kiel, Germany
e-mail: hkoenig@math.uni-kiel.de

V. Milman
Sackler Faculty of Exact Sciences, Department of Mathematics, Tel Aviv University, Ramat Aviv, 50 69978 Tel Aviv, Israel
e-mail: milman@post.tau.ac.il

B. Klartag et al. (eds.), *Geometric Aspects of Functional Analysis*, Lecture Notes in Mathematics 2050, DOI 10.1007/978-3-642-29849-3_16,
© Springer-Verlag Berlin Heidelberg 2012

T and A may be of the following three types

$$Tf = \tfrac{1}{2}d^2 f'' \qquad\qquad , \; Af = d\, f'$$
$$Tf = \tfrac{1}{2}d^2 f(\ln|f|)^2 \quad , \; Af = d\, f \ln|f|$$
$$Tf = d^2 f(\varepsilon|f|^p - 1) \;, \; Af = d\, f(\varepsilon|f|^p - 1)$$

for suitable continuous functions d, c and p and where ε is either 1 or sgn f and $p \geq -1$. The last operator solution is degenerate in the sense that T is a multiple of A. We also determine all solutions of (1) if T and A operate only on positive $C^2(\mathbb{R})$-functions or $C^2(\mathbb{R})$-functions which are nowhere zero.

1 Introduction and Results

In previous papers we showed that the derivative and other differentiation operators are characterized by simple operator equations like the chain rule or the Leibniz rule plus some natural initial conditions [3, 5, 6]. In this paper we consider an operator equation generalizing the second order Leibniz rule for C^2-functions on $\mathbb{R}$,

$$(f \cdot g)'' = f'' \cdot g + f \cdot g'' + 2\, f' \cdot g' \;; \; f, g \in C^2(\mathbb{R}) \;:$$

Suppose $T : C^2(\mathbb{R}) \to C(\mathbb{R})$ and $A : C^1(\mathbb{R}) \to C(\mathbb{R})$ are operators satisfying the operator equation

$$T(f \cdot g) = Tf \cdot g + f \cdot Tg + Af \cdot Ag \;; \; f, g \in C^2(\mathbb{R}) \;. \tag{1}$$

So in the case of the second derivative $Tf = f''$ we have that $Af = \sqrt{2}f'$. We determine all operators T and A verifying (1) for all $C^2(\mathbb{R})$-functions and show that T and A have to be of a very restricted type. This allows a characterization of the second derivative $Tf = f''$ by (1) and some natural smoothness and initial conditions. No regularity or linearity assumptions will be made on T and A besides a mild non-degeneracy and a weak continuity condition imposed on A. As a corollary we also determine the analogues of the second derivative operation on $C^1(\mathbb{R})$ and $C(\mathbb{R})$.

All operators $S : C^k(\mathbb{R}) \to C(\mathbb{R})$ satisfying the Leibniz rule derivation equation

$$S(f \cdot g) = Sf \cdot g + f \cdot Sg \;; \; f, g \in C^k(\mathbb{R}) \tag{2}$$

were determined in [6] for $k \in \mathbb{N}$: There are continuous functions $a, b \in C(\mathbb{R})$ such that any such S is of the form

$$(Sf)(x) = b(x) f'(x) + a(x) f(x) \ln|f(x)| \;, \tag{3}$$

without any continuity assumption on S. Linearity is not assumed and, in fact, is not present in the case of the entropy type solution. For $k = 0$, the operators satisfying (2) had been found by Goldmann and Šemrl [4]: then $b(x) = 0$ so that S is defined on $C(\mathbb{R})$ and given by the entropy function. Clearly, (2) is a homogeneous version of equation (1). The operators S satisfy (1) with $A = 0$, and the solutions (3) of the "homogeneous" equation (2) may be added to any solution T of the "inhomogeneous" equation (1), yielding another map fulfilling (1) with the same operator A. We prove that only very few types of operators T and A will be suitable for (1) to hold.

We consider $A : C^1(\mathbb{R}) \to C(\mathbb{R})$ to be "of lower order", not depending on second derivatives. To prove a localization property for A for $C^1(\mathbb{R})$-functions while (1) is assumed only for $C^2(\mathbb{R})$-functions, we need the following mild continuity assumption on A:

Definition. An operator $A : C^1(\mathbb{R}) \to C(\mathbb{R})$ is *pointwise C^1-continuous* provided that for any $f \in C^1(\mathbb{R})$ and any sequence $(f_n)_{n \in \mathbb{N}}$ in $C^2(\mathbb{R})$ such that $f_n \to f$ and $f_n' \to f'$ converge uniformly on $\mathbb{R}$, we have for all $x \in \mathbb{R}$ that $\lim_{n \to \infty} (Af_n)(x) = (Af)(x)$ holds.

To prove localization for $C^2(\mathbb{R})$-functions, we need a non-degeneracy assumption as well:

Definition. An operator $A : C^1(\mathbb{R}) \to C(\mathbb{R})$ is *non-degenerate* if for any open interval $J \subset \mathbb{R}$ and any $x \in J$ there are functions $g_1, g_2 \in C^2(\mathbb{R})$ with support in J such that the two vectors $(g_i(x), Ag_i(x)) \in \mathbb{R}^2$, $i \in \{1, 2\}$ are linearly independent.

This means that A should be essentially different from a multiple of the identity. Our main result gives all operators T and A verifying the second order Leibniz rule (1) on $C^2(\mathbb{R})$.

Theorem 1. *Let $T : C^2(\mathbb{R}) \to C(\mathbb{R})$ and $A : C^1(\mathbb{R}) \to C(\mathbb{R})$ be operators such that the equation*

$$T(f \cdot g)(x) = (Tf)(x) \cdot g(x) + f(x) \cdot (Tg)(x) + (Af)(x) \cdot (Ag)(x) \quad (1)$$

holds for all functions $f, g \in C^2(\mathbb{R})$ and any $x \in \mathbb{R}$. Assume that A is non-degenerate and pointwise C^1-continuous. Then there are continuous functions $a, b, d, e, p \in C(\mathbb{R})$ such that

$$(Tf)(x) = (T_1 f)(x) + (Sf)(x),$$

where

$$(Sf)(x) = b(x) f'(x) + a(x) f(x) \ln |f(x)|. \quad (3)$$

solves the "homogeneous" equation and where T_1 and A are of one of the following three forms

$$(T_1 f)(x) = \frac{d(x)^2}{2} f''(x), \qquad\qquad (Af)(x) = d(x) f'(x)$$

or

$$(T_1 f)(x) = \tfrac{e(x)^2}{2} f(x)(\ln |f(x)|)^2 , \qquad (Af)(x) = e(x) f(x) \ln |f(x)|$$

or

$$(T_1 f)(x) = e(x)^2 f(x)(\{\operatorname{sgn} f(x)\} |f(x)|^{p(x)} - 1),$$
$$(Af)(x) = e(x) f(x)(\{\operatorname{sgn} f(x)\} |f(x)|^{p(x)} - 1) .$$

Here $p(x) \geq -1$ and the bracket $\{\operatorname{sgn} f(x)\}$ means that there are two solutions, one without this term and one with the sgn-*term. The formulas for T hold for all $f \in C^2(\mathbb{R})$ and the ones for A for all $f \in C^1(\mathbb{R})$. The first operator T_1 is the only one that cannot be extended to $C^1(\mathbb{R})$. Conversely, these operators (T, A) satisfy* (1).

Remarks. 1. After this paper was finished, we reproved Theorem 1 in [7] using a different method which allowed to avoid the assumption of pointwise C^1-continuity of A. However, Theorem 3 below, which is the analogue of Theorem 1 for functions which never vanish, requires the assumption of pointwise C^1-continuity of A and is not considered in [7]. In addition, the proof given here explains clearer and more easily why the dependence on f'' in the first and main case is *linear*.
2. Let $x_0 \in \mathbb{R}$. If $d(x_0) = 0$ and $e(x_0) = 0$ or $e(x_0) = 0$ for two different of the three solutions, one might think that the corresponding operators might be mixed: taking for $x < x_0$ one formula and for $x > x_0$ the second formula: on both sides of x_0, equation (1) is satisfied with $Tf, Af \in C(\mathbb{R})$. However, the condition of non-degeneration of A would not be satisfied in x_0, so this is not a valid solution under the assumptions of Theorem 1.
3. The last operator (T_1, A) may be also considered degenerate since in these cases T_1 and A are proportional.
4. Starting with the derivations $Sf = f'$ and $Sf = f \ln |f|$ solving (2) in C^1 and C, the first two operators T_1 might be considered "second iterated derivations". In particular, $T_1 f = \tfrac{1}{2} f(\ln |f|)^2$ corresponds to the derivation $Sf = f \ln |f|$ on C.
5. The last two solutions for (T, A) extend to operators $T : C(\mathbb{R}) \to C(\mathbb{R})$ and $A : C(\mathbb{R}) \to C(\mathbb{R})$ satisfying the operator functional equation (1) for all $f, g \in C(\mathbb{R})$, if the $b(x) f'(x)$ term in Sf is omitted.

Theorem 1 yields a characterization of the second derivative by the operator equation (1) together with a natural smoothness and an initial condition:

Proposition 2. *Assume $T : C^2(\mathbb{R}) \to C(\mathbb{R})$ and $A : C^1(\mathbb{R}) \to C(\mathbb{R})$ satisfy the functional equation* (1) *on $C^2(\mathbb{R})$ and that A is non-degenerate and pointwise C^1-continuous. If T is zero on the affine functions on $\mathbb{R}$, it is the second derivative up to some continuous function, namely there exists $d \in C(\mathbb{R})$ such that $Tf = \tfrac{d}{2} f''$ and $Af = d f'$.*

We also determine the larger class of operators satisfying (1) on the set of *nowhere zero C^2-functions. Let $C_{\neq 0}^k(\mathbb{R}) := \{f \in C^k(\mathbb{R}) | f(x) \neq 0$ for all*

$x \in \mathbb{R}$}, $k \in \mathbb{N}$. In the case of the "homogeneous" equation (2), all $C^2_{\neq 0}(\mathbb{R})$-solutions are of the type

$$(S'f)(x) = c(x)(f''(x) - \frac{f'(x)^2}{f(x)}) + b(x)f'(x) + a(x)f(x)\ln|f(x)|, \quad (4)$$

where a, b and c are suitable continuous functions on $\mathbb{R}$, cf. [6]. Adding S' to any solution T of (1) again yields a solution of (1) for $C^2_{\neq 0}(\mathbb{R})$-functions with the same operator A. In this situation, we prove:

Theorem 3. *Let* $T : C^2_{\neq 0}(\mathbb{R}) \to C(\mathbb{R})$ *and* $A : C^1_{\neq 0}(\mathbb{R}) \to C(\mathbb{R})$ *be operators such that the functional equation*

$$T(f \cdot g)(x) = (Tf)(x) \cdot g(x) + f(x) \cdot (Tg)(x) + (Af)(x) \cdot (Ag)(x) \quad (1)$$

holds for all functions $f, g \in C^2_{\neq 0}(\mathbb{R})$ *and any* $x \in \mathbb{R}$. *Assume that* A *is non-degenerate and pointwise* C^1*-continuous restricted to the nowhere zero functions. Then there are continuous functions* $a, b, c, d, e, p \in C(\mathbb{R})$ *such that* T *and* A *have the form*

$$Tf(x) = (T'f)(x) + (S'f)(x)$$

where

$$(S'f)(x) = c(x)(f''(x) - \frac{f'(x)^2}{f(x)}) + b(x)f'(x) + a(x)f(x)\ln|f(x)| \quad (5)$$

solves the "homogeneous" equation and where T' *and* A *are of one of the following two forms*

$$(T'f)(x) = \frac{d(x)^2}{2}\frac{f'(x)^2}{f(x)} + \frac{e(x)^2}{2}f(x)(\ln|f(x)|)^2 + d(x)e(x)f'(x)\ln|f(x)|$$

$$(Af)(x) = d(x)f'(x) + e(x)f(x)\ln|f(x)|$$

or

$$(T'f)(x) = e(x)^2 f(x)(\{\text{sgn } f(x)\}|f(x)|^{p(x)}\exp(d(x)f'(x)/f(x)) - 1)$$

$$(Af)(x) = e(x)\ f(x)(\{\text{sgn } f(x)\}|f(x)|^{p(x)}\exp(d(x)f'(x)/f(x)) - 1)$$

In the second case the term $\{\text{sgn } f(x)\}$ *may be present both in* T' *and* A *or not. Conversely, these operators* (T, A) *satisfy equation* (1).

Remarks. (a) The second case is trivial in the sense that T' and A are proportional. The first solution combines the first two solutions in Theorem 1.

(b) Notice that a term $c(x)f''(x)$ involving the second derivative may show up in all solutions due to the fact that it may be present in $S'f$.

(c) The proof uses localization on open intervals $J \subset \mathbb{R}$, and Theorems 1 and 3 are true just as well for maps $T : C^2(J) \to C(J)$ and $A : C^1(J) \to C(J)$ satisfying (1) for $x \in J$. In the case of Theorem 3, the functions should be non-zero everywhere on J.

Acknowledgements We would like to thank the referee for valuable suggestions and remarks, in particular for pointing out a gap in the original localization argument. This led us to the example in the following section.

2 Localization

The first step in the Proof of Theorems 1 and 3 is to show that any operators T and A verifying the functional equation (1) are local operators, i.e. defined pointwise in the sense that $(Tf)(x)$ only depends on x, $f(x)$, $f'(x)$ and $f''(x)$ and that $(Af)(x)$ is a function of x, $f(x)$ and $f'(x)$. Afterwards, we will analyze this dependence in a second step. We start with a "localization on intervals" result.

Proposition 4. *Assume* $T : C^2(\mathbb{R}) \to C(\mathbb{R})$ *and* $A : C^1(\mathbb{R}) \to C(\mathbb{R})$ *satisfy the functional equation* (1) *on* $C^2(\mathbb{R})$ *and that* A *is non-degenerate. Then:*

(a) $T\mathbb{1} = A\mathbb{1} = 0$.
(b) *For any open interval* $J \subset \mathbb{R}$ *and* $f_1, f_2 \in C(\mathbb{R})$ *with* $f_1|_J = f_2|_J$, *we have* $(Tf_1)|_J = (Tf_2)|_J$ *and* $(Af_1)|_J = (Af_2)|_J$. *This is also true in the case of* $C^2_{\neq 0}(\mathbb{R})$-*functions, if* (1) *holds for* $f, g \in C^2_{\neq 0}(\mathbb{R})$.

Proof. (a) By the functional equation (1), we have for any $g \in C^2(\mathbb{R})$ and $x \in \mathbb{R}$

$$0 = T(\mathbb{1}(x) \cdot g(x)) - \mathbb{1}(x) \cdot T(g)(x) = T(\mathbb{1})(x) \cdot g(x) + A(\mathbb{1})(x) \cdot Ag(x) .$$

Since A is non-degenerate, we may choose functions $g_1, g_2 \in C^2(\mathbb{R})$ such that $(g_i(x), Ag_i(x))$, $i \in \{1, 2\}$ are linearly independent. Applying the previous equality to g_1, g_2, we find that $T(\mathbb{1})(x) = A(\mathbb{1})(x) = 0$. Therefore $T(\mathbb{1}) = A(\mathbb{1}) = 0$ on $\mathbb{R}$.

(b) Assume $J \subset \mathbb{R}$ is an open interval and $f_1, f_2 \in C^2(\mathbb{R})$ are such that $f_1|_J = f_2|_J$. Then for any $g \in C^2(\mathbb{R})$ with support in J, we have $f_1 \cdot g = f_2 \cdot g$ and hence by (1)

$$(Tf_1) \cdot g + f_1 \cdot (Tg) + (Af_1) \cdot (Ag) = T(f_1 \cdot g) = T(f_2 \cdot g)$$
$$= (Tf_2) \cdot g + f_2 \cdot (Tg) + (Af_2) \cdot (Ag) ,$$

and hence for any $x \in J$, with $f_1(x) = f_2(x)$,

$$((Tf_1)(x) - (Tf_2)(x)) \cdot g(x) + ((Af_1)(x) - (Af_2)(x)) \cdot (Ag)(x) = 0 . \quad (6)$$

Since A is assumed to be non-degenerate, we may choose two functions $g_1, g_2 \in C^2(\mathbb{R})$ with support in J such that $(g_1(x), (Ag_1)(x))$ and $(g_2(x), (Ag_2)(x))$ are linearly independent in $\mathbb{R}^2$. Applying (6) for $g = g_1$ and $g = g_2$ then also yields $(Tf_1)(x) = (Tf_2)(x)$ and $(Af_1)(x) = (Af_2)(x)$. Hence $Tf_1|_J = Tf_2|_J$ and $Af_1|_J = Af_2|_J$ in both cases. The argument for $C^2_{\neq 0}(\mathbb{R})$-functions is identical. $\square$

For non-degenerate A, the previous result implies as in [5] that T and A are pointwise defined operators:

Proposition 5. *Assume $T : C^2(\mathbb{R}) \to C(\mathbb{R})$ and $A : C^1(\mathbb{R}) \to C(\mathbb{R})$ satisfy the functional equation (1) on $C^2(\mathbb{R})$ and that A is non-degenerate and pointwise C^1-continuous. Then there are functions $F : \mathbb{R}^4 \to \mathbb{R}$ and $B : \mathbb{R}^3 \to \mathbb{R}$ such that*

$$(Tf)(x) = F_x(f(x), f'(x), f''(x)),$$
$$(Af)(x) = B_x(f(x), f'(x)); \quad f \in C^2(\mathbb{R}), \ x \in \mathbb{R}.$$

For convenience, we write the dependence on the first variable $x \in \mathbb{R}$ as index. A corresponding statement holds in the $C^2_{\neq 0}(\mathbb{R})$-case.

Proof. (a) For $x_0 \in \mathbb{R}$ and $f \in C^2(\mathbb{R})$, let g be the quadratic approximation to f in x_0,

$$g(x) := f(x_0) + f'(x_0)(x - x_0) + \frac{1}{2}f''(x_0)(x - x_0)^2,$$

and put

$$h(x) := \begin{cases} f(x) & x < x_0 \\ g(x) & x \geq x_0 \end{cases}.$$

Then $h \in C^2(\mathbb{R})$ and with $J_1 = (-\infty, x_0)$, $J_2 = (x_0, \infty)$ we have $f|_{J_1} = h|_{J_1}$, $h|_{J_2} = g|_{J_2}$. By Proposition 4, $(Tf)|_{J_1} = (Th)|_{J_1}$ and $(Th)|_{J_2} = (Tg)|_{J_2}$. Since Tf, Th and Tg are continuous in $x_0 \in \overline{J_1} \cap \overline{J_2}$, we find $(Tf)(x_0) = (Th)(x_0) = (Tg)(x_0)$. Since g only depends on x_0, $f(x_0)$, $f'(x_0)$ and $f''(x_0)$, $(Tf)(x_0) = F_{x_0}(f(x_0), f'(x_0), f''(x_0))$ for a suitable function $F : \mathbb{R}^4 \to \mathbb{R}$. In the case of $C^2_{\neq 0}(\mathbb{R})$ and $f(x_0) \neq 0$, there is an open interval $J \subset \mathbb{R}$ with $x_0 \in J$ and $g(x) \neq 0$ for any $x \in J$. By Proposition 4, $(Tf)(x_0)$ is determined by $g|_J$. Therefore again $(Tf)(x_0) = F_{x_0}(f(x_0), f'(x_0), f''(x_0))$ with $F : \mathbb{R} \times \mathbb{R}_{\neq 0} \times \mathbb{R}^2 \to \mathbb{R}$.

(b) In the case of $A : C^1(\mathbb{R}) \to C(\mathbb{R})$, consider the tangential approximation to $f \in C^2(\mathbb{R})$ in x_0,

$$g(x) := f(x_0) + f'(x_0)(x - x_0).$$

Then h, defined as above, is only in $C^1(\mathbb{R})$ and (1) is not directly applicable to h. However, as in [5], we find $C^2(\mathbb{R})$-functions g_n such that $g_n \to g$ and $g'_n \to g'$ converge uniformly and such that

$$h_n(x) := \begin{cases} f(x) & x < x_0 \\ g_n(x) & x \geq x_0 \end{cases}.$$

defines a $C^2(\mathbb{R})$-function. Using Proposition 4 in the same way as in (a) gives

$$(Af)(x_0) = (Ah_n)(x_0) = (Ag_n)(x_0),$$

and the pointwise C^1-continuity assumption on A implies

$$(Af)(x_0) = \lim_n (Ag_n)(x_0) = (Ag)(x_0).$$

Therefore $(Af)(x_0)$ only depends on the parameter x_0, $f(x_0)$ and $f'(x_0)$ defining the tangent g to f at x_0, i.e. $Af(x_0) = B_{x_0}(f(x_0), f'(x_0))$.   □

Example. By itself, without some assumption of non-degeneracy, the operator equation (1) does not yield a local operation in the sense of being determined completely by the values of x, $f(x)$, $f'(x)$ and $f''(x)$: Consider the operators $T, A : C^2(\mathbb{R}) \to C(\mathbb{R})$ defined by

$$(Tf)(x) = -f(x) + f(x+1), \quad (Af)(x) = f(x) - f(x+1).$$

One quickly checks that (1) is satisfied by T and A. On intervals J of length < 1 with $x \in J$ and functions f with support in J, $(Af)(x) = f(x)$. Therefore A is degenerate. T and A are not locally defined in one point x, depending also on $x+1$.

3   Two Functional Equations on $\mathbb{R}$

A function $c : \mathbb{R} \to \mathbb{R}$ is *additive* if $c(x + y) = c(x) + c(y)$ for all $x, y \in \mathbb{R}$. As well-known, additive functions are linear, i.e. $c(x) = \gamma x$, if they are measurable, cf. [1]. In the following, we write $c[x]$ for additive functions which at that stage are not known to be linear. We also do so for additive functions on $\mathbb{R}_+$, $c : \mathbb{R}_+ \to \mathbb{R}$.

To determine the specific form of the functions F and B representing T and A according to Proposition 5, we need two Lemmas on functional equations which show up analyzing the dependence on the function and the derivative variable in F and B.

Lemma 6. *Assume that $F, B : \mathbb{R} \to \mathbb{R}$ are functions such that for any $x, y \in \mathbb{R}$*

$$F(x + y) = F(x) + F(y) + B(x)B(y). \tag{7}$$

Then there are additive functions $c, d : \mathbb{R} \to \mathbb{R}$ and there is some $\gamma \in \mathbb{R}$ such that F and B are of one of the following three forms

(a) $F(x) = -\gamma^2 + d[x]$, $B(x) = \gamma$

(b) $F(x) = \frac{1}{2}(c[x])^2 + d[x]$, $B(x) = c[x]$

(c) $F(x) = \gamma^2(e^{c[x]} - 1) + d[x]$, $B(x) = \gamma(e^{c[x]} - 1)$.

Remark. Case (c) with $d = 0$ and $\gamma = 1$ yields a solution for (7) with $B = F$.

Proof. There are related functional equations in Sect. 3.1.3 in [1] and in Chap. 15, Theorem 1 in [2] to which the Lemma could be reduced. We prefer to give a direct proof.

(i) If $B = 0$, F is additive and we are in case (a) with $\gamma = 0$. Thus assume $B \neq 0$ and choose $a \in \mathbb{R}$ with $B(a) \neq 0$. Let

$$f(x) := F(x+a) - F(x) - F(a), \ b(x) := B(x+a) - B(x).$$

Equation (7) then is transformed into

$$f(x+y) = f(x) + b(x)B(y). \tag{8}$$

Hence for $x = 0$, $f(y) = f(0) + b(0)B(y)$. Replacing here y by $x + y$ and by x, respectively, and inserting this back into (8), we find that

$$b(0)(B(x+y) - B(x)) = b(x)B(y). \tag{9}$$

If $b(0) = 0$, (9) yields $b = 0$ since $B(a) \neq 0$ and that $f = f(0)$ is constant. Note that by (7), $f(x) = B(a)B(x)$. Thus also B is constant, $B = f(0)/B(a) =: \gamma$. Let $d(x) := F(x) + \gamma^2$. Then by (7)

$$d(x+y) = F(x+y) + \gamma^2 = (F(x) + F(y) + \gamma^2) + \gamma^2 = d(x) + d(y),$$

i.e. d is additive and F and B are of the form given in (a).

(ii) Assume now that $b(0) \neq 0$. Then putting $x = 0$ in (9), we find that $B(0) = 0$. Moreover,

$$B(x+y) = B(x) + \frac{b(x)}{b(0)}B(y). \tag{10}$$

We first consider the case that $b = b(0)$ is a constant function. Then $B(x) = c[x]$ is additive and $G(x) := F(x) - \frac{1}{2}(c[x])^2$ satisfies

$$G(x+y) = F(x+y) - \frac{1}{2}(c[x] + c[y])^2$$

$$= (F(x) + F(y) + B(x)B(y)) - \frac{1}{2}(c[x])^2 - \frac{1}{2}(c[y])^2 - c[x]c[y]$$

$$= G(x) + G(y).$$

Hence $G(x) = d[x]$ is additive on $\mathbb{R}$ and F and B are of the form in (b),

$$F(x) = \frac{1}{2}(c[x])^2 + d[x] , \quad B(x) = c[x] .$$

(iii) Now assume that $b(0) \neq 0$ and that b is not constant. Choose $x_0 \in \mathbb{R}$ with $b(x_0) \neq b(0)$. Since the left side of (10) is symmetric in x and y,

$$B(x) + \frac{b(x)}{b(0)} B(y) = B(y) + \frac{b(y)}{b(0)} B(x) .$$

Hence for $y = x_0$, $B(x) = \frac{B(x_0)}{b(x_0) - b(0)}(b(x) - b(0))$ and by (8)

$$f(x) - f(0) = b(0)B(x) = \gamma(b(x) - b(0)) \tag{11}$$

where $\gamma := b(0)B(x_0)/(b(x_0) - b(0))$. For $\gamma = 0$, i.e. $B(x_0) = 0$, we again are in case (a). So we may assume that $\gamma \neq 0$. By (8) and (11)

$$\begin{aligned}
\gamma(b(x + y) - b(0)) &= f(x + y) - f(0) \\
&= f(x) - f(0) + b(x)B(y) \\
&= \gamma(b(x) - b(0)) + \frac{b(x)}{b(0)}\gamma(b(y) - b(0)) \\
&= \gamma\left(\frac{b(x)b(y)}{b(0)} - b(0)\right) .
\end{aligned}$$

Hence $\tilde{b}(x) := b(x)/b(0)$ satisfies $\tilde{b}(x + y) = \tilde{b}(x)\,\tilde{b}(y)$. Hence $\tilde{b}(x) = \tilde{b}\left(\frac{x}{2}\right)^2 \geq 0$. Therefore $c(x) := \ln \tilde{b}(x)$ is additive and

$$b(x) = b(0)e^{c[x]} .$$

The formula for $B(x)$ before (11) then gives

$$B(x) = \gamma\left(e^{c[x]} - 1\right) . \tag{12}$$

Put similarly as above $G(x) := F(x) - \gamma^2(e^{c[x]} - 1)$. Then (7) and (12) imply that $G(x + y) = G(x) + G(y)$, i.e. G is additive, $G(x) = d[x]$. Therefore $F(x) = \gamma^2(e^{c[x]} - 1) + d[x]$ and this case yields case (c) of the Lemma. $\square$

We also need a multiplicative analogue of Lemma 6.

Lemma 7. *Assume that $F, B : \mathbb{R} \to \mathbb{R}$ are functions such that for any $\alpha, \beta \in \mathbb{R}$*

$$F(\alpha\beta) = F(\alpha)\beta + F(\beta)\alpha + B(\alpha)\,B(\beta) , \tag{13}$$

Then there are additive functions $c, d : \mathbb{R} \to \mathbb{R}$ and there is $\gamma \in \mathbb{R}$ such that F and B are of one of the following three forms

(a) $F(\alpha) = \alpha \left(c[\ln |\alpha|] - \gamma^2 \right)$, $B(\alpha) = \gamma \alpha$

(b) $F(\alpha) = \alpha \left(\frac{1}{2} c[\ln |\alpha|]^2 + d[\ln |\alpha|] \right)$, $B(\alpha) = \alpha \, c[\ln |\alpha|]$

(c) $F(\alpha) = \alpha \left(\gamma^2 (\{\operatorname{sgn} \alpha\} e^{c[\ln |\alpha|]} - 1) + d[\ln |\alpha|] \right)$,

$B(\alpha) = \alpha \, \gamma (\{\operatorname{sgn} \alpha\} e^{c[\ln |\alpha|]} - 1)$.

In (c), there are two possibilities, with $\operatorname{sgn} \alpha$ present in both B and F or not present in both. If the $\operatorname{sgn} \alpha$-term is present, B and F are not odd functions; in all other cases they are odd functions.

Proof. (i) For $\alpha \neq 0 \neq \beta$, (10) yields

$$\frac{F(\alpha \beta)}{\alpha \beta} = \frac{F(\alpha)}{\alpha} + \frac{F(\beta)}{\beta} + \frac{B(\alpha)}{\alpha} \frac{B(\beta)}{\beta} .$$

For $x, y \in \mathbb{R}$, let $\alpha = e^x$, $\beta = e^y$ and put

$$\tilde{F}(x) := F(e^x)/e^x , \quad \tilde{B}(x) := B(e^x)/e^x .$$

Then

$$\tilde{F}(x + y) = \tilde{F}(x) + \tilde{F}(y) + \tilde{B}(x) \tilde{B}(y)$$

is of the form considered in Lemma 6. Hence for positive $\alpha, \beta > 0$, $F(\alpha)$ equals α times one of the formulas (a), (b) or (c) in Lemma 6, with $x = \ln \alpha$, yielding (a), (b) or (c) of Lemma 7 in this case.

Choose $\beta = -1$ in (13) and use the symmetry in $(\alpha, -\alpha)$ to find that

$$F(\alpha) + F(-\alpha) = F(-1)\alpha + B(-1)B(\alpha) = -F(-1)\alpha + B(-1)B(-\alpha) ,$$

$$B(-1) \, B(-\alpha) = B(-1)B(\alpha) + 2F(-1)\alpha .$$

For $\alpha = 1$ we get that $B(-1)^2 = B(-1)B(1) + 2F(-1)$, hence

$$B(-1)B(-\alpha) = B(-1)[B(\alpha) + (B(-1) - B(1))\alpha] \tag{14}$$

and

$$F(\alpha) + F(-\alpha) = B(-1)[B(\alpha) + \frac{1}{2}(B(-1) - B(1))\alpha] . \tag{15}$$

(ii) If $B(-1) = 0$, also $F(-1) = 0$ and (15) implies for $\beta = -1$ that $F(-\alpha) = -F(\alpha)$ and hence

$$B(-\alpha)B(\beta) = F(-\alpha\beta) - F(-\alpha)\beta + F(\beta)\alpha$$

$$= -F(\alpha\beta) + F(\alpha)\beta + F(\beta)\alpha = -B(\alpha)B(\beta)$$

i.e. both F and B are odd functions.

(iii) If $B(-1) \neq 0$, (14) yields that $B(-\alpha) = B(\alpha) + (B(-1) - B(1))\alpha$. In the case of (a), for $\alpha > 0$, $B(-\alpha) = \gamma\alpha + (B(-1) - \gamma)\alpha = B(-1)\alpha$ and by (15)

$$F(\alpha) + F(-\alpha) = B(-1)(\gamma\alpha + \frac{1}{2}(B(-1) - \gamma)\alpha) = \frac{B(-1)}{2}(\gamma + B(-1))\alpha .$$
(16)

Using this for $\alpha\beta$ and again (13), we have

$$\frac{B(-1)}{2}(\gamma + B(-1))\alpha\beta = F(\alpha\beta) + F(-\alpha\beta)$$

$$= (F(\alpha) + F(-\alpha))\beta + (B(\alpha) + B(-\alpha))B(\beta)$$

$$= \frac{B(-1)}{2}(\gamma + B(-1))\alpha\beta + (\gamma + B(-1))\alpha B(\beta)$$

which implies that $B(-1) = -\gamma$, $B(-\alpha) = -\gamma\alpha = -B(\alpha)$ and by (14) $F(-\alpha) = -F(\alpha)$, i.e. B and F are odd functions.

(iv) In the case of (b) and (c) we know that $B(1) = F(1) = 0$. We only have to consider the case of $B(-1) \neq 0$. By (14) and (15)

$$B(-\alpha) = B(\alpha) + B(-1)\alpha$$

$$F(\alpha) + F(-\alpha) = B(-1)[B(\alpha) + \frac{B(-1)}{2}\alpha] .$$
(17)

We apply the last equation to $\alpha\beta$ and use (13) to find

$$B(-1)[B(\alpha\beta) + \frac{B(-1)}{2}\alpha\beta] = F(\alpha\beta) + F(-\alpha\beta)$$

$$= (F(\alpha) + F(-\alpha))\beta + (B(\alpha) + B(-\alpha))B(\beta)$$

$$= B(-1)[B(\alpha)\beta + \frac{B(-1)}{2}\alpha\beta] + 2B(\alpha)B(\beta) + B(-1)B(\beta)\alpha$$

which may be restated as

$$\frac{B(-1)}{2}(B(\alpha\beta) + \frac{B(-1)}{2}\alpha\beta) = (B(\alpha) + \frac{B(-1)}{2}\alpha)(B(\beta) + \frac{B(-1)}{2}\beta) .$$

Therefore the function $\varphi(\alpha) := \frac{2}{B(-1)}B(\alpha) + \alpha$ is multiplicative, $\varphi(\alpha\beta) = \varphi(\alpha)\varphi(\beta)$ and $\tilde{c}(x) = \ln\varphi(e^x)$ is additive. Therefore $\varphi(\alpha) = \{\text{sgn }\alpha\}e^{\tilde{c}[\ln|\alpha|]}$, and for any $\alpha \in \mathbb{R}$,

$$B(\alpha) = \frac{B(-1)}{2}(\{\text{sgn }\alpha\}e^{\tilde{c}[\ln|\alpha|]} - \alpha) = \frac{B(-1)}{2}\alpha(\{\text{sgn }\alpha\}e^{c[\ln|\alpha|]} - 1)$$

where $c[x] = \tilde{c}[x] - x$, i.e. we are in case (c) with $B(-1) = 2\gamma$ where we know $F(\alpha)$ for $\alpha > 0$. For $\alpha < 0$, by (17) with $-\alpha > 0$, $\operatorname{sgn}(-\alpha) = 1$,

$$F(\alpha) = -F(-\alpha) + B(-1)[B(\alpha) + \frac{B(-1)}{2}\alpha]$$

$$= \alpha(\gamma^2(e^{c[\ln|\alpha|]} - 1) + d[\ln|\alpha|]) + 2\gamma[-\gamma\alpha(e^{c[\ln|\alpha|]} + 1) + \gamma\alpha]$$

$$= -\alpha\gamma^2(e^{c[\ln|\alpha|]} + 1) + \alpha d[\ln|\alpha|].$$

This is the case in (c) where the term $(\operatorname{sgn}\alpha)$ appears and where F and B are not odd functions of α. In the cases (a) and (b), F and B are odd. This ends the proof of Lemma 7. $\qquad\square$

The following Proposition of Faifman is proved in [6].

Proposition 8. *Let* $H_j : \mathbb{R}^2 \to \mathbb{R}$ *for* $j = 1, \ldots, d$ *be a family of functions additive in the second variable,*

$$H_j(x, \alpha + \beta) = H_j(x, \alpha) + H_j(x, \beta) \quad ; \quad x, \alpha, \beta \in \mathbb{R}, \ j = 1, \ldots, d$$

such that

$$H_1(x, f(x)) + \cdots + H_d(x, f^{(d-1)}(x))$$

is continuous in x for any $f \in C^\infty(\mathbb{R})$. Then $H_j(x, \alpha) = c_j(x)\alpha$ is linear in the second variable α with continuous coefficients $c_j \in C(\mathbb{R})$.

4   Proofs of Theorems 1 and 3

We return to the functional equation

$$T(fg) = Tf \cdot g + f \cdot Tg + Af \cdot Ag \tag{1}$$

By Proposition 5, T and A are local in the sense that there are functions $F : \mathbb{R}^4 \to \mathbb{R}$ and $B : \mathbb{R}^3 \to \mathbb{R}$ such that for any $f \in C^2(\mathbb{R})$ and $x \in \mathbb{R}$

$$Tf(x) = F_x(f(x), \ f'(x), \ f''(x)), \ Af(x) = B_x(f(x), f'(x))$$

We now analyze the structure of F and B and prove Theorems 1 and 3.

Proof of Theorem 3.

(i) For any $\alpha_0, \alpha_1, \alpha_2, \beta_0, \beta_1, \beta_2 \in \mathbb{R}$ and $x \in \mathbb{R}$ there are functions $f, g \in C^2(\mathbb{R})$ such that

$$f^{(j)}(x) = \alpha_j, \ g^{(j)}(x) = \beta_j ; \ j = 0, 1, 2.$$

Therefore the functional equation (1) translates into the following equation for F and B ,

$$F_x(\alpha_0\beta_0, \ \alpha_0\beta_1 + \alpha_1\beta_0, \ \alpha_0\beta_2 + \alpha_2\beta_0 + 2\alpha_1\beta_1)$$

$$= F_x(\alpha_0, \alpha_1, \alpha_2)\beta_0 + F_x(\beta_0, \beta_1, \beta_2)\alpha_0 + B_x(\alpha_0, \alpha_1)B_x(\beta_0, \beta_1). \quad (18)$$

We first determine the general forms of F and B if $\alpha_0 \neq 0$ and $\beta_0 \neq 0$ are assumed, i.e. for functions f and g which are non-zero everywhere. By Proposition 4, we know that $A(\mathbb{1})(x) = 0$. Therefore, choosing $\beta_0 = 1$ and $\beta_1 = 0$ in (18) we conclude that

$$F_x(\alpha_0, \alpha_1, \alpha_2 + \alpha_0\beta_2) = F_x(\alpha_0, \alpha_1, \alpha_2) + F_x(1, 0, \beta_2)\alpha_0 \ .$$

For $\alpha_0 = 1$, $\alpha_1 = 0$ this means that $F_x(1, 0. \cdot)$ is additive,

$$F_x(1, 0, \alpha_2 + \beta_2) = F_x(1, 0, \alpha_2) + F_x(1, 0, \beta_2) \ .$$

We write

$$F_x(1, 0, \alpha_2) = c_x[\alpha_2] \quad (19)$$

for this additive function. We will later see that it is linear, i.e. that $F_x(1, 0, \alpha_2) = c_x\alpha_2$ with $c_x = F_x(1, 0, 1)$. Choosing $\alpha_2 = 0$ above, we get

$$F_x(\alpha_0, \alpha_1, \alpha_0\beta_2) = F_x(\alpha_0, \alpha_1, 0) + F_x(1, 0, \beta_2)\alpha_0 \ .$$

Since $\alpha_0 \neq 0$, $\alpha_0\beta_2$ may attain arbitrary values; this yields that

$$F_x(\alpha_0, \alpha_1, \alpha_2) = F_x(\alpha_0, \alpha_1, 0) + c_x\left[\frac{\alpha_2}{\alpha_0}\right] \cdot \alpha_0 \ , \quad (20)$$

i.e. for $\alpha_0 \neq 0$ we separated the second order variable α_2 in an additive way.

Choose $\beta_0 = 1$, $\alpha_2 = \beta_2 = 0$ in (18) and use (20) to get

$$F_x(\alpha_0, \alpha_1 + \alpha_0\beta_1, 0) + 2c_x\left[\frac{\alpha_1\beta_1}{\alpha_0}\right]\alpha_0$$

$$= F_x(\alpha_0, \alpha_1, 0) + F_x(1, \beta_1, 0)\alpha_0 + B_x(\alpha_0, \alpha_1)B_x(1, \beta_1) \ .$$

Since $\alpha_0 \neq 0$, $\alpha_0\beta_1$ may attain arbitrary values. We may therefore replace $\alpha_0\beta_1$ by β_1 and use the symmetry in (α_1, β_1) to find

$$F_x(\alpha_0, \alpha_1 + \beta_1, 0) + 2c_x\left[\frac{\alpha_1\beta_1}{\alpha_0^2}\right]\alpha_0$$

$$= F_x(\alpha_0, \alpha_1, 0) + F_x(1, \frac{\beta_1}{\alpha_0}, 0)\alpha_0 + B_x(\alpha_0, \alpha_1) \cdot B_x(1, \frac{\beta_1}{\alpha_0})$$

$$= F_x(\alpha_0, \beta_1, 0) + F_x(1, \frac{\alpha_1}{\alpha_0}, 0)\alpha_0 + B_x(\alpha_0, \beta_1) \cdot B_x(1, \frac{\alpha_1}{\alpha_0}) \ .$$

We know from Proposition 4 (a) that $B_x(1,0) = 0$ and $F_x(1,0,0) = 0$. Using this, the second equality yields for $\beta_1 \Leftarrow 0$ that

$$F_x(\alpha_0, \alpha_1, 0) = F_x(\alpha_0, 0, 0) + F_x\left(1, \frac{\alpha_1}{\alpha_0}, 0\right)\alpha_0 + B_x(\alpha_0, 0) \cdot B_x\left(1, \frac{\alpha_1}{\alpha_0}\right).$$

(21)

(ii) Equations (20) and (21) determine $F_x(\alpha_0, \alpha_1, \alpha_2)$ once we know the form of $F_x(\alpha_0, 0, 0)$, $B_x(\alpha_0, 0)$ and $F_x(1, \alpha_1, 0)$, $B_x(1, \alpha_1)$. Equation (18) implies for $\alpha_1 = \alpha_2 = \beta_1 = \beta_2 = 0$ that

$$F_x(\alpha_0\beta_0, 0, 0) = F_x(\alpha_0, 0, 0)\beta_0 + F_x(\beta_0, 0, 0)\alpha_0 + B_x(\alpha_0, 0)B_x(\beta_0, 0).$$ (22)

By Lemma 7, there are suitable additive functions $a_x, e_x, p_x : \mathbb{R} \to \mathbb{R}$ and constants γ_x such that, introducing functions $G_x : \mathbb{R} \to \mathbb{R}$ by

$$G_x(\alpha_0) := F_x(\alpha_0, 0, 0) - \alpha_0 a_x[\ln|\alpha_0|],$$

any solution of (22) is of one of the following three corresponding forms

$$G_x(\alpha_0) = \begin{cases} -\gamma_x^2\alpha_0 \\ \frac{\alpha_0}{2}(e_x[\ln|\alpha_0|])^2 \\ \gamma_x^2\alpha_0(\{\operatorname{sgn}\alpha_0\}e^{p_x[\ln|\alpha_0|]} - 1) \end{cases},$$

(23)

$$B_x(\alpha_0, 0) = \begin{cases} \gamma_x\alpha_0 \\ \alpha_0 e_x[\ln|\alpha_0|] \\ \gamma_x\alpha_0(\{\operatorname{sgn}\alpha_0\}e^{p_x[\ln|\alpha_0|]} - 1) \end{cases}.$$

Since $B_x(1,0) = 0$, the first solution pair is actually zero. The choice of $\alpha_0 = \beta_0 = 1$ and $\alpha_2 = \beta_2 = 0$ in (18) yields, using also (20)

$$F_x(1, \alpha_1 + \beta_1, 0) + 2c_x[\alpha_1\beta_1]$$
$$= F_x(1, \alpha_1, 0) + F_x(1, \beta_1, 0) + B_x(1, \alpha_1)B_x(1, \beta_1).$$

Let $\tilde{H}_x(\alpha_1) := F_x(1, \alpha_1, 0) + c_x[\alpha_1^2]$. Since c_x is additive, the previous equation means

$$\tilde{H}_x(\alpha_1 + \beta_1) = \tilde{H}_x(\alpha_1) + \tilde{H}_x(\beta_1) + B_x(1, \alpha_1)B_x(1, \beta_1).$$ (24)

By Lemma 6, there are additive functions $d_x, b_x : \mathbb{R} \to \mathbb{R}$ and constants δ_x such that with

$$H_x(\alpha_1) := F_x(1, \alpha_1, 0) + c_x[\alpha_1^2] - b_x[\alpha_1]$$

any solution of (24) may be written in one of the following forms

$$H_x(\alpha_1) = \begin{Bmatrix} -\delta_x^2 \\ \frac{1}{2}(d_x[\alpha_1])^2 \\ \delta_x^2(e^{d_x[\alpha_1]} - 1) \end{Bmatrix} , \quad B_x(1,\alpha_1) = \begin{Bmatrix} \delta_x \\ d_x[\alpha_1] \\ \delta_x(e^{d_x[\alpha_1]} - 1) \end{Bmatrix} . \tag{25}$$

Since $B_x(1,0) = 0$, again the first solution pair is zero. By (20), (21), (23) and (25) a term occurring in any case in $F_x(\alpha_0, \alpha_1, \alpha_2)$ is

$$S_x(\alpha_0, \alpha_1, \alpha_2) := \alpha_0 (c_x[\frac{\alpha_2}{\alpha_0} - \frac{\alpha_1^2}{\alpha_0^2}] + b_x[\frac{\alpha_1}{\alpha_0}] + \alpha_x[\ln |\alpha_o|]) . \tag{26}$$

This corresponds to the general solution of the homogeneous functional equation (1) with $A = 0$, i.e. $Tf = f(c[\frac{f''}{f} - \frac{f'^2}{f^2}] + b[\frac{f'}{f}] + a[\ln |f|])$. Moreover, put

$$K_x(\alpha_0, \alpha_1) := G_x(\alpha_0) + \alpha_0 H_x(\frac{\alpha_1}{\alpha_0}) + B_x(\alpha_0, 0) B_x(1, \frac{\alpha_1}{\alpha_0}) . \tag{27}$$

Then by (20), (21), (23) and (25)

$$F_x(\alpha_0, \alpha_1, \alpha_2) = S_x(\alpha_0, \alpha_1, \alpha_2) + K_x(\alpha_0, \alpha_1) . \tag{28}$$

Since S satisfies the homogeneous functional equation, (18) and (20) imply that K_x and B_x are related by

$$K_x(\alpha_0 \beta_0, \alpha_1 \beta_0 + \alpha_0 \beta_1)$$
$$= K_x(\alpha_0, \alpha_1) \beta_0 + K_x(\beta_0, \beta_1) \alpha_0 + B_x(\alpha_0, \alpha_1) B_x(\beta_0, \beta_1) . \tag{29}$$

(iii) Equations (26)–(28) determine the structure of F_x, once we know which combinations of the different cases for G_x and H_x are possible. We use (29) for answering this question and hence need to describe the general form of $B_x(\alpha_0, \alpha_1)$. To do so, we isolate the product of B's in (29) on the left side for $\alpha_0 \neq 0 \neq \beta_0$ and use equations (27) and (29) in the following calculation

$$B_x(\alpha_0, \alpha_1) B_x(\beta_0, \beta_1)$$
$$= K_x(\alpha_0 \beta_0, \alpha_1 \beta_0 + \alpha_0 \beta_1) - K_x(\alpha_0, \alpha_1) \beta_0 - K_x(\beta_0, \beta_1) \alpha_0$$
$$= (G_x(\alpha_0 \beta_0) - G_x'(\alpha_0) \beta_0 - G_x(\beta_0) \alpha_0)$$
$$+ \alpha_0 \beta_0 \left(H_x \left(\frac{\alpha_1}{\alpha_0} + \frac{\beta_1}{\beta_0} \right) - H_x \left(\frac{\alpha_1}{\alpha_0} \right) - H_x \left(\frac{\beta_1}{\beta_0} \right) \right)$$
$$+ B_x(\alpha_0 \beta_0, 0) B_x \left(1, \frac{\alpha_1}{\alpha_0} + \frac{\beta_1}{\beta_0} \right) - B_x(\alpha_0, 0) B_x \left(1, \frac{\alpha_1}{\alpha_0} \right) \beta_0$$
$$- B_x(\beta_0, 0) B_x \left(1, \frac{\beta_1}{\beta_0} \right) \alpha_0 .$$

Rewriting equations (22) and (24) as

$$G_x(\alpha_0\beta_0) = G_x(\alpha_0)\beta_0 + G_x(\beta_0)\alpha_0 + B_x(\alpha_0, 0)B_x(\beta_0, 0),$$

$$H_x(\frac{\alpha_1}{\alpha_0} + \frac{\beta_1}{\beta_0}) = H_x(\frac{\alpha_1}{\alpha_0}) + H_x(\frac{\beta_1}{\beta_0}) + B_x(1, \frac{\alpha_1}{\alpha_0})B_x(1, \frac{\beta_1}{\beta_0}),$$

we find that

$$B_x(\alpha_0, \alpha_1)B_x(\beta_0, \beta_1) = B_x(\alpha_0, 0)B_x(\beta_0, 0) + \alpha_0\beta_0 \, B_x\left(1, \frac{\alpha_1}{\alpha_0}\right) B_x\left(1, \frac{\beta_1}{\beta_0}\right)$$

$$+ B_x(\alpha_0\beta_0, 0)B_x\left(1, \frac{\alpha_1}{\alpha_0} + \frac{\beta_1}{\beta_0}\right) - B_x(\alpha_0, 0)B_x\left(1, \frac{\alpha_1}{\alpha_0}\right)\beta_0$$

$$- B_x(\beta_0, 0)B_x\left(1, \frac{\beta_1}{\beta_0}\right)\alpha_0. \tag{30}$$

Hence, if $B_x(1, \cdot) = 0$ identically, we find for $\alpha_0 = \beta_0$ and $\alpha_1 = \beta_1$ that $B_x(\alpha_0, \alpha_1)^2 = B_x(\alpha_0, 0)^2$ and hence $B_x(\alpha_0, \alpha_1) = B_x(\alpha_0, 0)$ does not depend on α_1. If $B_x(1, \cdot) \neq 0$, choose $\beta_1 \in \mathbb{R}$ with $B_x(1, \beta_1) \neq 0$. Then (30) with $\beta_0 = 1$ yields, after dividing by $B_x(1, \beta_1)$,

$$B_x(\alpha_0, \alpha_1) = \frac{B_x(\alpha_0, 0)}{B_x(1, \beta_1)}(B_x(1, \frac{\alpha_1}{\alpha_0} + \beta_1) - B_x(1, \frac{\alpha_1}{\alpha_0})) + \alpha_0 B_x(1, \frac{\alpha_1}{\alpha_0}). \tag{31}$$

(iv) We now determine the possible combinations of $H_x, B_x(1, \cdot)$ from (25) and $G_x, B_x(\cdot, 0)$ from (23). The first solutions were zero in our case. It turns out that the last two possibilities of (25) and (23) may be combined with one another. In the following, we consider these four cases.

In the second case of (25),

$$B_x(1, \alpha_1) = d_x[\alpha_1], \quad H_x(\alpha_1) = \frac{1}{2}(d_x[\alpha_1])^2 \tag{32}$$

where d_x is additive. Equation (31) yields in this case that

$$B_x(\alpha_0, \alpha_1) = B_x(\alpha_0, 0) + \alpha_0 \, d_x\left[\frac{\alpha_1}{\alpha_0}\right]. \tag{33}$$

We have to examine the second and third possibility in (23): The second case in (23) and equation (27) yield

$$B_x(\alpha_0, \alpha_1) = \alpha_0 \left(d_x\left[\frac{\alpha_1}{\alpha_0}\right] + e_x[\ln |\alpha_0|]\right)$$

$$K_x(\alpha_0, \alpha_1) = \frac{\alpha_0}{2} \left\{(d_x\left[\frac{\alpha_1}{\alpha_0}\right])^2 + (e_x[\ln |\alpha_0|])^2\right\} + \alpha_0 \, d_x\left[\frac{\alpha_1}{\alpha_0}\right] e_x[\ln |\alpha_0|]$$

$$\tag{34}$$

Calculation shows that (29) and hence (18) is satisfied for (34). This will give the first solution in Theorem 3.

Combining the second case of (25) with the third case of (23) again yield formulas for $B_x(\alpha_0, \alpha_1)$ and $K_x(\alpha_0, \alpha_1)$ by using (33) and (27). Inserting these into (29) shows, however, that they provide a solution of (29) and (18) only if $\gamma_x \cdot d_x = 0$, γ_x from (23). The case $d_x \neq 0$ requires $\gamma_x = 0$, $B_x(\alpha_0, 0) = 0$ and is a special case of (33) before. If $\gamma_x \neq 0$, $d_x = 0$ and

$$B_x(\alpha_0, \alpha_1) = \gamma_x \alpha_0(\{\operatorname{sgn} \alpha_0\} e^{p_x[\ln |\alpha_0|]} - 1), \quad K_x(\alpha_0, \alpha_1) = \gamma_x \, B_x(\alpha_0, \alpha_1).$$

This will be a special case of a more general solution below.

(v) Last, we consider the third case in (25), combined with one of the two last possibilities in (23),

$$B_x(1, \alpha_1) = \delta_x(e^{d_x[\alpha_1]} - 1), \quad H_x(\alpha_1) = \delta_x \, B_x(1, \alpha_1).$$

Using this, (31) and (27) yield

$$B_x(\alpha_0, \alpha_1) = e^{d_x\left[\frac{\alpha_1}{\alpha_0}\right]} B_x(\alpha_0, 0) + \delta_x \, \alpha_0(e^{d_x\left[\frac{\alpha_1}{\alpha_0}\right]} - 1),$$

$$K_x(\alpha_0, \alpha_1) = G_x(\alpha_0) + \delta_x(B_x(\alpha_0, 0) + \delta_x \, \alpha_0)(e^{d_x\left[\frac{\alpha_1}{\alpha_0}\right]} - 1). \tag{35}$$

To check whether (35) provides another solution of (18), we insert the formulas for B_x and K_x into (29). Calculation and reordering terms shows that (29) is satisfied if and only if

$$\delta_x(e^{d_x\left[\frac{\alpha_1}{\alpha_0} + \frac{\beta_1}{\beta_0}\right]} - 1)\varphi(\alpha_0, \beta_0) = e^{d_x\left[\frac{\alpha_1}{\alpha_0} + \frac{\beta_1}{\beta_0}\right]} B_x(\alpha_0, 0) \, B_x(\beta_0, 0) - \psi(\alpha_0, \beta_0), \tag{36}$$

where

$$\varphi(\alpha_0, \beta_0) = B_x(\alpha_0 \beta_0, 0) - \alpha_0 B_x(\beta_0, 0) - \beta_0 B_x(\alpha_0, 0),$$

$$\psi(\alpha_0, \beta_0) = G_x(\alpha_0 \beta_0) - \alpha_0 G_x(\beta_0) - \beta_0 G_x(\alpha_0).$$

In the third case of (23), $G_x(\alpha_0) = \gamma_x B_x(\alpha_0, 0)$ and hence $\psi(\alpha_0, \beta_0) = \gamma_x \varphi(\alpha_0, \beta_0)$. Moreover, in this case $B_x(\alpha_0, 0) B_x(\beta_0, 0) = \psi(\alpha_0, \beta_0)$. Hence (36) means

$$(\delta_x - \gamma_x)(e^{d_x\left[\frac{\alpha_1}{\alpha_0} + \frac{\beta_1}{\beta_0}\right]} - 1)\varphi(\alpha_0, \beta_0) = 0.$$

This requirement means that either $\gamma_x = \delta_x$ or $d_x = 0$ or $\varphi = 0$. For $\delta_x = \gamma_x$ we get

$$B_x(\alpha_0, \alpha_1) = \delta_x \alpha_0(\{\operatorname{sgn} \alpha_0\} e^{p_x[\ln |\alpha_0|]} e^{d_x\left[\frac{\alpha_1}{\alpha_0}\right]} - 1)$$

$$K_x(\alpha_0, \alpha_1) = \delta_x^2 \alpha_0(\{\operatorname{sgn} \alpha_0\} e^{p_x[\ln |\alpha_0|]} e^{d_x\left[\frac{\alpha_1}{\alpha_0}\right]} - 1) \tag{37}$$

as another solution of (18) which will yield the second solution in Theorem 3. If $d_x = 0$, $B_x(1, \cdot) = 0$ and $H_x = 0$. If $\varphi = 0$, $B_x(\cdot, 0) = 0$ and $G_x = 0$. Both cases yield special cases of (37) with $d_x = 0$ or $p_x = 0$ there.

In the second case of (23), $\varphi = 0$ and

$$\psi(\alpha_0, \beta_0) = B_x(\alpha_0, 0) B_x(\beta_0, 0) = \alpha_0 \beta_0 \, e_x[\ln |\alpha_0|] \, e_x[\ln |\beta_0|] \, .$$

Therefore equation (36) is equivalent to

$$(e^{d_x[\frac{\alpha_1}{\alpha_0} + \frac{\beta_1}{\beta_0}]} - 1)\psi(\alpha_0, \beta_0) = 0$$

which requires that $d_x = 0$, i.e. $H_x = 0$. The solution of (18), coming from this choice is a special case of the solution in (34), when $d_x = 0$ there.

We now know that any solution of (18) is of one of the forms in (34) or (37), where definitions (26) and (28) are used.

(vi) We now prove that the additive functions occurring in formulas (34) and (37) are linear and continuous in x. By these formulas and Proposition 5, we have for any $f \in C^2_{\neq 0}(\mathbb{R})$ that one of the following cases occurs

$$(Af)(x) = \begin{cases} f(x) \, d_x[f'(x)/f(x)] + e_x[\ln |f(x)|] \\ \delta_x \, f(x)(\{\operatorname{sgn} f(x)\} e^{p_x[\ln |f(x)|]} e^{d_x[f'(x)/f(x)]} - 1) \, . \end{cases}$$

For any $g \in C^2(\mathbb{R})$, let $f(x) = \exp(g(x))$. Then $f \in C^2_{\neq 0}(\mathbb{R})$ and since A maps into $C(\mathbb{R})$, we know that for any $g \in C^2(\mathbb{R})$

$$(Sg)(x) := \frac{(Af)(x)}{f(x)} = \begin{cases} d_x[g'(x)] + e_x[g(x)] \\ \delta_x(e^{p_x[g(x)] + d_x[g'(x)]} - 1) \end{cases}$$

is a continuous function of $x \in \mathbb{R}$. In the second case, if $p_x \neq 0$, choose g to be an appropriate constant function and, if $p_x = 0$ and $d_x \neq 0$, choose g to be a constant multiple of the identity to conclude that $e(x) := \delta_x$ is continuous in x. For those x with $\delta_x \neq 0$ in the second case

$$p_x[g(x)] + d_x[g'(x)] = \ln((Sg)(x)/\delta_x + 1)$$

is continuous for any $g \in C^2(\mathbb{R})$, and similarly $e_x[g(x)] + d_x[g'(x)]$ in the first case. Proposition 8 yields that p_x, d_x, e_x are linear and continuous in x, e.g. $d_x(\alpha_1) = d(x)\alpha_1$ for $d \in C(\mathbb{R})$ and similarly for p_x, e_x. If $\delta_x = 0$ in the second case but for a sequence x_n tending to x we have $\delta_{x_n} \neq 0$, we get the continuity of $p[g] + d[g']$ in x from the C^1-pointwise continuity of A. Hence A has the form given in Theorem 3. As a consequence, the form of T' is

$$(T'f)(x) = \begin{cases} \frac{d(x)^2}{2} \frac{f'(x)^2}{f(x)} + \frac{e(x)^2}{2} f(x) \, (\ln |f(x)|)^2 + d(x) \, e(x) \, f'(x) \ln |f(x)| \\ e(x)^2 \, f(x)(\{\operatorname{sgn} f(x)\}|f(x)|^{p(x)} e^{d(x)f'(x)/f(x)} - 1) \end{cases}$$

with the same continuous functions d, e and p as in the representation of A. Since Tf is continuous for any $f \in C^2_{\neq 0}(\mathbb{R})$, so is $S'f = Tf - T'f$. For $g \in C^2(\mathbb{R})$ and $f = \exp(g)$, we thus know that $S'f(x) = c_x[g''(x)] + b_x[g'(x)] + a_x[g(x)]$ is continuous in $x \in \mathbb{R}$. Again, Proposition 8 implies the linearity and continuity in x of c_x, b_x, a_x; i.e. $c_x[g''(x)] = c(x) \cdot g''(x)$ for $c \in C(\mathbb{R})$. This ends the Proof of Theorem 3. Remark (c) after the statement of Theorem 3 follows from the local nature (in terms of open intervals J) of the proof given.   □

Proof of Theorem 1. Since $C^2_{\neq 0}(\mathbb{R})$ is a subset of $C^2(\mathbb{R})$, T and A are given by one of the formulas in Theorem 3, with the additional requirement that they need to be extendable to all $f \in C^2(\mathbb{R})$ and $x \in \mathbb{R}$ with $f(x) = 0$ and with (Tf) being continuous. As a consequence of this requirement, we eliminate some solutions and show that the remaining ones are uniquely extendable to $C^2(\mathbb{R})$.

For the first solution in Theorem 3, this requires that $c(x) = d(x)^2/2$ and that $d(x)e(x) = 0$ since $1/f(x)$ and $\ln|f(x)|$ are not defined if $f(x) = 0$. Thus either $e(x) = 0$ or $c(x) = d(x) = 0$ which yields the first two solutions for T and A in Theorem 1. In both cases the "homogeneous" part in (5) is $Sf(x) = b(x)f'(x) + a(x)\ln|f(x)|$ and the "non-homogeneous" part is given by one of the first two formulas for $T_1 f$ in Theorem 1. We remark that this is the general solution of the homogeneous equation (2) when the domain is a linear space contained in $C^1(\mathbb{R})$ and containing $C^k(\mathbb{R})$ for some $k \in \mathbb{N}$. In the second solution of Theorem 3, $d(x) = 0$ and $p(x) \geq -1$ are required to guarantee that $(Tf)(x)$ is defined if $f(x) = 0$. This yields the third solution in Theorem 1. The last solution in both Theorems is identical.

We now show that these maps T are uniquely extended from $C^2_{\neq 0}(\mathbb{R})$ to $C^2(\mathbb{R})$. For $x_0 \in \mathbb{R}$ and $f \in C^2(\mathbb{R})$ with $\alpha_0 = f(x_0) = 0$, $\alpha_1 = f'(x_0)$, $\alpha_2 = f''(x_0)$ and $(\alpha_1, \alpha_2) \neq 0$, f is non-zero in $J \setminus \{x_0\}$ for a suitable open interval $J \subset \mathbb{R}$ with $x_0 \in J$. Therefore, using Remark (c) after the statement of Theorem 3, we have for any $x \in J \setminus \{x_0\}$ by Theorem 3 and the preceeding arguments

$$(Tf)(x) = F_x(f(x), f'(x), f''(x))$$

where e.g. in the first case the form of F_x is given by

$$F_x(\alpha_0, \alpha_1, \alpha_2) = \frac{1}{2}d(x)^2\alpha_2 + b(x)\alpha_1 + a(x)\alpha_0 \ln|\alpha_0|$$

with $d, b, a \in C(\mathbb{R})$. Therefore F is continuous in $x \in \mathbb{R}$ and $(\alpha_0, \alpha_1, \alpha_2) \in \mathbb{R}^3$ which implies that

$$(Tf)(x_0) = \lim_{x \to x_0}(Tf)(x) = \lim_{x \to x_0}F_x(f(x), f'(x), f''(x))$$

$$= \frac{1}{2}d(x_0)^2 f''(x_0) + b(x_0)f'(x_0),$$

so that solution formula for Tf also holds if $f(x_0) = 0$, with the same representing functions $d, b, a \in C(\mathbb{R})$. In the same way, one shows that the solution formulas remain valid for $f(x_0) = 0$ also in the other cases, and also for $(Af)(x_0)$.

To prove that the formula for $Af(x_0)$ also holds for $C^1(\mathbb{R})$-functions f which are not in $C^2(\mathbb{R})$ and $x_0 \in \mathbb{R}$, we choose a compact neighborhood I of x_0 and a sequence of $C^2(\mathbb{R})$-functions f_n such that $g = \lim_n f_n$, $g' = \lim_n f_n'$ exist uniformly on $\mathbb{R}$ and $g|_I = f|_I$. Using localization on I (Proposition 4(b)) and the C^1-pointwise continuity of A, we conclude that $Af(x_0) = \lim_n Af_n(x_0)$ and since the coefficients in B are continuous, the formulas for $Af(x_0)$ also holds for the $C^1(\mathbb{R})$-function f. This ends the proof of Theorem 1. □

Remarks. (a) The previous argument shows that the formulas for $(Af)(x)$ hold for all $C^1(\mathbb{R})$-functions f also in the case of Theorem 3.

(b) Instead of the continuity argument for $f(x_0) = 0$, we could also perform a direct analysis of $F_x(0, \alpha_0, \alpha_1)$ using (18) for $\alpha_0 = 0$ or $\beta_0 = 0$ similar as in part (i) of the proof of Theorem 3.

Proof of Proposition 2. By Theorem 1, any operator T satisfying (1) has the form

$$Tf(x) = b(x)f'(x) + a(x)f(x)\ln|f(x)| + T_1 f(x).$$

The last three forms of $T_1 f(x)$ only involve $f(x)$ but neither $f'(x)$ nor $f''(x)$. Choosing different constant functions for f, we get that the different terms, which are functions of $f(x)$ only, must be zero, i.e. $a(x) = 0, e(x) = 0$ or $p(x) = 0$. Then choosing $f(x) = x$, we also find that $b(x) = 0$. □

References

1. J. Aczél, in *Lectures on Functional Equations and Their Applications*. Mathematics in Science and Engineering, vol. 19 (Academic, New York, 1966)
2. J. Aczél, J. Dhombres, in *Functional Equations in Several Variables*. Encyclopedia of Mathematics and Its Applications, vol. 31 (Cambridge University Press, Cambridge, 1989)
3. S. Artstein-Avidan, H. König, V. Milman, The chain rule as a functional equation. J. Funct. Anal. **259**, 2999–3024 (2010)
4. H. Goldmann, P. Šemrl, Multiplicative derivations on $C(X)$. Monatsh. Math. **121**, 189–197 (1996)
5. H. König, V. Milman, A functional equation characerizing the second derivative. J. Funct. Anal. **261**, 876–896 (2011)
6. H. König, V. Milman, Characterizing the derivative and the entropy function by the Leibniz rule, with an appendix by D. Faifman. J. Funct. Anal. **261**, 1325–1344 (2011)
7. H. König, V. Milman, An operator equation characterizing the Laplacian. Algebra and Analysis (to appear)

Moments of Unconditional Logarithmically Concave Vectors

Rafał Latała

Abstract We derive two-sided bounds for moments of linear combinations of coordinates of unconditional log-concave vectors. We also investigate how well moments of such combinations may be approximated by moments of Gaussian random variables.

1 Introduction

The aim of this paper is to study moments of linear combinations of coordinates of unconditional, log-concave vectors $X = (X_1, \ldots, X_n)$. A nondegenerate random vector X is *log-concave* if it has a density of the form $g = e^{-h}$, where $h \colon \mathbb{R} \to (-\infty, \infty]$ is a convex function. We say that a random vector X is *unconditional* if the distribution of $(\eta_1 X_1, \ldots, \eta_n X_n)$ is the same as X for any choice of signs $\eta_1, \ldots, \eta_n$.

A typical example of an unconditional log-concave vector is a vector distributed uniformly in an unconditional convex body K, i.e. such convex body that $(\pm x_1, \ldots, \pm x_n) \in K$ whenever $(x_1, \ldots, x_n) \in K$.

A random vector X is called *isotropic* if it has identity covariance matrix, i.e. $\mathrm{Cov}(X_i, X_j) = \delta_{i,j}$. Notice that unconditional vector X is isotropic if and only if its coordinates have variance one, in particular if X is unconditional with nondegenerate coordinates then the vector $(X_1 / \mathrm{Var}^{1/2}(X_1), \ldots, X_n / \mathrm{Var}^{1/2}(X_n))$ is isotropic and unconditional.

R. Latała (✉)
Institute of Mathematics, University of Warsaw, Banacha 2, 02-097 Warszawa, Poland

Institute of Mathematics, Polish Academy of Sciences, ul. Śniadeckich 8, 00-956 Warszawa, Poland
e-mail: rlatala@mimuw.edu.pl

B. Klartag et al. (eds.), *Geometric Aspects of Functional Analysis*, Lecture Notes in Mathematics 2050, DOI 10.1007/978-3-642-29849-3_17,
© Springer-Verlag Berlin Heidelberg 2012

In [3] Gluskin and Kwapień derived two-sided estimates for moments of $\sum_{i=1}^{n} a_i X_i$ if X_i are independent, symmetric random variables with log-concave tails (coordinates of log-concave vector have log-concave tails). In Sect. 2 we derive similar result for arbitrary unconditional log-concave vectors X.

In [8] Klartag obtained powerful Berry-Essen type estimates for isotropic, unconditional, log-concave vectors X, showing in particular that if $\sum_i a_i^2 = 1$ and all a_i's are small then the distribution of $S = \sum_{i=1}^{n} a_i X_i$ is close to the standard Gaussian distribution $\mathcal{N}(0, 1)$. In Sect. 3 we investigate how well moments of S may be approximated by moments of $\mathcal{N}(0, 1)$.

Notation. By $\varepsilon_1, \varepsilon_2, \ldots$ we denote a Bernoulli sequence, i.e. a sequence of independent symmetric random variables taking values ± 1. We assume that the sequence (ε_i) is independent of other random variables.

For a random variable Y and $p > 0$ we write $\|Y\|_p = (\mathbb{E}|Y|^p)^{1/p}$. For a sequence (a_i) and $1 \leq q < \infty$, $\|a\|_q = (\sum_i |a_i|^q)^{1/p}$ and $\|a\|_\infty = \max_i |a_i|$. We set $B_q^n = \{a \in \mathbb{R}^n : \|a\|_q \leq 1\}$, $1 \leq q \leq \infty$. By $(a_i^*)_{1 \leq i \leq n}$ we denote the nonincreasing rearrangement of $(|a_i|)_{1 \leq i \leq n}$.

We use letter C (resp. $C(\alpha)$) for universal constants (resp. constants depending only on parameter α). Value of a constant C may differ at each occurence. Whenever we want to fix the value of an absolute constant we will use letters $C_1, C_2, \ldots$. For two functions f and g we write $f \sim g$ to signify that $\frac{1}{C} f \leq g \leq C f$.

2　Estimation of Moments

It is well known and easy to show (using e.g. Brunn's principle—see Lemma 4.1 in [17]) that if X has a uniform distribution over a symmetric convex body K in $\mathbb{R}^n$ then for any $p \geq n$, $\|\sum_{i \leq n} a_i X_i\|_p \sim \|a\|_{K^o} = \sup\{|\sum_{i \leq n} a_i x_i| : x \in K\}$. Our first proposition generalizes this statement to arbitrary log-concave symmetric distributions.

Proposition 1. *Suppose that X has a symmetric n-dimensional log-concave distribution with the density g. Then for any $p \geq n$ we have*

$$\left\| \sum_{i=1}^{n} a_i X_i \right\|_p \sim \|a\|_{K_p^o},$$

where

$$K_p := \{x : g(x) \geq e^{-p} g(0)\} \quad and \quad \|a\|_{K_p^o} = \sup\left\{ \sum_{i=1}^{n} a_i x_i : x \in K_p \right\}.$$

Proof. First notice that there exists an absolute constant C_1 such that

$$\Pr(X \in C_1 K_p) \geq 1 - e^{-p} \geq \frac{1}{2}.$$

For $n \leq p \leq 2n$ this follows by Corollary 2.4 and Lemma 2.2 in [9]. For $p \geq 2n$ we may either adjust arguments from [9] or take any log-concave symmetric $m = \lfloor p \rfloor - n$ dimensional vector Y independent of X with density g' and consider the set $K' = \{(x, y) \in \mathbb{R}^n \times \mathbb{R}^m : g(x)g'(y) \geq e^{-p}g(0)g'(0)\}$. Then K_p is a central n-dimensional section of K', hence $\Pr(X \in C_1 K_p) \geq \Pr((X, Y) \in C_1 K') \geq 1 - e^{-p}$.

Observe that for any $z \in K_p$,

$$\left| \left\{ x \in K_p : \left| \sum_{i=1}^n a_i x_i \right| \geq \frac{1}{2} \sum_{i=1}^n a_i z_i \right\} \right| \geq 2^{-n} |K_p| \geq (2C_1)^{-n} \Pr(X \in C_1 K_p)/g(0),$$

therefore choosing z such that $\sum_{i=1}^n a_i z_i = \|a\|_{K_p^o}$ we get

$$\left\| \sum_{i=1}^n a_i X_i \right\|_p \geq 2^{-1/p} \|a\|_{K_p^o} e^{-1} g(0)^{1/p} \left| \left\{ x \in K_p : \left| \sum_{i=1}^n a_i x_i \right| \geq \frac{1}{2} \sum_{i=1}^n a_i z_i \right\} \right|^{1/p}$$

$$\geq 2^{-1/p} \|a\|_{K_p^o} e^{-1} (2C_1)^{-n/p} \Pr(X \in C_1 K_p)^{1/p} \geq \frac{1}{8eC_1} \|a\|_{K_p^o}.$$

To get the upper estimate notice that

$$\Pr \left(\left| \sum_{i=1}^n a_i X_i \right| > C_1 \|a\|_{K_p^o} \right) \leq \Pr(X \notin C_1 K_p) \leq e^{-p}.$$

Together with the symmetry and log-concavity of $\sum_{i=1}^n a_i X_i$ this gives

$$\Pr \left(\left| \sum_{i=1}^n a_i X_i \right| > C_1 t \|a\|_{K_p^o} \right) \leq e^{-tp} \text{ for } t \geq 1.$$

Integration by parts yields $\left\| \sum_{i=1}^n a_i X_i \right\|_p \leq C \|a\|_{K_p^o}$. $\square$

Remark. The same argument as above shows that for $\alpha \geq e$ and $p \geq n$,

$$\left\| \sum_{i=1}^n a_i X_i \right\|_p \geq \frac{1}{8\alpha C_1} \sup \left\{ \sum_{i=1}^n a_i x_i : g(x) \geq \alpha^{-p} g(0) \right\}.$$

From now on till the end of this section we assume that a random vector X is unconditional, log-concave and isotropic. Jensen's inequality and Hitczenko estimates for moments of Rademacher sums [5] (see also [15]) imply that for $p \geq 2$,

$$\left\| \sum_i a_i X_i \right\|_p = \left\| \sum_i a_i \varepsilon_i |X_i| \right\|_p \geq \left\| \sum_i a_i \varepsilon_i \mathbb{E}|X_i| \right\|_p$$

$$\geq \frac{1}{C} \Big(\sum_{i \leq p} a_i^* + \sqrt{p} \Big(\sum_{i > p} |a_i^*|^2 \Big)^{1/2} \Big). \tag{1}$$

The result of Bobkov and Nazarov [2] yields for $p \geq 2$,

$$\left\| \sum_i a_i X_i \right\|_p \leq C \left\| \sum_i a_i E_i \right\|_p \leq C \Big(p \max_i |a_i| + \sqrt{p} \Big(\sum_i a_i^2 \Big)^{1/2} \Big), \tag{2}$$

where (E_i) is a sequence of independent symmetric exponential random variables with variance 1 and to get the second inequality we used the result of Gluskin and Kwapień [3].

Estimates (1) and (2) together with Proposition 1 give

$$\frac{1}{C} (\sqrt{p} B_2^n \cap B_\infty^n) \subset \left\{ x : g(x) \geq e^{-p} g(0) \right\} \subset C(\sqrt{p} B_2^n + p B_1^n) \quad \text{for } p \geq n. \tag{3}$$

Corollary 2. *Let* $X = (X_1, \ldots, X_n)$ *be an unconditional log-concave isotropic random vector with the density g. Then for any* $p \geq n$ *we have*

$$\left\| \sum_{i=1}^n a_i X_i \right\|_p \sim \sup \left\{ \sum_{i=1}^n a_i x_i : g(x) \geq e^{-p} g(0) \right\}$$

$$\sim \sup \left\{ \sum_{i=1}^n a_i x_i : g(x) \geq e^{-5p/2} \right\}$$

$$\sim \sup \left\{ \sum_{i=1}^n |a_i| t_i : \Pr(|X_1| \geq t_1, \ldots, |X_n| \geq t_n) \geq e^{-p} \right\}.$$

Proof. We have $g(0) = L_X^n$, where L_X is the isotropic constant of the vector X. Unconditionality of X implies boundedness of L_X, thus

$$e^{-3n/2} \leq (2\pi e)^{-n/2} \leq g(0) \leq C_2^n,$$

where C_2 is an absolute constant (see for example [2]). Hence

$$\{x : g(x) \geq e^{-p} g(0)\} \subset \{x : g(x) \geq e^{-5p/2}\} \subset \{x : g(x) \geq (e^{5/2} C_2)^{-p} g(0)\} \tag{4}$$

and first two estimates on moments follows by Proposition 1 (see also remark after it).

For any $t_1, \ldots, t_n \geq 0$,

$$\mathbb{E}\left|\sum_{i=1}^{n} a_i X_i\right|^p \geq \left(\sum_{i=1}^{n} |a_i| t_i\right)^p 2^{-n} \Pr(|X_1| \geq t_1, \ldots, |X_n| \geq t_n),$$

therefore

$$\left\|\sum_{i=1}^{n} a_i X_i\right\|_p \geq \frac{1}{2e} \sup\left\{\sum_{i=1}^{n} |a_i| t_i : \Pr(|X_1| \geq t_1, \ldots, |X_n| \geq t_n) \geq e^{-p}\right\}.$$

To prove the opposite estimate we use the already proven bound and take x such that $g(x) \geq e^{-5p/2}$ and $\sum_{i=1}^{n} a_i x_i \geq \frac{1}{C_3} \|\sum_{i=1}^{n} a_i X_i\|_p$. By the unconditionality without loss of generality we may assume that all a_i's and x_i's are nonnegative. Notice that by (3) and (4) we have $g(1/C_4, \ldots, 1/C_4) \geq e^{-5p/2}$. Hence by log-concavity of g we also have $g(y) \geq e^{-5p/2}$ for $y_i = (x_i + 1/C_4)/2$. Notice that g is coordinate increasing on $\mathbb{R}_+^n$, therefore

$$\Pr\left(X_1 \geq \frac{y_1}{2}, \ldots, X_n \geq \frac{y_n}{2}\right) \geq g(y) \prod_{i=1}^{n} \frac{y_i}{2} \geq e^{-5p/2}(4C_4)^{-n} \geq (4e^{5/2}C_4)^{-p}.$$

The function $F(s_1, \ldots, s_n) := -\ln \Pr(X_1 \geq s_1, \ldots, X_n \geq s_n)$ is convex on $\mathbb{R}_+^n$, $F(0) = n \ln 2$, therefore

$$\Pr\left(|X_1| \geq \frac{y_1}{C_5}, \ldots, |X_n| \geq \frac{y_n}{C_5}\right) = 2^n \Pr\left(X_1 \geq \frac{y_1}{C_5}, \ldots, X_n \geq \frac{y_n}{C_5}\right) \geq e^{-p}$$

for sufficiently large C_5. To conclude it is enough to notice that

$$\sum_{i=1}^{n} a_i \frac{y_i}{C_5} \geq \frac{1}{2C_5} \sum_{i=1}^{n} a_i x_i \geq \frac{1}{2C_3 C_5} \left\|\sum_{i=1}^{n} a_i X_i\right\|_p.$$

$\square$

Theorem 3. *Suppose that X is an unconditional log-concave isotropic random vector in $\mathbb{R}^n$. Then for any $p \geq 2$,*

$$\left\|\sum_{i=1}^{n} a_i X_i\right\|_p \sim \sup\left\{\sum_{i \in I_p} a_i x_i : g_{I_p}(x) \geq e^{-p} g_{I_p}(0)\right\} + \sqrt{p}\left(\sum_{i \notin I_p} a_i^2\right)^{1/2},$$

$$\sim \sup\left\{\sum_{i \in I_p} a_i x_i : g_{I_p}(x) \geq e^{-5p/2}\right\} + \sqrt{p}\left(\sum_{i \notin I_p} a_i^2\right)^{1/2}$$

$$\sim \sup\left\{\sum_{i \in I_p} |a_i| t_i : \Pr\left(\forall_{i \in I_p} |X_i| \geq t_i\right) \geq e^{-p}\right\} + \sqrt{p}\left(\sum_{i \notin I_p} a_i^2\right)^{1/2},$$

where g_{I_p} is the density of $(X_i)_{i \in I_p}$ and I_p is the set of indices of $\min\{\lceil p \rceil, n\}$ largest values of $|a_i|$'s.

Proof. By Corollary 2 it is enough to show that

$$\frac{1}{C}\left(\left\|\sum_{i \in I_p} a_i X_i\right\|_p + \sqrt{p}\left(\sum_{i \notin I_p} a_i^2\right)^{1/2}\right) \le \left\|\sum_{i=1}^n a_i X_i\right\|_p$$

$$\le C\left(\left\|\sum_{i \in I_p} a_i X_i\right\|_p + \sqrt{p}\left(\sum_{i \notin I_p} a_i^2\right)^{1/2}\right).$$

$$(5)$$

Observe also that $\sum_{i \notin I_p} a_i^2 = \sum_{i > p} |a_i^*|^2$.

Unconditionality of X_i implies that $\left\|\sum_{i=1}^n a_i X_i\right\|_p \ge \left\|\sum_{i \in I_p} a_i X_i\right\|_p$. Hence the lower estimate in (5) follows by (1).

Obviously we have

$$\left\|\sum_{i=1}^n a_i X_i\right\|_p \le \left\|\sum_{i \in I_p} a_i X_i\right\|_p + \left\|\sum_{i \notin I_p} a_i X_i\right\|_p.$$

Estimates (1) and (2) imply

$$\left\|\sum_{i \notin I_p} a_i X_i\right\|_p \le C\left(p \max_{i \notin I_p} |a_i| + \sqrt{p}\left(\sum_{i \notin I_p} a_i^2\right)^{1/2}\right)$$

$$\le C\left(\left\|\sum_{i \in I_p} a_i X_i\right\|_p + \sqrt{p}\left(\sum_{i \notin I_p} a_i^2\right)^{1/2}\right)$$

and the upper bound in (5) follows.   □

Example 1. Let X_i be independent symmetric log-concave r.v.'s. Define $N_i(t) := -\ln \Pr(|X_i| \ge t)$, then $\Pr(|X_i| \ge t_i \text{ for } i \in I_p) = \exp(-\sum_{i \in I_p} N_i(t_i))$ and Theorem 3 yields the Gluskin-Kwapień estimate

$$\left\|\sum_{i=1}^n a_i X_i\right\|_p \sim \sup\left\{\sum_{i \in I_p} |a_i| t_i : \sum_{i \in I_p} N_i(t_i) \le p\right\} + \sqrt{p}\left(\sum_{i \notin I_p} a_i^2\right)^{1/2}.$$

Example 2. Let X be uniformly distriputed on $r_{n,q} B_q^n$ with $1 \le q < \infty$, where $r_{n,q}$ is chosen in such a way that X is isotropic. Then it is easy to check that $r_{n,q} \sim n^{1/q}$. Since all k-dimensional sections of B_q^n are homogenous we immediately obtain that for $I \subset \{1, \dots, n\}$ and $x \in \mathbb{R}^I$, $g_I(x)/g_I(0) = (1 - (\|x\|_q/r_{n,q})^q)^{(n-|I|)/q}$. Hence for $1 \le p \le n/2$ we get that

$$\sup\left\{\sum_{i\in I_p} a_i x_i : g_{I_p}(x) \geq e^{-p} g_{I_p}(0)\right\} \sim \sup\left\{\sum_{i\in I_p} a_i x_i : \|x\|_q \leq p^{1/q}\right\}.$$

Since for $p \geq n/2$, $\|\sum_{i=1}^n a_i X_i\|_p \sim \|\sum_{i=1}^n a_i X_i\|_{n/2}$, we recover the result from [1] and show that for $p \geq 2$,

$$\left\|\sum_{i=1}^n a_i X_i\right\|_p \sim \min\{p,n\}^{1/q}\left(\sum_{i\leq p} |a_i^*|^{q'}\right)^{1/q'} + \sqrt{p}\left(\sum_{i>p} |a_i^*|^2\right)^{1/2},$$

where $1/q' + 1/q = 1$.

Remark. In the case of vector coefficients the following conjecture seems reasonable. For any isotropic unconditional log-concave vector $X = (X_1, \ldots, X_n)$, any vectors $v_1, \ldots, v_n$ in a normed space $(F, \|\cdot\|)$ and $p \geq 1$,

$$\left(\mathbb{E}\left\|\sum_{i=1}^n v_i X_i\right\|^p\right)^{1/p} \sim \left(\mathbb{E}\left\|\sum_{i=1}^n v_i X_i\right\| + \sup_{\|\varphi\|_* \leq 1}\left(\mathbb{E}\left|\sum_{i=1}^n \varphi(v_i) X_i\right|^p\right)^{1/p}\right).$$

The nontrivial part is the upper bound for $(\mathbb{E}\|\sum_{i=1}^n v_i X_i\|^p)^{1/p}$. It is known that the above conjecture holds if the space $(F, \|\cdot\|)$ has a nontrivial cotype – see [11] for this and some related results.

Remark. Let $S = \sum_{i=1}^n a_i X_i$, where X is as in Theorem 3. Then $\Pr(|S| \geq e\|S\|_p) \leq e^{-p}$ by the Chebyshev's inequality. Moreover $\|S\|_{2p} \leq C\|S\|_p$ for $p \geq 2$, hence by the Paley-Zygmund inequality, $\Pr(|S| \geq \|S\|_p/C) \geq \min\{1/C, e^{-p}\}$. This way Theorem 3 may be also used to get two-sided estimates for tails of S.

3 Gaussian Approximation of Moments

Let $\gamma_p = \|\mathcal{N}(0,1)\|_p = 2^{p/2}\Gamma(\frac{p+1}{2})/\sqrt{\pi}$. In [10] it was shown that for independent symmetric random variables $X_1, \ldots, X_n$ with log-concave tails (notice that log-concave symmetric random variables have log-concave tails) and variance 1,

$$\left|\left\|\sum_{i=1}^n a_i X_i\right\|_p - \gamma_p\|a\|_2\right| \leq p\|a\|_\infty \quad \text{for } a \in \mathbb{R}^n, \ p \geq 3 \qquad (6)$$

(see also [13] for $p \in [2,3)$). The purpose of this section is to discuss similar statements for general log-concave isotropic vectors X.

The lower estimate of moments is easy. In fact it holds for more general class of unconditional vectors with bounded fourth moments.

Proposition 4. *Suppose that X is an isotropic unconditional n-dimensional vector with finite fourth moment. Then for any nonzero $a \in \mathbb{R}^n$ and $p \geq 2$,*

$$\Big\| \sum_{i=1}^{n} a_i X_i \Big\|_p \geq \gamma_p \|a\|_2 - \frac{p}{\sqrt{2}\|a\|_2} \Big(\sum_{i=1}^{n} a_i^4 \mathbb{E} X_i^4 \Big)^{1/2}$$

$$\geq \gamma_p \|a\|_2 - \frac{p}{\sqrt{2}} \max_i (\mathbb{E} X_i^4)^{1/2} \|a\|_\infty.$$

Proof. Let us fix $p \geq 2$. By the homogenity we may and will assume that $\|a\|_2 = 1$. Corollary 1 in [10] gives

$$\Big\| \sum_{i=1}^{n} b_i \varepsilon_i \Big\|_p \geq \gamma_p \Big(\sum_{i \geq \lceil p/2 \rceil} |b_i^*|^2 \Big)^{1/2} \qquad \text{for } b \in \mathbb{R}^n,$$

where (b_i^*) denotes the nonicreasing rearrangement of $(|b_i|)_{i \leq n}$. Therefore

$$\Big\| \sum_{i=1}^{n} a_i X_i \Big\|_p^p = \mathbb{E} \Big| \sum_{i=1}^{n} a_i \varepsilon_i X_i \Big|^p \geq \gamma_p^p \mathbb{E} \Big(\sum_{i=1}^{n} a_i^2 X_i^2 - \max_{\#I < p/2} \sum_{i \in I} a_i^2 X_i^2 \Big)^{p/2}$$

$$\geq \gamma_p^p \Big(\mathbb{E} \Big(\sum_{i=1}^{n} a_i^2 X_i^2 - \max_{\#I < p/2} \sum_{i \in I} a_i^2 X_i^2 \Big) \Big)^{p/2} = \gamma_p^p \Big(1 - \mathbb{E} \max_{\#I < p/2} \sum_{i \in I} a_i^2 X_i^2 \Big)^{p/2}.$$

We have

$$\mathbb{E} \max_{\#I < p/2} \sum_{i \in I} a_i^2 X_i^2 \leq \mathbb{E} \max_{\#I < p/2} \sqrt{\#I} \Big(\sum_{i \in I} a_i^4 X_i^4 \Big)^{1/2} \leq \sqrt{\frac{p}{2}} \mathbb{E} \Big(\sum_{i=1}^{n} a_i^4 X_i^4 \Big)^{1/2}$$

$$\leq \sqrt{\frac{p}{2}} \Big(\sum_{i=1}^{n} a_i^4 \mathbb{E} X_i^4 \Big)^{1/2}.$$

Since $\sqrt{1-x} \geq 1 - x$ for $x \geq 0$ and $\gamma_p \leq \sqrt{p}$ the assertion easily follows. □

Since $\mathbb{E} Y^4 \leq 6$ for symmetric log-concave random variables Y we immediately get the following.

Corollary 5. *Let X be an isotropic unconditional n-dimensional log-concave vector. Then for any $a \in \mathbb{R}^n \setminus \{0\}$ and $p \geq 2$,*

$$\Big\| \sum_{i=1}^{n} a_i X_i \Big\|_p \geq \gamma_p \|a\|_2 - \frac{p}{\|a\|_2} \Big(3 \sum_{i=1}^{n} a_i^4 \Big)^{1/2} \geq \gamma_p \|a\|_2 - \sqrt{3} p \|a\|_\infty.$$

Now we turn our attention to the upper bound. Notice that for unconditional vectors X and $p \geq 2$,

$$\Big\| \sum_{i=1}^{n} a_i X_i \Big\|_p = \Big\| \sum_{i=1}^{n} a_i \varepsilon_i X_i \Big\|_p \leq \gamma_p \Big\| \Big(\sum_{i=1}^{n} a_i^2 X_i^2 \Big)^{1/2} \Big\|_p, \tag{7}$$

where the last inequality follows by the Khintchine inequality with the optimal constant [4]. First we will bound moments of $(\sum_{i=1}^{n} a_i^2 X_i^2)^{1/2}$ using the result of Klartag [8].

Proposition 6. *For any isotropic unconditional n-dimensional log-concave vector X, $p \geq 2$ and $a \in \mathbb{R} \setminus \{0\}$ we have*

$$\left\| \sum_{i=1}^{n} a_i X_i \right\|_p - \gamma_p \|a\|_2 \leq C p^{5/2} \frac{1}{\|a\|_2} \left(\sum_{i=1}^{n} |a_i|^4 \right)^{1/2} \leq C p^{5/2} \|a\|_\infty.$$

Proof. By the homogenity we may assume that $\|a\|_2 = 1$. We have

$$\left\| \left(\sum_{i=1}^{n} a_i^2 X_i^2 \right)^{1/2} \right\|_p \leq 1 + \left\| \left(\left(\sum_{i=1}^{n} a_i^2 X_i^2 \right)^{1/2} - 1 \right)_+ \right\|_p.$$

Notice that

$$\sum_{i=1}^{n} a_i^2 (X_i^2 - 1) = \left(\left(\sum_{i=1}^{n} a_i^2 X_i^2 \right)^{1/2} - 1 \right) \left(\left(\sum_{i=1}^{n} a_i^2 X_i^2 \right)^{1/2} + 1 \right),$$

thus

$$\left\| \left(\left(\sum_{i=1}^{n} a_i^2 X_i^2 \right)^{1/2} - 1 \right)_+ \right\|_p \leq \left\| \sum_{i=1}^{n} a_i^2 (X_i^2 - 1) \right\|_p.$$

Lemma 4 in [8] gives

$$\left\| \sum_{i=1}^{n} a_i^2 (X_i^2 - 1) \right\|_2^2 = \mathrm{Var} \left(\sum_{i=1}^{n} a_i^2 X_i^2 \right) \leq \frac{8}{3} \sum_{i=1}^{n} a_i^4 \mathbb{E} X_i^4 \leq 16 \sum_{i=1}^{n} a_i^4.$$

Comparison of moments of polynomials with respect to log-concave distributions [16] implies

$$\left\| \sum_{i=1}^{n} a_i^2 (X_i^2 - 1) \right\|_p \leq (Cp)^2 \left\| \sum_{i=1}^{n} a_i^2 (X_i^2 - 1) \right\|_2 \leq C p^2 \left(\sum_{i=1}^{n} a_i^4 \right)^{1/2}.$$

$\square$

We may improve $p^{5/2}$ term if we assume some concentration properties of a vector X. We say that a random vector X satisfies *exponential concentration* with constant κ if

$$\Pr(X \in A) \geq \frac{1}{2} \implies \Pr(X \in A + \kappa t B_2^n) \geq 1 - e^{-t} \quad \text{for } t \geq 0.$$

For log-concave vectors exponential concentration is equivalent to several other important functional and concentration inequalities including Poincaré and

Cheeger [14]. The strong conjecture due to Kannan, Lovász and Simonovits [7] states that every isotropic log-concave vector satisfies Cheeger's (and therefore also exponential) inequality with a uniform constant. The conjecture is wide open— however a recent result of Klartag [8] shows that unconditional isotropic vectors satisfy exponential concentration with $\kappa = C \log n$ (see also [6] for examples of log-concave measures that satisfy Poincaré inequalities with uniform constants).

Proposition 7. *Let X be an isotropic unconditional vector that satisfies exponential concentration with constant κ. Then for any $p \geq 2$ and $a \in \mathbb{R}^n$,*

$$\Big\| \sum_{i=1}^n a_i X_i \Big\|_p \leq \gamma_p \|a\|_2 + C\kappa p^{3/2} \|a\|_\infty.$$

Proof. Let $M := \mathrm{Med}((\sum_{i=1}^n a_i^2 X_i^2)^{1/2})$. Notice that

$$\sup \Big\{ \Big(\sum_{i=1}^n a_i^2 y_i^2 \Big)^{1/2} : y \in t B_2^n \Big\} = t \|a\|_\infty,$$

therefore exponential concentration applied to the set $A := \{ (\sum_{i=1}^n a_i^2 x_i^2)^{1/2} \leq M \}$ gives

$$\Pr \Big(\Big(\sum_{i=1}^n a_i^2 X_i^2 \Big)^{1/2} \leq M + \kappa t \|a\|_\infty \Big) \geq 1 - e^{-t}.$$

Integration by parts gives for $p \geq 2$,

$$\Big\| \Big(\sum_{i=1}^n a_i^2 X_i^2 \Big)^{1/2} \Big\|_p \leq M + C\kappa p \|a\|_\infty.$$

Using exponential concentration for the set $A := \{ (\sum_{i=1}^n a_i^2 x_i^2)^{1/2} \geq M \}$ we get

$$\Pr \Big(\Big(\sum_{i=1}^n a_i^2 X_i^2 \Big)^{1/2} \geq M - \kappa t \|a\|_\infty \Big) \geq 1 - e^{-t},$$

hence

$$\|a\|_2 = \Big\| \Big(\sum_{i=1}^n a_i^2 X_i^2 \Big)^{1/2} \Big\|_2 \geq M - C\kappa \|a\|_\infty.$$

Thus by (7) we get for $p \geq 2$,

$$\Big\| \sum_{i=1}^n a_i X_i \Big\|_p \leq \gamma_p \Big\| \Big(\sum_{i=1}^n a_i^2 X_i^2 \Big)^{1/2} \Big\|_p \leq \gamma_p (\|a\|_2 + C\kappa p \|a\|_\infty).$$

$\square$

Since by the result of Klartag [8] unconditional log-concave vectors satisfy exponential concentration with constant $C \log n$ we get

Corollary 8. *Let X be an isotropic unconditional log-concave vector. Then for any $p \geq 2$ and $a \in \mathbb{R}^n$,*

$$\left\| \sum_{i=1}^n a_i X_i \right\|_p \leq \gamma_p \|a\|_2 + C p^{3/2} \log n \|a\|_\infty.$$

To get the factor p instead of $p^{3/2}$ we need a stronger notion than exponential concentration. We say that a random vector X satisfies *two level concentration with constant κ* if

$$\Pr(X \in A) \geq \frac{1}{2} \implies \Pr(X \in A + \kappa(\sqrt{t} B_2^n + t B_1^n)) \geq 1 - e^{-t} \quad \text{for } t \geq 0.$$

Since it is enough to consider $t \geq 1$ two level concentration is indeed stronger than exponential concentration.

Proposition 9. *Suppose that X is an isotropic unconditional vector that satisfies two level concentration with constant κ. Then for any $p \geq 2$ and $a \in \mathbb{R}^n$,*

$$\left\| \sum_{i=1}^n a_i X_i \right\|_p \leq \gamma_p \|a\|_2 + C \kappa p \|a\|_\infty.$$

Proof. For $p \geq 2$ define a norm $\|\| \cdot \|\|_p$ on $\mathbb{R}^n$ by $\|\|x\|\|_p = \| \sum_{i=1}^n x_i \varepsilon_i \|_p$. Notice that $\|\|x\|\|_p \leq \gamma_p \|x\|_2$, hence

$$\mathbb{E} \|\|(a_i X_i)\|\|_p^2 \leq \gamma_p^2 \|a\|_2^2.$$

Observe also that

$$\sup\{\|\|(a_i x_i)\|\|_p : x \in \sqrt{t} B_2^n + t B_1^n\}$$

$$\leq \sqrt{t} \sup\{\|\|(a_i x_i)\|\|_p : x \in B_2^n\} + t \sup_{j \leq n} \|\|(a_i \delta_{i,j})\|\|_p$$

$$\leq \sqrt{t} \gamma_p \sup\{\|\|(a_i x_i)\|\|_2 : x \in B_2^n\} + t \|a\|_\infty = (\sqrt{t} \gamma_p + t) \|a\|_\infty.$$

Let $M_p = \text{Med}(\|\|(a_i X_i)\|\|_p)$, two level concentration (applied twice to sets $A = \{\|\|(a_i x_i)\|\|_p \leq M_p\}$ and $A = \{\|\|(a_i x_i)\|\|_p \geq M_p\}$) implies that

$$\Pr\left(\left| \|\|(a_i X_i)\|\|_p - M_p \right| \geq \kappa(\sqrt{t} \gamma_p + t) \|a\|_\infty \right) \leq 2 \exp(-t).$$

Integrating by parts this gives for $p \geq q \geq 2$,

$$\left\| \|(a_i X_i)\|_p - M_p \right\|_q \leq C\kappa(\sqrt{q}\gamma_p + q)\|a\|_\infty \leq C\kappa p\|a\|_\infty.$$

Hence

$$\left\| \sum_{i=1}^n a_i X_i \right\|_p = \left\| \|(a_i X_i)\|_p \right\|_p \leq \left\| \|(a_i X_i)\|_p \right\|_2 + C\kappa p\|a\|_\infty \leq \gamma_p\|a\|_2 + C\kappa p\|a\|_\infty.$$

$\square$

Unfortunately we do not know many examples of random vectors satisfying two level concentration with a good constant. Using estimate (2) it is not hard to see that infimum convolution inequality investigated in [12] implies two level concentration. In particular isotropic log-concave unconditional vectors with independent coordinates and isotropic vectors uniformly distributed on the (suitably rescaled) B_q^n balls satisfy two level concentration with an absolute constant.

The last approach to the problem of Gaussian approximation of moments we will discuss is based on the notion of negative association. We say that random variables $(Y_1, \ldots, Y_n)$ are *negatively associated* if for any disjoint sets I_1, I_2 in $\{1, \ldots, n\}$ and any bounded functions $f_i : \mathbb{R}^{I_i} \to \mathbb{R}$, $i = 1, 2$ that are coordinate nondecreasing we have

$$\mathrm{Cov}\left(f_1((Y_i)_{i \in I_1}), f_2((Y_i)_{i \in I_2}) \right) \leq 0.$$

Our next result is an unconditional version of Theorem 1 in [19].

Theorem 10. *Suppose that $X = (X_1, \ldots, X_n)$ is an unconditional random vector with finite second moment and random variables $(|X_i|)_{i=1}^n$ are negatively associated. Let $X_1^*, \ldots, X_n^*$ be independent random variables such that X_i^* has the same distribution as X_i. Then for any nonnegative function f on $\mathbb{R}$ such that f'' is convex and any $a_1, \ldots, a_n$ we have*

$$\mathbb{E}f\left(\sum_{i=1}^n a_i X_i \right) \leq \mathbb{E}f\left(\sum_{i=1}^n a_i X_i^* \right). \tag{8}$$

In particular

$$\mathbb{E}\left| \sum_{i=1}^n a_i X_i \right|^p \leq \mathbb{E}\left| \sum_{i=1}^n a_i X_i^* \right|^p \quad \text{for } p \geq 3.$$

Proof. Since random variables $|a_i X_i|$ are also negatively associated, it is enough to consider the case when $a_i = 1$ for all i. We may also assume that variables X_i^* are independent of X. Assume first that random variables X_i are bounded.

Let $Y = (Y_1, \ldots, Y_n)$ be independent copy of X and $2 \leq k \leq n$. To shorten the notation put for $1 \leq l \leq n$, $S_l = \sum_{i=1}^l \varepsilon_i |X_i|$ and $\tilde{S}_l = \sum_{i=1}^l \varepsilon_i |Y_i|$ (recall that ε_i denotes a Bernoulli sequence independent of other variables).

We have

$$f(S_k) + f(\tilde{S}_k) - f(S_{k-1} + \varepsilon_k|Y_k|) - f(\tilde{S}_{k-1} + \varepsilon_k|X_k|)$$

$$= \int_{|Y_k|}^{|X_k|} \varepsilon_k(f'(S_{k-1} + \varepsilon_k t) - f'(\tilde{S}_{k-1} + \varepsilon_k t))dt$$

$$= \int_0^\infty \varepsilon_k(f'(S_{k-1} + \varepsilon_k t) - f'(\tilde{S}_{k-1} + \varepsilon_k t))(\mathbf{I}_{\{|X_k|\geq t\}} - \mathbf{I}_{\{|Y_k|\geq t\}})dt. \quad (9)$$

Define for $t > 0$, $g_t(x) = \mathbb{E}\varepsilon_k f'(x + \varepsilon_k t) = (f'(x+t) - f'(x-t))/2$ and

$$h_t(|x_1|,\ldots,|x_{k-1}|) = \mathbb{E}_\varepsilon\varepsilon_k f'\Big(\sum_{i=1}^{k-1} \varepsilon_i|x_i| + \varepsilon_k t\Big) = \mathbb{E}g_t\Big(\sum_{i=1}^{k-1} \varepsilon_i|x_i|\Big).$$

Taking the expectation in (9) and using the unconditionality we get

$$2\Big(\mathbb{E}f\Big(\sum_{i=1}^{k} X_i\Big) - \mathbb{E}f\Big(\sum_{i=1}^{k-1} X_i + X_k^*\Big)\Big)$$

$$= \mathbb{E}\int_0^\infty \varepsilon_k(f'(S_{k-1} + \varepsilon_k t) - f'(\tilde{S}_{k-1} + \varepsilon_k t))(\mathbf{I}_{\{|X_k|\geq t\}} - \mathbf{I}_{\{|Y_k|\geq t\}})dt$$

$$= \int_0^\infty \mathbb{E}\big[\big(h_t(|X_1|,\ldots,|X_{k-1}|) - h_t(|Y_1|,\ldots,|Y_{k-1}|)\big)\big(\mathbf{I}_{\{|X_k|\geq t\}} - \mathbf{I}_{\{|Y_k|\geq t\}}\big)\big]dt$$

$$= \int_0^\infty \mathrm{Cov}\big(h_t(|X_1|,\ldots,|X_{k-1}|), \mathbf{I}_{\{|X_k|\geq t\}}\big)dt.$$

Convexity of f'' implies that the function g_t is convex on $\mathbb{R}$, therefore the function h_t is coordinate increasing on $\mathbb{R}_+^{k-1}$. So by the negative association we get

$$\mathbb{E}f\Big(\sum_{i=1}^{k} X_i\Big) \leq \mathbb{E}f\Big(\sum_{i=1}^{k-1} X_i + X_k^*\Big) \quad (10)$$

The same inequality holds if we change the function f into the function $f(\cdot + h)$ for any $h \in \mathbb{R}$. Therefore applying (10) conditionally we get

$$\mathbb{E}f\Big(\sum_{i=1}^{k} X_i + \sum_{i=k+1}^{n} X_i^*\Big) \leq \mathbb{E}f\Big(\sum_{i=1}^{k-1} X_i + \sum_{i=k}^{n} X_i^*\Big)$$

and inequality (8) easily follows in the bounded case.

To settle the unbounded case first notice that random variables $|X_i| \wedge m$ are bounded and negatively associated for any $m > 0$. Hence we know that

$$\mathbb{E}f\left(\sum_{i=1}^{n}\varepsilon_i(|X_i| \wedge m)\right) \leq \mathbb{E}f\left(\sum_{i=1}^{n}\varepsilon_i(|X_i^*| \wedge m)\right).$$

We have $\liminf_{m\to\infty} \mathbb{E}f(\sum_{i=1}^{n}\varepsilon_i(|X_i| \wedge m)) \geq \mathbb{E}f(\sum_{i=1}^{n}\varepsilon_i|X_i|)$, so it is enough to show that, $\liminf_{m\to\infty} \mathbb{E}f(\sum_{i=1}^{n}\varepsilon_i(|X_i^*| \wedge m)) \leq \mathbb{E}f(\sum_{i=1}^{n}\varepsilon_i|X_i^*|)$.

Let us define $u(x) = f(x) - \frac{1}{2}f''(0)x^2$, the function u'' is convex and $u''(0) = 0$. Since $\mathbb{E}|X_i|^2 = \mathbb{E}|X_i^*|^2 < \infty$ it is enough to show that for any $m > 0$,

$$\mathbb{E}u\left(\sum_{i=1}^{n}\varepsilon_i(|X_i^*| \wedge m)\right) \leq \mathbb{E}u\left(\sum_{i=1}^{n}\varepsilon_i|X_i^*|\right). \tag{11}$$

Let for $s \in \mathbb{R}$, $v_s(t) := \mathbb{E}u(\varepsilon_1 s + \varepsilon_2 t)$. Then $v_s''(t) = \mathbb{E}u''(\varepsilon_1 s + \varepsilon_2 t) \geq u''(\mathbb{E}(\varepsilon_1 s + \varepsilon_2 t)) = 0$ and $v_s'(0) = 0$, hence v_s is nondecreasing on $[0, \infty)$. Thus for any $x \in \mathbb{R}^n$,

$$\mathbb{E}_\varepsilon u\left(\sum_{i=1}^{n}\varepsilon_i(|x_i| \wedge m)\right) \leq \mathbb{E}_\varepsilon u\left(\sum_{i=1}^{n}\varepsilon_i|x_i|\right)$$

and (11) immediately follows. □

Corollary 11. *Suppose that X is an isotropic unconditional n-dimensional log-concave vector such that variables $|X_i|$ are negatively associated. Then for any $a_1, \ldots, a_n$ and $p \geq 3$,*

$$-\sqrt{3}p\|a\|_\infty \leq \left\|\sum_{i=1}^{n}a_i X_i\right\|_p - \gamma_p\|a\|_2 \leq p\|a\|_\infty.$$

In particular the above inequality holds if X has a uniform distribution on a (suitably rescaled) Orlicz ball.

Proof. First inequality follows by Corollary 5, second by Theorem 10 and (6). The last part of the statement is a consequence of the result of Pilipczuk and Wojtaszczyk [18] (see also [20] for a simpler proof and a slightly more general class of unconditional log-concave measures with negatively associated absolute values of coordinates). □

Acknowledgements Part of this work was done at the Newton institute for Mathematical Sciences in Cambridge (UK) during the program Discrete Analysis. Research partially supported by MNiSW Grant no. N N201 397437 and the Foundation for Polish Science.

References

1. F. Barthe, O. Guédon, S. Mendelson, A. Naor, A probabilistic approach to the geometry of the l_p^n-ball. Ann. Probab. **33**, 480–513 (2005)
2. S.G. Bobkov, F.L. Nazarov, in *On Convex Bodies and Log-Concave Probability Measures with Unconditional Basis*. Geometric Aspects of Functional Analysis. Lecture Notes in Math., vol. 1807 (Springer, Berlin, 2003), pp. 53–69
3. E.D. Gluskin and S. Kwapień, Tail and moment estimates for sums of independent random variables with logarithmically concave tails. Studia Math. **114**, 303–309 (1995)
4. U. Haagerup, The best constants in the Khintchine inequality. Studia Math. **70**, 231–283 (1982)
5. P. Hitczenko, Domination inequality for martingale transforms of a Rademacher sequence. Israel J. Math. **84**, 161–178 (1993)
6. N. Huet, Spectral gap for some invariant log-concave probability measures. Mathematika **57**, 51–62 (2011)
7. R. Kannan, L. Lovász, M. Simonovits, Isoperimetric problems for convex bodies and a localization lemma. Discrete Comput. Geom. **13**, 541–559 (1995)
8. B. Klartag, A Berry-Esseen type inequality for convex bodies with an unconditional basis. Probab. Theor. Relat. Fields **145**, 1–33 (2009)
9. B. Klartag, V. Milman, Geometry of log-concave functions and measures. Geom. Dedicata **112**, 169–182 (2005)
10. R. Latała, Gaussian approximation of moments of sums of independent symmetric random variables with logarithmically concave tails. IMS Collect. **5**, 37–42 (2009)
11. R. Latała, Weak and strong moments of random vectors. Banach Center Publ. **95**, 115–121 (2011)
12. R. Latała, J.O. Wojtaszczyk, On the infimum convolution inequality. Studia Math. **189**, 147–187 (2008)
13. M. Lis, Gaussian approximation of moments of sums of independent random variables. Bull. Polish Acad. Sci. Math. **60**, 77–90 (2012)
14. E. Milman, On the role of convexity in isoperimetry, spectral gap and concentration. Invent. Math. **177**, 1–43 (2009)
15. S.J. Montgomery-Smith, The distribution of Rademacher sums. Proc. Am. Math. Soc. **109**, 517–522 (1990)
16. F. Nazarov, M. Sodin, A. Volberg, The geometric KLS lemma, dimension-free estimates for the distribution of values of polynomials, and distribution of zeroes of random analytic functions (Russian). Algebra i Analiz **14**, 214–234 (2002); Translation in St. Petersburg Math. J. **14**, 351–366 (2003)
17. G. Paouris, in Ψ_2-*Estimates for Linear Functionals on Zonoids*. Geometric Aspects of Functional Analysis. Lecture Notes in Math., vol. 1807 (Springer, Berlin, 2003), pp. 211–222
18. M. Pilipczuk, J.O. Wojtaszczyk, Negative association property for absolute values of random variable equidistributed on generalized Orlicz balls. Positivity **12**, 421–474 (2008)
19. Q.-M. Shao, A comparison theorem on moment inequalities between negatively associated and independent random variables. J. Theoret. Probab. **13**, 343–356 (2000)
20. J.O. Wojtaszczyk, A simpler proof of the negative association property for absolute values of measures tied to generalized Orlicz balls. Bull. Pol. Acad. Sci. Math. **57**, 41–56 (2009)

Projections of Probability Distributions: A Measure-Theoretic Dvoretzky Theorem

Elizabeth Meckes

Abstract Many authors have studied the phenomenon of typically Gaussian marginals of high-dimensional random vectors; e.g., for a probability measure on $\mathbb{R}^d$, under mild conditions, most one-dimensional marginals are approximately Gaussian if d is large. In earlier work, the author used entropy techniques and Stein's method to show that this phenomenon persists in the bounded-Lipschitz distance for k-dimensional marginals of d-dimensional distributions, if $k = o(\sqrt{\log(d)})$. In this paper, a somewhat different approach is used to show that the phenomenon persists if $k < \frac{2\log(d)}{\log(\log(d))}$, and that this estimate is best possible.

1 Introduction

The explicit study of typical behavior of the margins of high-dimensional probability measures goes back to Sudakov [15], although some of the central ideas appeared much earlier; e.g., the 1906 monograph [2] of Borel, which contains the first rigorous proof that projections of uniform measure on the n-dimensional sphere are approximately Gaussian for large n. Subsequent major contributions were made by Diaconis and Freedman [3], von Weizsäcker [18], Bobkov [1], and Klartag [8], among others. The objects of study are a random vector $X \in \mathbb{R}^d$ and its projections onto subspaces; the central problem here is to show that for most subspaces, the resulting distributions are about the same, approximately Gaussian, and moreover to determine how large the dimension k of the subspace may be relative to d for this phenomenon to persist. This aspect in particular of the problem was addressed in earlier work [10] of the author. In this paper, a different approach is presented to

E. Meckes (✉)
Department of Mathematics, Case Western Reserve University, 10900 Euclid Ave., Cleveland, OH 44106, USA
e-mail: ese3@cwru.edu

B. Klartag et al. (eds.), *Geometric Aspects of Functional Analysis*, Lecture Notes in Mathematics 2050, DOI 10.1007/978-3-642-29849-3_18,
© Springer-Verlag Berlin Heidelberg 2012

proving the main result of [10], which, in addition to being technically simpler and perhaps more geometrically natural, also gives a noticable quantiative improvement. The result shows that the phenomenon of typical Gaussian marginals persists under mild conditions for $k < \frac{2\log(d)}{\log(\log(d))}$, as opposed to the results of [10], which requires $k = o(\sqrt{\log(d)})$ (note that a misprint in the abstract of that paper claimed that $k = o(\log(d))$ was sufficient).

The fact that typical k-dimensional projections of probability measures on $\mathbb{R}^d$ are approximately Gaussian when $k < \frac{2\log(d)}{\log(\log(d))}$ can be viewed as a measure-theoretic version of a famous theorem of Dvoretzky [5], Milman's proof of which [12] shows that for $\epsilon > 0$ fixed and $\mathcal{X}$ a d-dimensional Banach space, typical k-dimensional subspaces $E \subseteq \mathcal{X}$ are $(1 + \epsilon)$-isomorphic to a Hilbert space, if $k \leq C(\epsilon)\log(d)$. (This is the usual formulation, although one can give a dual formulation in terms of projections and quotient norms rather than subspaces.) These results should be viewed as analogous, in the following sense: in both cases, an additional structure is imposed on $\mathbb{R}^n$ (a norm in the case of Dvoretzky's theorem; a probability measure in the present context); in either case, there is a particularly nice way to do this (the Euclidean norm and the Gaussian distribution, respectively). The question is then: if one projects an arbitrary norm or probability measure onto lower dimensional subspaces, does it tend to resemble this nice structure? If so, by how much must one reduce the dimension in order to see this phenomenon?

Aside from the philosophical similarity of these results, they are also similar in that additional natural geometric assumptions lead to better behavior under projections. The main result of Klartag [9] shows that if the random vector $X \in \mathbb{R}^d$ is assumed to have a log-concave distribution, then typical marginals of the distribution of X are approximately Gaussian even when $k = d^\epsilon$ (for a specific universal constant $\epsilon \in (0, 1)$). This should be compared in the context of Dvoretzky's theorem to, for example, the result of Figiel et al. [6] showing that if a d-dimensional Banach space $\mathcal{X}$ has cotype $q \in [2, \infty)$, then $\mathcal{X}$ has subspaces of dimension of the order $d^{\frac{2}{q}}$ which are approximately Euclidean; or the result of Szarek [16] showing that if $\mathcal{X}$ has bounded volume ratio, then $\mathcal{X}$ has nearly Euclidean subspaces of dimension $\frac{d}{2}$. One interesting difference in the measure-theoretic context from the classical context is that, for measures, it is possible to determine *which* subspaces have approximately Gaussian projections under symmetry assumptions on the measure (see [11]); there is no known method to find explicit almost Euclidean subspaces of Banach spaces, even under natural geometric assumptions such as symmetry properties.

Following the statements of the main results below, an example is given to show that the estimate $k < \frac{2\log(d)}{\log(\log(d))}$ is best possible in the metric used here.

Before formally stating the results, some notation and context are needed. The Stiefel manifold $\mathfrak{W}_{d,k}$ is defined by

$$\mathfrak{W}_{d,k} := \{\theta = (\theta_1, \ldots, \theta_k) : \theta_i \in \mathbb{R}^d, \langle \theta_i, \theta_j \rangle = \delta_{ij} \ \forall \ 1 \leq i, j \leq k\},$$

with metric $\rho(\theta, \theta') = \left[\sum_{j=1}^{k} |\theta_j - \theta'_j|^2\right]^{1/2}$. The manifold $\mathfrak{W}_{d,k}$ posseses a rotation-invariant (Haar) probability measure.

Let X be a random vector in $\mathbb{R}^d$ and let $\theta \in \mathfrak{W}_{d,k}$. Let

$$X_\theta := \left(\langle X, \theta_1 \rangle, \ldots, \langle X, \theta_k \rangle \right);$$

that is, X_θ is the projection of X onto the span of θ. Consider also the "annealed" version X_Θ for $\Theta \in \mathfrak{W}_{d,k}$ distributed according to Haar measure and independent of X. The notation $e_X[\cdot]$ is used to denote expectation with respect to X only; that is, $e_X[f(X, \Theta)] = e\left[f(X, \Theta)|\Theta\right]$. When X_Θ is being thought of as conditioned on Θ with randomness coming from X only, it is written X_θ. The following results describe the behavior of the random variables X_θ and X_Θ. In what follows, c and C are used to denote universal constants which need not be the same in every appearance.

Theorem 1. *Let X be a random vector in $\mathbb{R}^n$, with $eX = 0$, $e\left[|X|^2\right] = \sigma^2 d$, and let $A := e\left||X|^2\sigma^{-2} - d\right|$. If Θ is a random point of $\mathfrak{W}_{d,k}$, X_Θ is defined as above, and Z is a standard Gaussian random vector, then*

$$d_{BL}(X_\Theta, \sigma Z) \leq \frac{\sigma[\sqrt{k}(A+1) + k]}{d-1}.$$

Theorem 2. *Let Z be a standard Gaussian random vector. Let*

$$B := \sup_{\xi \in \mathbb{S}^{d-1}} e\langle X, \xi \rangle^2.$$

For $\theta \in \mathfrak{W}_{d,k}$, let

$$d_{BL}(X_\theta, \sigma Z)$$

$$= \sup_{\max(\|f\|_\infty, |f|_L) \leq 1} \left| e\left[f(\langle X, \theta_1 \rangle, \ldots, \langle X, \theta_k \rangle)) | \theta\right] - ef(\sigma Z_1, \ldots, \sigma Z_k) \right|;$$

that is, $d_{BL}(X_\theta, \sigma Z)$ is the conditional bounded-Lipschitz distance from X_Θ to σZ, conditioned on Θ. Then if $\mathbb{P}_{d,k}$ denotes the Haar measure on $\mathfrak{W}_{d,k}$,

$$\mathbb{P}_{d,k}\left[\theta : \left|d_{BL}(X_\theta, \sigma Z) - ed_{BL}(X_\theta, \sigma Z)\right| > \epsilon\right] \leq Ce^{-\frac{cd\epsilon^2}{B}}.$$

Theorem 3. *With notation as in the previous theorems,*

$$ed_{BL}(X_\theta, \sigma Z) \leq C\left[\frac{(kB + B\log(d))B^{\frac{2}{9k+12}}}{(kB)^{\frac{2}{3}} d^{\frac{2}{3k+4}}} + \frac{\sigma[\sqrt{k}(A+1) + k]}{d-1}\right].$$

In particular, under the additional assumptions that $A \leq C'\sqrt{d}$ and $B = 1$, then

$$\mathrm{ed}_{BL}(X_\theta, \sigma Z) \leq C \frac{k + \log(d)}{k^{\frac{2}{3}} d^{\frac{2}{3k+4}}}.$$

Remark. The assumption that $B = 1$ is automatically satisfied if the covariance matrix of X is the identity; in the language of convex geometry, this is simply the case that the vector X is isotropic. The assumption that $A = O(\sqrt{d})$ is a geometrically natural one which arises, for example, if X is distributed uniformly on the isotropic dilate of the ℓ_1 ball in $\mathbb{R}^d$.

Together, Theorems 2 and 3 give the following.

Corollary 4. *Let X be a random vector in $\mathbb{R}^d$ satisfying*

$$\mathrm{e}|X|^2 = \sigma^2 d \qquad \mathrm{e}||X|^2\sigma^{-2} - d| \leq L\sqrt{d} \qquad \sup_{\xi \in \mathbb{S}^{d-1}} \mathrm{e}\langle\xi, X\rangle^2 \leq 1.$$

Let X_θ denote the projection of X onto the span of θ, for $\theta \in \mathfrak{W}_{d,k}$. Fix $a > 0$ and $b < 2$ and suppose that $k = \delta \frac{\log(d)}{\log(\log(d))}$ with $a \leq \delta \leq b$. Then there is a $c > 0$ depending only on a and b such that for

$$\epsilon = 2\exp\left[-c\frac{\log(\log(d))}{\delta}\right],$$

there is a subset $\mathfrak{T} \subseteq \mathfrak{W}_{d,k}$ with $\mathbb{P}_{d,k}[\mathfrak{T}] \geq 1 - C\exp\left(-c'd\epsilon^2\right)$, such that for all $\theta \in \mathfrak{T}$,

$$d_{BL}(X_\theta, \sigma Z) \leq C'\epsilon.$$

Remark. For the bound on $\mathrm{ed}_{BL}(X_\theta, \sigma Z)$ given in [10] to tend to zero as $d \to \infty$, it is necessary that $k = o(\sqrt{\log(d)})$, whereas Theorem 3 gives a similar result if $k = \delta\left(\frac{\log(d)}{\log(\log(d))}\right)$ for $\delta < 2$. Moreover, the following example shows that the bound above is best possible in our metric.

1.1 Sharpness

In the presence of log-concavity of the distribution of X, Klartag [9] proved a stronger result than Corollary 4 above; namely, that the typical total variation distance between X_θ and the corresponding Gaussian distribution is small even when $\theta \in \mathfrak{W}_{d,k}$ and $k = d^\epsilon$ (for a specific universal constant $\epsilon \in (0, 1)$). The result above allows k to grow only a bit more slowly than logarithmically with d. However, as the following example shows, either the log-concavity or some other additional assumption is necessary; with only the assumptions here, logarithmic-type growth of k in d is best possible for the bounded-Lipschitz metric. (It should be noted that the specific constants appearing in the results above are almost certainly non-optimal.)

Let X be distributed uniformly among $\{\pm\sqrt{d}e_1,\ldots,\pm\sqrt{d}e_d\}$, where the e_i are the standard basis vectors of $\mathbb{R}^d$. That is, X is uniformly distributed on the vertices of a cross-polytope. Then $e[X] = 0$, $|X|^2 \equiv d$, and given $\xi \in \mathbb{S}^{d-1}$, $e\langle X, \xi\rangle^2 = 1$; Theorems 1–3 apply with $\Sigma^2 = 1$, $A = 0$ and $B = 1$.

Consider a projection of $\{\pm\sqrt{d}e_1,\ldots,\pm\sqrt{d}e_d\}$ onto a random subspace E of dimension k, and define the Lipschitz function $f : E \to \mathbb{R}$ by $f(x) := (1 - d(x, S_E))_+$, where S_E is the image of $\{\pm\sqrt{d}e_1,\ldots,\pm\sqrt{d}e_d\}$ under projection onto E and $d(x, S_E)$ denotes the (Euclidean) distance from the point x to the set S_E. Then if μ_{S_E} denotes the probability measure putting equal mass at each of the points of S_E, $\int f d\mu_{S_E} = 1$. On the other hand, it is classical (see, e.g., [7]) that the volume ω_k of the unit ball in $\mathbb{R}^k$ is asymptotically given by $\frac{\sqrt{2}}{\sqrt{k\pi}}\left[\frac{2\pi e}{k}\right]^{\frac{k}{2}}$ for large k, in the sense that the ratio tends to one as k tends to infinity. It follows that the standard Gaussian measure of a ball of radius 1 in $\mathbb{R}^k$ is bounded by $\frac{1}{(2\pi)^{k/2}}\omega_k \sim \frac{\sqrt{2}}{\sqrt{k\pi}}\left[\frac{e}{k}\right]^{\frac{k}{2}}$. If γ_k denotes the standard Gaussian measure in $\mathbb{R}^k$, then this estimate means that $\int f d\gamma_k \leq \frac{2\sqrt{2}d}{\sqrt{k\pi}}\left[\frac{e}{k}\right]^{\frac{k}{2}}$. Now, if $k = \frac{c\log(d)}{\log(\log(d))}$ for $c > 2$, then this bound tends to zero, and thus $d_{BL}(\mu_{S_E}, \gamma_k)$ is close to 1 for any choice of the subspace E; the measures μ_{S_E} are far from Gaussian in this regime.

Taken together with Corollary 4, this shows that the phenomenon of typically Gaussian marginals persists for $k = \frac{c\log(d)}{\log(\log(d))}$ for $c < 2$, but fails in general if $k = \frac{c\log(d)}{\log(\log(d))}$ for $c > 2$.

Continuing the analogy with Dvoretzky's theorem, it is worth noting here that, for the projection formulation of Dvoretzky's theorem (the dual viewpoint to the slicing version discussed above), the worst case behavior is achieved for the ℓ_1 ball, that is, for the convex hull of the points considered above.

Acknowledgements The author thanks Mark Meckes for many useful discussions, without which this paper may never have been completed. Thanks also to Michel Talagrand, who pointed out a simplification in the proof of the main theorem, and to Richard Dudley for clarifying the history of "Dudley's entropy bound". Research supported by an American Institute of Mathematics 5-year Fellowship and NSF grant DMS-0852898.

2 Proofs

Theorems 1 and 2 were proved in [10], and their proofs will not be reproduced.

This section is mainly devoted to the proof of Theorem 3, but first some more definitions and notation are needed. Firstly, a comment on distance: as is clear from the statement of Theorems 2 and 3, the metric on random variables used here is the bounded-Lipschitz distance, defined by $d_{BL}(X, Y) := \sup_f |ef(X) - ef(Y)|$, where the supremum is taken over functions f with $\|f\|_{BL} := \max\{\|f\|_\infty, |f|_L\} \leq 1$ ($|f|_L$ is the Lipschitz constant of f).

A centered stochastic process $\{X_t\}_{t \in T}$ indexed by a space T with a metric d is said to satisfy a *sub-Gaussian increment condition* if there is a constant C such that, for all $\epsilon > 0$,

$$\mathbb{P}[|X_s - X_t| \geq \epsilon] \leq C \exp\left(-\frac{\epsilon^2}{2d^2(s,t)}\right). \tag{1}$$

A crucial point for the proof of Theorem 3 is that in the presence of a sub-Gaussian increment condition, there are powerful tools availabe to bound the expected supremum of a stochastic process; the one used here is what is usually called Dudley's entropy bound, formulated in terms of entropy numbers *à la* Talagrand [17]. For $n \geq 1$, the *entropy number* $e_n(T, d)$ is defined by

$$e_n(T, d) := \inf\{\sup_t d(t, T_n) : T_n \subseteq T, |T_n| \leq 2^{2^n}\}.$$

Dudley's entropy bound is the following.

Theorem 5. *If $\{X_t\}_{t \in T}$ is a centered stochastic process satisfying the sub-Gaussian increment condition* (1), *then there is a constant L such that*

$$e\left[\sup_{t \in T} X_t\right] \leq L \sum_{n=0}^{\infty} 2^{n/2} e_n(T, d). \tag{2}$$

Although the bound above is usually attributed to Dudley [4], it appears to have first appeared in print in a more general formulation due to Pisier [14, Theorem 1.1].

We now give the proof of the main theorem.

Proof of Theorem 3. As in [10], the key initial step is to view the distance as the supremum of a stochastic process: let $X_f = X_f(\theta) := e_X f(X_\theta) - e f(X_\Theta)$. Then $\{X_f\}_f$ is a centered stochastic process indexed by the unit ball of $\|\cdot\|_{BL}$, and $d_{BL}(X_\theta, X_\Theta) = \sup_{\|f\|_{BL} \leq 1} X_f$. The fact that Haar measure on $\mathfrak{W}_{d,k}$ has a measure-concentration property for Lipschitz functions (see [13]) implies that X_f is a sub-Gaussian process, as follows.

Let $f : \mathbb{R}^k \to \mathbb{R}$ be Lipschitz with Lipschitz constant L and consider the function $G = G_f$ defined on $\mathfrak{W}_{d,k}$ by

$$G(\theta_1, \ldots, \theta_k) = e_X f(X_\theta) = e\left[f(\langle\theta_1, X\rangle, \ldots, \langle\theta_k, X\rangle)|\theta\right].$$

Then

$$\left|G(\theta) - G(\theta')\right| = \left|e\left[f(\langle X, \theta_1'\rangle, \ldots, \langle X, \theta_k'\rangle) - f(\langle X, \theta_1\rangle, \ldots, \langle X, \theta_k\rangle)|\theta, \theta'\right]\right|$$

$$\leq Le\left[\left|(\langle X, \theta_1' - \theta_1\rangle, \ldots, \langle X, \theta_k' - \theta_k\rangle)\right||\theta, \theta'\right]$$

$$\leq L \sqrt{\sum_{j=1}^{k} |\theta'_j - \theta_j|^2 e \left\langle X, \frac{\theta'_j - \theta_j}{|\theta'_j - \theta_j|} \right\rangle^2}$$

$$\leq L\rho(\theta, \theta')\sqrt{B},$$

thus $G(\theta)$ is a Lipschitz function on $\mathfrak{W}_{k,d}$, with Lipschitz constant $L\sqrt{B}$. It follows immediately from Theorem 6.6 and Remark 6.7.1 of [13] that

$$\mathbb{P}_{d,k}\left[|G(\theta) - M_G| > \epsilon\right] \leq \sqrt{\frac{\pi}{2}} e^{-\frac{d\epsilon^2}{8L^2B}},$$

where M_G is the median of G with respect to Haar measure on $\mathfrak{W}_{d,k}$. It is then a straightforward exercise to show that for some universal constant C,

$$\mathbb{P}\left[|G(\theta) - eG(\theta)| > \epsilon\right] \leq C e^{-\frac{d\epsilon^2}{32L^2B}}. \tag{3}$$

Observe that, for Θ a Haar-distributed random point of $\mathfrak{W}_{d,k}$, $eG(\Theta) = ef(X_\Theta)$, and so (3) can be restated as $\mathbb{P}\left[|X_f| > \epsilon\right] \leq C \exp\left[-cd\epsilon^2\right]$.

Note that $X_f - X_g = X_{f-g}$, thus for $|f - g|_L$ the Lipschitz constant of $f - g$ and $\|f - g\|_{BL}$ the bounded-Lipschitz norm of $f - g$,

$$\mathbb{P}\left[|X_f - X_g| > \epsilon\right] \leq C \exp\left[\frac{-cd\epsilon^2}{2|f-g|_L^2}\right] \leq C \exp\left[\frac{-cd\epsilon^2}{2\|f-g\|_{BL}^2}\right].$$

The process $\{X_f\}$ therefore satisfies the sub-Gaussian increment condition in the metric $d^*(f,g) := \frac{1}{\sqrt{cd}}\|f - g\|_{BL}$; in particular, the entropy bound (2) applies. We will not be able to apply it directly, but rather use a sequence of approximations to arrive at a bound.

The first step is to truncate the indexing functions. Let

$$\varphi_R(x) = \begin{cases} 1 & |x| \leq R, \\ R+1-|x| & R \leq |x| \leq R+1, \\ 0 & R+1 \leq |x|, \end{cases}$$

and define $f_R := f \cdot \varphi_R$. It is easy to see that if $\|f\|_{BL} \leq 1$, then $\|f_R\|_{BL} \leq 2$. Since $|f(x) - f_R(x)| = 0$ if $x \in B_R$ and $|f(x) - f_R(x)| \leq 1$ for all $x \in \mathbb{R}^k$,

$$\left|e_X f(X_\theta) - e_X f_R(X_\theta)\right| \leq \mathbb{P}\left[|X_\theta| > R|\theta\right] \leq \frac{1}{R^2} \sum_{i=1}^{k} e\left[\langle X, \theta_i \rangle^2\right] \leq \frac{Bk}{R^2},$$

and the same holds if e_X is replaced by e. It follows that $|X_f - X_{f_R}| \le \frac{2Bk}{R^2}$. Consider therefore the process X_f indexed by $BL_{2,R+1}$ (with norm $\| \cdot \|_{BL}$), for some choice of R to be determined, where

$$BL_{2,R+1} := \{f : \mathbb{R}^k \to \mathbb{R} : \|f\|_{BL} \le 2; f(x) = 0 \text{ if } |x| > R+1\};$$

what has been shown is that

$$e\left[\sup_{\|f\|_{BL}\le 1} X_f\right] \le e\left[\sup_{f\in BL_{2,R+1}} X_f\right] + \frac{2Bk}{R^2}. \tag{4}$$

The next step is to approximate functions in $BL_{2,R+1}$ by "piecewise linear" functions. Specifically, consider a cubic lattice of edge length ϵ in $\mathbb{R}^k$. Triangulate each cube of the lattice into simplices inductively as follows: in $\mathbb{R}^2$, add an extra vertex in the center of each square to divide the square into four triangles. To triangulate the cube of $\mathbb{R}^k$, first triangulate each facet as was described in the previous stage of the induction. Then add a new vertex at the center of the cube; connecting it to each of the vertices of each of the facets gives a triangulation into simplices. Observe that when this procedure is carried out, each new vertex added is on a cubic lattice of edge length $\frac{\epsilon}{2}$. Let $\mathcal{L}$ denote the supplemented lattice comprised of the original cubic lattice, together with the additional vertices needed for the triangulation. The number of sites of $\mathcal{L}$ within the ball of radius $R + 1$ is then bounded by, e.g., $c\left(\frac{3R}{\epsilon}\right)^k \omega_k$, where ω_k is the volume of the unit ball in $\mathbb{R}^k$.

Now approximate $f \in BL_{2,R+1}$ by the function $\tilde{f}$ defined such that $\tilde{f}(x) = f(x)$ for $x \in \mathcal{L}$, and the graph of $\tilde{f}$ is determined by taking the convex hull of the vertices of the image under f of each k-dimensional simplex determined by $\mathcal{L}$. The resulting function $\tilde{f}$ still has $\|\tilde{f}\|_{BL} \le 2$, and $\|f - \tilde{f}\|_\infty \le \frac{\epsilon\sqrt{k}}{2}$, since the distance between points in the same simplex is bounded by $\epsilon\sqrt{k}$. Moreover, $\|\tilde{f}\|_{BL} = \sup_{x\in\mathcal{L}} |f(x)| + \sup_{x\sim y} \frac{|f(x)-f(y)|}{|x-y|}$, where $x \sim y$ if $x, y \in \mathcal{L}$ and x and y are part of the same triangulating simplex. Observe that, for a given $x \in \mathcal{L}$, those vertices which are part of a triangulating simplex with x are all contained in a cube centered at x of edge length ϵ; the number of such points is thus bounded by 3^k, and the number of differences which must be considered in order to compute the Lipschitz constant of $\tilde{f}$ is therefore bounded by $c\left(\frac{9R}{\epsilon}\right)^k \omega_k$. Recall that $\omega_k \sim \frac{2}{\sqrt{k\pi}}\left[\frac{2\pi e}{k}\right]^{\frac{k}{2}}$ for large k, and so the number of differences determining the Lipschitz constant of $\tilde{f}$ is bounded by $\frac{c}{\sqrt{k}}\left(\frac{c'R}{\epsilon\sqrt{k}}\right)^k$, for some absolute constants c, c'. It follows that

$$e\left[\sup_{f\in BL_{2,R+1}} X_f\right] \le e\left[\sup_{f\in BL_{2,R+1}} X_{\tilde{f}}\right] + \epsilon\sqrt{k}, \tag{5}$$

that the process $\{X_{\tilde{f}}\}_{f \in BL_{2,R+1}}$ is sub-Gaussian with respect to $\frac{1}{\sqrt{cd}} \| \cdot \|_{BL}$, and that the values of $\tilde{f}$ for $f \in BL_{2,R+1}$ are determined by a point of the ball $2B_\infty^M$ of ℓ_∞^M, where

$$M = \frac{c}{\sqrt{k}} \left(\frac{c'R}{\epsilon\sqrt{k}} \right)^k. \tag{6}$$

The virtue of this approximation is that it replaces a sub-Gaussian process indexed by a ball in an infinite-dimensional space with one indexed by a ball in a finite-dimensional space, where Dudley's bound is finally to be applied. Let $T := \left\{ \tilde{f} : f \in BL_{2,R+1} \right\} \subseteq 2B_\infty^M$; the covering numbers of the unit ball B of a finite-dimensional normed space $(X, \| \cdot \|)$ of dimension M are known (see Lemma 2.6 of [13]) to be bounded as $\mathcal{N}(B, \| \cdot \|, \epsilon) \leq \exp\left[M \log\left(\frac{3}{\epsilon}\right)\right]$. This implies that

$$\mathcal{N}(B_\infty^M, \rho, \epsilon) \leq \exp\left[M \log\left(\frac{3}{\epsilon\sqrt{cd}}\right)\right],$$

which in turn implies that

$$e_n(2B_\infty^M, \rho) \leq \frac{24\sqrt{B}}{\sqrt{d}} 2^{-\frac{2^n}{M}}.$$

Applying Theorem 5 now yields

$$e\left[\sup_{f \in BL_{2,R+1}} X_{\tilde{f}}\right] \leq L \sum_{n \geq 0} \left(\frac{24\sqrt{B}}{\sqrt{d}} 2^{\left(\frac{n}{2} - \frac{2^n}{M}\right)} \right). \tag{7}$$

Now, for the terms in the sum with $\log(M) \leq (n+1)\log(2) - 3\log(n)$, the summands are bounded above by 2^{-n}, contributing only a constant to the upper bound. On the other hand, the summand is maximized for $2^n = \frac{M}{2}\log(2)$, and is therefore bounded by $\sqrt{M}$. Taken together, these estimates show that the sum on the right-hand side of (7) is bounded by $L \log(M) \sqrt{\frac{MB}{d}}$.

Putting all the pieces together,

$$e\left[\sup_{\|f\|_{BL} \leq 1} \left(e\left[f(X_\Theta) | \Theta\right] - ef(X_\Theta)\right)\right] \leq \frac{9kB}{R^2} + 2\epsilon\sqrt{k} + L \log(M)\sqrt{\frac{MB}{d}}.$$

Choosing $\epsilon = \frac{\sqrt{k}B}{2R^2}$ and using the value of M in terms of R yields

$$e\left[\sup_{\|f\|_{BL} \leq 1} \left(e\left[f(X_\Theta) | \Theta\right] - ef(X_\Theta)\right)\right]$$

$$\leq \frac{10kB}{R^2} + Lk \log\left(\frac{c'R^3}{kB}\right) \frac{c}{k^{1/4}} \left[\frac{c'R^3}{kB}\right]^{\frac{k}{2}} \sqrt{\frac{B}{d}}.$$

Now choosing $R = cd^{\frac{1}{3k+4}} k^{\frac{2k+1}{6k+8}} B^{\frac{k+1}{3k+4}}$ yields

$$\mathrm{e}\left[\sup_{\|f\|_{BL}\leq 1} \left(\mathrm{e}\left[f(X_\Theta)|\Theta\right] - \mathrm{e}f(X_\Theta)\right)\right] \leq L\frac{kB + B\log(d)}{d^{\frac{2}{3k+4}} k^{\frac{2k+1}{3k+4}} B^{\frac{2k+2}{3k+4}}}.$$

This completes the proof of the first statement of the theorem. The second follows immediately using that $B = 1$ and observing that, under the assumption that $A \leq C'\sqrt{d}$, the bound above is always worse than the error $\frac{\sigma[\sqrt{k}(A+1)+k]}{d-1}$ coming from Theorem 1. □

The proof of Corollary 4 is essentially immediate from Theorems 2 and 3.

References

1. S.G. Bobkov, On concentration of distributions of random weighted sums. Ann. Probab. **31**(1), 195–215 (2003)
2. E. Borel, Sur les principes de la theorie cinétique des gaz. Annales de l'ecole normale sup. **23**, 9–32 (1906)
3. P. Diaconis, D. Freedman, Asymptotics of graphical projection pursuit. Ann. Statist. **12**(3), 793–815 (1984)
4. R.M. Dudley, The sizes of compact subsets of Hilbert space and continuity of Gaussian processes. J. Funct. Anal. **1**, 290–330 (1967)
5. A. Dvoretzky, in *Some Results on Convex Bodies and Banach Spaces*. Proc. Internat. Sympos. Linear Spaces, Jerusalem, 1960 (Jerusalem Academic Press, Jerusalem, 1961), pp. 123–160
6. T. Figiel, J. Lindenstrauss, V.D. Milman, The dimension of almost spherical sections of convex bodies. Acta Math. **139**(1–2), 53–94 (1977)
7. G.B. Folland, How to integrate a polynomial over a sphere. Am. Math. Mon. **108**(5), 446–448 (2001)
8. B. Klartag, A central limit theorem for convex sets. Invent. Math. **168**(1), 91–131 (2007)
9. B. Klartag, Power-law estimates for the central limit theorem for convex sets. J. Funct. Anal. **245**(1), 284–310 (2007)
10. E. Meckes, *Approximation of projections of random vectors*. J. Theoret. Probab. **25**(2), (2012)
11. M.W. Meckes, Gaussian marginals of convex bodies with symmetries. Beiträge Algebra Geom. **50**(1), 101–118 (2009)
12. V.D. Milman, A new proof of A. Dvoretzky's theorem on cross-sections of convex bodies. Funkcional. Anal. i Priložen. **5**(4), 28–37 (1971)
13. V.D. Milman, G. Schechtman, in *Asymptotic Theory of Finite-Dimensional Normed Spaces*, with an appendix by M. Gromov. Lecture Notes in Mathematics, vol. 1200 (Springer, Berlin, 1986)
14. G. Pisier, in *Some Applications of the Metric Entropy Condition to Harmonic Analysis*. Banach Spaces, Harmonic Analysis, and Probability Theory. Lecture Notes in Mathematics, vol. 995 (Springer, Berlin, 1983)
15. V.N. Sudakov, Typical distributions of linear functionals in finite-dimensional spaces of high dimension. Dokl. Akad. Nauk SSSR **243**(6), 1402–1405 (1978)
16. S. Szarek, On Kashin's almost Euclidean orthogonal decomposition of l_n^1. Bull. Acad. Polon. Sci. Sér. Sci. Math. Astronom. Phys. **26**(8), 691–694 (1978)
17. M. Talagrand, *The Generic Chaining. Upper and Lower Bounds of Stochastic Processes*. Springer Monographs in Mathematics (Springer, Berlin, 2005)
18. H. von Weizsäckerm Sudakov's typical marginals, random linear functionals and a conditional central limit theorem. Probab. Theor. Relat. Fields **107**(3), 313–324 (1997)

On a Loomis–Whitney Type Inequality for Permutationally Invariant Unconditional Convex Bodies

Piotr Nayar and Tomasz Tkocz

Abstract For a permutationally invariant unconditional convex body K in $\mathbb{R}^n$ we define a finite sequence $(K_j)_{j=1}^n$ of projections of the body K to the space spanned by first j vectors of the standard basis of $\mathbb{R}^n$. We prove that the sequence of volumes $(|K_j|)_{j=1}^n$ is log-concave.

1 Introduction

The main interest in convex geometry is the examination of sections and projections of sets. Some introduction can be found in a monograph by Gardner [5]. We are interested in a class $\mathcal{PU}_n$ of convex bodies in $\mathbb{R}^n$ which are unconditional and permutationally invariant.

Let us briefly recall some definitions. A convex body K in $\mathbb{R}^n$ is called *unconditional* if for every point $(x_1, \ldots, x_n) \in K$ and every choice of signs $\epsilon_1, \ldots, \epsilon_n \in \{-1, 1\}$ the point $(\epsilon_1 x_1, \ldots, \epsilon_n x_n)$ also belongs to K. A convex body K in $\mathbb{R}^n$ is called *permutationally invariant* if for every point $(x_1, \ldots, x_n) \in K$ and every permutation $\pi : \{1, \ldots, n\} \longrightarrow \{1, \ldots, n\}$ the point $(x_{\pi(1)}, \ldots, x_{\pi(n)})$ is also in K. A sequence $(a_i)_{i=1}^n$ of positive real numbers is called *log-concave* if $a_i^2 \geq a_{i-1} a_{i+1}$, for $i = 2, \ldots, n-1$.

The main result of this paper reads as follows.

Theorem 1. *Let $n \geq 3$ and let $K \in \mathcal{PU}_n$. For each $i = 1, \ldots, n$ we define a convex body $K_i \in \mathcal{PU}_i$ as an orthogonal projection of K to the subspace $\{(x_1, \ldots, x_n) \in \mathbb{R}^n \mid x_{i+1} = \ldots = x_n = 0\}$. Then the sequence of volumes $(|K_i|)_{i=1}^n$ is log-concave. In particular*

$$|K_{n-1}|^2 \geq |K_n| \cdot |K_{n-2}|. \tag{1}$$

P. Nayar (✉) • T. Tkocz
Institute of Mathematics, University of Warsaw, Banacha 2, 02-097 Warszawa, Poland
e-mail: nayar@mimuw.edu.pl; tkocz@mimuw.edu.pl

B. Klartag et al. (eds.), *Geometric Aspects of Functional Analysis*, Lecture Notes in Mathematics 2050, DOI 10.1007/978-3-642-29849-3_19,
© Springer-Verlag Berlin Heidelberg 2012

Inequality (1) is related to the problem of negative correlation of coordinate functions on $K \in \mathcal{PU}_n$, i.e. the question whether for every $t_1, \ldots, t_n \geq 0$

$$\mu_K \left(\bigcap_{i=1}^n \{|x_i| \geq t_i\} \right) \leq \prod_{i=1}^n \mu_K \left(|x_i| \geq t_i \right), \tag{2}$$

where μ_K is normalized Lebesgue measure on K. Indeed, the Taylor expansion of the function $h(t) = \mu_K(|x_1| \geq t)\mu_K(|x_2| \geq t) - \mu_K(|x_1| \geq t, |x_2| \geq t)$ at $t = 0$ contains

$$\frac{1}{|K_n|^2} \left(|K_{n-1}|^2 - |K_{n-2}| \cdot |K_n| \right) t^2,$$

cf. (1), as a leading term. The Property (2), the so-called concentration hypothesis and the central limit theorem for convex bodies are closely related, see [1]. The last theorem has been recently proved by Klartag [8].

The negative correlation property in the case of generalized Orlicz balls was originally investigated by Wojtaszczyk in [11]. A generalized Orlicz ball is a set

$$B = \left\{ (x_1, \ldots, x_n) \in \mathbb{R}^n \mid \sum_{i=1}^n f_i(|x_i|) \leq n \right\},$$

where $f_1, \ldots, f_n$ are some Young functions (see [11] for the definition). In probabilistic terms Pilipczuk and Wojtaszczyk (see [10]) have shown that the random variable $X = (X_1, \ldots, X_n)$ uniformly distributed on B satisfies the inequality

$$\mathrm{Cov}(f(|X_{i_1}|, \ldots, |X_{i_k}|), g(|X_{j_1}|, \ldots, |X_{j_l}|)) \leq 0$$

for any bounded coordinate-wise increasing functions $f : \mathbb{R}^k \longrightarrow \mathbb{R}$, $g : \mathbb{R}^l \longrightarrow \mathbb{R}$ and any disjoint subsets $\{i_1, \ldots, i_k\}$ and $\{j_1, \ldots, j_l\}$ of $\{1, \ldots, n\}$. In the case of generalized isotropic Orlicz balls this result implies the inequality

$$\mathrm{Var}|X|^p \leq \frac{Cp^2}{n} \mathbb{E}|X|^{2p}, \qquad p \geq 2,$$

from which some reverse Hölder inequalities can be deduced (see [3]).

One may ask about an example of a nice class of Borel probability measures on $\mathbb{R}^n$ for which the negative correlation inequality hold. Considering the example of the measure with the density

$$p(x_1, \ldots, x_n) = \exp\left(-2(n!)^{1/n} \max\{|x_1|, \ldots, |x_n|\} \right),$$

which was mentioned by Bobkov and Nazarov in a different context (see [2, Lemma 3.1]), we certainly see that the class of unconditional and permutationally invariant log-concave measures would not be the answer. Nevertheless, it remains still open whether the negative correlation of coordinate functions holds for measures uniformly distributed on the bodies from the class $\mathcal{PU}_n$.

We should remark that our inequality (1) is similar to some auxiliary result by Giannopoulos, Hartzoulaki and Paouris, see [6, Lemma 4.1]. They proved that a version of inequality (1) holds, up to the multiplicative constant $\frac{n}{2(n-1)}$, for an arbitrary convex body.

The paper is organised as follows. In Sect. 2 we give the proof of Theorem 1. Section 3 is devoted to some remarks. Several examples are there provided as well.

2 Proof of the Main Result

Here we deal with the proof of Theorem 1. We start with an elementary lemma.

Lemma 1. Let $f : [0, L] \longrightarrow [0, \infty)$ be a nonincreasing concave function such that $f(0) = 1$. Then

$$\frac{n-1}{n} \left(\int_0^L f(x)^{n-2} dx \right)^2 \geq \int_0^L x f(x)^{n-2} dx, \qquad n \geq 3. \tag{3}$$

Proof. By a linear change of a variable one can assume that $L = 1$. Since f is concave and nonincreasing, we have $1 - x \leq f(x) \leq 1$ for $x \in [0, 1]$. Therefore, there exists a real number $\alpha \in [0, 1]$ such that for $g(x) = 1 - \alpha x$ we have

$$\int_0^1 f(x)^{n-2} dx = \int_0^1 g(x)^{n-2} dx.$$

Clearly, we can find a number $c \in [0, 1]$ such that $f(c) = g(c)$. Since f is concave and g is affine, we have $f(x) \geq g(x)$ for $x \in [0, c]$ and $f(x) \leq g(x)$ for $x \in [c, 1]$. Hence,

$$\int_0^1 x(f(x)^{n-2} - g(x)^{n-2}) dx \leq \int_0^c c(f(x)^{n-2} - g(x)^{n-2}) dx$$

$$+ \int_c^1 c(f(x)^{n-2} - g(x)^{n-2}) dx = 0.$$

We conclude that it suffices to prove (3) for the function g, which is by simple computation equivalent to

$$\frac{1}{\alpha^2 n(n-1)} \left(1 - (1-\alpha)^{n-1} \right)^2 \geq \frac{1}{\alpha^2} \left(\frac{1}{n-1} \left(1 - (1-\alpha)^{n-1} \right) \right.$$

$$\left. - \frac{1}{n} \left(1 - (1-\alpha)^n \right) \right).$$

To finish the proof one has to perform a short calculation and use Bernoulli's inequality.   □

Remark 1 (Added in Proofs). A slightly more general form of this lemma appeared in [7] and, as it is pointed out in that paper, the lemma is a particular case of a result of [9, p. 182]. Only after the paper was written we heard about these references from Prof. A. Zvavitch, for whom we are thankful. Our proof differs only in a few details, yet it is provided for the convenience of the reader.

Proof of Theorem 1. Due to an inductive argument it is enough to prove inequality (1).

Let $g: \mathbb{R}^{n-1} \longrightarrow \{0, 1\}$ be a characteristic function of the set K_{n-1}. Then, by permutational invariance and unconditionality, we have

$$|K_{n-1}| = 2^{n-1}(n-1)! \int_{x_1 \geq \ldots \geq x_{n-1} \geq 0} g(x_1, \ldots, x_{n-1}) dx_1 \ldots dx_{n-1}, \qquad (4)$$

and similarly

$$|K_{n-2}| = 2^{n-2}(n-2)! \int_{x_1 \geq \ldots \geq x_{n-2} \geq 0} g(x_1, \ldots, x_{n-2}, 0) dx_1 \ldots dx_{n-2}. \qquad (5)$$

Moreover, permutational invariance and the definition of a projection imply

$$\mathbf{1}_{K_n}(x_1, \ldots, x_n) \leq \prod_{i=1}^{n} g(x_1, \ldots, \hat{x}_i, \ldots, x_n). \qquad (6)$$

Thus

$$|K_n| \leq 2^n n! \int_{x_1 \geq \ldots \geq x_n \geq 0} \prod_{i=1}^{n} g(x_1, \ldots, \hat{x}_i, \ldots, x_n) dx_1 \ldots dx_n$$

$$= 2^n n! \int_{x_1 \geq \ldots \geq x_n \geq 0} g(x_1, \ldots, x_{n-1}) dx_1 \ldots dx_n \qquad (7)$$

$$= 2^n n! \int_{x_1 \geq \ldots \geq x_{n-1} \geq 0} x_{n-1} g(x_1, \ldots, x_{n-1}) dx_1 \ldots dx_{n-1},$$

where the first equality follows from the monotonicity of the function g for nonnegative arguments with respect to each coordinate. We define a function $F: [0, \infty) \longrightarrow [0, \infty)$ by the equation

$$F(x) = \frac{\int_{x_1 \geq \ldots \geq x_{n-2} \geq x} g(x_1, \ldots, x_{n-2}, x) dx_1 \ldots dx_{n-2}}{\int_{x_1 \geq \ldots \geq x_{n-2} \geq 0} g(x_1, \ldots, x_{n-2}, 0) dx_1 \ldots dx_{n-2}}.$$

One can notice that

1. $F(0) = 1$.
2. The function F is nonincreasing as so is the function

$$x \mapsto g(x_1, \ldots, x_{n-2}, x)\mathbf{1}_{\{x_1 \geq \ldots \geq x_{n-2} \geq x\}}.$$

3. The function $F^{1/(n-2)}$ is concave on its support $[0, L]$ since $F(x)$ multiplied by some constant equals the volume of the intersection of the convex set $K_{n-1} \cap \{x_1 \geq \ldots \geq x_{n-1} \geq 0\}$ with the hyperplane $\{x_{n-1} = x\}$. This is a simple consequence of the Brunn–Minkowski inequality, see for instance [4, p. 361].

By the definition of the function F and equations (4), (5) we obtain

$$\int_0^L F(x)\mathrm{d}x = \frac{\frac{1}{2^{n-1}(n-1)!}|K_{n-1}|}{\frac{1}{2^{n-2}(n-2)!}|K_{n-2}|} = \frac{1}{2(n-1)} \cdot \frac{|K_{n-1}|}{|K_{n-2}|},$$

and using inequality (7)

$$\int_0^L xF(x)\mathrm{d}x \geq \frac{\frac{1}{2^n n!}|K_n|}{\frac{1}{2^{n-2}(n-2)!}|K_{n-2}|} = \frac{1}{2^2 n(n-1)} \cdot \frac{|K_n|}{|K_{n-2}|}.$$

Therefore it is enough to show that

$$\frac{n-1}{n}\left(\int_0^L F(x)\mathrm{d}x\right)^2 \geq \int_0^L xF(x)\mathrm{d}x.$$

This inequality follows from Lemma 1.   □

3 Some Remarks

In this section we give some remarks concerning Theorem 1.

Remark 2. Apart from the trivial example of the B_∞^n ball, there are many other examples of bodies for which equality in (1) is attained. Indeed, analysing the proof, we observe that for the equality in (1) the equality in Lemma 1 is needed. Therefore, the function $F^{1/(n-2)}$ has to be linear and equal to $1 - x$. Taking into account the equality conditions in the Brunn–Minkowski inequality (consult [4, p. 363]), this is the case if and only if the set $K_{n-1} \cap \{x_1 \geq \ldots \geq x_{n-1} \geq 0\}$ is a cone C with the base $(K_{n-2} \cap \{x_1 \geq \ldots \geq x_{n-2} \geq 0\}) \times \{0\} \subset \mathbb{R}^{n-1}$ and the vertex $(z_0, \ldots, z_0) \in \mathbb{R}^{n-1}$. Thus if for a convex body $K \in \mathcal{PU}_n$ we have the equality in (1), then this body K is constructed in the following manner. Take an arbitrary $K_{n-2} \in \mathcal{PU}_{n-2}$. Define the set K_{n-1} as the smallest permutationally invariant unconditional body containing C. For z_0 from some interval the set K_{n-1} is convex. For the characteristic function of the body K we then set $\prod_{i=1}^{n} \mathbf{1}_{K_{n-1}}(x_1, \ldots, \hat{x_i}, \ldots, x_n)$.

A one more natural question to ask is when a sequence $(|K_i|)_{i=1}^{n}$ is geometric. Bearing in mind what has been said above for $i = 2, 3, \ldots, n-1$ we find that a sequence $(|K_i|)_{i=1}^{n}$ is geometric if and only if

$$K = [-L, L]^n \cup \bigcup_{i \in \{1, \ldots, n\}, \epsilon \in \{-1, 1\}} \operatorname{conv}\{\epsilon a e_i, \{x_i = \epsilon L, |x_k| \le L, k \ne i\}\},$$

for some positive parameters a and L satisfying $L < a < 2L$, where $e_1, \ldots, e_n$ stand for the standard orthonormal basis in $\mathbb{R}^n$. One can easily check that $|K_i| = 2^i L^{i-1} a$.

Remark 3. Suppose we have a sequence of convex bodies $K_n \in \mathcal{PU}_n$, for $n \ge 1$, such that $K_n = \pi_n(K_{n+1})$, where by $\pi_n: \mathbb{R}^{n+1} \longrightarrow \mathbb{R}^n$ we denote the projection $\pi_n(x_1, \ldots, x_n, x_{n+1}) = (x_1, \ldots, x_n)$. Since Theorem 1 implies that the sequence $(|K_n|)_{n=1}^{\infty}$ is log-concave we deduce the existence of the limits

$$\lim_{n \to \infty} \frac{|K_{n+1}|}{|K_n|}, \qquad \lim_{n \to \infty} \sqrt[n]{|K_n|}.$$

We can obtain this kind of sequences as finite dimensional projections of an Orlicz ball in ℓ_∞.

Acknowledgements The authors would like to thank Prof. K. Oleszkiewicz for a valuable comment regarding the equality conditions in Theorem 1 as well as Prof. R. Latała for a stimulating discussion. Research of the First named author partially supported by NCN Grant no. 2011/01/N/ST1/01839. Research of the second named author partially supported by NCN Grant no. 2011/01/N/ST1/05960.

References

1. M. Anttila, K. Ball, I. Perissinaki, The central limit problem for convex bodies. Trans. Am. Math. Soc. **355**, 4723–4735 (2003)
2. S.G. Bobkov, F.L. Nazarov, in *On Convex Bodies and Log-Concave Probability Measures with Unconditional Basis*. Geometric Aspects of Functional Analysis. Lecture Notes in Math., vol. 1807 (Springer, Berlin, 2003), pp. 53–69
3. B. Fleury, Between Paouris concentration inequality and variance conjecture. Ann. Inst. Henri Poincar Probab. Stat. **46**(2), 299–312 (2010)
4. R.J. Gardner, The Brunn-Minkowski inequality. Bull. Am. Math. Soc. **39**, 355–405 (2002)
5. R.J. Gardner, in *Geometric Tomography*. Encyclopedia of Mathematics, vol. 58 (Cambridge University Press, London, 2006)
6. A. Giannopoulos, M. Hartzoulaki, G. Paouris, On a local version of the Aleksandrov-Fenchel inequalities for the quermassintegrals of a convex body. Proc. Am. Math. Soc. **130**, 2403–2412 (2002)

7. Y. Gordon, M. Meyer, S. Reisner, Zonoids with minimal volume-product – a new proof. Proc. Am. Math. Soc. **104**(1), 273–276 (1988)
8. B. Klartag, A central limit theorem for convex sets. Invent. Math. **168**, 91–131 (2007)
9. A.W. Marshall, I. Olkin, F. Proschan, in *Monotonicity of Ratios of Means and Other Applications of Majorization*, ed. by O. Shisha. Inequalities (Academic, New York, 1967), pp. 177–190
10. M. Pilipczuk, J.O. Wojtaszczyk, The negative association property for the absolute values of random variables equidistributed on a generalized Orlicz ball. Positivity **12**(3), 421–474 (2008)
11. J.O. Wojtaszczyk, in *The Square Negative Correlation Property for Generalized Orlicz Balls*. Geometric Aspects of Functional Analysis. Lecture Notes in Math., vol. 1910 (Springer, Berlin, 2007), pp. 305–313

The Hörmander Proof of the Bourgain–Milman Theorem

Fedor Nazarov

Abstract We give a proof of the Bourgain–Milman theorem based on Hörmander's Existence Theorem for solutions of the $\bar{\partial}$-problem.

1 Introduction

The formal aim of this paper is to present a complex-analytic proof of the Bourgain–Milman estimate

$$v_n(K)v_n(K^\circ) \geq c^n \frac{4^n}{n!}$$

where K is an origin-symmetric bounded convex body in $\mathbb{R}^n$, $K^\circ = \{t \in \mathbb{R}^n : \langle x, t \rangle \leq 1\}$ is its polar body, v_n stands for the n-dimensional volume measure in $\mathbb{R}^n$, and $c > 0$ is a numeric constant (see [2]).

The best value of c I could get on this way is $\left(\frac{\pi}{4}\right)^3$, which is 3 times worse (on the logarithmic scale) than the current record $c = \frac{\pi}{4}$ due to Kuperberg [7]. Still, I hope that this approach may be of some interest to those who enjoy fancy interplays between convex geometry and Fourier analysis.

Since the title line and the author line of this article contradict each other, I should, probably, clarify that, like in many other cases, my personal contribution to the proof below was merely to combine the ideas of other, greater, minds in a way they just didn't have enough time to think of and to prepare this write-up, i.e., to do something that any other qualified mathematician could do equally well, if not better, under favorable circumstances.

F. Nazarov (✉)
Department of Mathematics, University of Wisconsin-Madison, 480 Lincoln Drive, Madison, WI 53706, USA
e-mail: nazarov@math.wisc.edu

B. Klartag et al. (eds.), *Geometric Aspects of Functional Analysis*, Lecture Notes in Mathematics 2050, DOI 10.1007/978-3-642-29849-3_20,
© Springer-Verlag Berlin Heidelberg 2012

Artem Zvavich and Dmitry Ryabogin attracted my attention to this problem. Mikhail Sodin suggested using the Hörmander theorem in the construction and, in parallel with Greg Kuperberg, read the original draft and has made many pertinent remarks about it, which resulted in eliminating many misprints it contained and improving the presentation in general. I'm most grateful to them as well as to numerous other friends, relatives, and colleagues of mine who made my work on this project possible.

With all that said, let's turn to mathematics now.

2 The Main Idea

We shall start with recasting the question into the language of Hilbert spaces of analytic functions of several complex variables. The optimal way to restate the problem is to use the Paley–Wiener theorem, which asserts that the following two classes of functions are the same:

1. The class of all entire finctions $f : \mathbb{C}^n \to \mathbb{C}$ of finite exponential type (i.e., satisfying the bound $f(z) \leq Ce^{C|z|}$ for all $z \in \mathbb{C}^n$ with some $C > 0$) such that their restriction to $\mathbb{R}^n$ belongs to L^2 and such that $|f(iy)| \leq Ce^{\rho_K(y)}$ with some $C > 0$ for all $y \in \mathbb{R}^n$ where $\rho_K(x) = \inf\{\beta > 0 : x \in \beta K\}$.
2. The class of the Fourier transforms $f(z) = \int_{K^\circ} g(t)e^{-i\langle z,t\rangle}\, dv_n(t)$ of L^2-functions g supported on K°.

We shall denote the class given by any of these conditions by PW(K). If $f \in$ PW(K) is the Fourier transform of g, then, by Plancherel's formula, $\|f\|^2_{L^2(\mathbb{R}^n)} = (2\pi)^n \|g\|^2_{L^2(K^\circ)}$.

Note now that, by the Cauchy–Schwarz inequality, we have

$$|f(0)|^2 = \left|\int_{K^\circ} g\, dv_n\right|^2 \leq v_n(K^\circ)\|g\|^2_{L^2(K^\circ)} = \frac{1}{(2\pi)^n}v_n(K^\circ)\|f\|^2_{L^2(\mathbb{R}^n)}$$

and that the equality sign is attained when $g = 1$ in K°. Thus,

$$v_n(K^\circ) = (2\pi)^n \sup_{f \in \mathrm{PW}(K)} |f(0)|^2 \cdot \|f\|^{-2}_{L^2(\mathbb{R}^n)}.$$

Note that the quantity on the right does not include any metric characteristics of the polar body K° and that the problem of *proving* a lower bound for $v_n(K^\circ)$ has been thus transformed into the problem of *finding an example* of an entire function $f \in$ PW(K) that has not too small value at the origin and not too large $L^2(\mathbb{R}^n)$-norm.

Unfortunately, constructing fast decaying on $\mathbb{R}^n$ analytic functions of several complex variables is quite a non-trivial task by itself and this approach would look rather hopeless if not for the remarkable theorem of Hörmander that allows one to conjure up such functions in Bergman type spaces $L^2(\mathbb{C}^n, e^{-\varphi} dv_{2n})$ with plurisubharmonic φ.

The whole point now is to approximate the Paley–Wiener space by some weighted Bergman space with a Hörmander type weight and to carry out the relevant computations in that space. There is a lot of freedom in what Bergman space to choose. It turns out that almost any decent approximation works and gives the desired inequality with its own exponential factor c^n. The particular choice below was made just because it gives the best constant c among all spaces I tried but I do not guarantee its optimality.

The reader should be aware though that there may be no ideal approximation of the Paley–Wiener space by a Bergman–Hörmander one and, in order to get the Mahler conjecture itself on this way, one would have to work directly with the Paley–Wiener space by either finding a good analogue of the Hörmander theorem allowing to control the Paley–Wiener norm of the solution, or by finding some novel way to construct decaying analytic functions of several variables.

Now it is time to present the formal argument. We shall start with

3 The Rothaus–Korányi–Hsin Formula for the Reproducing Kernel in a Tube Domain

Let K be any (strictly) convex open subset of $\mathbb{R}^n$ and let $T_K = \{x + iy : x \in \mathbb{R}^n, y \in K\} \subset \mathbb{C}^n$ be the corresponding tube domain. Let $A^2(T_K)$ be the Bergman space of all analytic in T_K functions for which

$$\|f\|^2_{A^2(T_K)} = \int_{T_K} |f|^2 \, dv_{2n} < +\infty.$$

In [5] Hsin presented the following nice formula for the reproducing kernel $\mathcal{K}(z, w)$ associated with the Hilbert space $A^2(T_K)$:

$$\mathcal{K}(z, w) = \frac{1}{(2\pi)^n} \int_{\mathbb{R}^n} \frac{e^{i\langle z - \overline{w}, t\rangle}}{J_K(t)} \, dv_n(t)$$

where

$$J_K(t) = \int_K e^{-2\langle x, t\rangle} \, dv_n(x).$$

The idea of this formula can be traced back to Rothaus' dissertation [8]. Korányi [6] seems to be the first to publish it in 1962 for the very similar to our case situation of a tube domain constructed on a convex cone instead of a bounded convex set. In 1968, Rothaus published a paper [9] that, among other things, contained a formula analogous to Hsin's but for the *vertical* tube $x \in K, y \in \mathbb{R}^n$, not for the horizontal one we need here.

Anyway, what concerns us at this point is the following simple corollary. If $0 \in K$, then

$$\mathcal{K}(0,0) = \frac{1}{(2\pi)^n} \int_{\mathbb{R}^n} \frac{dv_n(t)}{J_K(t)}.$$

Suppose now that K is origin-symmetric. Since $\frac{x+y}{2} \in K$ for all $x, y \in K$ and since the function $x \mapsto e^{-\langle x,t \rangle}$ is convex, we can write

$$J_K(t) \geq 2^{-n} \int_K e^{-\langle x,t \rangle - \langle y,t \rangle} \, dv_n(x) \geq 2^{-n} v_n(K) e^{-\langle y,t \rangle}$$

for all $y \in K$. Maximizing this quantity over y, we get

$$J_K(t) \geq 2^{-n} v_n(K) e^{\rho_{K^\circ}(-t)}$$

where $\rho_{K^\circ}(t) = \inf\{\beta > 0 : t \in \beta K^\circ\}$. The immediate corollary of this estimate is the inequality

$$\int_{\mathbb{R}^n} \frac{dv_n(t)}{J_K(t)} \leq 2^n v_n(K)^{-1} \int_{\mathbb{R}^n} e^{-\rho_{K^\circ}} \, dv_n = 2^n n! v_n(K^\circ) v_n(K)^{-1}.$$

Now, one of the key properties of the reproducing kernel $\mathcal{K}(z,w)$ is the inequality

$$|f(0)|^2 = \left| \int_{T_K} \mathcal{K}(0,w) f(w) \, dv_{2n}(w) \right|^2$$

$$\leq \left[\int_{T_K} |\mathcal{K}(0,w)|^2 \, dv_{2n}(w) \right] \cdot \left[\int_{T_K} |f(w)|^2 \, dv_{2n}(w) \right] = \mathcal{K}(0,0) \|f\|_{A^2(T_K)}^2$$

valid for all $f \in A^2(T_K)$. Thus, in order to estimate $\mathcal{K}(0,0)$ (and, thereby, $v_n(K^\circ)$) from below, it will suffice to construct a function $f \in A^2(T_K)$ with $|f(0)|$ not too small compared to $\|f\|_{A^2(T_K)}$. For this construction, we shall need the celebrated

4 Hörmander's Existence Theorem for Solutions of the $\bar{\partial}$-Problem

In [3] Hörmander proved the following statement. Let $\Omega \subset \mathbb{C}^n$ be any open pseudoconvex domain. Let $\varphi : \Omega \to \mathbb{R}$ be any plurisubharmonic function in Ω satisfying the inequality

$$\sum_{i,j=1}^n \frac{\partial^2 \varphi}{\partial z_i \, \partial \bar{z}_j} w_i \bar{w}_j \geq \tau |w|^2$$

for all $w \in \mathbb{C}^n$ at every point of Ω with some $\tau > 0$. Then, for every $(0, 1)$-form ω in Ω satisfying $\bar{\partial}\omega = 0$, we can find a solution g of the equation $\bar{\partial}g = \omega$ in Ω such that

$$\int_\Omega |g|^2 e^{-\varphi} \, dv_{2n} \le \tau^{-1} \int_\Omega |\omega|^2 e^{-\varphi} \, dv_{2n}$$

where, as usual, $|\omega|^2 = \sum_{i=1}^n |a_j|^2$ if $\omega = \sum_{i=1}^n a_i(z) d\bar{z}_i$.

This amazing theorem has become the main tool for constructing analytic functions in $\mathbb{C}^n$ with good growth/decay estimates. It has essentially wiped out all previous ad-hoc procedures based on power series, Cauchy integrals, and such. The details of how to use this remarkable tool vary slightly from proof to proof. The particular way we apply it below can be traced back to Hörmander himself (see, for example, the subsection "The construction of analytic functions with prescribed zeros" on p. 349 in [4]). We shall start with

5 The Construction of φ

For technical reasons, it will be convenient to assume that K is not too wild. Since all the quantities we will use change in a very simple way under linear transformations of $\mathbb{R}^n$, by John's theorem, we can replace K by its suitable affine image and ensure that $v_n(K) = 1$ and that K contains the ball of radius r and is contained in the ball of radius R centered at the origin with the ratio of radii $\frac{R}{r} \le \sqrt{n}$ (see [1], for example).

Now, for every $t \in K^\circ$, the mapping $z \mapsto \langle z, t \rangle$ sends T_K to the horizontal unit strip $|\operatorname{Im}\zeta| < 1$. Let

$$\Phi(\zeta) = \frac{4}{\pi} \cdot \frac{e^{\frac{\pi}{2}\zeta} - 1}{e^{\frac{\pi}{2}\zeta} + 1}$$

be the standard conformal mapping of this strip to the disk of radius $\frac{4}{\pi}$ centered at the origin. Note that $\Phi(0) = 0$ and $\Phi'(0) = 1$.

The function $\log|\Phi|$ is subharmonic in the strip $|\operatorname{Im}\zeta| < 1$ and satisfies $\left|\log|\Phi(\zeta)| - \log|\zeta|\right| \le C|\zeta|$ when $|\zeta| \le \frac{1}{2}$ with some finite $C \ge 1$.

Define

$$\varphi(z) = R^{-2}|y|^2 + 2n \log \sup_{t \in K^\circ} |\Phi(\langle z, t \rangle)|.$$

(as usual, we write $z = x + iy$ for $z \in \mathbb{C}^n$)

Note that the second term is plurisubharmonic as a supremum of a family of plurisubharmonic functions and therefore φ satisfies the conditions of the Hörmander existence theorem with $\tau = \frac{1}{4}R^{-2}$. Also,

$$\varphi(z) \le 2n \log \frac{4}{\pi} + R^{-2}|y|^2 \le 2n \log \frac{4}{\pi} + 1$$

in T_K. At last, $e^{-\varphi}$ is comparable to $|z|^{-2n}$ near the origin, so $e^{-\varphi}$ is not locally integrable at 0.

Now we turn to

6   The Construction of the Analytic Function f

Fix two parameters $\sigma \in (1, 2)$ and $\delta \in (0, \frac{1}{4})$. Let

$$K_{\mathbb{C}} = \{z \in \mathbb{C}^n : |\langle z, t \rangle| \leq 1 \text{ for all } t \in K^\circ\} \subset K \times K.$$

Note first of all that

$$\varphi(z) \geq 2n(\log \delta - 2C\delta) = 2n \log \delta - 4Cn\delta$$

in $(\sigma\delta K_{\mathbb{C}}) \setminus (\delta K_{\mathbb{C}})$.

Now note that $K_{\mathbb{C}}$ is convex and contains $\frac{1}{\sqrt{2}}(K \times K)$, which, in turn, contains the ball of radius $\frac{r}{\sqrt{2}}$ centered at the origin. Thus, we can construct a smooth function $g : \mathbb{C}^n \to [0, 1]$ such that $g = 1$ in $\delta K_{\mathbb{C}}$, $g = 0$ outside $\sigma\delta K_{\mathbb{C}}$, and

$$|\bar{\partial}g| = \frac{1}{2}|\nabla g| \leq \sqrt{2}\, r^{-1}[\delta(\sigma - 1)]^{-1}.$$

This function will satisfy

$$\int_{T_K} |\bar{\partial}g|^2 e^{-\varphi}\, dv_{2n} \leq C(\sigma, \delta) r^{-2} \sigma^{2n} \delta^{2n} v_{2n}(K_{\mathbb{C}}) e^{-2n \log \delta + 4Cn\delta}$$

$$= C(\sigma, \delta) r^{-2} e^{2n(\log \sigma + 2C\delta)} v_{2n}(K_{\mathbb{C}})$$

with some $C(\sigma, \delta)$ that does not depend on n.

Now, by Hörmander's theorem, there exists a solution h of the equation $\bar{\partial}h = -\bar{\partial}g$ in T_K such that

$$\int_{T_K} |h|^2 e^{-\varphi}\, dv_{2n} \leq C(\sigma, \delta) 4 \left(\frac{R}{r}\right)^2 e^{2n(\log \sigma + 2C\delta)} v_{2n}(K_{\mathbb{C}})$$

$$\leq C(\sigma, \delta) 4n\, e^{2n(\log \sigma + 2C\delta)} v_{2n}(K_{\mathbb{C}}).$$

Note that $\bar{\partial}h = 0$ in $\delta K_{\mathbb{C}}$, so h is analytic and, thereby, continuous in $\delta K_{\mathbb{C}}$. Since $e^{-\varphi}$ is not locally integrable at the origin, the integral $\int_{T_K} |h|^2 e^{-\varphi}\, dv_{2n}$ can be finite only if $h(0) = 0$.

Thus, the analytic in T_K function $f = g + h$ satisfies $f(0) = g(0) = 1$. On the other hand,

$$\int_{T_K} |f|^2 \, dv_{2n} \le 2 \int_{T_K} |g|^2 \, dv_{2n} + 2 \int_{T_K} |h|^2 \, dv_{2n}$$

The first integral does not exceed $(\sigma\delta)^{2n} v_{2n}(K_{\mathbb{C}}) \le v_{2n}(K_{\mathbb{C}})$. The second one can be bounded by

$$e^{2n \log \frac{4}{\pi} + 1} \int_{T_K} |h|^2 e^{-\varphi} \, dv_{2n} \le 4neC(\sigma, \delta) e^{2n(\log \sigma + 2C\delta)} \left(\frac{4}{\pi} \right)^{2n} v_{2n}(K_{\mathbb{C}}).$$

Thus

$$\| f \|_{A^2(T_K)}^2 \le 2[4neC(\sigma, \delta) + 1] e^{2n(\log \sigma + 2C\delta)} \left(\frac{4}{\pi} \right)^{2n} v_{2n}(K_{\mathbb{C}}),$$

say, while

$$|f(0)|^2 = 1.$$

Therefore

$$\mathcal{K}(0, 0) \ge c(\sigma, \delta) n^{-1} e^{-2n(\log \sigma + 2C\delta)} \left(\frac{\pi}{4} \right)^{2n} v_{2n}(K_{\mathbb{C}})^{-1}.$$

Now observe that we can choose δ very small and σ very close to 1 to make $\log \sigma + 2C\delta$ as small as we wish. Recalling that $\mathcal{K}(0, 0) = \frac{1}{(2\pi)^n} \int_{\mathbb{R}^n} \frac{dv_n(t)}{J_K(t)}$, we get the inequality

$$v_{2n}(K_{\mathbb{C}}) \int_{\mathbb{R}^n} \frac{dv_n(t)}{J_K(t)} \ge e^{-o(n)} \left(\frac{\pi^3}{8} \right)^n$$

as $n \to \infty$.

To get rid of the $e^{-o(n)}$ factor, we use

7 The Tensor Power Trick

Fix $n \ge 1$ and the body $K \in \mathbb{R}^n$. Choose a very big number m and consider $K' = \underbrace{K \times \cdots \times K}_{m} \subset \mathbb{R}^{mn}$. Note that $K'_{\mathbb{C}} = K_{\mathbb{C}} \times \cdots \times K_{\mathbb{C}}$ and $J_{K'}(t_1, \dots, t_m) = J_K(t_1) \cdot \ldots \cdot J_K(t_m)$. Applying the above inequality to K' instead of K and raising both parts to the power $\frac{1}{m}$, we get

$$v_{2n}(K_{\mathbb{C}}) \int_{\mathbb{R}^n} \frac{dv_n(t)}{J_K(t)} \ge e^{-o(mn)/m} \left(\frac{\pi^3}{8} \right)^n$$

as $m \to \infty$. Since $o(mn)/m \to 0$ as $m \to \infty$ and everything else does not depend on m at all, we get the clean estimate

$$v_{2n}(K_{\mathbb{C}}) \int_{\mathbb{R}^n} \frac{dv_n(t)}{J_K(t)} \geq \left(\frac{\pi^3}{8}\right)^n$$

valid for all origin-symmetric convex bodies K of volume 1 in $\mathbb{R}^n$.

Though it doesn't relate directly to our story, it is worth mentioning here that this estimate is sharp. If $n = 1$ and $K = (-\frac{1}{2}, \frac{1}{2})$, we have $v_2(K_{\mathbb{C}}) = \frac{\pi}{4}$, $J_K(t) = \frac{e^t - e^{-t}}{2t}$, and

$$\int_{\mathbb{R}} \frac{dt}{J_K(t)} = 4 \int_0^\infty \frac{t}{e^t - e^{-t}} = 4 \int_0^\infty t \left(\sum_{k \geq 1, k \text{ odd}} e^{-kt} \right) dt$$

$$= 4 \sum_{k \geq 1, k \text{ odd}} \int_0^\infty t e^{-kt} \, dt = 4 \sum_{k \geq 1, k \text{ odd}} \frac{1}{k^2} = \frac{\pi^2}{2}.$$

Thus, for the interval in $\mathbb{R}^1$ (and, thereby, for the cube in every dimension), the equality sign is attained.

It is now time to finish with the

8   Derivation of the Bourgain–Milman Theorem

Recalling that $K_{\mathbb{C}} \subset K \times K$, we see that $v_{2n}(K_{\mathbb{C}}) \leq v_n(K)^2$. Also, as we have seen earlier,

$$\int_{\mathbb{R}^n} \frac{dv_n(t)}{J_K(t)} \leq 2^n n! v_n(K^\circ) v_n(K)^{-1}.$$

Plugging these estimates in, we get

$$v_n(K) v_n(K^\circ) \geq \left(\frac{\pi}{4}\right)^{3n} \frac{4^n}{n!}$$

as promised.

It is, probably, worth mentioning that the same technique with minor modifications can be applied to the non-symmetric case as well. If somebody wants to follow this way, he should note first that it suffices to consider the case when 0 is the center of mass of K, after which the whole argument can be repeated almost verbatim using the lower half-plane Im $\zeta < 1$ instead of the unit strip. Since the conformal radius of this half-plane with respect to the origin is 2, the final estimate will change to

$$v_n(K) v_n(K^\circ) \geq \left(\frac{\pi}{16}\right)^n \frac{4^n}{n!} = \left(\frac{\pi}{4e}\right)^n \frac{e^n}{n!}.$$

Unfortunately, this is well below the bound you can get by the symmetrization trick.

References

1. K. Ball, in *Flavors of Geometry*, ed. by S. Levy. Math. Sci. Res. Inst. Publ., vol. 31 (Cambridge University Press, Cambridge, 1997)
2. J. Bourgain, V.D. Milman, New volume ratio properties for convex symmetric bodies in $\mathbb{R}^n$. Invent. Math. **88**(2), 319–340 (1987)
3. L. Hörmander, L^2 estimates and existence theorems for the $\bar{\partial}$ operator. Acta Math. **113**, 89–152 (1965)
4. L. Hörmander, A history of existence theorems for the Cauchy-Riemann complex in L^2 spaces. J. Geom. Anal. **13**(2), 329–357 (2003)
5. C.-I. Hsin, The Bergman kernel on tube domains. Rev. Unión Mat. Argentina **46**(1), 23–29 (2005)
6. A. Korányi, The Bergman kernel function for tubes over convex cones. Pac. J. Math. **12**(4), 1355–1359 (1962)
7. G. Kuperberg, *From the Mahler conjecture to Gauss linking integrals*. Geom. Funct. Anal., **18**, 870–892 (2008)
8. O.S. Rothaus, Domains of positivity. Abh. Math. Semin. Hamburg **24**, 189–235 (1960)
9. O.S. Rothaus, Some properties of Laplace transforms of measures. Trans. Am. Math. Soc. **131**(1), 163–169 (1968)

On Some Extension of Feige's Inequality

Krzysztof Oleszkiewicz

Abstract An extension of the Feige inequality [Feige, SIAM J. Comput. **35**, 964–984 (2006)] is formulated and proved in a relatively simple way.

1 Main Theorem

The main result of this note is the following theorem:

Theorem 1.1. *Let $t_0, M > 0$ and let $X_1, X_2, \ldots, X_n$ be independent zero-mean real random variables. Assume also that for $i = 1, 2, \ldots, n$ and for every $t > t_0$ there is*

$$\mathbb{E} X_i 1_{X_i \geq t} \geq \mathbb{E} |X_i| 1_{X_i \leq -Mt}. \tag{1}$$

Then for every $\delta > 0$ we have

$$\mathbb{P}(X_1 + \ldots + X_n \leq \delta) \geq \varepsilon(\delta, t_0, M),$$

where $\varepsilon = \varepsilon(\delta, t_0, M)$ is strictly positive and does not depend on the number n and distribution of the random variables $X_1, X_2, \ldots, X_n$.

The proof yields $\varepsilon(\delta, t_0, M)$ such that $\liminf_{\delta \to 0^+} \varepsilon(\delta, t_0, M)/\delta > 0$ and it is easy to see that even for $n = 1$ one cannot in general expect better asymptotics with respect to δ. Theorem 1.1 is an extension of the following result of Uriel Feige.

K. Oleszkiewicz (✉)

Institute of Mathematics, University of Warsaw, ul. Banacha 2, 02-097 Warsaw, Poland

Institute of Mathematics, Polish Academy of Sciences, ul. Śniadeckich 8, 00-956 Warsaw, Poland
e-mail: koles@mimuw.edu.pl

B. Klartag et al. (eds.), *Geometric Aspects of Functional Analysis*, Lecture Notes in Mathematics 2050, DOI 10.1007/978-3-642-29849-3_21,
© Springer-Verlag Berlin Heidelberg 2012

Theorem 1.2 (Feige [1]). *For every $\delta > 0$ there exists some $\varepsilon = \varepsilon(\delta) > 0$ such that for any positive integer n and any sequence of independent non-negative random variables $Y_1, Y_2, \ldots, Y_n$ with $\mathbb{E}Y_i \leq 1$ for $i = 1, 2, \ldots, n$ there is*

$$\mathbb{P}(S \leq \mathbb{E}S + \delta) \geq \varepsilon(\delta),$$

where $S = Y_1 + Y_2 + \ldots + Y_n$.

Indeed, by setting $X_i = Y_i - \mathbb{E}Y_i$ one immediately reduces Theorem 1.2 to a special case of Theorem 1.1 with $M = t_0 = 1$, the inequality (1) being then trivially satisfied since its right hand side is equal to zero.

Actually, the proof given in [1] yields $\varepsilon(\delta) = \frac{\delta}{1+\delta}$ for δ close enough to zero, and this bound cannot be improved as indicated by an example of $n = 1$ and $\mathbb{P}(Y_1 = 0) = 1 - \mathbb{P}(Y_1 = 1 + \delta) = (\delta + \eta)/(1 + \delta)$ with $\eta \to 0^+$. Neither the present note nor the recent paper by He et al. [3] recovers this optimal estimate for small values of δ. However, it seems that both of them offer some better understanding of probabilistic phenomena related to Feige's inequality than the (elementary but very complicated) proof contained in [1]. While He et al. work hard to obtain as good as possible value of $\varepsilon(1)$ in Theorem 1.2, managing to improve it from Feige's $1/13$ to $1/8$, we will not care much about constants, trying to keep the proof as simple and transparent as possible.

It should be explained here that, at the time the results of this note were proved, the present author was not aware of [3]. Nevertheless, as must be obvious to a careful reader, methods used in both papers are quite similar.

2   Proof of the Main Theorem

2.1   Reduction to Two-Point Distributions

We will start by showing that it suffices to prove Theorem 1.1 under additional assumption that each of X_i's takes on exactly two values—a similar initial reduction occured already in [1]. The proof given there relied on an approximation argument which, if extended to our setting, would have to become quite technical and complicated since the inequality (1) is less "stable" under approximation than assumptions of the Feige theorem. Hence we present a different approach, closer to functional analysis (extreme point theory) or stochastic ordering theory while still quite explicit and elementary.

Given $M, t_0 > 0$ we will say that a real random variable X is (M, t_0)-controlled (meaning the control of its lower tail by its upper tail) if it is zero-mean and for all $t > t_0$ we have

$$\mathbb{E}X 1_{X \geq t} \geq \mathbb{E}|X| 1_{X \leq -Mt}.$$

We will say that a probability measure μ on $\mathbb{R}$ is (M, t_0)-controlled if a random variable with distribution μ is (M, t_0)-controlled. Assume that an (M, t_0)-controlled X is non-trivial, i.e. $q = \mathbb{P}(X \neq 0) > 0$. It is a well known fact that there exists a non-decreasing right-continuous function $f : (0, 1) \to \mathbb{R}$ with the same distribution as X, i.e. $\mathbb{P}(X \in A) = \lambda(f^{-1}(A))$ for every Borel $A \subset \mathbb{R}$, where λ denotes the Lebesgue measure on $(0, 1)$. Let $\alpha = \sup f^{-1}((-\infty, 0))$ and $\beta = \inf f^{-1}((0, \infty))$. Obviously, $0 < \alpha \leq \beta < 1$. Set $\rho = \frac{\mathbb{E}|X|}{2q}$. Let us consider two continuous functions, $a : (0, q) \to (0, \alpha]$ and $b : (0, q) \to [\beta, 1)$, first of them increasing and the second one decreasing, defined by

$$-\int_0^{a(x)} f(s)\,ds = \int_{b(x)}^1 f(s)\,ds = \rho x$$

for $x \in (0, q)$. Furthermore, define a weight function w on $(0, 1)$ by

$$w(x) = \frac{\rho}{f(b(x))} - \frac{\rho}{f(a(x))}$$

for $x \in (0, q)$ and $w(x) = 1$ for $x \in [q, 1)$. Note that $f(a(x))$ is negative, so $w \geq 0$ on $(0, 1)$. Let us define a family of probabilistic measures $(\mu^x)_{x \in (0,1)}$ by setting $\mu^x = \delta_0$ for $x \in [q, 1)$ and

$$\mu^x = \frac{f(b(x))\delta_{f(a(x))} - f(a(x))\delta_{f(b(x))}}{f(b(x)) - f(a(x))}$$

for $x \in (0, q)$. Since both a and b are locally Lipschitz, and there is $a'(x) = -\rho/f(a(x))$ and $b'(x) = -\rho/f(b(x))$ for all except countably many points $x \in (0, q)$, it is easy to check that for any Borel $A \subset \mathbb{R}$ we have

$$\int_0^q w(x)\mu^x(A)\,dx = -\rho \int_0^q \frac{1_A(f(a(x)))}{f(a(x))}\,dx + \rho \int_0^q \frac{1_A(f(b(x)))}{f(b(x))}\,dx$$

$$= \int_0^q 1_{f^{-1}(A)}(a(x))a'(x)\,dx - \int_0^q 1_{f^{-1}(A)}(b(x))b'(x)\,dx$$

$$= \int_0^\alpha 1_{f^{-1}(A)}(y)\,dy + \int_\beta^1 1_{f^{-1}(A)}(y)\,dy = \lambda(f^{-1}(A \setminus \{0\}))$$

$$= \mathbb{P}(X \in A \setminus \{0\}).$$

The above reasoning works for all Borel sets A; however the readers who feel uncertain about technicalities of the change of variables may find it useful to verify the equality for A being open intervals (which is simple) and then use the standard π- and λ-systems argument. We have proved that

$$\mathbb{P}(X \in A) = \int_0^1 w(x)\mu^x(A)\,dx$$

for every Borel set $A \subset \mathbb{R}$. For $A = \mathbb{R}$ we get $\int_0^1 w(x)\,dx = 1$.

Now notice that all the measures μ^x are (M, t_0)-controlled. Indeed, they are obviously zero-mean. Let $t > t_0$. We are to check that $\int_{(-\infty, -Mt]} |s|\,d\mu^x(s) \le \int_{[t,\infty)} s\,d\mu^x(s)$. If $f(a(x)) > -Mt$ then our assertion is trivial, whereas for $f(a(x)) \le -Mt$ we need only to prove that $f(b(x)) \ge t$:

$$\int_{b(x)}^1 f(s)\,ds = \int_0^{a(x)} |f(s)|\,ds \le \int_{f^{-1}((\infty, -Mt])} |f(s)|\,ds$$

$$= \mathbb{E}|X|1_{X \le -Mt} \le \mathbb{E}X1_{X \ge t} = \int_{f^{-1}([t,\infty))} f(s)\,ds,$$

so that $b(x) \ge \inf f^{-1}([t, \infty))$—recall that f is strictly positive on $(\beta, 1)$. Thus $f(b(x)) \ge t$ (here we use the right-continuity of f) and we are done.

Using the above procedure we may express distributions of the random variables $X_1, X_2, \ldots, X_n$ from Theorem 1.1 as integrals $\int_0^1 w_i(x)\mu_i^x\,dx$ $(i = 1, 2, \ldots, n$, respectively). Now it is easy to prove that

$$\mathbb{P}((X_1, X_2, \ldots, X_n) \in A) = \int_{(0,1)^n} \Big(\prod_{i=1}^n w_i(x_i)\Big)(\mu_1^{x_1} \otimes \mu_2^{x_2} \otimes \ldots \otimes \mu_n^{x_n})(A)\,dx$$

for every Borel set $A \subset \mathbb{R}^n$. This follows by the Fubini theorem for product sets A and then again by the standard π- and λ-systems argument extends to all Borel sets. By considering

$$A = \{t \in \mathbb{R}^n : \sum_{i=1}^n t_i \le \delta\}$$

we see that it suffices to prove Theorem 1.1 for X_i's distributed according to the measures $\mu_i^{x_i}$ and then the assertion for original variables X_i immediately follows from the fact that $\int_{(0,1)^n} \prod_{i=1}^n w_i = \prod_{i=1}^n \int_0^1 w_i = 1$.

Thus we may and will henceforth assume that each of the (M, t_0)-controlled random variables X_i in Theorem 1.1 takes on exactly two values (strictly speaking, the above reduction may have led to some of the random variables being identically zero but then it is trivial to get rid of those):

$$\mathbb{P}(X_i = z_i) = p_i, \mathbb{P}(X_i = y_i) = 1 - p_i, y_i < 0 < z_i, 0 < p_i < 1, s_i = z_i - y_i.$$

Without loss of generality we may and will assume that the sequence of spreads $(s_i)_{i=1}^n$ is non-increasing: $s_1 \ge s_2 \ge \ldots \ge s_n > 0$. Also, let $s_{n+1} = 0$.

2.2 Auxiliary Estimate

In the proof of the main theorem we will use the following auxiliary bound.

Proposition 2.1. *For every $C > 0$ there exists $\kappa(C) > 0$ such that for any positive integer n, any $K > 0$, and any sequence of independent random variables Z_1, $Z_2, \ldots, Z_n$, satisfying $\mathbb{E}Z_i = 0$ and $-K \leq Z_i \leq K$ a.s. for $i = 1, 2, \ldots, n$, we have*

$$\mathbb{P}(Z_1 + Z_2 + \ldots + Z_n \leq C \cdot K) \geq \kappa(C).$$

Moreover, $\liminf_{C \to 0+} \bar{\kappa}(C)/C > 0$, where $\bar{\kappa}(C)$ denotes the optimal (largest) value of $\kappa(C)$ for which the above assertion holds true for given $C > 0$.

Since the above estimate is quite simple and standard we will postpone its proof until next section. Obviously, it is enough to prove it in the case $K = 1$ and then use the homogeneity to deduce the result for general $K > 0$—the formulation above was chosen only for its convenience of use.

Now we are in position to prove the main theorem (the main trick being quite similar to the one in [3] although discovered independently).

2.3 Main Argument

Let k be the least index i such that $p_1 s_1 + \ldots + p_i s_i \geq s_{i+1}/2$. So, $p_1 s_1 + \ldots + p_k s_k \geq s_{k+1}/2$ but $p_1 s_1 + \ldots + p_{k-1} s_{k-1} < s_k/2$ and hence $p_1 + \ldots + p_{k-1} < 1/2$ (recall that the spreads form a non-increasing sequence).

Let us consider two cases:

2.3.1 Case $s_k < (M + 1)t_0$

Then we have also $s_{k+1}, \ldots, s_n \leq (M + 1)t_0$, so that $X_k, X_{k+1}, \ldots, X_n$ are independent mean-zero random variables with values in $[-(M + 1)t_0, (M + 1)t_0]$. Now

$$\mathbb{P}(X_1 + \ldots + X_n \leq \delta) \geq \mathbb{P}(X_1 = y_1, \ldots, X_{k-1} = y_{k-1}, X_k + \ldots + X_n \leq \delta) =$$

$$(1 - p_1) \ldots (1 - p_{k-1})\mathbb{P}(X_k + \ldots + X_n \leq \delta) \geq$$

$$\left(1 - (p_1 + \ldots + p_{k-1})\right)\kappa\left(\frac{\delta}{(M + 1)t_0}\right) \geq \frac{1}{2}\kappa\left(\frac{\delta}{(M + 1)t_0}\right).$$

We have used Proposition 2.1 for $C = \delta/((M + 1)t_0)$ and $K = (M + 1)t_0$.

2.3.2   Case $s_k \geq (M+1)t_0$

Recall that $p_i z_i + (1 - p_i)y_i = \mathbb{E}X_i = 0$, so that $p_i s_i = |y_i|$ and $p_i = |y_i|/(z_i + |y_i|)$ for $i = 1, 2, \ldots, n$.

Assume that $p_k > M/(M+1)$. Then

$$|y_k| = p_k s_k > \frac{M}{M+1} \cdot (M+1)t_0 = Mt_0.$$

Set $t = |y_k|/M > t_0$. Since X_k is (M, t_0)-controlled we have

$$(1 - p_k)|y_k| = \mathbb{E}|X_k|1_{X_k \leq -Mt} \leq \mathbb{E}X_k 1_{X_k \geq t} = p_k z_k 1_{z_k \geq t}.$$

Hence $z_k \geq t = |y_k|/M$ and thus $p_k = |y_k|/(z_k + |y_k|) \leq M/(M+1)$, contradicting our assumption. We have proved that $p_k \leq M/(M+1)$. Now,

$$\mathbb{P}(X_1 + \ldots + X_n \leq \delta) \geq \mathbb{P}(X_1 + \ldots + X_n \leq 0)$$

$$\geq \mathbb{P}(X_1 = y_1, \ldots, X_k = y_k, X_{k+1} + \ldots + X_n \leq |y_1| + \ldots + |y_k|)$$

$$= (1 - p_1) \ldots (1 - p_{k-1})(1 - p_k)\mathbb{P}(X_{k+1} + \ldots + X_n \leq p_1 s_1 + \ldots + p_k s_k)$$

$$\geq \left(1 - (p_1 + \ldots + p_{k-1})\right) \cdot \frac{1}{M+1} \cdot \mathbb{P}(X_{k+1} + \ldots + X_n \leq s_{k+1}/2)$$

$$\geq \frac{1}{2} \cdot \frac{1}{M+1} \cdot \kappa(1/2) = \frac{\kappa(1/2)}{2(M+1)},$$

where we have used Proposition 2.1 for $C = 1/2$ and $K = s_{k+1}$ (if $k < n$).

Putting together both cases we finish the proof of Theorem 1.1 with

$$\varepsilon(\delta, t_0, M) = \min\left(\frac{\kappa(1/2)}{2(M+1)}, \kappa\left(\frac{\delta}{(M+1)t_0}\right)/2\right).$$

The second assertion of Theorem 1.1 follows from the second assertion of Proposition 2.1.

3   Proof of Proposition 2.1

With a slight abuse of notation, for $C > 0$ let us denote by $\bar{\kappa}(C)$ the largest real number κ for which the inequality

$$\mathbb{P}(Z_1 + Z_2 + \ldots + Z_n \leq C \cdot K) \geq \kappa$$

holds true under assumptions of Proposition 2.1 (note that a priori this $\bar{\kappa}(C)$ may be equal to zero). Obviously, $\bar{\kappa}(C)$ is non-decreasing and by considering symmetric ± 1 random variables and $n \to \infty$ we see that $\bar{\kappa}(C) \leq 1/2$ for all $C > 0$. First we will prove that $\bar{\kappa}(C) > 0$ for **some** $C > 0$ and then by using the standard "amplifier trick" we will infer that $\bar{\kappa}(C) > 0$ for **all** $C > 0$ and that $\liminf_{C \to 0+} \bar{\kappa}(C)/C > 0$. Then we will use another approach to prove that in fact $\bar{\kappa}(C) = 1/2$ for C large enough.

3.1 Fourth Moment Method

Let $S = Z_1 + Z_2 + \ldots + Z_n$ and $\sigma^2 = \mathbb{E}S^2$. Note that

$$\mathbb{E}S^4 = \sum_{i=1}^{n} \mathbb{E}Z_i^4 + 6 \sum_{1 \leq i < j \leq n} \mathbb{E}Z_i^2 \cdot \mathbb{E}Z_j^2 \leq K^2 \sum_{i=1}^{n} \mathbb{E}Z_i^2 + 3(\sum_{i=1}^{n} \mathbb{E}Z_i^2)^2 = K^2\sigma^2 + 3\sigma^4.$$

By Hölder's inequality $\mathbb{E}S^2 = \mathbb{E}(|S|^{2/3} \cdot |S|^{4/3}) \leq (\mathbb{E}|S|)^{2/3}(\mathbb{E}S^4)^{1/3}$, so

$$(\mathbb{E}|S|)^2/\mathbb{E}S^2 \geq (\mathbb{E}S^2)^2/\mathbb{E}S^4 \geq \sigma^4(K^2\sigma^2 + 3\sigma^4)^{-1} = 1/(3 + K^2\sigma^{-2}).$$

Now we use the classical Paley-Zygmund type bound: $\mathbb{E}S = 0$, so that

$$\mathbb{E}|S|/2 = \mathbb{E}|S|1_{S<0} \leq (\mathbb{E}S^2)^{1/2} \cdot \left(\mathbb{P}(S < 0)\right)^{1/2}$$

and therefore

$$\mathbb{P}(S \leq C \cdot K) \geq \mathbb{P}(S < 0) \geq \frac{(\mathbb{E}|S|)^2}{4\,\mathbb{E}S^2} \geq \frac{1}{4(3 + K^2\sigma^{-2})}.$$

Thus $\mathbb{P}(S \leq C \cdot K) \geq 1/16$ if $\sigma \geq K$, whereas for $\sigma \leq K$ by Chebyshev's inequality we get

$$\mathbb{P}(S > C \cdot K) \leq \frac{\sigma^2}{C^2K^2} \leq C^{-2},$$

in particular $\mathbb{P}(S \leq 2K) \geq 3/4 > 1/16$. We have proved $\bar{\kappa}(2) \geq 1/16 > 0$. The amplifier trick will do the rest.

3.2 Amplifier Trick

Let m be a natural number. Let $\left(Z_i^{(j)}\right)_{i=1,\ldots,n}^{j=1,\ldots,m}$ be independent random variables such that $Z_i^{(j)}$ has the same distribution as Z_i for $i \leq n$ and $j \leq m$. Furthermore, let $S_j = Z_1^{(j)} + Z_2^{(j)} + \ldots + Z_n^{(j)}$, so that $S_1, S_2, \ldots, S_m$ are independent copies of S. Then

$$\mathbb{P}(S_1 + S_2 + \ldots + S_m \le C \cdot K) \ge \bar{\kappa}(C)$$

since $S_1 + S_2 + \ldots + S_m$ is a sum of mn independent zero-mean random variables with values in $[-K, K]$ a.s. On the other hand, we have

$$\mathbb{P}(S_1 + S_2 + \ldots + S_m \le C \cdot K) \le \sum_{j=1}^{m} \mathbb{P}\left(S_j \le \frac{C}{m}K\right) = m \cdot \mathbb{P}\left(S \le \frac{C}{m}K\right).$$

Thus we have proved that for Z_i's satisfying assumptions of Proposition 2.1 there is $\mathbb{P}(S \le \frac{C}{m}K) \ge \bar{\kappa}(C)/m$, so that $\bar{\kappa}(C/m) \ge \bar{\kappa}(C)/m$ for all integer $m \ge 1$. Hence $\bar{\kappa}(C) > 0$ for all $C > 0$, and $\liminf_{C \to 0+} \bar{\kappa}(C)/C > 0$. The proof of Proposition 2.1 is complete.

Remark 3.1. The amplifier trick may be easily modified to yield a proof that $\bar{\kappa}$ is subadditive on $(0, \infty)$. Also, note that the amplifier trick may be as well, after minor modifications, applied directly to Theorem 1.1 (instead of Proposition 2.1) thus allowing its simple reduction to the case $\delta = 1$.

Remark 3.2. It is easy to see that the above proof of Proposition 2.1 remains valid (up to an obvious modification of the amplifier trick) if we replace in Proposition 2.1 the assumption $|Z_i| \le K$ a.s. by a much weaker condition $\mathbb{E}Z_i^4 \le K^2 \mathbb{E}Z_i^2 < \infty$ (for $i = 1, 2, \ldots, n$).

3.3 Berry-Esseen Inequality Approach

Let $Z_1, Z_2, \ldots, Z_n$ be as in Proposition 2.1 and let $S = Z_1 + Z_2 + \ldots + Z_n$, and $\sigma = \left(\sum_{i=1}^{n} \mathbb{E}Z_i^2\right)^{1/2}$. First let us observe that by Chebyshev's inequality $\mathbb{P}(S \le CK) \ge 1/2$ if $\sigma \le CK/\sqrt{2}$. Now let us assume $\sigma > CK/\sqrt{2}$.

The classical Berry-Esseen inequality (in the form which can be found for example in [5]) states that there exists a universal positive constant B such that for any integer n and independent zero-mean random variables $\xi_1, \xi_2, \ldots, \xi_n$ satisfying $\sum_{i=1}^{n} \mathbb{E}\xi_i^2 = 1$, there is

$$\sup_{s \in \mathbb{R}} |\mathbb{P}(\sum_{i=1}^{n} \xi_i \le s) - \Phi(s)| \le B \cdot \sum_{i=1}^{n} \mathbb{E}|\xi_i|^3,$$

where Φ denotes the standard normal distribution function. Using the above quantitative version of the CLT for random variables $\xi_i = Z_i/\sigma$ we arrive at

$$\mathbb{P}(S \leq CK) = \mathbb{P}(\sum_{i=1}^{n} \xi_i \leq CK/\sigma) \geq \Phi(CK/\sigma) - B\sigma^{-3}\sum_{i=1}^{n}\mathbb{E}|Z_i|^3$$

$$\geq \Phi(CK/\sigma) - B\sigma^{-3}\sum_{i=1}^{n}K\mathbb{E}Z_i^2$$

$$= 1/2 + (2\pi)^{-1/2}\int_0^{CK/\sigma} e^{-t^2/2}\,dt - BK/\sigma.$$

Recall that $\sigma > CK/\sqrt{2}$, so that $e^{-t^2/2} \geq 1/e$ on $[0, CK/\sigma]$ and we have

$$(2\pi)^{-1/2}\int_0^{CK/\sigma} e^{-t^2/2}\,dt \geq (2\pi)^{-1/2}e^{-1}CK/\sigma \geq BK/\sigma$$

whenever $C \geq e\sqrt{2\pi}B$. Thus $\bar{\kappa}(C) \geq 1/2$ for $C \geq e\sqrt{2\pi}B$. A similar reasoning may be found in [4]—we have reproduced it here for reader's convenience (the reader may also like to check [2] for interesting related considerations).

Remark 3.3. As can be easily seen, the above argument remains valid if we replace in Proposition 2.1 the assumption $|Z_i| \leq K$ a.s. by a much weaker condition $\mathbb{E}|Z_i|^3 \leq K\mathbb{E}Z_i^2 < \infty$ (for $i = 1, 2, \ldots, n$). Note that the Schwarz inequality implies $\mathbb{E}Z^4/\mathbb{E}Z^2 \geq (\mathbb{E}|Z|^3/\mathbb{E}Z^2)^2$ for any square-integrable random variable Z, therefore this approach yields a proof of Proposition 2.1 under assumptions slightly weaker than those needed for the fourth moment method to work.

Acknowledgements I would like to thank Franck Barthe for a fruitful discussion which inspired my work on Feige's inequality during my visit to Institut de Mathématiques at Université Paul Sabatier in Toulouse in May 2010. It is a pleasure to acknowledge their kind hospitality. Research partially supported by Polish MNiSzW Grant N N201 397437.

References

1. U. Feige, On sums of independent random variables with unbounded variance and estimating the average degree in a graph. SIAM J. Comput. **35**, 964–984 (2006)
2. M.G. Hahn, M.J. Klass, Uniform local probability approximations: Improvements on Berry-Esseen. Ann. Probab. **23**, 446–463 (1995)
3. S. He, J. Zhang, S. Zhang, Bounding probability of small deviation: A fourth moment approach. Math. Oper. Res. **35**, 208–232 (2010)
4. K. Oleszkiewicz, Concentration of capital – the product form of the law of large numbers in L_1. Statist. Probab. Lett. **55**, 159–162 (2001)
5. V.V. Petrov, in *Sums of Independent Random Variables*. Ergeb. Math. Grenzgeb., vol. 82 (Springer, Berlin, 1975)

On the Mean Width of Log-Concave Functions

Liran Rotem

Abstract In this work we present a new, natural, definition for the mean width of log-concave functions. We show that the new definition coincides with a previous one by B. Klartag and V. Milman, and deduce some properties of the mean width, including an Urysohn type inequality. Finally, we prove a functional version of the finite volume ratio estimate and the low-M^* estimate.

1 Introduction and Definitions

This paper is another step in the "geometrization of probability" plan, a term coined by V. Milman. The main idea is to extend notions and results about convex bodies into the realm of log-concave functions. Such extensions serve two purposes: Firstly, the new functional results can be interesting on their own right. Secondly, and perhaps more importantly, the techniques developed can be used to prove new results about convex bodies. For a survey of results in this area see [11].

A function $f : \mathbb{R}^n \to [0, \infty)$ is called *log-concave* if it is of the form $f = e^{-\phi}$, where $\phi : \mathbb{R}^n \to (-\infty, \infty]$ is a convex function. For us, the definition will also include the technical assumptions that f is upper semi-continuous and f is not identically 0. Whenever we discuss f and ϕ simultaneously, we will always assume they satisfy the relation $f = e^{-\phi}$. Similar relation will be assumed for $\widetilde{f}$ and $\widetilde{\phi}$, f_k and ϕ_k, etc. The class of log-concave functions naturally extends the class of convex bodies: if $\emptyset \neq K \subseteq \mathbb{R}^n$ is a closed, convex set, then its characteristic function $\mathbf{1}_K$ is a log-concave function.

L. Rotem (✉)
School of Mathematical Sciences, Sackler Faculty of Exact Sciences, Tel Aviv University, Ramat Aviv, Tel Aviv 69978, Israel
e-mail: liranro1@post.tau.ac.il

B. Klartag et al. (eds.), *Geometric Aspects of Functional Analysis*, Lecture Notes in Mathematics 2050, DOI 10.1007/978-3-642-29849-3_22,
© Springer-Verlag Berlin Heidelberg 2012

On the class of convex bodies there are two important operations. If K and T are convex bodies then their *Minkowski sum* is $K + T = \{k+t : k \in K, t \in T\}$. If in addition $\lambda > 0$, then the λ-*homothety* of K is $\lambda \cdot K = \{\lambda k : k \in K\}$. These operations extend to log-concave functions: If f and g are log-concave we define their *Asplund product* (or *sup-convolution*), to be

$$(f \star g)(x) = \sup_{x_1+x_2=x} f(x_1)g(x_2).$$

If in addition $\lambda > 0$ we define the λ-*homothety* of f to be

$$(\lambda \cdot f)(x) = f\left(\frac{x}{\lambda}\right)^\lambda.$$

It is easy to see that these operations extend the classical operations, in the sense that $1_K \star 1_T = 1_{K+T}$ and $\lambda \cdot 1_K = 1_{\lambda K}$ for every convex bodies K, T and every $\lambda > 0$. It is also useful to notice that if f is log-concave and $\alpha, \beta > 0$ then $(\alpha \cdot f) \star (\beta \cdot f) = (\alpha + \beta) \cdot f$. In particular, $f \star f = 2 \cdot f$.

The main goal of this paper is to define the notion of *mean width* for log-concave functions. For convex bodies, this notion requires we fix an Euclidean structure on $\mathbb{R}^n$. Once we fix such a structure we define the *support function* of a body K to be $h_K(x) = \sup_{y \in K} \langle x, y \rangle$. The function $h_K : \mathbb{R}^n \to (-\infty, \infty]$ is convex and 1-homogeneous. The mean width of K is defined to be

$$M^*(K) = \int_{S^{n-1}} h_K(\theta)d\sigma(\theta), \tag{1}$$

where σ is the normalized Haar measure on the unit sphere $S^{n-1} = \{x \in \mathbb{R}^n : |x| = 1\}$.

The correspondence between convex bodies and support functions is linear, in the sense that $h_{\lambda K+T} = \lambda h_K + h_T$ for every convex bodies K and T and every $\lambda > 0$. It immediately follows that the mean width is linear as well. It is also easy to check that M^* is translation and rotation invariant, so $M^*(uK) = M^*(K)$ for every isometry $u : \mathbb{R}^n \to \mathbb{R}^n$.

We will also need the equivalent definition of mean width as a *quermassintegrals*: Let $D \subseteq \mathbb{R}^n$ denote the euclidean ball. If $K \subseteq \mathbb{R}^n$ is any convex body then the n-dimensional volume $|K + tD|$ is a polynomial in t of degree n, known as the *Steiner polynomial*. More explicitly, one can write

$$|K + tD| = \sum_{i=0}^{n} \binom{n}{i} V_{n-i}(K)t^i,$$

and the coefficients $V_i(K)$ are known as the quermassintegrals of K. One can also give explicit definitions for the V_i's, and it follows that $V_1(K) = |D| \cdot M^*(K)$

(more information and proofs can be found for example in [9] or [14]). From this it's not hard to prove the equivalent definition

$$M^*(K) = \frac{1}{n\,|D|} \cdot \lim_{\epsilon \to 0^+} \frac{|D + \epsilon K| - |D|}{\epsilon}. \tag{2}$$

This last definition is less geometric in nature, but it suits some purposes extremely well. For example, using the Brunn–Minkowski theorem (again, check [9] or [14]), one can easily deduce the *Urysohn inequality*:

$$M^*(K) \geq \left(\frac{|K|}{|D|}\right)^{\frac{1}{n}}$$

for every convex body K.

In [6], Klartag and Milman give a definition for the mean width of a log-concave function, based on definition (2). The role of the volume is played by Lebesgue integral (which makes sense because $\int \mathbf{1}_K dx = |K|$), and the euclidean ball D is replaced by a Gaussian $G(x) = e^{-\frac{|x|^2}{2}}$. The result is the following definition:

Definition 1.1. The mean width of a log-concave function f is

$$\widetilde{M}^*(f) = c_n \lim_{\epsilon \to 0^+} \frac{\int G \star (\epsilon \cdot f) - \int G}{\epsilon}.$$

Here $c_n = \frac{2}{n(2\pi)^{\frac{n}{2}}}$ is a normalization constant, chosen to have $\widetilde{M}^*(G) = 1$.

Some properties of $\widetilde{M}^*$ are not hard to prove. For example, it is easy to see that $\widetilde{M}^*$ is rotation and translation invariant. It is also not hard to prove a functional Urysohn inequality:

Proposition. *If f is log-concave and $\int f = \int G$, then $\widetilde{M}^*(f) \geq \widetilde{M}^*(G) = 1$.*

The proof that appears in [6] is similar to the standard proof for convex bodies. Instead of the Brunn–Minkowski theorem one uses its functional version, known as the Prékopa–Leindler inequality (see, e.g. [13]). For other applications, however, this definition is rather cumbersome to work with. For example, by looking at the definition it is not at all obvious that $\widetilde{M}^*$ is a linear functional. It is proven in [6] that indeed

$$\widetilde{M}^* ((\lambda \cdot f) \star g) = \lambda \widetilde{M}^*(f) + \widetilde{M}^*(g),$$

but only for sufficiently regular log-concave functions f and g. These difficulties, and the fact that the definition has no clear geometric intuition, made V. Milman raise the questions of whether Definition 1.1 is the "right" definition for mean width of log-concave functions.

We would like to give an alternative definition for mean width, based on the original definition (1). To do so, we first need to explain what is the support function of a log-concave function, following a series of papers by S. Artstein-Avidan and

V. Milman. To state their result, assume that $\mathcal{T}$ maps every (upper semi-continuous) log-concave function to its support function which is lower semi-continuous and convex. It is natural to assume that $\mathcal{T}$ is a bijection, so a log-concave function can be completely recovered from its support function. It is equally natural to assume that $\mathcal{T}$ is order preserving, that is $\mathcal{T}f \geq \mathcal{T}g$ if and only if $f \geq g$—this is definitely the case for the standard support function defined on convex bodies. In [2] it is shown that such a $\mathcal{T}$ must be of the form

$$(\mathcal{T}f)(x) = C_1 \cdot [\mathcal{L}(-\log f)](Bx + v_0) + \langle x, v_1 \rangle + C_0$$

for constants $C_0, C_1 \in \mathbb{R}$, vectors $v_0, v_1 \in \mathbb{R}^n$ and a transformation $B \in GL_n$. Here $\mathcal{L}$ is the classical *Legendre transform*, defined by

$$(\mathcal{L}\phi)(x) = \sup_{y \in \mathbb{R}^n} (\langle x, y \rangle - \phi(y)).$$

We of course also want $\mathcal{T}$ to extend the standard support function. This significantly reduces the number of choices and we get that $(\mathcal{T}f)(x) = \frac{1}{C}[\mathcal{L}(-\log f)](Cx)$ for some $C > 0$. The exact choice of C is not very important, and we will choose the convenient $C = 1$. In other words, we define the support function h_f of a log-concave function f to be $\mathcal{L}(-\log f)$. Notice that the support function interacts well with the operations we defined on log-concave functions: it is easy to check that $h_{(\lambda \cdot f) \star g} = \lambda h_f + h_g$ for every log-concave functions f and g and every $\lambda > 0$ (in fact this property also completely characterizes the support function—see [1]).

We would like to define the mean width of a log-concave function as the integral of its support function with respect to some measure on $\mathbb{R}^n$. In (1) the measure being used is the Haar measure on S^{n-1}, but since h_K is always 1-homogeneous this is completely arbitrary: for *every* rotationally invariant probability measure μ on $\mathbb{R}^n$ one can find a constant $C_\mu > 0$ such that

$$M^*(K) = C_\mu \int_{\mathbb{R}^n} h_K(x) d\mu(x)$$

for every convex body $K \subseteq \mathbb{R}^n$. We choose to work with Gaussians:

Definition 1.2. The mean width of log-concave function f is

$$M^*(f) = \frac{2}{n} \int_{\mathbb{R}^n} h_f(x) d\gamma_n(x),$$

where γ_n is the standard Gaussian probability measure on $\mathbb{R}^n$ ($d\gamma_n = (2\pi)^{-\frac{n}{2}} e^{-\frac{|x|^2}{2}} dx$).

The main result of Sect. 2 is the fact that the two definitions given above are, in fact, the same:

Theorem 1.3. $M^*(f) = \widetilde{M}^*(f)$ *for every log-concave function* f.

This theorem gives strong indication that our definition for mean width is the "right" one.

In Sect. 3 we present some basic properties of the functional mean width. The highlight of this section is a new proof of the functional Urysohn inequality, based on Definition 1.2. Since this definition involves no limit procedure, it is also possible to characterize the equality case:

Theorem 1.4. *For any log-concave* f

$$M^*(f) \geq 2 \log \left(\frac{\int f}{\int G} \right)^{\frac{1}{n}} + 1,$$

with equality if and only if $\int f = \infty$ *or* $f(x) = Ce^{-\frac{|x-a|^2}{2}}$ *for some* $C > 0$ *and* $a \in \mathbb{R}^n$.

Finally, in Sect. 4, we prove a functional version of the classical low-M^* estimate (see, e.g. [10]). All of the necessary background information will be presented there, so for now we settle on presenting the main result:

Theorem 1.5. *For every* $\varepsilon < M$, *every large enough* $n \in \mathbb{N}$, *every* $f : \mathbb{R}^n \to [0, \infty)$ *such that* $f(0) = 1$ *and* $M^*(f) \leq 1$ *and every* $0 < \lambda < 1$ *one can find a subspace* $E \hookrightarrow \mathbb{R}^n$ *such that* $\dim E \geq \lambda n$ *with the following property: for every* $x \in E$ *such that* $e^{-\varepsilon n} \geq (f \star G)(x) \geq e^{-Mn}$ *one have*

$$f(x) \leq \left(C(\varepsilon, M)^{\frac{1}{1-\lambda}} \cdot G \right)(x).$$

In fact, one can take

$$C(\varepsilon, M) = C \max \left(\frac{1}{\varepsilon}, M \right).$$

Acknowledgements I would like to thank my advisor, Vitali Milman, for raising most of the questions in this paper, and helping me tremendously in finding the answers.

2 Equivalence of the Definitions

Our first goal is to prove that $M^*(f) = \widetilde{M}^*(f)$ for every log-concave function f. We'll start by proving it under some technical assumptions:

Lemma 2.1. *Let* $f : \mathbb{R}^n \to [0, \infty)$ *be a compactly supported, bounded, log-concave function, and assume that* $f(0) > 0$. *Then* $M^*(f) = \widetilde{M}^*(f)$.

Proof. We'll begin by noticing that

$$[G \star (\varepsilon \cdot f)](x) = \sup_y G(x - y) \cdot f\left(\frac{y}{\varepsilon}\right)^\varepsilon = \sup_y \exp\left(-\frac{|x - y|^2}{2} - \epsilon\phi\left(\frac{y}{\epsilon}\right)\right)$$

$$= \sup_y \exp\left(-\frac{|x|^2}{2} + \langle x, y\rangle - \frac{|y|^2}{2} - \epsilon\phi\left(\frac{y}{\epsilon}\right)\right)$$

$$= e^{-\frac{|x|^2}{2}} \exp\left(\sup_z \left(\langle x, \epsilon z\rangle - \frac{|\epsilon z|^2}{2} - \epsilon\phi(z)\right)\right)$$

$$= e^{-\frac{|x|^2}{2} + \epsilon H(x, \epsilon)},$$

where

$$H(x, \epsilon) = \sup_z \left(\langle x, z\rangle - \phi(z) - \epsilon\frac{|z|^2}{2}\right) = \mathcal{L}\left(\phi(x) + \epsilon\frac{|x|^2}{2}\right).$$

Since the functions $\phi(x) + \epsilon\frac{|x|^2}{2}$ converge pointwise to ϕ as $\epsilon \to 0$, it follows that $H(x, \epsilon) \to (\mathcal{L}\phi)(x)$ for every x in the interior of $A = \{x : (\mathcal{L}\phi)(x) < \infty\}$ (see for example Lemma 3.2 (3) in [3]).

To find A, notice the following: since f is bounded there exists an $M \in \mathbb{R}$ such that $\phi(x) > -M$ for all x. Since f is compactly supported there exists an $R > 0$ such that $\phi(x) = \infty$ if $|x| > R$. It follows that for every x

$$(\mathcal{L}\phi)(x) = \sup_y (\langle x, y\rangle - \phi(y)) = \sup_{|y| \le R} (\langle x, y\rangle - \phi(y))$$

$$\le \sup_{|y| \le R} (|x| |y| - \phi(y)) \le R |x| + M < \infty.$$

Therefore $A = \mathbb{R}^n$ and $H(x, \epsilon) \to (\mathcal{L}\phi)(x)$ for all x.

We wish to calculate

$$\widetilde{M}^*(f) = c_n \lim_{\epsilon \to 0^+} \frac{\int e^{-\frac{|x|^2}{2} + \epsilon H(x, \epsilon)} dx - \int e^{-\frac{|x|^2}{2}} dx}{\epsilon}$$

$$= c_n \lim_{\epsilon \to 0^+} \int \frac{e^{\epsilon H(x, \epsilon)} - 1}{\epsilon} \cdot e^{-\frac{|x|^2}{2}} dx,$$

and to do so we would like to justify the use of the dominated convergence theorem. Notice that for every fixed t, the function $\frac{\exp(\epsilon t) - 1}{\epsilon}$ is increasing in ϵ. By substituting $z = 0$ we also see that for every $\epsilon > 0$

$$(\mathcal{L}\phi)(x) \geq H(x,\epsilon) = \sup_{z}\left(\langle x,z\rangle - \phi(z) - \epsilon\frac{|z|^2}{2}\right) \geq -\phi(0).$$

Therefore on the one hand we get that for every $\epsilon > 0$

$$\frac{e^{\epsilon H(x,\epsilon)} - 1}{\epsilon} \geq \frac{e^{-\epsilon\phi(0)} - 1}{\epsilon} \geq \lim_{\epsilon \to 0^+}\frac{e^{-\epsilon\phi(0)} - 1}{\epsilon} = -\phi(0) > -\infty,$$

and on the other hand we get that for every $0 < \epsilon < 1$

$$\frac{e^{\epsilon H(x,\epsilon)} - 1}{\epsilon} \leq \frac{e^{\epsilon(\mathcal{L}\phi)(x)} - 1}{\epsilon} \leq e^{(\mathcal{L}\phi)(x)} - 1 \leq e^{R|x|+M}.$$

Since the functions $-\phi(0)$ and $e^{R|x|+M}$ are both integrable with respect to the Gaussian measure the conditions of the dominated convergence theorem apply, so we can write

$$\widetilde{M}^*(f) = c_n \int \lim_{\epsilon \to 0^+}\frac{e^{\epsilon H(x,\epsilon)} - 1}{\epsilon}\cdot e^{-\frac{|x|^2}{2}}dx.$$

To finish the proof we calculate

$$\lim_{\epsilon \to 0^+}\frac{e^{\epsilon H(x,\epsilon)} - 1}{\epsilon} = \lim_{\epsilon \to 0^+}\frac{e^{\epsilon H(x,\epsilon)} - 1}{\epsilon H(x,\epsilon)}\cdot \lim_{\epsilon \to 0^+} H(x,\epsilon)$$

$$= \lim_{\eta \to 0^+}\frac{e^\eta - 1}{\eta}\cdot \lim_{\epsilon \to 0^+} H(x,\epsilon) = (\mathcal{L}\phi)(x) = h_f(x).$$

Therefore

$$\widetilde{M}^*(f) = c_n \int h_f(x)e^{-\frac{|x|^2}{2}}dx = \frac{2}{n}\int h_f(x)d\gamma_n(x) = M^*(f)$$

like we wanted.   □

In order to prove Theorem 1.3 in its full generality, we first need to eliminate one extreme case: usually we think of $\widetilde{M}^*(f)$ as the differentiation with respect to ϵ of $\int G \star (\epsilon \cdot f)$. However, this is not always the case, since it is quite possible that $\int G \star (\epsilon \cdot f) \not\to \int G$ as $\epsilon \to 0^+$ (for example this happens for $f(x) = e^{-|x|}$). The next lemma characterizes this case completely:

Lemma 2.2. *The following are equivalent for a log-concave function f:*

(i) $(\mathcal{L}\phi)(x) < \infty$ *for every x.*
(ii) $\int G \star [\epsilon \cdot f] \to \int G$ *as $\epsilon \to 0^+$.*

Proof. First, notice that both conditions are translation invariant: if we define $\tilde{f} = f(x - a)$ then it's easy to check that

$$(\mathcal{L}\tilde{\phi})\,(x) = (\mathcal{L}\phi)\,(x) + \langle x, a \rangle \tag{3}$$

and

$$\int \left(G \star \left[\epsilon \cdot \tilde{f} \right] \right)(x)dx = \int (G \star [\epsilon \cdot f])\,(x - a\epsilon)dx = \int (G \star [\epsilon \cdot f])\,(x)dx. \tag{4}$$

Therefore, since we assumed $f \not\equiv 0$, we can translate f and assume without loss of generality that $f(0) > 0$ (or $\phi(0) < \infty$).

Assume first that condition (i) holds. In the proof of Lemma 2.1 we saw that

$$[G \star (\varepsilon \cdot f)]\,(x) = e^{-\frac{|x|^2}{2} + \epsilon H(x,\epsilon)},$$

and that if $(\mathcal{L}\phi)\,(x) < \infty$ for every x then $H(x, \epsilon) \to (\mathcal{L}\phi)\,(x)$ as $\epsilon \to 0^+$. It follows that

$$\lim_{\epsilon \to 0^+} [G \star (\varepsilon \cdot f)]\,(x) = e^{-\frac{|x|^2}{2} + 0 \cdot (\mathcal{L}\phi)(x)} = G(x)$$

for every x. Since the functions $G \star (\varepsilon \cdot f)$ are log-concave, we get that $\int G \star [\epsilon \cdot f] \to \int G$ like we wanted (See Lemma 3.2 (1) in [3]).

Now assume that (i) doesn't hold. Since the set $A = \{x : (\mathcal{L}\phi)\,(x) < \infty\}$ is convex, we must have $A \subseteq H$ for some half-space

$$H = \{x : \langle x, \theta \rangle \leq a\}$$

(here $\theta \in S^{n-1}$ and $a > 0$). It follows that for every $t > 0$

$$\phi(t\theta) = (\mathcal{L}\mathcal{L}\phi)\,(t\theta) = \sup_{y \in H} [\langle y, t\theta \rangle - (\mathcal{L}\phi)\,(y)].$$

But for every y we know that

$$(\mathcal{L}\phi)\,(y) = \sup_z (\langle y, z \rangle - \phi(z)) \geq -\phi(0),$$

so

$$\phi(t\theta) \leq at + b$$

where $b = \phi(0)$. Therefore

$$H(x, \epsilon) \geq \sup_{t > 0} \left(\langle x, t\theta \rangle - \phi(t\theta) - \epsilon \frac{|t\theta|^2}{2} \right) \geq \sup_{t > 0} \left(t \langle x, \theta \rangle - at - b - \frac{\epsilon t^2}{2} \right)$$

$$= \frac{(\langle x, \theta \rangle - a)^2}{2\epsilon} - b,$$

and then

$$\int [G \star (\varepsilon \cdot f)](x)dx \geq e^{-b\epsilon} \int e^{-\frac{|x|^2}{2} + \frac{((x,\theta)-a)^2}{2}} dx \to \int e^{-\frac{|x|^2}{2} + \frac{((x,\theta)-a)^2}{2}} dx$$

$$> \int G.$$

It follows that we can't have convergence in (ii) and we are done. □

The last ingredient we need is a monotone convergence result which may be interesting on its own right:

Proposition 2.3. Let f be a log-concave function such that $(\mathcal{L}\phi)(x) < \infty$ for all x. Assume that (f_k) is a sequence of log-concave functions such that for every x

$$f_1(x) \leq f_2(x) \leq f_3(x) \leq \cdots$$

and $f_k(x) \to f(x)$. Then:

(i) $M^*(f_k) \to M^*(f)$.
(ii) $\widetilde{M}^*(f_k) \to \widetilde{M}^*(f)$.

Proof. (i) By our assumption $\phi_k(x) \to \phi(x)$ pointwise. Since we assumed that $(\mathcal{L}\phi)(x) < \infty$ it follows that $\mathcal{L}\phi_k$ converges pointwise to $\mathcal{L}\phi$ (again, Lemma 3.2 (3) in [3]). Now one can apply the monotone convergence theorem and get that

$$M^*(f_k) = \frac{2}{n} \int (\mathcal{L}\phi_k)(x)d\gamma_n(x) \to \frac{2}{n} \int (\mathcal{L}\phi)(x)d\gamma_n(x) = M^*(f),$$

like we wanted.

(ii) For $\epsilon > 0$ define

$$F_k(\epsilon) = \int G \star [\epsilon \cdot f_k]$$

and

$$F(\epsilon) = \int G \star [\epsilon \cdot f].$$

It was observed already in [6] that F_k and F are log-concave. By our assumption on f and Lemma 2.2, F_k and F will be (right) continuous at $\epsilon = 0$ if we define $F_k(0) = F(0) = \int G$. We would first like the show that F_k converges pointwise to F. Because all of the functions involved are log-concave, it is enough to prove that for a fixed $\epsilon > 0$ and $x \in \mathbb{R}^n$

$$(G \star [\epsilon \cdot f_k])(x) \to (G \star [\epsilon \cdot f])(x)$$

(Lemma 3.2 (1) in [3]). Since $f_k \leq f$ for all k it is obvious that $\lim (G \star [\epsilon \cdot f_k])(x) \leq (G \star [\epsilon \cdot f])(x)$. For the other direction, choose $\delta > 0$. There exists $y_\delta \in \mathbb{R}^n$ such that

$$(G \star [\epsilon \cdot f]) (x) \leq G(x - y_\delta) f \left(\frac{y_\delta}{\epsilon}\right)^\epsilon + \delta$$

$$= \lim_{k \to \infty} G(x - y_\delta) f_k \left(\frac{y_\delta}{\epsilon}\right)^\epsilon + \delta \leq \lim_{k \to \infty} (G \star [\epsilon \cdot f_k]) (x) + \delta.$$

Finally taking $\delta \to 0$ we obtain the result.

We are interested in calculating $\widetilde{M}^*(f) = c_n F'(0)$ (the derivative here is right-derivative, but it won't matter anywhere in the proof). Since F is log-concave, it will be easier for us to compute $(\log F)'(0) = \frac{F'(0)}{\int G}$. Indeed, notice that

$$(\log F)'(0) = \sup_{\epsilon > 0} \frac{(\log F)(\epsilon) - (\log F)(0)}{\epsilon} = \sup_{\epsilon > 0} \sup_k \frac{(\log F_k)(\epsilon) - (\log F_k)(0)}{\epsilon}$$

$$= \sup_k \sup_{\epsilon > 0} \frac{(\log F_k)(\epsilon) - (\log F_k)(0)}{\epsilon} = \sup_k (\log F_k)'(0)$$

$$= \sup_k \frac{F_k'(0)}{\int G}.$$

Since the sequence $F_k'(0)$ is monotone increasing we get that

$$\widetilde{M}^*(f) = c_n \int G \cdot (\log F)'(0) = \lim_{k \to \infty} c_n F_k'(0) = \lim_{k \to \infty} \widetilde{M}^*(f_k)$$

like we wanted. □

Now that we have all of the ingredients, it is fairly straightforward to prove the main result of this section:

Theorem 1.3. $M^*(f) = \widetilde{M}^*(f)$ *for every log-concave function* f.

Proof. Let $f : \mathbb{R}^n \to [0, \infty)$ be a log-concave function. By equations (3) and (4) we see that both M^* and $\widetilde{M}^*$ are translation invariant. Hence we can translate f and assume without loss of generality that $f(0) > 0$.

If there exists a point x_0 such that $(\mathcal{L}\phi)(x_0) = \infty$, then $\mathcal{L}\phi = \infty$ on an entire half-space, so $M^*(f) = \infty$. By Lemma 2.2 we know that $\int G \star [\epsilon \cdot f] \not\to \int G$, and then $\widetilde{M}^*(f) = \infty$ as well and we get an equality.

If $(\mathcal{L}\phi)(x) < \infty$ for all x we define a sequence of functions $\{f_k\}_{k=1}^\infty$ as

$$f_k = \min(f \cdot \mathbf{1}_{|x| \leq k}, k).$$

Every f_k is log-concave, compactly supported, bounded and satisfies

$$f_k(0) = \min(f(0), k) > 0.$$

Therefore we can apply Lemma 2.1 and conclude that $M^*(f_k) = \widetilde{M}^*(f_k)$. Since the sequence $\{f_k\}$ is monotone and converges pointwise to f we can apply proposition 2.3 and get that

$$M^*(f) = \lim_{k \to \infty} M^*(f_k) = \lim_{k \to \infty} \widetilde{M}^*(f_k) = \widetilde{M}^*(f),$$

so we are done. $\qquad\square$

3 Properties of the Mean Width

We start by listing some basic properties of the mean width, all of which are almost immediate from the definition:

Proposition 3.1. *(i)* $M^*(f) > -\infty$ *for every log-concave function* f.
(ii) If there exists a point $x_0 \in \mathbb{R}^n$ *such that* $f(x_0) \geq 1$, *then* $M^*(f) \geq 0$.
(iii) M^* *is linear: for every log-concave functions* f, g *and every* $\lambda > 0$

$$M^* ((\lambda \cdot f) \star g) = \lambda M^*(f) + M^*(g).$$

(iv) M^* *in rotation and translation invariant.*
(v) If f *is a log-concave function and* $a > 0$ *define* $f_a(x) = a \cdot f(x)$. *Then*

$$M^*(f_a) = M^*(f) + \frac{2}{n} \log a$$

Proof. For (i), remember we explicitly assumed that $f \not\equiv 0$, so there exists a point $x_0 \in \mathbb{R}^n$ such that $f(x_0) > 0$. Hence

$$h_f(x) = \sup_y (\langle x, y \rangle - \phi(y)) \geq \langle x, x_0 \rangle - \phi(x_0),$$

and then

$$M^*(f) = \frac{2}{n} \int h_f(x) d\gamma_n(x) \geq \frac{2}{n} \left[\int \langle x, x_0 \rangle \, d\gamma_n(x) - \phi(x_0) \right] = -\frac{2}{n} \phi(x_0)$$
$$> -\infty$$

like we wanted. For (ii) we know that $\phi(x_0) < 0$, and we simply repeat the argument.

(iii) follows from the easily verified fact that the support function has the same property. In other words, if f, g are log-concave and $\lambda > 0$ then

$$h_{(\lambda \cdot f) \star g}(x) = \lambda h_f(x) + h_g(x)$$

for every x. Integrating over x we get the result.

For (iv), we already saw in the proof of Theorem 1.3 that M^* is translation invariant. For rotation invariance, notice that if u is any linear operator then

$$h_{f \circ u}(x) = \sup_y \left[\langle x, y \rangle - \phi(u(y)) \right] = \sup_z \left[\langle x, u^{-1}z \rangle - \phi(z) \right] =$$

$$= \sup_z \left[\langle (u^{-1})^* x, z \rangle - \phi(z) \right] = h_f \left((u^{-1})^* x \right).$$

In particular if u is orthogonal then $h_{f \circ u}(x) = h_f(ux)$, and the result follows since γ_n is rotation invariant.

Finally for (v), notice that $\phi_a = \phi - \log a$. Therefore

$$h_{f_a} = \mathcal{L}(\phi - \log a) = \mathcal{L}\phi + \log a = h_f + \log a,$$

and the result follows.   □

Remark. A comment in [6] states that $M^*(f)$ is always positive. This is not the case: from (v) we see that if f is any log-concave function with $M^*(f) < \infty$ then $M^*(f_a) \to -\infty$ as $a \to 0^+$. (ii) gives one condition that guarantees that $M^*(f) \geq 0$, and another condition can be deduced from Theorem 1.4.

We now turn our focus to the proof of Theorem 1.4, the functional Urysohn inequality. The main ingredient of the proof is the functional Santaló inequality, proven in [4] for the even case and in [3] for the general case. The result can be stated as follows:

Proposition. *Let $\phi : \mathbb{R}^n \to (-\infty, \infty]$ be any function such that $0 < \int e^{-\phi} < \infty$. Then, there exists $x_0 \in \mathbb{R}^n$ such that for $\tilde{\phi}(x) = \phi(x - x_0)$ one has*

$$\int e^{-\tilde{\phi}} \cdot \int e^{-\mathcal{L}\tilde{\phi}} \leq (2\pi)^n$$

We will also need the following corollary of Jensen's inequality, sometimes known as Shannon's inequality:

Proposition. *For measurable functions $p, q : \mathbb{R}^n \to \mathbb{R}$, assume the following:*

(i) $p(x) > 0$ *for all $x \in \mathbb{R}^n$ and $\int_{\mathbb{R}^n} p(x)dx = 1$*
(ii) $q(x) \geq 0$ *for all $x \in \mathbb{R}^n$*

Then

$$\int p \log \frac{1}{p} \leq \int p \log \frac{1}{q} + \log \int q,$$

with equality if and only if $q(x) = \alpha \cdot p(x)$ almost everywhere.

For a proof of this result see, e.g. Theorem B.1 in [7] (the result is stated for $n = 1$, but the proof is completely general). Using these propositions we can now prove:

Theorem 1.4. *For any log-concave function* f

$$M^*(f) \geq 2 \log \left(\frac{\int f}{\int G} \right)^{\frac{1}{n}} + 1,$$

with equality if and only if $\int f = \infty$ or $f(x) = C e^{-\frac{|x-a|^2}{2}}$ for some $C > 0$ and $a \in \mathbb{R}^n$.

Proof. If $\int f = 0$ there is nothing to prove. Assume first that $\int f < \infty$. We start by applying Shannon's inequality with $p = \frac{d\gamma_n}{dx} = (2\pi)^{-\frac{n}{2}} e^{-\frac{|x|^2}{2}}$ and $q = e^{-h_f}$:

$$M^*(f) = \frac{2}{n} \int h_f(x) d\gamma_n(x) = \frac{2}{n} \int p \log \frac{1}{q} \geq \frac{2}{n} \left[\int p \log \frac{1}{p} - \log \int q \right]$$

$$= \frac{2}{n} \left[\int \left(\frac{|x|^2}{2} + \frac{n}{2} \log(2\pi) \right) d\gamma_n(x) - \log \left(\int e^{-h_f} \right) \right]$$

$$= \int x_1^2 d\gamma_n(x) + \log(2\pi) - \frac{2}{n} \log \left(\int e^{-h_f} \right)$$

$$= 1 + \log(2\pi) - \frac{2}{n} \log \left(\int e^{-h_f} \right).$$

Now we wish to use the functional Santaló inequality. Since the inequality we need to prove is translation invariant, we can translate f and assume without loss of generality that $x_0 = 0$. Hence we get

$$\int f \cdot \int e^{-h_f} \leq (2\pi)^n.$$

Substituting back it follows that

$$M^*(f) \geq 1 + \log(2\pi) - \frac{2}{n} \log \left(\frac{(2\pi)^n}{\int f} \right)$$

$$= 1 + \frac{2}{n} \log \left(\frac{\int f}{\int G} \right),$$

which is what we wanted to prove.

From the proof we also see that equality in Urysohn inequality implies equality in Shannon's inequality. Hence for equality we must have $q(x) = \alpha \cdot p(x)$ for some constant α, or $h_f = \frac{|x|^2}{2} + a$ for some constant a. This implies that

$$\phi = \mathcal{L}(\mathcal{L}\phi) = \mathcal{L} \left(\frac{|x|^2}{2} + a \right) = \frac{|x|^2}{2} - a,$$

so $f(x) = Ce^{-\frac{|x|^2}{2}}$ for $C = e^{-a}$. Since we allowed translations of f in the proof, the general equality case is $f(x) = Ce^{-\frac{|x-a|^2}{2}}$ for some $C > 0$ and $a \in \mathbb{R}^n$.

Finally, we need to handle the case that $\int f = \infty$. Like in Theorem 1.3, we choose a sequence of compactly supported, bounded functions f_k such that $f_k \uparrow f$. It follows that

$$M^*(f) \geq M^*(f_k) \geq 2 \log \left(\frac{\int f_k}{\int G} \right)^{\frac{1}{n}} + 1 \xrightarrow{k \to \infty} \infty,$$

so $M^*(f) = \infty$ and we are done.   □

4   Low-M^* Estimate

Remember the following important result, known as the low-M^* estimate:

Theorem. *There exists a function $f : (0, 1) \to \mathbb{R}^+$ such that for every convex body $K \subseteq \mathbb{R}^n$ and every $\lambda \in (0, 1)$ one can find a subspace $E \hookrightarrow \mathbb{R}^n$ such that $\dim E \geq \lambda n$ and*

$$K \cap E \subseteq f(\lambda) \cdot M^*(K) \cdot D_E$$

This result was first proven by Milman in [8] with $f(\lambda) = C^{\frac{1}{1-\lambda}}$ for some universal constant C. Many other proofs were later found, most of which give sharper bounds on $f(\lambda)$ as $\lambda \to 1^-$ (an incomplete list includes [10, 12], and [5]).

The original proof of the low-M^* estimate passes through another result, known as the finite volume ratio estimate. Remember that if K is a convex body, then the volume ratio of K is

$$V(K) = \inf \left(\frac{|K|}{|\mathcal{E}|} \right)^{\frac{1}{n}},$$

where the infimum is over all ellipsoids $\mathcal{E}$ such that $\mathcal{E} \subseteq K$. In order to state the finite volume ratio estimate it is convenient to assume without loss of generality that this maximizing ellipsoid is the euclidean ball D. The finite volume ratio estimate [15, 16] then reads:

Theorem. *Assume $D \subseteq K$ and $\left(\frac{|K|}{|D|} \right)^{\frac{1}{n}} \leq A$. Then for every $\lambda \in (0, 1)$ one can find a subspace $E \hookrightarrow \mathbb{R}^n$ such that $\dim E \geq \lambda n$ and*

$$K \cap E \subseteq (C \cdot A)^{\frac{1}{1-\lambda}} \cdot (D \cap E)$$

for some universal constant C. In fact, a random subspace will have the desired property with probability $\geq 1 - 2^{-n}$.

We would like to state and prove functional versions of these results. For simplicity, we will only define the functional volume ratio of a log-concave function f when $f \geq G$:

Definition 4.1. Let f be a log-concave function and assume that $f(x) \geq G(x)$ for every x. We define the relative volume ratio of f with respect to G as

$$V(f) = \left(\frac{\int f}{\int G} \right)^{\frac{1}{n}} = \frac{1}{\sqrt{2\pi}} \left(\int f \right)^{\frac{1}{n}}.$$

Theorem 4.2. For every $\epsilon < 1 < M$, every large enough $n \in \mathbb{N}$, every log-concave $f : \mathbb{R}^n \to [0, \infty)$ such that $f \geq G$ and every $0 < \lambda < 1$ one can find a subspace $E \hookrightarrow \mathbb{R}^n$ such that $\dim E \geq \lambda n$ with the following property: for every $x \in E$ such that $e^{-\epsilon n} \geq f(x) \geq e^{-Mn}$ one have

$$f(x) \leq \left([C(\epsilon, M) \cdot V(f)]^{\frac{2}{1-\lambda}} \cdot G \right)(x).$$

Here $C(\epsilon, M)$ is a constant depending only on ϵ and M, and in fact we can take

$$C(\varepsilon, M)^2 = C \max \left(\frac{1}{\varepsilon}, M \right).$$

Proof. For any $\beta > 0$ define

$$K_{f,\beta} = \left\{ x \in \mathbb{R}^n \mid f(x) \geq e^{-\beta n} \right\}.$$

We will bound the volume ratio of $K_{f,\beta}$ in terms of $V(f)$. Because $f \geq G$ we get

$$K_{f,\beta} \supseteq K_{G,\beta} = \left\{ x \in \mathbb{R}^n \mid e^{-\frac{|x|^2}{2}} \geq e^{-\beta n} \right\} = \sqrt{2\beta n} D.$$

We will prove a simple upper bound for the volume of $K_{f,\beta}$. Since f is log-concave one get that for every $\beta_1 \leq \beta_2$

$$K_{f,\beta_1} \subseteq K_{f,\beta_2} \subseteq \frac{\beta_2}{\beta_1} K_{f,\beta_1}.$$

In particular, we can conclude that for every $\beta > 0$

$$K_{f,\beta} \subseteq \max(1, \beta) \cdot K_{f,1}.$$

However, a simple calculation tells us that

$$\int f \geq \int_{K_{f,1}} f \geq |K_{f,1}| \cdot e^{-n},$$

so

$$|K_{f,\beta}| \leq \max(1,\beta)^n |K_{f,1}| \leq [e \cdot \max(1,\beta)]^n \int f.$$

Putting everything together we can bound the volume ratio for $K_{f,\beta}$ with respect to the ball $\sqrt{2\beta n}\,D$:

$$V(K_{f,\beta}) = \left(\frac{|K_{f,\beta}|}{\left|\sqrt{2\beta n}D\right|} \right)^{\frac{1}{n}} \leq \frac{e \cdot \max(1,\beta)}{\sqrt{2\beta n}} \cdot \left(\frac{\int f}{\int G} \right)^{\frac{1}{n}} \cdot \left(\frac{\int G}{|D|} \right)^{\frac{1}{n}}$$

$$\leq C \max(\frac{1}{\sqrt{\beta}}, \sqrt{\beta}) \cdot V(f).$$

Now we pick a one dimensional net $\epsilon = \beta_0 < \beta_1 < \ldots < \beta_{N-1} < \beta_N = M$ such that $\frac{\beta_{i+1}}{\beta_i} \leq 2$. Using the standard finite volume ratio theorem for convex bodies we find a subspace $E \subseteq \mathbb{R}^n$ such that

$$K_{f,\beta_i} \cap E \subseteq \left[C \max(\frac{1}{\sqrt{\beta_i}}, \sqrt{\beta_i}) \cdot V(f) \right]^{\frac{1}{1-\lambda}} \sqrt{2\beta_i n}\,D$$

$$\subseteq \left[C \max(\frac{1}{\sqrt{\epsilon}}, \sqrt{M}) \cdot V(f) \right]^{\frac{1}{1-\lambda}} \sqrt{2\beta_i n}\,D.$$

for every $0 \leq i \leq N$ (This will be possible for large enough n. In fact, it's enough to take $n \geq \log\log \frac{M}{\epsilon}$).

For every $x \in E$ such that $e^{-\epsilon n} \geq f(x) \geq e^{-Mn}$ pick the smallest i such that $e^{-\beta_i n} \leq f(x)$. Then $x \in K_{f,\beta_i} \cap E$, and therefore

$$|x| \leq \left[C \max(\frac{1}{\sqrt{\epsilon}}, \sqrt{M}) \cdot V(f) \right]^{\frac{1}{1-\lambda}} \sqrt{2\beta_i n},$$

or

$$G(x) = e^{-\frac{|x|^2}{2}} \geq \exp\left(-(\beta_i n) \cdot \left[C \max(\frac{1}{\sqrt{\epsilon}}, \sqrt{M}) \cdot V(f) \right]^{\frac{2}{1-\lambda}} \right)$$

$$\geq \exp\left(-(\beta_{i-1} n) \cdot \left[C' \max(\frac{1}{\sqrt{\epsilon}}, \sqrt{M}) \cdot V(f) \right]^{\frac{2}{1-\lambda}} \right).$$

This is equivalent to

$$\left(\left[C' \max(\frac{1}{\sqrt{\epsilon}}, \sqrt{M}) \cdot V(f) \right]^{\frac{2}{1-\lambda}} \cdot G \right)(x) \geq e^{-\beta_{i-1} n} > f(x)$$

which is exactly what we wanted.   □

Remark. The role of ϵ and M in the above theorem might seem a bit artificial, as the condition $e^{-\epsilon n} \geq f(x) \geq e^{-Mn}$ has no analog in the classical theorem. This condition is necessary however, as some simple examples show. For example, consider $f(x) = e^{-\phi(|x|)}$ where

$$\phi(x) = \begin{cases} 0 & x < \sqrt{n} \\ 2\sqrt{n}x - 2n & \sqrt{n} \leq x \leq 2\sqrt{n} \\ \frac{x^2}{2} & 2\sqrt{n} \leq x. \end{cases}$$

To explain the origin of this example notice that f is the log-concave envelope of $\max(G, 1_{\sqrt{n}D})$. It is easy to check that $f \geq G$ and $V(f)$ is bounded from above by a universal constant independent of n. Since f is rotationally invariant the role of the subspace E in the theorem is redundant, and one easily checks that $f(x) \geq e^{-\epsilon n}$ if and only if

$$f(x) \leq \left(\frac{(\epsilon + 2)^2}{8\epsilon} \cdot G\right)(x) \sim \left(\frac{1}{2\epsilon} \cdot G\right)(x).$$

This shows that not only does $C(\epsilon, M)$ must depend on ϵ, but the dependence we showed is essentially sharp as $\epsilon \to 0$. Similar examples show that the same is true for the dependence in M.

Using Theorem 4.2 we can easily prove Theorem 1.5:

Theorem 1.5. *For every $\varepsilon < M$, every large enough $n \in \mathbb{N}$, every $f : \mathbb{R}^n \to [0, \infty)$ such that $f(0) = 1$ and $M^*(f) \leq 1$ and every $0 < \lambda < 1$ one can find a subspace $E \hookrightarrow \mathbb{R}^n$ such that $\dim E \geq \lambda n$ with the following property: for every $x \in E$ such that $e^{-\varepsilon n} \geq (f \star G)(x) \geq e^{-Mn}$ one have*

$$f(x) \leq \left(C(\varepsilon, M)^{\frac{1}{1-\lambda}} \cdot G\right)(x).$$

In fact, one can take

$$C(\varepsilon, M) = C \max\left(\frac{1}{\varepsilon}, M\right).$$

Proof. Define $h = f \star G$. Since $f(0) = 1$ it follows that

$$(f \star G)(x) = \sup_{x_1 + x_2 = x} f(x_1)G(x_2) \geq f(0)G(x) = G(x).$$

Since M^* is linear $M^*(h) = M^*(f) + M^*(G) \leq 2$, so by Theorem 1.4 we get that $V(h) \leq \sqrt{e}$. Applying Theorem 4.2 for h, and noticing that $f(x) \leq h(x)$ for all x, we get the result. $\qquad \square$

References

1. S. Artstein-Avidan, V. Milman, A characterization of the support map. Adv. Math. **223**(1), 379–391 (2010)
2. S. Artstein-Avidan, V. Milman, Hidden structures in the class of convex functions and a new duality transform. J. Eur. Math. Soc. (JEMS) **13**(4), 975–1004 (2011)
3. S. Artstein-Avidan, B. Klartag, V. Milman, The santaló point of a function, and a functional form of the santaló inequality. Mathematika **51**(1–2), 33–48 (2004)
4. K. Ball, Isometric problems in l_p and sections of convex sets. PhD thesis, Cambridge University (1986)
5. Y. Gordon, in *On Milman's Inequality and Random Subspaces which Escape Through a Mesh in $\mathbb{R}^n$*, ed. by J. Lindenstrauss, V. Milman. Geometric Aspects of Functional Analysis, vol. 1317 (Springer, Berlin, 1988), pp. 84–106
6. B. Klartag, V. Milman, Geometry of log-concave functions and measures. Geometriae Dedicata **112**(1), 169–182 (2005)
7. R. McEliece, in *The Theory of Information and Coding*. Encyclopedia of Mathematics and Its Applications, vol. 86, 2nd edn. (Cambridge University Press, London, 2002)
8. V. Milman, Almost euclidean quotient spaces of subspaces of a Finite-Dimensional normed space. Proc. Am. Math. Soc. **94**(3), 445–449 (1985)
9. V. Milman, Geometrical inequalities and mixed volumes in the local theory of banach spaces. (Colloquium in honor of Laurent Schwartz, vol. 1 (Palaiseau, 1983). Astérisque **131**, 373–400 (1985)
10. V. Milman, in *Random Subspaces of Proportional Dimension of Finite Dimensional Normed Spaces: Approach Through the Isoperimetric Inequality*, ed. by N. Kalton, E. Saab. Banach Spaces (Columbia, MO, 1984). Lecture Notes in Math., vol. 1166 (Springer, Berlin, 1985), pp. 106–115
11. V. Milman, in *Geometrization of Probability*, ed. by M. Kapranov, Y. Manin, P. Moree, S. Kolyada, L. Potyagailo. Geometry and Dynamics of Groups and Spaces, vol. 265 (Birkhäuser, Basel, 2008), pp. 647–667
12. A. Pajor, N. Tomczak-Jaegermann, Subspaces of small codimension of Finite-Dimensional banach spaces. Proc. Am. Math. Soc. **97**(4), 637–642 (1986)
13. G. Pisier, in *The Volume of Convex Bodies and Banach Space Geometry*. Cambridge Tracts in Mathematics, vol. 94 (Cambridge University Press, Cambridge, 1989)
14. R. Schneider, in *Convex Bodies: The Brunn-Minkowski Theory*, Encyclopedia of Mathematics and Its Applications, vol. 44 (Cambridge University Press, Cambridge, 1993)
15. S. Szarek, On kashin's almost euclidean orthogonal decomposition of l_n^1. Bulletin de l'Académie Polonaise des Sciences. Série des Sciences Mathématiques, Astronomiques et Physiques **26**(8), 691–694 (1978)
16. S. Szarek, N. Tomczak-Jaegermann, On nearly euclidean decomposition for some classes of banach spaces. Compos. Math. **40**(3), 367–385 (1980)

Approximate Gaussian Isoperimetry for k Sets

Gideon Schechtman

Abstract Given $2 \leq k \leq n$, the minimal $(n-1)$-dimensional Gaussian measure of the union of the boundaries of k disjoint sets of equal Gaussian measure in $\mathbb{R}^n$ whose union is $\mathbb{R}^n$ is of order $\sqrt{\log k}$. A similar results holds also for partitions of the sphere S^{n-1} into k sets of equal Haar measure.

1 Introduction

Consider the canonical Gaussian measure on $\mathbb{R}^n$, γ_n. Given $k \in \mathbb{N}$ and k disjoint measurable subsets of $\mathbb{R}^n$ each of γ_n measure $1/k$ we can compute the $(n-1)$-dimensional Gaussian measure of the union of the boundaries of these k sets. Below (see Definition 1) we shall make clear what exactly we mean by the $(n-1)$-dimensional Gaussian measure but in particular our normalization will be such that the $(n-1)$-dimensional Gaussian measure of a hyperplane at distance t from the origin will be $e^{-t^2/2}$ (and not $\frac{1}{\sqrt{2\pi}} e^{-t^2/2}$ which is also a natural choice). The question we are interested in is what is the minimal value that this quantity can take when ranging over all such partitions of $\mathbb{R}^n$. As is well known, the Gaussian isoperimetric inequality [1, 4] implies that, for $k = 2$, the answer is 1 and is attained when the two sets are half spaces. The answer is also known for $k = 3$ and $n \geq 2$ and is given by 3 $2\pi/3$-sectors in $\mathbb{R}^2$ (product with R^{n-2}) [2]. The value in question is then $3/2$. If the k sets are nice enough (for example if, with respect to the $(n-1)$-dimensional Gaussian measure, almost every point in the union of the boundaries of the k sets belongs to the boundary of only two of the sets) then the quantity in question is bounded from below by $c\sqrt{log k}$ for some absolute $c > 0$. This was pointed out to us by Elchanan Mossel. Indeed, by the Gaussian isoperimetric

G. Schechtman
Weizmann Institute of Science, Department of Mathematics, Rehovot, Israel
e-mail: gideon.schechtman@weizmann.ac.il

B. Klartag et al. (eds.), *Geometric Aspects of Functional Analysis*, Lecture Notes in Mathematics 2050, DOI 10.1007/978-3-642-29849-3_23,
© Springer-Verlag Berlin Heidelberg 2012

inequality, the boundary of each of the sets has measure at least $e^{-t^2/2}$ where t is such that $\frac{1}{\sqrt{2\pi}} \int_t^\infty e^{-s^2/2} ds = 1/k$. If k is large enough t satisfies

$$\frac{e^{-t^2/2}}{\sqrt{2\pi}2t} < \frac{1}{k} < \frac{e^{-t^2/2}}{\sqrt{2\pi}t}$$

which implies $\sqrt{\log k} \le t \le \sqrt{2 \log k}$ and so the boundary of each of the k sets has $(n-1)$-dimensional Gaussian measure at least $e^{-t^2/2} \ge \sqrt{2\pi}t/k \ge \sqrt{2\pi \log k}/k$. Under the assumption that the sets are nice we then get a lower bound of order $\sqrt{2\pi \log k}$ to the quantity we are after.[1]

Of course the minimality of the boundary of each of the k sets cannot occur simultaneously for even 3 of the k sets (as the minimal configuration is a set bounded by an affine hyperplane) so it may come as a surprise that one can actually achieve a partition with that order of the size of the boundary. To show this is the main purpose of this note. It is natural to conjecture that, for $k - 1 \le n$ the minimal configuration is that given by the Voronoi cells of the k vertices of a simplex centered at the origin of $\mathbb{R}^n$. So it would be nice to compute or at least estimate well what one gets in this situation. This seems an unpleasant computation to do. However, in Corollary 1 below we compute such an estimate for a similar configuration - for even k with $k/2 \le n$, we look at the k cells obtained as the Voronoi cells of $\pm e_i, i = 1, \ldots, k/2$ and show that the order of the $(n-1)$-dimensional Gaussian measure of the boundary is of order $\sqrt{\log k}$ and we deduce the main result of this note:

Main Result. *Given even k with $k \le 2n$, the minimal $(n - 1)$-dimensional Gaussian measure of the union of the boundaries of k disjoint sets of equal Gaussian measure in $\mathbb{R}^n$ whose union is $\mathbb{R}^n$ is of order $\sqrt{\log k}$.*

In Corollary 2 we deduce analogue estimates for the Haar measure on the sphere S^{n-1}.

This note benefitted from discussions with Elchanan Mossel and Robi Krauthgamer. I first began to think of the subject after Elchanan and I spent some time trying (alas in vain) to use symmetrization techniques to gain information on the (say, Gaussian) "k-bubble" conjecture and some variant of it (see Conjecture 1.4 in [3]). Robi asked me specifically the question that is solved here, with some possible applications to designing some algorithm in mind (but apparently the solution turned out to be no good for that purpose). I thank Elchanan and Robi also for several remarks on a draft of this note. I had also a third motivation to deal with this question. It is related to the computation of the dependence on ε in (the probabilistic version of) Dvoretzky's theorem. It is too long to explain here, especially since it does not seem to lead to any specific result.

---

[1]One may think that the right quantity should be $\sqrt{2\pi \log k}/2$ since (almost) every boundary point is counted twice but our Definition 1 is such that almost every boundary point is counted with multiplicity of the number of sets in the partition it is on the boundary of. In any case, absolute constants do not play a significant role here.

2 Approximate Isoperimetry for k Sets

We begin with the formal definition of the $(n-1)$-dimensional Gaussian measure of the boundary of a partition of $\mathbb{R}^n$ into k sets.

Definition 1. Let $A_1, A_2, \ldots, A_k$ be a partition of $\mathbb{R}^n$ into k measurable sets. Put $A = \{A_1, A_2, \ldots, A_k\}$ and denote

$$\partial_\varepsilon A = \cup_{i=1}^k ((\cup_{j \neq i} A_j)_\varepsilon \setminus \cup_{j \neq i} A_j)$$

(where B_ε denotes the ε-neighborhood of the set B). We shall call $\partial_\varepsilon A$ the *ε-boundary of A. The $(n-1)$-dimensional Gaussian measure of the boundary of A* will be defined and denoted by

$$\gamma_{n-1}(\partial A) = \liminf_{\varepsilon \to 0} \frac{\gamma_n(\partial_\varepsilon A) - \gamma_n(A)}{\sqrt{2/\pi}\varepsilon}.$$

The reason we are using the above definition of $\partial_\varepsilon A$ rather than what might look more natural, $\cup_{i=1}^k ((A_i)_\varepsilon \setminus A_i)$, is that in the former the sets in the union are disjoint, making the computation of the measure of the union easier. Note that we do not define the boundary of the partition, only the measure of the boundary. However, in simple cases when the boundary and its $(n-1)$-dimensional Gaussian measure are well understood, this definition coincides with the classical one (modulo normalization by absolute constant). In particular notice that if the partition is into two sets which are separated by a hyperplane at distance t from the origin the definition says that the $(n-1)$-dimensional Gaussian measure of the boundary is $e^{-t^2/2}$ and in particular when $t = 0$ the measure is 1 which coincides with what we understand as the classical γ_{n-1} measure of a hyperplane through 0. This is why the factor $\sqrt{2/\pi}$ is present in the definition above.

The main technical tool here is Proposition 1 below for its proof we need the following simple inequality.

Lemma 1. *For all $\varepsilon > 0$ if C is large enough (depending on ε) then for all $k \in \mathbb{N}$*

$$\frac{1}{\sqrt{2\pi}} \int_{\sqrt{2 \log \frac{k}{C}} - 1}^{\sqrt{2 \log Ck}} \left(\frac{1}{\sqrt{2\pi}} \int_{-s}^s e^{-t^2/2} dt \right)^{k-1} e^{-s^2/2} ds \geq \frac{1}{(2+\varepsilon)k}$$

Proof. Let $g_1, g_2, \ldots, g_k$ be independent identically distributed $N(0,1)$ variables. Then by symmetry,

$$\frac{1}{\sqrt{2\pi}} \int_0^\infty \left(\frac{1}{\sqrt{2\pi}} \int_{-s}^s e^{-t^2/2} dt \right)^{k-1} e^{-s^2/2} ds = P(g_1 \geq |g_2|, \ldots, |g_k|) = \frac{1}{2k}.$$

$$(1)$$

Also,

$$\frac{1}{\sqrt{2\pi}} \int_0^{\sqrt{2\log\frac{k}{C}}-1} \left(\frac{1}{\sqrt{2\pi}} \int_{-s}^s e^{-t^2/2}dt\right)^{k-1} e^{-s^2/2}ds$$

$$= \frac{1}{2k}\left(\frac{2}{\sqrt{2\pi}} \int_0^s e^{-t^2/2}\right)^k \Big]_{s=0}^{\sqrt{2\log\frac{k}{C}}-1} \tag{2}$$

$$\leq \frac{1}{2k}\left(1 - \frac{2}{\sqrt{2\pi}}e^{-\log\frac{k}{C}}\right)^k \leq \frac{1}{2k}e^{-\frac{2C}{\sqrt{2\pi}}},$$

and, for C large enough,

$$\frac{1}{\sqrt{2\pi}} \int_{\sqrt{2\log Ck}}^{\infty} \left(\frac{1}{\sqrt{2\pi}} \int_{-s}^s e^{-t^2/2}dt\right)^{k-1} e^{-s^2/2}ds$$

$$= \frac{1}{2k}\left(\frac{2}{\sqrt{2\pi}} \int_0^s e^{-t^2/2}\right)^k \Big]_{s=\sqrt{2\log Ck}}^{\infty} \tag{3}$$

$$\leq \frac{1}{2k}\left(1 - \left(1 - \frac{2}{\sqrt{2\pi}} \int_{\sqrt{2\log Ck}}^{\infty} e^{-s^2/2}\right)^k\right) \leq \frac{1}{2k}(1 - e^{-1/C}).$$

The Lemma now follows from (1)–(3).   □

The next proposition is the main technical tool of this note. The statement involves the (appropriately normalized) $(k-1)$-dimensional Gaussian measure of a certain subset of $\mathbb{R}^k$. The set is the common boundary of the Voronoi cells corresponding to e_1 and e_2 in the partition of $\mathbb{R}^k$ obtained by the Voronoi cells of $\pm e_i, i = 1, \ldots, k/2$. This set is a subset of a hyperplane (through the origin of $\mathbb{R}^k$) and for such a set the measure in question coincides with the canonical Gaussian measure associated with this subspace (appropriately normalized).

Proposition 1. *For each $\varepsilon > 0$ there is a C such that for all $k \geq 2$, the $(k-1)$-dimensional Gaussian measure of the set $\{(t_1, t_2, \ldots, t_k); t_1 = t_2 \geq |t_3|, \ldots, |t_k|\}$ is bounded between $\frac{\sqrt{\pi \log\frac{k}{C}}-1}{(1+\varepsilon)2k(k-1)}$ and $\frac{(1+\varepsilon)\sqrt{\pi \log Ck}}{2k(k-1)}$.*

Proof. The measure in question is

$$\frac{1}{\sqrt{2\pi}} \int_0^{\infty} \left(\frac{1}{\sqrt{2\pi}} \int_{-s}^s e^{-t^2/2}dt\right)^{k-2} e^{-s^2}ds.$$

Integration by parts (with parts $\left(\frac{2}{\sqrt{2\pi}} \int_0^s e^{-t^2/2}dt\right)^{k-2} e^{-s^2/2}$ and $e^{-s^2/2}$) gives that this it is equal to

$$\frac{1}{2(k-1)} \int_0^{\infty} \left(\frac{2}{\sqrt{2\pi}} \int_0^s e^{-t^2/2}dt\right)^{k-1} se^{-s^2/2}ds. \tag{4}$$

Now,

$$\int_{\sqrt{2\log Ck}}^{\infty} \left(\frac{2}{\sqrt{2\pi}}\int_0^s e^{-t^2/2}dt\right)^{k-1} se^{-s^2/2}ds \tag{5}$$

$$= -\int_s^{\infty}\left(\frac{2}{\sqrt{2\pi}}\int_0^u e^{-t^2/2}dt\right)^{k-1}e^{-u^2/2}du\bigg]_{s=\sqrt{2\log Ck}}^{\infty}$$

$$+ \int_{\sqrt{2\log Ck}}^{\infty}\int_s^{\infty}\left(\frac{2}{\sqrt{2\pi}}\int_0^u e^{-t^2/2}dt\right)^{k-1}e^{-u^2/2}duds$$

$$\leq \frac{\sqrt{2\pi}}{2k}(1-e^{-1/C})\sqrt{2\log Ck} + \int_{\sqrt{2\log Ck}}^{\infty}\frac{\sqrt{2\pi}}{2k}(1-e^{-ke^{-s^2/2}})ds, \tag{6}$$

where the estimate for the first term in (6) follows from (3) and of the second term follows from a similar computation to (3). Now (6) is at most

$$\frac{\sqrt{2\pi}}{2Ck}\sqrt{2\log Ck} + \int_{\sqrt{2\log Ck}}^{\infty}\frac{\sqrt{2\pi}}{2}e^{-s^2/2}ds \leq \frac{\sqrt{2\pi}(\sqrt{2\log Ck}+1)}{2Ck} \tag{7}$$

and we conclude that

$$\int_{\sqrt{2\log Ck}}^{\infty}\left(\frac{2}{\sqrt{2\pi}}\int_0^s e^{-t^2/2}dt\right)^{k-1}se^{-s^2/2}ds \leq \frac{\sqrt{2\pi}(\sqrt{2\log Ck}+1)}{2Ck}. \tag{8}$$

On the other hand

$$\int_0^{\sqrt{2\log Ck}}\left(\frac{2}{\sqrt{2\pi}}\int_0^s e^{-t^2/2}dt\right)^{k-1}se^{-s^2/2}ds \tag{9}$$

$$\leq \sqrt{2\log Ck}\int_0^{\infty}\left(\frac{2}{\sqrt{2\pi}}\int_0^s e^{-t^2/2}dt\right)^{k-1}e^{-s^2/2}ds = \frac{\sqrt{2\pi}\sqrt{2\log Ck}}{2k}.$$

Now, (4), (8) and (9) gives the required upper bound. The lower bound (which also follows from the Gaussian isoperimetric inequality) is easier. By Lemma 1

$$\frac{1}{2(k-1)}\int_0^{\infty}\left(\frac{2}{\sqrt{2\pi}}\int_0^s e^{-t^2/2}dt\right)^{k-1}se^{-s^2/2}ds \tag{10}$$

$$\geq \frac{1}{2(k-1)}\int_{\sqrt{2\log\frac{k}{C}-1}}^{\sqrt{2\log Ck}}\left(\frac{2}{\sqrt{2\pi}}\int_0^s e^{-t^2/2}dt\right)^{k-1}se^{-s^2/2}ds \tag{11}$$

$$\geq \frac{\sqrt{2\log\frac{k}{C}}-1}{2(k-1)} \int_{\sqrt{2\log\frac{k}{C}}-1}^{\sqrt{2\log Ck}} \left(\frac{2}{\sqrt{2\pi}} \int_0^s e^{-t^2/2}dt\right)^{k-1} e^{-s^2/2}ds \qquad (12)$$

$$\geq \frac{\sqrt{\pi\log\frac{k}{C}}-1}{(1+\varepsilon)2k(k-1)}. \qquad (13)$$

□

The next Corollary is the main result here, in the most general setting we need the definition of the $(n-1)$-dimensional Gaussian measure of the boundary of a partition of $\mathbb{R}^n$ into k sets given in Definition 1 above. Note that this is the first time we use the full details of the definition; Until now we dealt only with subsets of hyperplanes for which simpler definitions could suffice.

Corollary 1. *For some universal constants* $0 < c < C < \infty$ *and all* $k = 2, 3, \ldots,$

(1) If $A = \{A_1, A_2, \ldots, A_k\}$ *is a partition of* $\mathbb{R}^n$ *into k measurable sets each of γ_n measure $1/k$. Then* $\gamma_{n-1}(\partial A) \geq c\sqrt{\log k}$.
(2) If $k \leq n$, *there is a partition* $A = \{A_1, A_2, \ldots, A_{2k}\}$ *of* $\mathbb{R}^n$ *into $2k$ measurable sets each of γ_n measure $1/2k$ such that* $\gamma_{n-1}(\partial A) \leq C\sqrt{\log k}$.

(1) follows very similarly to the argument in the introduction, except that there is no need for the boundary to be nice anymore: By the Gaussian isoperimetric inequality, for each $\varepsilon > 0$ and each $i = 1, \ldots, k$,

$$\gamma_n((\cup_{j\neq i} A_j)_\varepsilon \setminus \cup_{j\neq i} A_j) \geq \frac{1}{\sqrt{2\pi}} \int_t^{t+\varepsilon} e^{-s^2/2}ds,$$

where t is such that $\frac{1}{\sqrt{2\pi}} \int_t^\infty e^{-s^2/2}ds = 1/k$. If ε is small enough, the argument in the introduction gives that the integral in question is of order $\varepsilon\frac{\sqrt{\log k}}{k}$. Since the k sets $(\cup_{j\neq i} A_j)_\varepsilon \setminus \cup_{j\neq i} A_j$ are disjoint, we deduce (1). (2) follows directly from Proposition 1 since the boundary of the partition into the Voronoi cells corresponding to $\{\pm e_i\}_{i=1}^k$ is contained in the union of $k(k-1)$ hyperplans through zero and thus $\gamma_{n-1}(\partial A)$ coincide with the classical $\gamma_{n-1}(\partial A)$ which is what is estimated in Proposition 1.

A similar result to Corollary 1 holds on the n-dimensional sphere, S^{n-1} with its normalized Haar measure σ_n. One defines the ε-boundary of a partition A of the sphere in a similar way to the first part of Definition 1 (using, say, the geodesic distance to define the ε-neighborhood of a set). Then one defines the $(n-1)$-dimensional Haar measure of the boundary of A by

$$\sigma_{n-1}(\partial A) = \liminf_{\varepsilon\to 0} \frac{\sigma_n(\partial_\varepsilon A) - \sigma_n(A)}{\sqrt{2n/\pi\varepsilon}}.$$

The choice of the normalization constant $\sqrt{2n/\pi}$ was made so that if the partition is into two sets separated by a hyperplane then the measure of the boundary (which "is" S^{n-2}) will be 1. The proof of the upper bound ((2) of Corollary 2) can be obtained from that of Corollary 1 by a standard reduction, using the fact that if $(g_1, \ldots, g_n)$ is a standard Gaussian vector then the distribution of $(\sum g_i^2)^{-1/2}(g_1, \ldots, g_n)$ is σ_n. Note that we deal only with subsets of centered hyperplanes here so there is no problem with the reduction. The lower bound (1) can also be achieved using this reduction although one needs to be a bit more careful. Alternatively, a similar argument to that of the Gaussian case (given in the introduction), replacing the Gaussian isoperimetric inequality with the spherical isoperimetric inequality can be used.

Corollary 2. *For some universal constants $0 < c < C < \infty$ and all $k = 2, 3, \ldots,$*

(1) If $A = \{A_1, A_2, \ldots, A_k\}$ is a partition of S^{n-1} into k measurable sets each of σ_n measure $1/k$. Then $\sigma_{n-1}(\partial A) \geq c \sqrt{\log k}$.

(2) If $k \leq n$, there is a partition $A = \{A_1, A_2, \ldots, A_{2k}\}$ of S^{n-1} into $2k$ measurable sets each of σ_n measure $1/2k$ such that $\sigma_{n-1}(\partial A) \leq C \sqrt{\log k}$.

Remark 1. It may be interesting to investigate what happens when $k \gg n$. In particular, if $k = 2^n$ then the partition of R^n into its $k = 2^n$ quadrants satisfy that the γ_{n-1} measure of its boundary (consisting of the coordinates hyperplanes) is $n = \log k$. Is that the best (order) that can be achieved?

Acknowledgements Supported by the Israel Science Foundation.

References

1. C. Borell, Convex measures on locally convex spaces. Ark. Mat. **12**, 239–252 (1974)
2. J. Corneli, I. Corwin, S. Hurder, V. Sesum, Y. Xu, E. Adams, D. Davis, M. Lee, R. Visocchi, N. Hoffman, Double bubbles in Gauss space and spheres. Houston J. Math. **34**(1), 181–204 (2008)
3. M. Isaksson, E. Mossel, Maximally stable Gaussian partitions with discrete applications. Isaksson J. Math. to appear
4. V.N. Sudakov, B.S. Cirel'son, Extremal properties of half-spaces for spherically invariant measures (Russian). Problems in the Theory of Probability Distributions, vol. II. Zap. Naucn. Sem. Leningrad. Otdel. Mat. Inst. Steklov (LOMI) **41**(165), 14–24 (1974)

Remark on Stability of Brunn–Minkowski and Isoperimetric Inequalities for Convex Bodies

Alexander Segal

Abstract This paper is a note on the work of Figalli, Maggi and Pratelli, regarding the stability of Brunn–Minkowski and the isoperimetric inequalities. By a careful examination of the methods presented in the mentioned papers, we slightly improve the constants that appear in stability versions of these inequalities, which play an important role in asymptotic geometric analysis. In addition we discuss a stability version of Urysohn's inequality and the relation to Dar's conjecture.

1 Introduction

Assume that E and F are convex bodies in $\mathbb{R}^n$. Then we have the famous Brunn–Minkowski (B–M) inequality

$$|E + F|^{1/n} \geq |E|^{1/n} + |F|^{1/n}, \tag{1}$$

where $|\cdot|$ denotes the Lebesgue measure in $\mathbb{R}^n$. The Brunn–Minkowski inequality is a well known inequality that plays a central role in theory of convex bodies, as explained in great detail in Schneider's book [12]. It is easy to see that equality holds if and only if the bodies E and F are homothetic. This fact raises the natural questions: Given that the left-hand side and right-hand side of (1) differ only slightly, can we say that E and F are "close"? Is the Brunn–Minkowski inequality sensitive to small perturbations?

A. Segal (✉)
School of Mathematical Sciences, Sackler Faculty of Exact Sciences, Tel Aviv University, Ramat Aviv, Tel Aviv 69978, Israel
e-mail: segalale@gmail.com

B. Klartag et al. (eds.), *Geometric Aspects of Functional Analysis*, Lecture Notes in Mathematics 2050, DOI 10.1007/978-3-642-29849-3_24,
© Springer-Verlag Berlin Heidelberg 2012

Clearly, to answer these questions one needs to define what "close" means. Answers were given, when the distance between bodies is measured by Hausdorff distance [7], but a more natural way to measure how "close" two convex bodies are, is using volume. For example, we could ask how large the volume of the maximal intersection of E and homotheties of F is, compared to the volume of E. This question was answered by Figalli et al. in [9] and in [8] using mass transportation approach.

Let us state some notations used in the mentioned papers in order to discuss their results:

Given two convex bodies E and F, define their *relative asymmetry* to be:

$$A(E, F) := \inf_{x_0 \in \mathbb{R}^n} \frac{|E \Delta (x_0 + rF)|}{|E|}, \tag{2}$$

where $E \Delta F = (F \setminus E) \cup (E \setminus F)$ is the symmetric difference and $r^n |E| = |F|$.

Also, define the *Brunn–Minkowski deficit* of E and F to be:

$$\beta(E, F) = \frac{|E + F|^{1/n}}{|E|^{1/n} + |F|^{1/n}} - 1. \tag{3}$$

Using these notations, the authors of [9] showed a refined version of the Brunn–Minkowski inequality in case $|E| = |F|$:

$$A(E, F) \leq C(n) \sqrt{\beta(E, F)},$$

where $C(n) \approx n^7$.

Remark 1.1. In this paper, for convenience purposes, we only discuss the case where the bodies are of equal volume. The general case follows exactly in the same way, with an added scaling parameter.

In [8], the authors showed the same inequality with a much simpler proof, but the constant had an exponential dependence on n. Knowing the optimal behaviour of $C(n)$ is very important for certain problems in asymptotic geometrical analysis, and we would like to find the best possible estimate. Although we have not found the optimal behaviour, we would like to note that some improvement is possible just by careful examination of each step in [8, 9] and minor modification of some of them. We show the following statement:

Theorem 1.2. *Assume E and K are convex bodies in $\mathbb{R}^n$ with equal volume. The following inequality holds:*

$$A(E, K) \leq C n^{3.5} \sqrt{\beta(E, K)},$$

where C is some universal constant.

Acknowledgements I would like to thank my advisor, professor Vitali Milman, and Professors Figalli and Pratelli for discussions and their useful advices regarding the Brunn Minkowski stability problem. This research was supported by ISF grant 387/09 and BSF grant 200 6079.

2 Preliminaries and Notation

We denote by l_2^n the space $\mathbb{R}^n$ equipped with the standard euclidean norm. Also denote by d_E the Banach-Mazur distance of the body E from the unit ball of l_2^n. Recall that the Banach-Mazur distance between two finite dimensional normed spaces X, Y is defined as follows:

$$d(X,Y) = \inf_T \left\{ \|T\| \|T^{-1}\| : T : X \to Y \text{ is a linear isomorphism} \right\}.$$

The proof of the main theorem follows directly from a similar result regarding the stability of the isoperimetric inequality. We outline the details here.

First we state a modified Poincaré type inequality, proved in [8] and used for proving the main theorem:

Lemma 2.1. *Let E be a convex body such that $B_r \subset E \subset B_R$, for $0 < r < R$. Then, the following inequality holds:*

$$n \frac{\sqrt{2}}{\log 2} \frac{R}{r} \int_E |\nabla f| \geq \int_{\partial E} |f - M_f| d\mathcal{H}^{n-1}, \tag{4}$$

for every f smooth enough, where M_f denotes the median of function f in E.

Remark 2.2. The dependence on the dimension n, and on R/r is optimal in Lemma 2.1, as shown in [8].

Next we turn to basic definitions and statements regarding the isoperimetric inequality.

Denote by $h_K(x)$ the supporting functional of a convex body K in $\mathbb{R}^n$:

$$h_K(v) = \sup\{\langle x, v \rangle : x \in K\}.$$

Now, given a convex body E, we can define the *anisotropic perimeter* of E with respect to K:

$$P_K(E) = \int_{\partial E} h_K(v_E(x)) d\mathcal{H}^{n-1}, \tag{5}$$

where $v_E(x)$ is the outer unit normal vector on the boundary ∂E, and $\mathcal{H}^{n-1}$ is the $n-1$ dimensional Hausdorff measure on $\mathbb{R}^n$. Since we restrict our discussion to compact convex sets, the anisotropic perimeter can be written as

$$P_K(E) = \lim_{\epsilon \to 0^+} \frac{|E + \epsilon K| - |E|}{\epsilon}.$$

This fact follows from the definition of mixed volumes, mixed area measures and Minkowski's theorem on mixed volumes. For details see [10] (pp. 397–399). Actually, one can check this fact directly for polytopes and conclude the proof by approximation argument. Applying Brunn–Minkowski inequality to the expression under the limit we get:

$$\frac{|E + \epsilon K| - |E|}{\epsilon} \geq \frac{(|E|^{1/n} + \epsilon |K|^{1/n})^n - (|E|^{1/n})^n}{\epsilon}.$$

Clearly, when $\epsilon \to 0$ the right hand side converges to $n|K|^{1/n}|E|^{1/n'}$, where n' is the conjugate of n:

$$\frac{1}{n} + \frac{1}{n'} = 1.$$

Finally, we get the anisotropic isoperimetric inequality:

$$P_K(E) \geq n|K|^{1/n}|E|^{1/n'}. \tag{6}$$

Like in the Brunn–Minkowski inequality, the case of equality holds if and only if the bodies E and K are homothetic. Hence, the questions of stability is relevant again.

Define the *isoperimetric deficit* of a body E to be:

$$\delta(E, K) := \frac{P_K(E)}{n|K|^{1/n}|E|^{1/n'}} - 1.$$

We are interested in estimating the behavior of $A(E, K)$ with the deficit $\delta(E, K)$. An answer to this question was given in [9]:

$$A(E, K) \leq C_0(n)\sqrt{\delta(E, K)}, \tag{7}$$

where $C_0(n)$ behaves like n^7. The authors also proved that the result is sharp, in the sense that the power of $\delta(E, K)$ cannot be improved. The equivalent refinement for Brunn–Minkowski follows immediately, as will be shown in the text. In order to improve the constant in the Brunn–Minkowski inequality, we will prove the following theorem:

Theorem 2.3. *Let E and K be two convex bodies in $\mathbb{R}^n$. Then, the following inequality holds:*

$$A(E, K) \leq C_1 n^{3.5}\sqrt{\delta(E, K)},$$

where C_1 is a universal constant, $A(E, K)$ and $\delta(E, K)$ are the relative asymmetry and the isoperimetric deficit, defined above.

The proof is in the following sections.

3 Proof of Isoperimetric Inequality

From now on, for our convenience, we assume that E and K are smooth and uniform convex bodies that have equal volumes (scaling argument). We will use the method of mass transportation to show the isoperimetric inequality. For details regarding optimal transport we refer the read to Villani's book [13]. Brenier map theorem [2] states there exists a convex function $\phi : \mathbb{R}^n \to \mathbb{R}$, such that the map $T = \nabla \phi$ pushes forward the measure $|E|^{-1} 1_E(x)dx$ to $|K|^{-1} 1_K(x)dx$. A result by Caffarelli [4, 5] also states that under our assumptions the map T is smooth, i.e $T \in C^\infty(\bar{E}, \bar{K})$. Since T is measure preserving, we also get:

$$\det(Hess(\phi(x))) = \det(\nabla T(x)) = \frac{|K|}{|E|} = 1.$$

Since $\nabla T(x)$ is a symmetric positive definite matrix, we may assume that the eigenvalues are ordered $0 < \lambda_1(x) \leq \lambda_2(x) \cdots \leq \lambda_n(x)$. Denote the Arithmetic and Geometric means of the eigenvalues,

$$\lambda_A(x) = \frac{1}{n} \sum_{i=1}^{n} \lambda_i(x), \quad \lambda_G = \prod_{i=1}^{n} \lambda_i(x)^{1/n} = 1.$$

Now, we can prove the isoperimetric inequality using mass transportation. First denote by $|| \cdot ||$, the norm induced by the body K:

$$||x|| = \inf_{\lambda > 0} \left\{ \lambda : \frac{x}{\lambda} \in K \right\}.$$

Notice that whenever $x \in E$, we have $T(x) \in K$, thus $||T|| \leq 1$. Applying arithmetic-geometric mean inequality and the divergence theorem, we get

$$n|K|^{1/n}|E|^{1/n'} = n|E| = \int_E n(\det \nabla T(x))^{1/n}\, dx \leq \int_E div T(x)\, dx$$

$$= \int_{\partial E} <T, v_E>\, d\mathcal{H}^{n-1} \leq \int_{\partial E} ||T||h_K(v_E)d\mathcal{H}^{n-1} \leq P_K(E).$$

Where the last inequalities were obtained applying Cauchy–Schwarz inequality and $||T|| \leq 1$.

4 Estimating the Isoperimetric Deficit

From the proof of the isoperimetric inequality, we conclude that

$$|K|\delta(E, K) \geq \int_E \frac{div T(x)}{n} - (\det \nabla T(x))^{1/n}\, dx = \int_E (\lambda_A - \lambda_G)\, dx \quad (8)$$

Now, since $A(E, K)$ is controlled by $\int_{\partial E} |x - T(x)|\, d\mathcal{H}^{n-1}$, we would like to estimate it by using Poincaré inequality mentioned above. Thus, first we estimate the following expression:

$$\int_E |\nabla T(x) - Id|\, dx = \int_E \sqrt{\sum_{k=1}^{n} (\lambda_k - \lambda_G)^2}.$$

Using a refinement of the Arithmetic-Geometric mean inequality proved by Alzer in [1] we conclude:

$$\sum_{k=1}^{n} (\lambda_k - \lambda_G)^2 \leq 2n\lambda_n(\lambda_A - \lambda_G).$$

Applying this inequality we write:

$$\int_E |\nabla T(x) - Id|\, dx \leq \int_E \sqrt{2n\lambda_n(\lambda_A - \lambda_G)} \leq \sqrt{2n\|\lambda_n\|_{L^1(E)}|K|\delta(E, K)}, \quad (9)$$

where the last inequality was obtained by applying Hölder's inequality and using (8). Now we need an upper bound for $\|\lambda_n\|_{L^1(E)}$. Notice that

$$\|\lambda_n\|_{L^1(E)} \leq \int_E div T(x)\, dx \leq P_K(E) = n|K|(\delta(E, K) + 1) \leq 2n|K|,$$

where in the last inequality we assumed the isoperimetric deficit to be smaller than 1. This assumption is natural, as otherwise Theorem 2.3 holds in a trivial way. Substitue it in (9) and conclude:

$$\int_E |\nabla T(x) - Id|\, dx \leq \sqrt{4n^2|K|^2\delta(E, K)} = 2n|K|\sqrt{\delta(E, K)}. \quad (10)$$

Now, we can apply the mentioned Poincaré trace-type inequality (4) on (10) entrywise, to get the following (up to translation of K):

$$\int_E |\nabla T(x) - Id|\, dx \geq \frac{C}{n^{1.5} d_E} \int_{\partial E} |T(x) - x|\, d\mathcal{H}^{n-1}, \quad (11)$$

Where we replaced R/r with d_E, which is the Banach-Mazur distance of the body E from l_2^n, by choosing the correct position for E. Combining (10) and (11), we come to:

$$Cn^{2.5}d_E|K|\sqrt{\delta(E, K)} \geq \int_{\partial E} |T(x) - x|d\mathcal{H}^{n-1} \tag{12}$$

As shown in [8], the right hand side of (12) controls $A(E, K)$. This is done by replacing the Brenier map T, with the projection $P : \mathbb{R}^n \setminus K \to \partial K$. Clearly,

$$\int_{\partial E} |T(x) - x|d\mathcal{H}^{n-1} \geq \int_{\partial E \setminus K} |P(x) - x|d\mathcal{H}^{n-1}. \tag{13}$$

Now, consider the map $\Phi : (\partial E \setminus K) \times (0, 1) \to E \setminus K$, defined by:

$$\Phi(x, t) = tx + (1 - t)P(x). \tag{14}$$

Also, consider the tangent space to ∂E at x to be $\{\epsilon_k(x)\}_1^{n-1}$. Now, since Φ is a bijection, we can write:

$$|E \setminus K| = \int_0^1 dt \int_{\partial E \setminus K} |(x-P(x)| \wedge (\bigwedge_{k=1}^{n-1} (t\epsilon_k(x) + (1-t)dP_x(\epsilon_k(x))))d\mathcal{H}^{n-1}(x), \tag{15}$$

where dP_x denotes the differential of the projection P at x. Since P is a projection over a convex set, it decreases distances, which means that $|dP_x(e)| \leq 1$ for $e \in S^{n-1}$. Hence

$$|t\epsilon_k(x) + (1-t)dP_x(\epsilon_k(x))| \leq 1, \quad \forall 1 \leq k \leq n-1.$$

Combining it all together, we write:

$$A(E, K) \leq \frac{|E \Delta K|}{|K|} = 2\frac{|E \setminus K|}{|K|} \leq \frac{1}{|K|} \int_{\partial E \setminus K} |P(x) - x|d\mathcal{H}^{n-1}$$

$$\leq \frac{1}{|K|} \int_{\partial E} |T(x) - x|d\mathcal{H}^{n-1} \leq Cn^{2.5}d_E\sqrt{\delta(E, K)}.$$

By John's theorem [11] we know $d_E \leq n$, so we may write that $A(E, K) \leq Cn^{3.5}\sqrt{\delta(E, K)}$, as stated in Theorem 2.3.

Remark 4.1. Notice that in case E is a symmetric body, we get a slightly improved result, $A(E, K) \leq Cn^3\sqrt{\delta(E, K)}$, since $d_E \leq \sqrt{n}$.

4.1 Brunn–Minkowski Refinement

Now, when we have a refinement for the isoperimetric inequality, using Theorem 2.3 we can easily get the same refinement for the Brunn–Minkowski inequality, by repeating the argument in [9] (Sect. 4). We outline the proof for the sake of completeness. We show that given two convex bodies E and K,

$$A(E, K) \leq C n^{3.5} \sqrt{\beta(E, K)}, \tag{16}$$

where $\beta(E, K)$ is the Brunn–Minkowski deficit, defined in (3). To prove this, we use the following property of $P_K(E)$:

$$P_E(G) + P_K(G) = P_{E+K}(G), \tag{17}$$

which can be easily verified using the definition. Let us examine the case where $|E| = |K| = |G| = 1$, $A(E, G) = |E \Delta G|$ and $A(G, K) = |G \Delta K|$:

$$A(E, K) \leq |E \Delta K| \leq |E \Delta G| + |G \Delta K| = A(E, G) + A(G, L) \tag{18}$$

By the first part of Theorem 1.2 we may write:

$$P_E(E + K) \geq n |E|^{1/n} |E + K|^{1/n'} \left(1 + \left(\frac{A(E + K, E)}{C n^{3.5}} \right)^2 \right), \tag{19}$$

$$P_K(E + K) \geq n |K|^{1/n} |E + K|^{1/n'} \left(1 + \left(\frac{A(E + K, K)}{C n^{3.5}} \right)^2 \right). \tag{20}$$

Adding up the two inequalities, applying (17) and the fact that $n |E + K| = P_{E+K}(E + K)$, we find that

$$C n^{3.5} \sqrt{\beta(E, K)} \geq A(E, K),$$

which completes the proof of Theorem 1.2.

Remark 4.2. Notice that $n^{3.5}$ is an estimate of the worst case that is achieved only when the bodies have a large Banach-Mazur distance from l_2^n. The general estimate can be written as $C n^{2.5} \max\{d_E, d_K\} \sqrt{\beta(E, K)} \geq A(E, K)$.

5 Some Special Cases

We outline two special cases of the discussed inequalities which are of special interest in the field of high dimensional convex geometry.

5.1 Urysohn's Inequality Refinement

Denote by $h_K(\cdot)$ the support functional of convex body K with $0 \in int(K)$. Then, the width of K in some direction $\nu \in S^{n-1}$ is given by $h_K(\nu) + h_K(-\nu)$, and the mean width of K is given by

$$\omega(K) = \int_{S^{n-1}} h_K(\nu) + h_K(-\nu)d\sigma(\nu) = 2 \int_{S^{n-1}} h_K(\nu)d\sigma(\nu),$$

where σ is the normalized Lebesgue measure on the sphere S^{n-1}. Notice that by definition (5) of the anisotropic perimeter $P_K(B_n)$, where B_n is the n-dimensional Euclidean unit ball, we may write:

$$P_K(B_n) = \int_{S^{n-1}} h_K(\nu)d\mathcal{H}^{n-1} = \frac{\omega_{n-1}}{2}\omega(K), \qquad (21)$$

where ω_{n-1} is the surface area of S^{n-1}. Substituting the formula above in the isoperimetric inequality (6) for $P_K(B_m)$ we get:

$$\frac{\omega_{n-1}}{2}\omega(K) \geq n|K|^{1/n}|B_n|^{1-1/n}.$$

Using the fact that $\omega_{n-1} = n|B_n|$, we come to the well known Urysohn's inequality (see [3] pp. 93–94):

$$\frac{1}{2}\omega(K) \geq \left(\frac{|K|}{|B_n|}\right)^{\frac{1}{n}}. \qquad (22)$$

Notice that the result of Theorem 2.3 can be rewritten in the form:

$$\frac{P_K(E)}{n|E|} \geq \left(\frac{|K|}{|E|}\right)^{\frac{1}{n}}\left(1 + \left(\frac{A(E,K)}{Cn^{3.5}}\right)^2\right).$$

Thus, for the case where $E = B_n$ we get a refinement for Urysohn's inequality:

$$\frac{1}{2}\omega(K) \geq \left(\frac{|K|}{|B_n|}\right)^{\frac{1}{n}}\left(1 + \left(\frac{A(B_n,K)}{Cn^{3.5}}\right)^2\right). \qquad (23)$$

Actually, it is easy to notice that the trace type inequality in Lemma 2.1 holds with a constant n in our case since the Banach-Mazur distance d_{B_n} equals 1. So we may rewrite (23) with a slightly better constant:

$$\frac{1}{2}\omega(K) \geq \left(\frac{|K|}{|B_n|}\right)^{\frac{1}{n}}\left(1 + \left(\frac{A(B_n,K)}{Cn^{2.5}}\right)^2\right).$$

5.2   The Case of K and $-K$

Given two convex bodies K, L define their maximal intersection by:

$$M(K, L) = \max_{x \in \mathbb{R}^n} |K \cap (L + x_0)|. \tag{24}$$

Notice that whenever $|K| = |L|$ we have that $A(K, L)|K| + 2M(K, L) = |K| + |L| = 2|K|$. Thus, in this case,

$$A(K, L) = 2 \left(1 - \frac{M(K, L)}{|K|} \right) \tag{25}$$

Notice that the constant C which appears in Theorem 1.2, satisfies $C \leq \frac{4\sqrt{2}}{\log 2}$. This is easily verified by following the proof of the theorem.

Let us assume now that the Brunn–Minkowski deficit satisfies: $\beta(K, -K) \leq \alpha n^{-7}$, for some $\alpha > 0$. Then, using the result in Theorem 1.2 we may write:

$$A(K, -K) \leq C n^{3.5} \sqrt{\beta(K, -K)} \leq C \sqrt{\alpha},$$

or equivalently, using (25) and the upper bound for C:

$$M(K, -K) \geq \left(1 - \frac{C}{2} \sqrt{\alpha} \right) |K| \geq \left(1 - \frac{2\sqrt{2}}{\log 2} \sqrt{\alpha} \right) |K|.$$

The last inequality implies that given a convex body K that has a small enough Brunn–Minkowski deficit, then we can choose our origin in such way that $|K \cap (-K)|$ is equivalent to $|K|$. In other words, K must be close to a symmetric body. For example, if C^{-2} is an admissible value for α, then we have that $|K \cap (-K)| \geq \frac{1}{2}|K|$.

6   Relation to Dar's Conjecture

Another possible refinement of the Brunn–Minkowski inequality was proposed and proven to hold in some special cases by Dar in [6]. Given two convex bodies K, L the conjecture states the following:

$$|K + L|^{1/n} \geq M(K, L)^{1/n} + \frac{|K|^{1/n}|L|^{1/n}}{M(K, L)^{1/n}}, \tag{26}$$

where $M(K, L)$ is the volume of the maximal intersection, defined in (24). Let us check that this conjecture implies a stability result equivalent to Theorem 1.2

with better constants. We will assume, as always, that $|K| = |L| = 1$ and denote $\epsilon := \beta(K, L)$. Then, in this case, Dar's conjecture implies:

$$2(1 + \epsilon) \geq M(K, L)^{1/n} + \frac{1}{M(K, L)^{1/n}}.$$

Solving this inequality for small values of ϵ we get the stability result:

$$M(K, L)^{1/n} \geq 1 - C\sqrt{\beta(K, L)}.$$

Applying the fact that $M(K, L) = 1 - \frac{1}{2}A(K, L)$ together with Bernoulli's inequality, we get:

$$A(K, L) \leq Cn\sqrt{\beta(K, L)},$$

which is a stability result of the same spirit as in Theorem 1.2, except for an improved dimensional constant.

References

1. H. Alzer, A new refinement of the arithmetic mean-geometric mean inequality. Rocky Mt. J. Math. **27**(3), 663–667 (1997)
2. Y. Brenier, Polar factorization and monotone rearrangement of vector-valued functions. Comm. Pure Appl. Math. **44**(4), 375–417 (1991)
3. Yu.D. Burago, V.A. Zalgaller, *Geometric Inequalities*. A series of comprehensive studies in mathematics (Springer-Verlag, Berlin, 1980)
4. L.A. Caffarelli, The regularity of mappings with a convex potential. J. Am. Math. Soc. **5**(1), 99–104 (1992)
5. L.A. Caffarelli, Boundary regularity of maps with convex potentials. II. Ann. of Math. (2) **144**(3), 453–496 (1996)
6. S. Dar, A Brunn Minkowski type inequality. Geometria Dedicata **77**, 1–9 (1999)
7. V.I. Diskant, Stability of the solution of a Minkowski equation (Russian). Sibirsk. Mat. Z. **14**, 669–673, 696 (1973)
8. A. Figalli, F. Maggi, F. Pratelli, A refined Brunn-Minkowski inequality for convex sets. Ann. Inst. H. Poincaré Anal. Non Linaire **26**(6), 2511–2519 (2009)
9. A. Figalli, F. Maggi, F. Pratelli, A mass transportation approach to quantitative isoperimetric inequalities. Invent. Math. **182**(1), 167–211 (2010)
10. R. Gardner, *Geometric Tomography*, 2nd edn. (Cambridge University Press, London, 2006)
11. F. John, An inequality for convex bodies. Univ. Kentucky Res. Club Bull. **8**, 8–11 (1942)
12. R. Schneider, *Convex Bodies: The Brunn Minkowski Theory* (Cambridge University Press, London, 1993)
13. C. Villani, in *Topics in Optimal Transportation*. Graduate Studies in Mathematics, vol. 58 (American Mathematical Society, Providence, 2003)

On Contact Points of Convex Bodies

Nikhil Srivastava

Abstract We show that for every convex body K in $\mathbb{R}^n$, there is a convex body H such that

$$H \subset K \subset c \cdot H \qquad \text{with } c = 2.24$$

and H has at most $O(n)$ contact points with the minimal volume ellipsoid that contains it. When K is symmetric, we can obtain the same conclusion for every constant $c > 1$. We build on work of Rudelson [Israel J. Math. **101**(1), 92–124 (1997)], who showed the existence of H with $O(n \log n)$ contact points. The approximating body H is constructed using the "barrier" method of Batson, Spielman, and the author, which allows one to extract a small set of vectors with desirable spectral properties from any John's decomposition of the identity. The main technical contribution of this paper is a way of controlling the *mean* of the vectors produced by that method, which is necessary in the application to John's decompositions of nonsymmetric bodies.

1 Introduction

Let K be an arbitary convex body in $\mathbb{R}^n$ and let $\mathcal{E}$ be a minimal volume ellipsoid containing K. Then the *contact points* of K are the points of intersection of $\mathcal{E}$ and K. The ellipsoid $\mathcal{E}$ is unique and characterized by a celebrated theorem of John [1], which says that if K is embedded via an affine transformation in $\mathbb{R}^n$ so that

N. Srivastava (✉)
Mathematical Sciences Research Institute, 16 Gauss Way, Berkeley, CA 94720-5070, USA
e-mail: nikhils@math.ias.edu

B. Klartag et al. (eds.), *Geometric Aspects of Functional Analysis*, Lecture Notes in Mathematics 2050, DOI 10.1007/978-3-642-29849-3_25,
© Springer-Verlag Berlin Heidelberg 2012

$\mathcal{E}$ becomes the standard Euclidean ball B_2^n, then there are $m \leq N = n(n+3)/2$ contact points $x_1, \ldots, x_m \in K \cap B_2^n$ and nonnegative weights $c_1, \ldots c_m$ for which

$$\sum_i c_i x_i = 0 \qquad \text{(mean zero)} \qquad (1)$$

$$\sum_i c_i x_i x_i^T = I \qquad \text{(inertia matrix identity)}. \qquad (2)$$

Moreover, any convex body K' containing $x_1, \ldots, x_m$ must have B_2^n inside its John ellipsoid. We refer to a weighted collection of unit vectors $(c_i, x_i)_{i \leq m}$ which satisfies (1) and (2) as a *John's decomposition* of the identity.

The study of contact points has been fruitful in convex geometry, for instance in understanding the behavior of volume ratios of symmetric and nonsymmetric convex bodies [1] and in estimating distances between convex bodies and the cube or simplex [3,5]. In this work, we consider the number of contact points of a convex body. Define a distance d between two (not necessarily symmetric) convex bodies K and H in $\mathbb{R}^n$ as follows[1]:

$$d(K, H) = \inf_{T \in GL(n), u, v \in \mathbb{R}^n} \{c : H + u \subset T(K + v) \subset c(H + u)\}$$

and let $\mathcal{K}$ be the space of all convex bodies equipped with the topology induced by d. Gruber [4] proved that the set of K having fewer than $N = n(n+3)/2$ contact points is of the first Baire category in $\mathcal{K}$. However, Rudelson has shown that every K is arbitrarily close to a body which has a much smaller number of contact points.

Theorem 1 (Rudelson [6]). *Suppose K is a convex body in $\mathbb{R}^n$ and $\epsilon > 0$. Then there is a convex body H such that $d(H, K) \leq 1 + \epsilon$ and H has at most $m \leq Cn \log n/\epsilon^2$ contact points, where C is a universal constant.*

In this note, we show that the $\log n$ factor in Rudelson's theorem is unnecessary in many cases. For symmetric convex bodies, we obtain exactly the same distance guarantee $d(H, K) \leq 1 + \epsilon$ but with a much smaller number $m \leq 32n/\epsilon^2$ of contact points of H. For arbitrary convex bodies, we show a somewhat weaker result that only guarantees an H within constant distance $d(H, K) \leq 2.24$, with $m \leq Cn$ contact points for some universal C. Thus Rudelson's $O(n \log n)$ bound is still the best known in the regime $d(H, K) < 2.24$ for nonsymmetric bodies.

Our approach for constructing H is the same as Rudelson's, and consists of two steps:

1. Given a John's decomposition $(c_i, x_i)_{i \leq m}$ for K, extract a small subsequence of points x_i which are *approximately* a John's decomposition. To be precise,

---

[1] When K and H are symmetric then we can take $u = v = 0$ and d becomes the usual Banach-Mazur distance.

find a set of scalars b_i, at most $s = O(n)$ of which are nonzero, and a small 'recentering' vector u for which

$$\sum_i b_i (x_i + u) = 0 \qquad \left\| I - \sum_i b_i (x_i + u)(x_i + u)^T \right\| \le \epsilon.$$

2. Use the approximate John's decomposition $(b_i, x_i + u)_{i \le s}$ to construct an *exact* John's decomposition $(a_i, u_i)_{i \le s}$, and show that it characterizes the John Ellipsoid of a body H that is close to K.

The source of our improvement is a new method for extracting the approximate decomposition in step (1). Whereas the b_i were chosen by random methods in Rudelson's work, we now use a deterministic procedure which was introduced in recent work of Batson et al. on spectral sparsification of graphs [2]. The main theorem of [2] is the following:

Theorem 2 (Batson, Spielman, Srivastava). *Suppose $d > 1$ and $\mathbf{v}_1, \mathbf{v}_2, \ldots, \mathbf{v}_m$ are vectors in $\mathbb{R}^n$ with*

$$\sum_{i \le m} \mathbf{v}_i \mathbf{v}_i^T = I.$$

Then there exist scalars $s_i \ge 0$ with $|\{i : s_i \ne 0\}| \le dn$ so that

$$I \preceq \sum_{i \le m} s_i \mathbf{v}_i \mathbf{v}_i^T \preceq \left(\frac{\sqrt{d} + 1}{\sqrt{d} - 1} \right)^2 I.$$

A sharp result regarding the contact points of *symmetric* convex bodies can be derived as an immediate corollary of Theorem 2 and Rudelson's proof of Theorem 1 [5].

Corollary 3. *If K is a symmetric convex body in $\mathbb{R}^n$ and $\epsilon > 0$, then there exists a body H such that $H \subset K \subset (1 + \epsilon)H$ and H has at most $m \le 32n/\epsilon^2$ contact points with its John Ellipsoid.*

Proof. Suppose K is a symmetric convex body whose John ellipsoid is B_2^n, and let $X = \{x_1, \ldots x_m\}$ be contact points satisfying (1,2) with weights $c_1, \ldots c_m$. Since K is symmetric we can assume that $x_i \in X \iff -x_i \in X$, and that the corresponding weights c_i are equal.

We will extract an approximate John's decomposition from X. Apply Theorem 2 to the vectors $\mathbf{v}_i = \sqrt{c_i} x_i$ with parameter $d = 16/\epsilon^2$ to obtain scalars s_i, and let $Y \subset X$ be the set of x_i with nonzero s_i. We are now guaranteed that

$$I \preceq \sum_{x_i \in Y} s_i c_i x_i x_i^T \preceq \left(\frac{4/\epsilon + 1}{4/\epsilon - 1} \right)^2 I.$$

with $|Y| \leq 16n/\epsilon^2$. Notice that by an easy calculation

$$\left(\frac{4/\epsilon + 1}{4/\epsilon - 1}\right)^2 \leq 1 + \epsilon$$

for sufficiently small epsilon.

In order to obtain a John's decomposition from these vectors, we need to ensure the mean zero condition (1). This is achieved easily by taking a negative copy of each vector in Y and halving the scalars s_i, since

$$\sum_{x_i \in Y} (s_i/2) c_i x_i + (s_i/2) c_i (-x_i) = 0$$

and

$$\sum_{x_i \in Y} (s_i/2) c_i x_i x_i^T + (s_i/2) c_i (-x_i)(-x_i)^T = \sum_{x_i \in Y} s_i c_i x_i x_i^T$$

which we know is a good approximation to the identity. Thus the vectors in $Y \cup -Y$ with weights $b_i = s_i c_i/2$ on x_i and $-x_i$ constitute a $(1 + \epsilon)$-approximate John's decomposition with only $32n/\epsilon^2$ points. Substituting this fact in place of [5, Lemma 3.1] in the proof of [5, Theorem 1.1] gives the promised result.   □

When the body K is not symmetric, there is no immediate way to guarantee the mean zero condition. If we simply recenter the vectors produced by Theorem 2 to have mean zero by adding $u = -\frac{\sum_i b_i x_i}{\sum_i b_i}$ to each x_i, then the corresponding inertia matrix is

$$\sum_i b_i (x_i + u)(x_i + u)^T = \sum_i b_i x_i x_i^T - \left(\sum_i b_i\right) uu^T \qquad (3)$$

which no longer well-approximates the identity if $\| \left(\sum_i b_i\right) uu^T \|$ is large. This is the issue that we address here. In Sect. 2, we prove a variant of Theorem 2 which allows us to obtain very good control on the mean u at the cost of having a worse (at best factor 4, rather than $1 + \epsilon$) approximation of the inertia matrix to the identity. In Sect. 3, we show that this is still sufficient to carry out Rudelson's construction of the approximating body H. The end result is the following theorem.

Theorem 4. *For every $\epsilon > 0$ the following is true for n sufficiently large. If K is a convex body in $\mathbb{R}^n$, then there is a convex body H such that*

$$H \subset K \subset (\sqrt{5} + \epsilon)H$$

and H has at most $O_\epsilon(n)$ contact points with its John Ellipsoid.

2 Approximate John's Decompositions

In this section we will prove the following theorem.

Theorem 5. *Suppose we are given a John's decomposition of the identity, i.e., unit vectors $x_1, \ldots, x_m \in \mathbb{R}^n$ with nonnegative scalars c_i such that*

$$\sum_i c_i x_i = 0 \tag{4}$$

$$\sum_i c_i x_i x_i^T = I. \tag{5}$$

Then for every $\epsilon > 0$ there are scalars b_i, at most $O_\epsilon(n)$ nonzero, and a vector u such that

$$I \preceq \sum_i b_i (x_i + u)(x_i + u)^T \preceq (4 + \epsilon)I \tag{6}$$

$$\sum_i b_i (x_i + u) = 0 \tag{7}$$

$$\left(\sum_i b_i \right) \|u\|^2 \leq \epsilon. \tag{8}$$

We remark that the requirement (4) is necessary to allow a useful bound on u, since otherwise we can take $x_i = e_i$ with $c_i = 1$ for the canonical basis vectors $\{e_i\}_{i \leq n}$ and it is easily checked (for instance, using concavity of λ_{min}) that

$$\sum_i b_i e_i e_i^T - \frac{\left(\sum_i b_i e_i \right)\left(\sum_i b_i e_i \right)}{\sum_i b_i}$$

is singular for every choice of scalars b_i, which is worthless considering (3).

2.1 An Outline of the Proof

As in the proof of Theorem 2 [2], we will build the approximate John's decomposition $(b_i, x_i + u)$ by an iterative process which adds one vector at a time. At any step of the process, let

$$A = \sum_j b_j x_j x_j^T$$

denote the inertia matrix of the vectors that have already been added (i.e., the b_i's that have been set to some nonzero value), and let

$$z = \sum_j b_j x_j$$

denote their weighted sum. Initially both $A = 0$ and $z = 0$. We will take $s = O(n)$ steps, adding one $b_j v_j v_j^T$ in each step, in a way that guarantees that at the end the inertia matrix A approximates the identity and the sum z is close to zero. Since we only increment one b_i in each step, at most $O(n)$ will be nonzero at the end, as promised.

2.1.1 Barrier Functions

The choice of vector to add in each step will be guided by two "barrier" potential functions which we will use to maintain control on the eigenvalues of A. For real numbers $u, l \in \mathbb{R}$, which we will call the upper and lower barrier respectively, define:

$$\Phi^u(A) \overset{\text{def}}{=} \text{Tr}(uI - A)^{-1} = \sum_i \frac{1}{u - \lambda_i} \quad \text{(Upper potential)}.$$

$$\Phi_l(A) \overset{\text{def}}{=} \text{Tr}(A - lI)^{-1} = \sum_i \frac{1}{\lambda_i - l} \quad \text{(Lower potential)},$$

where $\lambda_1, \ldots, \lambda_n$ are the eigenvalues of A.

As long as $A \prec uI$ and $A \succ lI$ (i.e., $\lambda_{\max}(A) < u$ and $\lambda_{\min}(A) > l$), these potential functions measure how far the eigenvalues of A are from the barriers u and l. In particular, they blow up as any eigenvalue approaches a barrier, since then $uI - A$ (or $A - lI$) approaches a singular matrix. Thus if $\Phi_l(A)$ and $\Phi^u(A)$ are appropriately bounded, we can conclude that the eigenvalues of A are 'well-behaved' in that there is no accumulation near the barriers u and l. This will allow us to prove that the process never gets stuck.

For a thorough discussion of these potential functions, where they come from, and why they work, see [2].

2.1.2 Invariants

We will maintain three invariants throughout the process. Note that u, l, A and z vary from step to step, while P_U, P_L, and ϵ remain fixed.

- The eigenvalues of A lie strictly between l and u:

$$lI \prec A \prec lU. \tag{9}$$

- Both the upper and lower potentials are bounded by some fixed values P_L and P_U:

$$\Phi_l(A) \leq P_L \qquad \Phi^u(A) \leq P_U. \tag{10}$$

- The running sum z is appropriately small:

$$\|z\|^2 \le \epsilon \cdot \text{Tr}(A) \tag{11}$$

2.1.3 Initialization

At the beginning of the process we have $A = 0$ and $z = 0$, and the barriers at initial values

$$l = l_0 = -1 \quad \text{and} \quad u = u_0 = 1. \tag{12}$$

It is easy to see that (9)–(11) all hold with $P_U = P_L = n$ at this point.

2.1.4 Maintenance

The process will evolve in steps. Each step will consist of adding two vectors, tv and rw, where $t, r \ge 0$ and $v, w \in \{x_i\}_{i \le m}$. We will call these the main vector and the fix vector, respectively.

The main vector will allow us to move the upper and lower barriers forward by fixed amounts $\delta_l > 0$ and $\delta_u > 0$ while maintaining the invariants (9) and (10); in particular, we will choose it in a manner which satisfies:

$$\Phi_{l+\delta_l}(A + tvv^T) \le \Phi_l(A) \qquad \Phi^{u+\delta_u}(A + tvv^T) \le \Phi^u(A)$$
$$(l + \delta_l)I \prec A + tvv^T \prec (u + \delta_u)I. \tag{13}$$

The fix will correct any undesirable impact that the main has on the sum; specifically, we will choose rw in a way that guarantees

$$\langle z, tv + rw \rangle \le 0 \tag{14}$$

where z is the sum at the end of the previous step. Thus the net increase in the length of the sum in any step is given by

$$\|z + (tv + rw)\|^2 = \|z\|^2 + 2\langle z, tv + rw \rangle + \|tv + rw\|^2 \le \|z\|^2 + (t + r)^2, \tag{15}$$

since v and w are unit vectors. The corresponding increase in the trace is simply $t + r$, and so if we guarantee in addition that the steps are sufficiently small:

$$t + r \le \epsilon \tag{16}$$

then the invariant (11) can be maintained by induction as follows:

$$\frac{\|z + (tv + rw)\|^2}{\mathrm{Tr}(A + tvv^T + rww^T)} \leq \frac{\|z\|^2 + (t + r)^2}{\mathrm{Tr}(A) + (t + r)} \qquad \text{by (15)}$$

$$\leq \max\left\{\frac{\|z\|^2}{\mathrm{Tr}(A)}, \frac{(t + r)^2}{t + r}\right\}$$

$$\leq \epsilon \qquad \text{by (16).}$$

However, we need to make sure that adding the fix vector does not cause us to violate (9) or (10). To do this, the addition of rw will be accompanied by an appropriately large shift of $\delta_u^f > 0$ in the upper barrier. In particular, we will make sure that on top of satisfying (14), rw also satisfies

$$\Phi^{u+\delta_u+\delta_u^f}(A + tvv^T + rww^T) \leq \Phi^{u+\delta_u}(A + tvv^T)$$

$$\text{and} \quad A + tvv^T + rww^T \prec (u + \delta_u + \delta_u^f)I. \tag{17}$$

Since $A + tvv^T + A + rww^T \succeq A + tvv^T$, the analogous bound for the lower potential follows immediately without any additional lower shift:

$$\Phi_{l+\delta_l+0}(A + tvv^T + rww^T) \leq \Phi_{l+\delta_l}(A + tvv^T)$$

$$\text{with} \quad A + tvv^T + rww^T \succ (l + \delta_l)I.$$

Together with (13), these inequalities guarantee that

$$\Phi_{l+\delta_l}(A + tvv^T + rww^T) \leq \Phi_l(A) \leq P_L$$

and

$$\Phi^{u+\delta_u+\delta_u^f}(A + tvv^T + rww^T) \leq \Phi^u(A) \leq P_U,$$

thus maintaining both (9) and (10), as desired.

To summarize what has occured during the step: we have added two vectors tv and rw and shifted u and l forward by $\delta_u + \delta_u^f$ and δ_l, respectively, in a manner that our three invariants continue to hold. To show that such a step can actually be taken, we need to prove that as long as the invariants are maintained there must exist scalars $t, r \geq 0$ and vectors $v, w \in \{x_i\}_{i \leq m}$ which satisfy all of the conditions (13), (14), (16), and (17). We will do this in Lemma 6.

2.1.5 Termination

After s steps of the process, we have

$$(-1 + s\delta_l)I \prec A \prec (1 + s(\delta_u + \delta_u^f))I \qquad \text{by (9).}$$

If we take s sufficiently large, then we can make the ratio $\lambda_{\max}(A)/\lambda_{\min}(A)$ arbitrarily close to $\frac{\delta_u+\delta_u^f}{\delta_l}$. In the actual proof, which we will present shortly, we will show that the process can be realized with parameters $\delta_l, \delta_u, \delta_u^f$ for which this ratio can be made arbitrarily close to 4 in $s = O(n)$ steps.

As for the mean, we will set $u = -\frac{\sum_j b_j x_j}{\sum_j b_j} = -\frac{z}{\mathrm{Tr}(A)}$ at the end of the process, so that immediately

$$(\sum_j b_j)\|u\|^2 = \frac{\|z\|^2}{\mathrm{Tr}(A)} \leq \epsilon \qquad (18)$$

by (11) as desired.

2.2 Realizing the Proof

To complete the proof, we will identify a range of parameters $\delta_l, \delta_u, \delta_u^f, P_U, P_L$, and ϵ for which the above process can actually be sustained.

Lemma 6 (One Step). *Suppose $(c_i, x_i)_{i \leq m}$ is a John's decomposition and z is any vector. Let $A \succeq 0$ be a matrix satisfying the invariants (9) and (10). If*

$$\frac{1}{\delta_u} + \frac{1}{\delta_u^f} + 2P_U + P_L + \frac{4n}{\epsilon} \leq \frac{1}{\delta_l} \qquad (19)$$

Then there are scalars $t, r \geq 0$ and vectors $v, w \in \{x_i\}$ which satisfy (13), (14), (16), and (17).

To this end, we recall the following lemmas from [2], which characterize how much of a vector one can add to a matrix without increasing the upper and lower potentials.

Lemma 7 (Upper Barrier Shift). *Suppose $A \prec uI$ and $\delta_u > 0$. Then there is a positive definite matrix $\mathbb{U} = \mathbb{U}(A, u, \delta_u)$ so that if v is any vector which satisfies*

$$v^T \mathbb{U} v \geq \frac{1}{t}$$

then

$$\Phi^{u+\delta_u}(A + tvv^T) \leq \Phi^u(A) \quad \text{and} \quad \lambda_{\max}(A + tvv^T) < u + \delta_u.$$

That is, if we add t times vv^T to A and shift the upper barrier by δ_u, then we do not increase the upper potential.

Lemma 8 (Lower Barrier Shift). *Suppose $A \succ lI$, $\delta_l > 0$, and $\Phi_l(A) < 1/\delta_l$. Then there is a matrix $\mathbb{L} = \mathbb{L}(A, l, \delta_l)$ so that if v is any vector which satisfies*

$$v^T \mathbb{L} v \leq \frac{1}{t}$$

then
$$\Phi_{l+\delta_l}(A + t\mathbf{v}\mathbf{v}^T) \leq \Phi_l(A) \quad and \quad \lambda_{\min}(A + t\mathbf{v}\mathbf{v}^T) > l + \delta_l.$$

That is, if we add t times $\mathbf{v}\mathbf{v}^T$ to A and shift the lower barrier by δ_l, then we do not increase the lower potential.

We will prove that desirable vectors exist by taking averages of the quantities $\mathbf{v}^T\mathbb{U}\mathbf{v}$ and $\mathbf{v}^T\mathbb{L}\mathbf{v}$ over our contact points $\{x_i\}$ with weights $\{c_i\}$. Since

$$\sum_i c_i x_i^T \mathbb{U} x_i = \mathbb{U} \bullet \left(\sum_i c_i x_i x_i^T \right)$$

$$= \mathbb{U} \bullet I \qquad\qquad \text{by (2)}$$

$$= \text{Tr}(\mathbb{U})$$

(and similarly for $\mathbb{L}$), it will be useful to recall bounds on $\text{Tr}(\mathbb{U})$ and $\text{Tr}(\mathbb{L})$ from [2]. Crucially, these bounds do not depend on the matrix A or on u, l at all, but only on the shifts δ_u and δ_l and on the potentials.

Lemma 9 (Traces of $\mathbb{L}$ and $\mathbb{U}$). *If $lI \prec A \prec uL$ with $\Phi_l(A) \leq P_L$ and $\Phi^u(A) \leq P_U$ then*
$$\text{Tr}(\mathbb{U}) \leq \frac{1}{\delta_u} + P_U$$

and
$$\text{Tr}(\mathbb{L}) \geq \frac{1}{\delta_l} - P_L.$$

We are now in a position to prove Lemma 6.

Proof of Lemma 6. Let $\mathbb{L} = \mathbb{L}(A, l, \delta_l), \mathbb{U} = \mathbb{U}(A, u, \delta_u), \mathbb{U}^f = \mathbb{U}(A, u + \delta_u, \delta_u^f)$ be the matrices produced by Lemmas 7 and 8.

Let us focus on the main vector first. By Lemmas 7 and 8, we can add $t\mathbf{v}$ without increasing potentials if
$$v^T\mathbb{U}v \leq 1/t \leq v^T\mathbb{L}v.$$

In fact, we will insist on v for which

$$v^T\mathbb{U}v + 2/\epsilon \leq 1/t \leq v^T\mathbb{L}v$$

as this will ensure that we can take $t \leq \epsilon/2$.

Let $D(v) = v^T\mathbb{L}v - v^T\mathbb{U}v - 2/\epsilon$ and call $\mathcal{F} = \{x_i : D(x_i) \geq 0\}$ the set of *feasible* vectors. Let $\mathcal{P} = \{x_i : \langle x_i, z \rangle > 0\}$ be the set of vectors with positive inner product with z, and let $\mathcal{N} = \{x_i : \langle x_i, z \rangle \leq 0\}$ be the vectors in the complementary halfspace.

We will always add as little of a main vector can, so we can assume that we take $1/t = v^T\mathbb{L}v$ whenever $v \in \mathcal{F}$. Here is the rule for choosing which v to add: choose the feasible v for which $t \langle v, z \rangle$ is minimized. If this quantity is negative then there is no need for a fix vector, and taking $w = 0$ we are done. Otherwise let $\alpha := \min\{t\langle v, z\rangle : v \in \mathcal{F}\}$ and notice that $\mathcal{F} \subset \mathcal{P}$:j

$$\frac{\langle v, z\rangle}{\alpha} \geq \frac{1}{t} = v^T \mathbb{L} v \quad \forall v \in \mathcal{F}. \tag{20}$$

Taking a sum, we find that

$$\sum_{\mathcal{P}} c_i \frac{\langle x_i, z\rangle}{\alpha} \geq \sum_{\mathcal{F}} c_i \frac{\langle x_i, z\rangle}{\alpha} \quad \text{since } \mathcal{F} \subset \mathcal{P}$$

$$\geq \sum_{\mathcal{F}} c_i x_i^T \mathbb{L} x_i \quad \text{by (20)}$$

$$\geq \sum_{\mathcal{F}} c_i D(x_i) \quad \text{since } \mathbb{U} \succeq 0 \text{ implies that } D(x_i) \leq x_i^T \mathbb{L} x_i$$

$$\geq \sum_{i} c_i D(x_i) \quad \text{since } D < 0 \text{ outside } \mathcal{F}$$

However, since $\sum_i c_i x_i = 0$ this implies that

$$\sum_{\mathcal{N}} c_i \frac{\langle x_i, -z\rangle}{\alpha} = \sum_{\mathcal{P}} c_i \frac{\langle x_l, z\rangle}{\alpha} \geq \sum_{i} c_i D(x_i). \tag{21}$$

We will use (21) to show that a suitable fix vector w exists. We are interested in finding a $w \in \{x_i\}$ and $r \geq 0$ for which

$$r\langle w, -z\rangle \geq \alpha \quad \text{(sufficient to reverse } \alpha = t\langle v, z\rangle\text{)—for (14)}$$

$$w^T \mathbb{U}^f w + 2/\epsilon \leq 1/r \quad \text{(upper barrier feasible with shift } \delta_u^f \text{ and } r \leq \epsilon/2\text{)}$$

$$\text{—for (16,17)}$$

Thus it suffices to find a w for which

$$w^T \mathbb{U}^f w + 2/\epsilon \leq \frac{\langle w, -z\rangle}{\alpha},$$

and then we can squeeze $1/r$ in between. Taking a weighted sum over all vectors of interest, it will be sufficient to show that

$$\sum_{\mathcal{N}} c_i x_i^T \mathbb{U}^f x_i + 2c_i/\epsilon \leq \sum_{\mathcal{N}} c_i \frac{\langle x_i, -z\rangle}{\alpha}.$$

For the left hand side we use the crude estimate

$$\sum_{\mathcal{N}} c_i x_i^T \mathbb{U}^f x_i + 2c_i/\epsilon \le \sum_{\mathcal{N} \cup \mathcal{P}} c_i x_i^T \mathbb{U}^f x_i + 2c_i/\epsilon = \text{Tr}(\mathbb{U}^f) + 2n/\epsilon$$

and for the right hand side we consider that

$$\sum_{\mathcal{N}} c_i \frac{\langle x_i, -z \rangle}{\alpha} \ge \sum_i c_i D(x_i) \qquad \text{by (21)}$$

$$= \sum_i c_i x_i^T \mathbb{L} x_i - c_i x_i^T \mathbb{U} x_i - 2c_i/\epsilon$$

$$= \text{Tr}(\mathbb{L}) - \text{Tr}(\mathbb{U}) - 2n/\epsilon \qquad \text{since } \sum_i c_i x_i x_i^T = I$$

Thus it will be enough to have

$$\text{Tr}(\mathbb{U}^f) + 2n/\epsilon \le \text{Tr}(\mathbb{L}) - \text{Tr}(\mathbb{U}) - 2n/\epsilon$$

which follows from our hypothesis (19) and Lemma 9.   □

Proof of Theorem 5. If we start with $l_0 = -1$ and $u_0 = 1$ then we can take $P_L = P_U = n$. Setting $\delta_u = \delta_u^f = (2 + \epsilon)\delta_l$, (19) reduces to

$$\frac{2}{(2 + \epsilon)\delta_l} + 3n + \frac{4n}{\epsilon} \le \frac{1}{\delta_l},$$

which it is easy to check is satisfied for small enough δ_l, in particular for

$$\delta_l = \frac{\epsilon^2}{10n}.$$

At the end of s steps we take

$$u = -\frac{\sum_i b_i x_i}{\sum_i b_i} = -\frac{z}{\text{Tr}(A)},$$

immediately satisfying (7) and (8). To finish the proof, we notice that

$$(-1 + s\delta_l - \frac{\|z\|^2}{\text{Tr}(A)})I \preceq \sum_i b_i (x_i + u)(x_i + u)^T = A - \text{Tr}(A) u u^T$$

$$\preceq (1 + s \cdot 2(2 + \epsilon)\delta_l)I.$$

Setting $s = 100n/\epsilon^3$ and replacing ϵ by $\epsilon/3$ yields (6), as promised.   □

3 Construction of the Approximating Body

The contents of this section are very similar to Rudelson [5, Sect. 4] but the calculations are more delicate because we must work with a $(4 + \epsilon)$-approximate John's decomposition rather than a $(1 + \epsilon)$-approximate one.

The main technical contribution is a generic procedure for turning approximate John's decompositions into approximating bodies:

Lemma 10. *Suppose K has contact points $(c_i, x_i)_{i \leq m}$ and $(b_i, y_i = x_i + u)_{i \leq s}$ are the vectors produced by Theorem 5, with*

$$A = \sum_i b_i y_i y_i^T$$

satisfying

$$\kappa(A) = \|A\| \cdot \|A^{-1}\| = O(1).$$

Then for n sufficiently large, there is a body H with at most s contact points and

$$d(H, K) \leq (1 + o(1)) \left(\kappa(A) \left(1 + \frac{(\sqrt{\kappa(A)} - 1)^2}{4} \right) \right)^{1/2}. \tag{22}$$

This immediately yields a proof of Theorem 4:

Proof of Theorem 4. Since the condition number $\kappa(A)$ guaranteed by Theorem 5 can be made arbitrarily close to 4, the number in (22) can be made arbitrarily close to

$$\left(4 \left(1 + \frac{1}{4} \right) \right)^{1/2} = \sqrt{5}$$

as desired. $\qquad \square$

Let $(b_i, y_i = x_i + u)_{i \leq s}$ be the approximate John's decomposition guaranteed by Theorem 5, with

$$\sum_i b_i y_i = 0$$

and

$$\sum_i b_i y_i y_i^T = A.$$

There are two problems with this: the y_i are not unit vectors, and their moment ellipsoid $\mathcal{E} = A^{1/2} B_2^n$ is not the sphere. We will adjust the vectors in a manner that fixes both these problems, to obtain an exact John's decomposition $(\hat{a}_i, \hat{u}_i)_{i \leq s}$.

Add a small vector v, to be determined later, to each y_i to obtain vectors

$$\hat{y}_i := y_i + v$$

with inertia matrix

$$\hat{A} := \sum_i b_i \hat{y}_i \hat{y}_i^T = \sum_i b_i y_i y_i^T + \sum_i b_i v v^T = A + \text{Tr}(A) v v^T. \quad (23)$$

(we will use $\hat{}$ to denote vectors that depend on v.) Let $\hat{R} := \hat{A}^{1/2}$ and let $\hat{E} := \hat{R} B_2^n$ be the corresponding moment ellipsoid. If we rescale each $\hat{y}_i$ to lie on $\hat{E}$, taking

$$\hat{z}_i := \frac{\hat{y}_i}{\|\hat{y}_i\|_{\hat{E}}} \qquad \text{where } \|\hat{y}_i\|_{\hat{E}} = \|\hat{R}^{-1}\hat{y}_i\|, \quad (24)$$

and then apply the inverse transformation $\hat{R}^{-1}$ which maps $\hat{E}$ to B_2^n, we obtain unit vectors

$$\hat{u}_i := \hat{R}^{-1}\hat{z}_i.$$

Moreover, if these are given weights

$$\hat{a}_i := b_i \|\hat{y}_i\|_{\hat{E}}^2$$

then we have an exact decomposition of the identity since

$$\sum_i \hat{a}_i \hat{u}_i \hat{u}_i^T = \hat{R}^{-1} \left(\sum_i b_i \|\hat{y}_i\|_{\hat{E}}^2 \frac{\hat{y}_i \hat{y}_i^T}{\|\hat{y}_i\|_{\hat{E}}^2} \right) \hat{R}^{-1} = \hat{R}^{-1} \hat{A} \hat{R}^{-1} = I. \quad (25)$$

In the following lemma, we show that there must exist a small v for which the weighted sum

$$\sum_i \hat{a}_i \hat{u}_i = \sum_i b_i \|\hat{y}_i\|_{\hat{E}}^2 \hat{R}^{-1} \frac{\hat{y}_i}{\|\hat{y}_i\|_{\hat{E}}} = \hat{R}^{-1} \left(\sum_i b_i \|\hat{y}_i\|_{\hat{E}} \hat{y}_i \right)$$

is equal to zero. This will complete the construction of $(\hat{a}_i, \hat{u}_i)_{i \leq s}$.

Lemma 11. *Let b_i, y_i, A, etc. be as above and suppose $\text{Tr}(A) = \Omega(n)$. Then there is a vector v with*

$$\|v\| \leq v_A := \frac{1 + o(1)}{2} \left(\sqrt{\kappa(A)} - 1 \right) \sqrt{\frac{\|A\|}{\text{Tr}(A)}} \quad (26)$$

for which

$$\sum_i b_i \|\hat{y}_i\|_{\hat{E}} \hat{y}_i = 0.$$

Proof. We need to find a v for which

$$\sum_i b_i \sqrt{(y_i + v)^T \hat{A}^{-1}(y_i + v)}(y_i + v) = 0.$$

As in [5], we will do this using the Brouwer fixed point theorem. In particular it will suffice to show that the function

$$F(v) = -\frac{\sum_i b_i \beta_i^{(v)} y_i}{\sum_i b_i \beta_i^{(v)}} \quad \text{where } \beta_i^{(v)} = \sqrt{(y_i + v)^T (A + \mathrm{Tr}(A)vv^T)^{-1}(y_i + v)}$$

$$= -\frac{\sum_i b_i (\beta_i^{(v)} - \mu) y_i}{\sum_i b_i \beta_i^{(v)}} \quad \text{for any } \mu \in \mathbb{R} \text{ since } \sum_i b_i y_i = 0$$

maps $v_A B_2^n$ to itself. We begin with the preliminary bounds

$$\beta_i^{(0)} = \sqrt{(x_i + u)^T A^{-1}(x_i + u)}$$

$$\leq \|x_i + u\| \sqrt{\|A^{-1}\|}$$

$$\leq (1 + o(1)) \sqrt{\|A^{-1}\|} \quad \text{since } \|x_i\| = 1 \text{ and } \|u\| \leq O(\mathrm{Tr}(A)^{-1/2}), \quad (27)$$

and similarly

$$\beta_i^{(0)} \geq (1 - o(1)) \frac{1}{\|A\|^{1/2}}, \quad (28)$$

for all i. We now have the estimate:

$$\|F(v)\| = \max_{\|w\|=1} \frac{\sum_i b_i (\beta_i^{(v)} - \mu)\langle y_i, w\rangle}{\sum_i b_i \beta_i^{(v)}}$$

$$\leq \frac{(\sum_i b_i (\beta_i^{(v)} - \mu)^2)^{1/2}}{\sum_i b_i \beta_i^{(v)}} \cdot \max_{\|w\|=1}\left(\sum_i b_i \langle y_i, w\rangle^2\right)^{1/2} \quad \text{by Cauchy–Schwarz}$$

$$= \frac{(\sum_i b_i (\beta_i^{(v)} - \mu)^2)^{1/2}}{\sum_i b_i \beta_i^{(v)}} \cdot \|A\|^{1/2}$$

$$\leq \frac{1}{1 - o(1)} \frac{(\sum_i b_i (\beta_i^{(v)} - \mu)^2)^{1/2}}{\sum_i b_i \beta_i^{(0)}} \cdot \|A\|^{1/2} \quad \text{by Lemma 12}$$

$$\leq (1 + o(1)) \cdot \frac{(\sum_i b_i (\beta_i^{(v)} - \mu)^2)^{1/2}}{\sum_i b_i} \cdot \|A\|^{1/2} \cdot \|A\|^{1/2} \quad \text{by (28)}$$

Applying Lemma 13, we control the sum in the numerator as

$$\left(\sum_i b_i (\beta_i^{(v)} - \mu)^2\right)^{1/2} \leq \left(\mathrm{Tr}(A)^{1/2} + \sum_i b_i (\beta_i^{(0)} - \mu)^2\right)^{1/2}$$

$$\leq \left(\sum_i b_i (\beta_i^{(0)} - \mu)^2\right)^{1/2} + O(\mathrm{Tr}(A)^{1/4})$$

$$\text{using } \sqrt{a+b} \leq \sqrt{a} + \sqrt{b}$$

$$\leq \left(\sum_i b_i\right)^{1/2} \cdot \max_i |\beta_i^{(0)} - \mu| + O(\mathrm{Tr}(A)^{1/4})$$

$$= \mathrm{Tr}(A)^{1/2} \cdot \left(\frac{|\max_i \beta_i^{(0)} - \min_i \beta_i^{(0)}|}{2} + o(1)\right)$$

$$\text{setting } \mu = (\max_i \beta_i^{(0)} - \min_i \beta_i^{(0)})/2$$

$$\leq (1+o(1)) \cdot \frac{\mathrm{Tr}(A)^{1/2}}{2} \left(\|A^{-1}\|^{1/2} - \frac{1}{\|A\|^{1/2}}\right)$$

$$= (1+o(1)) \cdot \frac{\mathrm{Tr}(A)^{1/2}}{2\|A\|^{1/2}} \left(\sqrt{\kappa(A)} - 1\right).$$

Substituting into the previous bound gives

$$\|F(v)\| \leq (1+o(1)) \cdot \frac{\sqrt{\kappa(A)} - 1}{2} \cdot \sqrt{\frac{\|A\|}{\mathrm{Tr}(A)}},$$

as advertised in (26).   □

Lemma 12. *If* $\|v\| = O(\mathrm{Tr}(A)^{-1/2})$, *then*

$$\sum_i b_i \beta_i^{(v)} \geq \sum_i b_i \beta_i^{(0)} - O(\mathrm{Tr}(A)^{1/2}) \geq (1-o(1)) \cdot \sum_i b_i \beta_i^{(v)}.$$

Proof. We can lowerbound the individual terms as

$$\beta_i^{(v)} = \|\hat{R}^{-1}(y_i + v)\|$$

$$\geq \|\hat{R}^{-1} y_i\| - \|\hat{R}^{-1} v\|$$

$$= \left(y_i^T (A + \mathrm{Tr}(A)vv^T)^{-1} y_i\right)^{1/2} - \|\hat{R}^{-1} v\|$$

$$= \left(y_i^T \left(A^{-1} - \frac{A^{-1} \mathrm{Tr}(A)vv^T A^{-1}}{1 + \mathrm{Tr}(A)v^T A^{-1} v}\right) y_i\right)^{1/2} - \|\hat{R}^{-1} v\|$$

$$\text{by the Sherman-Morrison formula}$$

$$\geq \sqrt{y_i^T A^{-1} y_i} - \left(y_i^T \left(\frac{A^{-1} v v^T A^{-1}}{\mathrm{Tr}(A)^{-1} + v^T A^{-1} v} \right) y_i \right)^{1/2} - \| \hat{R}^{-1} v \|$$

$$\text{since } \sqrt{a - b} \geq \sqrt{a} - \sqrt{b}$$

Taking a sum, we observe the difference of the sums that we are interested in is bounded by

$$\sum_i b_i \beta_i^{(0)} - \sum_i b_i \beta_i^{(v)} \leq \sum_i b_i \left(y_i^T \left(\frac{A^{-1} v v^T A^{-1}}{\mathrm{Tr}(A)^{-1} + v^T A^{-1} v} \right) y_i \right)^{\frac{1}{2}} + \sum_i b_i \| \hat{R}^{-1} v \|.$$

$$(29)$$

The first sum is handled by Cauchy–Schwarz:

$$\sum_i b_i \left(y_i^T \left(\frac{A^{-1} v v^T A^{-1}}{\mathrm{Tr}(A)^{-1} + v^T A^{-1} v} \right) y_i \right)^{1/2}$$

$$\leq \left(\sum_i b_i \right)^{1/2} \left(\sum_i b_i y_i^T \left(\frac{A^{-1} v v^T A^{-1}}{\mathrm{Tr}(A)^{-1} + v^T A^{-1} v} \right) y_i \right)^{1/2}$$

$$= \mathrm{Tr}(A)^{1/2} \left(\frac{\mathrm{Tr}(A A^{-1} v v^T A^{-1})}{\mathrm{Tr}(A)^{-1} + v^T A^{-1} v} \right)^{1/2}$$

$$\text{since } \sum_i b_i y_i y_i^T = A$$

$$< \mathrm{Tr}(A)^{1/2}.$$

For the second, we observe crudely that

$$\sum_i b_i \| \hat{R}^{-1} v \| \leq \mathrm{Tr}(A) \| \hat{R}^{-1} \| \| v \| = O(\mathrm{Tr}(A)^{1/2})$$

since $\| \hat{R}^{-1} \| = O(1)$ and $\| v \| = O(\mathrm{Tr}(A)^{-1/2})$. Plugging these two bounds into (29), we obtain

$$\sum_i b_i \beta_i^{(v)} \geq \sum_i b_i \beta_i^{(0)} - O(\mathrm{Tr}(A)^{1/2})$$

$$\geq \sum_i b_i \beta_i^{(0)} \left(1 - \frac{O(\mathrm{Tr}(A)^{1/2})}{(\min_i \beta_i^{(0)}) \cdot \sum_i b_i} \right)$$

$$\geq (1 - o(1)) \sum_i b_i \beta_i^{(0)} \qquad \text{noting that } \min_i \beta_i^{(0)} = \Omega(1).$$

$$\square$$

Lemma 13. *If* $\|v\| = O(\text{Tr}(A)^{-1/2})$ *then*

$$\sum_i b_i (\beta_i^{(v)} - \mu)^2 \leq \sum_i b_i (\beta_i^{(0)} - \mu)^2 + O(\text{Tr}(A)^{1/2}).$$

Proof. Write

$$\sum_i b_i (\beta_i^{(v)} - \mu)^2 = \sum_i b_i (\beta_i^{(v)})^2 - 2b_i \beta_i^{(v)} + b_i \mu^2.$$

Then $(\beta_i^{(v)})^2 \leq (\beta_i^{(0)})^2$ by $A + \text{Tr}(A)vv^T \succeq A$, and

$$-2 \sum_i b_i \beta_i^{(v)} \leq -2 \sum_i b_i \beta_i^{(0)} + O(2\,\text{Tr}(A)^{1/2}) \quad \text{by Lemma 12,}$$

as desired.   □

Proof of Lemma 10. We are now in a position to construct the body H promised in Theorem 4. Let K be a convex body with B_2^n as its John ellipsoid and contact points $(c_i, x_i)_{i \leq m}$. Use Theorem 5 to obtain a subsequence $(b_i, y_i = x_i + u)_{i \leq s}$ with only $O(n)$ points. Let $v, \hat{A}, \{\hat{z}_i\}_{i \leq s}, \{\hat{y}_i\}_{i \leq s}, \hat{R}, \hat{E},$ and $(\hat{a}_i, \hat{u}_i)_{i \leq s}$ be as above and take

$$\hat{K} = K + u + v.$$

Let

$$\theta_{min} := \min_{\|y\|_{\hat{E}} = 1} \|y\|_2,$$

be the size of the largest copy of B_2^n which fits inside $\hat{E}$, let $\epsilon > 0$ be a small constant, and consider the body

$$H = \text{conv}\left(\frac{\theta_{min}}{1 + \epsilon} \hat{K}, \hat{z}_1, \ldots, \hat{z}_s \right), \tag{30}$$

which is as a shrunken version of $\hat{K}$ with "spikes" $\hat{z}_i$.

We first show that H has very few contact points with its John Ellipsoid, which is $\hat{E}$. Observe that each $\hat{z}_i \in \partial \hat{E}$ since $\|\hat{z}_i\|_{\hat{E}} = 1$ by construction. Moreover, as the John Ellipsoid of K is B_2^n we have the containments

$$\frac{\theta_{min}}{1 + \epsilon} \hat{K} \subset \frac{\theta_{min}}{1 + \epsilon} (B_2^n + u + v) \subset \frac{\theta_{min}(1 + o(1))}{1 + \epsilon} B_2^n,$$

since $\|u\|, \|v\| = O(n^{-1/2})$ and $\theta_{min} = \Omega(1)$, immediately implying that

$$\frac{\theta_{min}}{1 + \epsilon} \hat{K} \in \text{int}(\theta_{min} B_2^n) \subset \text{int}(\hat{E}).$$

Thus we must have $\hat{z}_i \notin \text{conv}\frac{\theta_{min}}{1+\epsilon}\hat{K}$ and $\hat{z}_i \in \partial H$; moreover, these are the *only* contact points of H with $\hat{E}$. To see that $\hat{E}$ is indeed the John Ellipsoid of H, apply the inverse transformation $\hat{R}^{-1}$, and note that $\hat{R}^{-1}H$ has contact points $\hat{u}_i = \hat{R}^{-1}\hat{z}_i$ with B_2^n, which we have already shown satisfy the conditions of John's theorem with weights $\hat{a}_i$ in (25) and Lemma 11.

It remains to bound the distance $d(K, H)$; we will show that

$$\frac{\theta_{min}}{1+\epsilon}\hat{K} \subset H \subset (1+\epsilon)\theta_{max}\hat{K},$$

for $\theta_{max} := \max_{\|y\|_{\hat{E}}=1}\|y\|_2 = \max_{\|y\|=1}\|y\|_{\hat{E}}^{-1}$. The first containment is obvious; for the second, observe that for each i we have

$$\|\hat{z}_i\|_{\hat{K}} = \frac{\|x_i + u + v\|_{\hat{K}}}{\|\hat{y}_i\|_{\hat{E}}}$$

$$= \frac{1}{\|\hat{y}_i\|_{\hat{E}}} \quad \text{since } \hat{y}_i = x_i + u + v \in \partial\hat{K}$$

$$\leq \|\hat{y}_i\| \cdot \theta_{max}$$

$$\leq (1 + o(1))\theta_{max} \quad \text{since } \hat{y}_i = x_i + u + v \text{ and } \|u\|, \|v\| = O(n^{-1/2}).$$

The distance between H and K is thus bounded by the ratio

$$(1+\epsilon)^2\frac{\theta_{max}}{\theta_{min}} \leq (1+\epsilon)^2\frac{\max_{\|y\|=1}\|y\|_{\hat{E}}}{\min_{\|y\|=1}\|y\|_{\hat{E}}} \leq (1+\epsilon)^2\sqrt{\kappa(\hat{A})}. \tag{31}$$

To complete the proof, we bound $\kappa(\hat{A})$ using Theorem 5 (6) and Lemma 11 as follows:

$$\kappa(\hat{A}) = \|\hat{A}\|\|\hat{A}^{-1}\|$$

$$\leq \|A + \text{Tr}(A)vv^T\|\|A^{-1}\| \quad \text{since } \hat{A} = A + \text{Tr}(A)vv^T \succeq A$$

$$\leq (\|A\| + \text{Tr}(A)\|v\|^2)\|A^{-1}\|$$

$$\leq \left(\|A\| + (\frac{1}{4} + o(1))(\sqrt{\kappa(A)} - 1)^2\|A\|\right)\|A^{-1}\| \quad \text{by (26)}$$

$$= \kappa(A)\left(1 + (\frac{1}{4} + o(1))(\sqrt{\kappa(A)} - 1)^2\right).$$

Combining this with (31) and taking $\epsilon = o(1)$ gives the required bound (22). $\square$

Acknowledgements The author wishes to thank Daniel Spielman for helpful comments as well as the Department of Computer Science at Yale University, where this work was done.

References

1. K. Ball, in *An Elementary Introduction to Modern Convex Geometry*. Flavors of Geometry (University Press, 1997), pp. 1–58
2. J.D. Batson, D.A. Spielman, N. Srivastava, in *Twice-Ramanujan Sparsifiers*. STOC '09: Proceedings of the 41st annual ACM symposium on Theory of computing (ACM, New York, 2009), pp. 255–262
3. A. Giannopoulos, V. Milman, Extremal problems and isotropic positions of convex bodies. Israel J. Math. **117**, 29–60 (2000)
4. P.M. Gruber, Minimal ellipsoids and their duals. Rendiconti del Circolo Matematico di Palermo **37**(1), 35–64 (1988)
5. M. Rudelson, Contact points of convex bodies. Israel J. Math. **101**(1), 92–124 (1997)
6. M. Rudelson, Random vectors in the isotropic position. J. Funct. Anal. **163**(1), 60–72 (1999)

Appendix A
Related Conference and Seminar Talks

Israel GAFA Seminar
Tel Aviv University, 2006–2011

Friday, November 10, 2006

Sasha Sodin (Tel Aviv) Non-backtracking walks and the spectrum of random matrices

Eyal Lubetzky (Tel – Aviv) Non-backtracking random walks mix faster (joint work with N. Alon, I. Benjamini, S. Sodin)

Friday, December 1, 2006

Semyon Alesker (Tel Aviv) A new proof of the Bourgain-Milman inequality (after G. Kuperberg)

Mark Rudelson (Columbia, MO) The smallest singular number of a random matrix

Friday, December 15, 2006

B. Mityagin (Columbus, OH) Asymptotics of instability zones of the Hill operator with trigonometric polynomial potentials

S. Mendelson (Haifa and Canberra) Subgaussian processes are well-balanced

Friday, December 29, 2006

James Lee (Seattle, Washington) Vertex cuts, random walks, and dimension reduction

Gady Kozma (Rehovot) The scaling limit of loop-erased random walk in three dimensions

Friday, January 12, 2007

William Johnson (College Station, TX) Eight 20+ year old problems in the geometry of Banach spaces

Ofer Zeitouni (Haifa and Minneapolis, MN) Markov chains on an infinite dimensional simplex, the symmetric group, and a conjecture of Vershik.

B. Klartag et al. (eds.), *Geometric Aspects of Functional Analysis*, Lecture Notes
in Mathematics 2050, DOI 10.1007/978-3-642-29849-3,
© Springer-Verlag Berlin Heidelberg 2012

Friday, March 23, 2007

Avi Wigderson (IAS, Princeton) The power and weakness of randomness in computation

Semyon Alesker (Tel Aviv) A Fourier type transform on translation invariant valuations on convex sets

Friday, April 27, 2007

Emanuel Milman (Rehovot) Isoperimetric inequalities for uniformly log-concave measures and uniformly convex bodies (Joint work with Sasha Sodin)

Victor Katsnelson (Rehovot) The Weyl and Minkowski polynomials

Wednesday, December 22, 2010

Yuval Peres (Microsoft) Cover times, blanket times and the Gaussian free field

Friday, December 31, 2010

Bo'az Klartag (Tel Aviv) The vector in subspace problem

Sergei Bobkov (Minneapolis) Concentration of information in data generated by log-concave probability distributions

Friday, January 7, 2011

Emanuel Milman (Haifa) Interpolating thin-shell and sharp large-deviation estimates for isotropic log-concave measures

Apostolos Giannopoulos (Athens) On the distribution of the ψ_2-norm of linear functionals on isotropic convex bodies

Friday, March 4, 2011

Semyon Alesker (Tel Aviv) A Radon type transform on valuations

Franz Schuster (Vienna) Towards a Hadwiger type theorem for Minkowski valuations

December 26, 2011

Yuval Peres (Microsoft) Mixing times are hitting times of large sets

Peter Mester (Jerusalem) A factor of iid with continuous marginals and infinite clusters spanned by identical labels

Friday, December 30, 2011

Yanir Rubinstein (Stanford) A priori estimates for complex Monge-Ampere equations

Zbigniew Blocki (Krakow) Suita conjecture and the Ohsawa-Takegoshi extension theorem

Asymptotic Geometric Analysis Seminar
Tel Aviv University, 2007–2011

December 3, 2007
Muli Safra   Extractors from polynomials

December 24, 2007
Sasha Sodin   Functional versions of isoperimetric inequalities

December 12, 2007
James Lee   Trees and Markov convexity

January 8, 2008
Zemer Kosalov   Random complex zeroes-asymptotic normality

January 15, 2008
Nir Shahaf The geometry of majorizing measures

January 22, 2008
Emanuel Milman   Almost sub-Gaussian marginals of convex bodies

March 18, 2008
Sasha Sodin   Simple bounds for the extreme eigenvalues of Gaussian random matrices, following Gordon, Szarek and Davidson

April 15, 2008
Jesús Suárez   Extending operators into Lindenstrauss spaces

April 18, 2008
M. Meyer Some inequalities for increasing functions on $\mathbb{R}^n$

April 22, 2008
Tal Weisblatt   Alexandrov inequalities for mixed descriminants
Daniel Pasternak   Power-law estimates for the central limit theorem for convex sets
April 29, 2008
Carla Peri   Geometric tomography: classical and recent results

May 16, 2008
Tal Weisblatt   Bergstorm's inequality for positive definite matrices, and a generalization due to Ky Fan

May 20, 2008

Omer Friedland   On Bollobás' paper titled "An extension of the isoperimetric inequality on the sphere"

Ronen Eldan   The volume distribution of convex bodies

May 27, 2008

Sasha Sodin   Hubbard–Stratonovich transform and the P. Levy concentration function.

May 29, 2008

Boaz Klartag   Unconditional convex bodies, Neumann laplacian, optimal transportation and all the rest of it

Olivier Guedon   Subspaces generated by orthogonal bounded systems

June 17, 2008

Tal Weisblatt   Bergstorm's inequality for mixed volumes with a Euclidean ball (after Giannopoulos et al.)

July 7, 2008

Orit Raz   Shadows of convex bodies

July 11, 2008

B. Rubin   Radon transforms and comparison of volumes

July 14, 2008

Dan Florentin   Cube slicing

July 18, 2008

Sasha Litvak   Vertex index of convex bodies and asymmetry of convex polytopes

August 1, 2008

Shiri Artstein-Avidan   Characterization of duality and the Legendre transform

August 8, 2008

Sasha Sodin   A bound on the P. Levy concentration function: the multidimensional case

August 15, 2008

Ronen Eldan   Pointwise estimates for the CLT for convex sets

August 18, 2008

Boaz Slomka   Characterization of duality for convex bodies

August 24,28, 2008

Apostolos Giannopoulos   Volume estimates for isotropic convex bodies

August 29, 2008

Jesús Suárez   Maurey's extension theorem

Sasha Sodin   Kwapień's proof of the Maurey-Pisier lemma

September 7, 2008

Ronen Eldan   Differential geometry of the 19th century—minimal surfaces of revolution

September 14, 2008

Sasha Sodin   A non-scary account of extremal problems related to Chebyshev, Markov and Stieltjes and to T_+ systems

September 21, 2008

Sasha Sodin   Extremal problems related to log-concave measures, and more

September 28, 2008

Bo'az Klartag   On a seven-line paragraph by Gromov

October 5, 2008

Mark Rudelson   Non-commutative Khinchine inequality for interpolation norms (a question)

October 12, 2008

Rolf Schneider   Mixed functionals and mixed bodies

October 19, 2008

Rolf Schneider   Inequalities for convex bodies applied to random mosaics

November 30, 2008

Ronen Eldan   Monotonicity of optimal transportation á la Caffarelli and also Kolesnikov

December 21, 2008

Boaz Slomka   Mass transport generated by a flow of Gauss maps, after Bogachev and Kolesnikov

December 28, 2008

Alex Segal   On Bokowski and Heil's inequalities

January 11, 2009

Omer Friedland   Remez inequalities (after Yomdin)

January 25 and February 2, 2009

Ronen Eldan   Polynomial number of random samples cannot determine the volume of a convex body

February 22, 2009

Omer Friedland   Turan's lemma for polynomials

March 15, 2009

Orit Raz   An introduction to Busemann-Petty type problems

March 22, 2009

Bo'az Klartag   What is convexity good for?

March 29, 2009

Bo'az Klartag   Geometric symmetrizations of convex bodies

April 19, 2009

Sasha Sodin   Poisson asymptotics for random projections of points on a high-dimensional sphere (joint work with Itai Benjamini and Oded Schramm)

April 26, 2009

Florin Avram   Asymptotic approximations for the stationary distribution of queuing networks

May 3, 2009

Dmitry Faifman   Muntz' theorem and some related inequalities

May 10, 2009

Alex Segal   On the paper "On the Volume ratio of two convex bodies" by Giannopoulos and Hartzoulaki.

May 24, 2009

Boaz Slomka   Uniformly convex functions on Banach spaces

May 28, 2009

Bo'az Klartag   On Nazarov's paper "The Hormander-Hsin proof of the Bourgain-Milman theorem"

June 6, 2009

Orit Raz   Covering n-space by convex bodies and its chromatic number after a paper by Furedi and Kang

July 3, 2009

Efim Gluskin   The diameter of the Banach Mazur compactum

June 21, 2009

Ronen Eldan   Brascamp-Lieb and Entropy sub-additivity are equivalent (after Carlen and Cordero)

Bo'az Klartag   The proof of Nazarov for the Bourgain-Milman inverse Santalo inequality

July 26, 2009

Daniel Dadush   Rapidly mixing random walks on convex bodies: Why isoperimetric inequalities are awesome

August 2, 2009

Orit Raz   A survey about successive minima and Minkowski's 1st and 2nd fundamental theorems (some theorems, some conjectured)

October 6, 2009

Dima Faifman   The Poisson formula and a characterisation of Fourier transform

Bo'az Klartag   Some classical results regarding the restriction of homogeneous polynomials of fixed (odd) degree and many variables to linear subspaces

October 11, 2009

Roy Wagner   Intersection of l_1^n and l_2^n spheres after a recent paper by Sourav Chatterjee

November 1, 2009

Dima Faifman   Every polynomial of odd degree, vanishes on large subspaces

November 15, 2009

Orit Raz Best known bounds for packing balls in Euclidean space, and more

November 22, 2009

Ronen Eldan Hilbert's proof of the Brunn-Minkowski inequality

January 10, 2010

Grigoris Paouris   A recent work with Nikos Dafnis, which in particular gives a new proof of $n^{1/4} \sqrt{(\log n)}$ bound for the isotropic constant

February 2, 2010

Omer Friedland   A recent result with Olivier Guedon, on embedding of ℓ_p^n into ℓ_r^N

March 14, 2010

Gideon Schechtman   Fine estimates in Dvoretzky's theorem

March 21, 2010

Ronen Eldan   A generalization of Caffarelli's contraction theorem via (reverse) heat flow (On the paper by E. Milman and Kim)

April 11, 2010

Sasha Sodin   Berry-Esseen inequality: history and more recent results

April 18, 2010

Boaz Slomka   Klartag's result regarding Minkowski symmetrizations using spherical harmonics

April 25, 2010

Bo'az Klartag   Approximately gaussian marginals and the hyperplane conjecture

May 2, 2010

Dima Faifman The algebraic Dvoretzky theorem and conjecture, and the recent (possibly unpublished) work of Dol'nikov and Karasev on the subject

May 9, 2010

Vitali Milman The mystique of duality

May 16, 2010

Dan Florentin Fractional linear maps and order isomorphisms

May 23, 2010

Liran Rotem Results concerning the Banach Mazur distance between a general convex body in R^n and the cube

June 6, 2010

Alex Segal Rotation invariant Minkowski classes of convex bodies (after a paper by Rolf Schneider and Franz E. Schuster)

June 21, 2010

Ronen Eldan A stochastic formula for the entropy due to Lehec, in the spirit of Borell

July 5, 2010

Uri Grupel Fleury's improvement of the thin shell bound for log-concave densities

July 12, 2010

Oded Badt Alexandrov's result concerning preservation of light cones

July 19, 2010

Ronen Eldan Kendall's work on domain monotonicity of the heat kernel in convex domains

July 26, 2010

Alexander Litvak On the symmetric average of a convex body (joint work with Olivier)

August 2, 2010

Edward G. Effros Quantum information theory in the context of quantized functional analysis

August 19, 2010

Dima Faifman Proof of the Gromov-Milman conjecture by Dol'nikov and Karasev

August 22, 2010

Sasha Sodin Wegner's estimate for random Schroedinger operators

September 6, 2010

Mark Kozdoba Euclidean Bernstein widths of some Lipschitz balls

September 20, 2010

Uri Grupel Sudakov's classical theorem on marginals of high-dimensional distributions

January 9, 2011

Yaron Ostrover Symplectic measurements and convex geometry

February 20, 2011

Shiri Artstein-Avidan Stability of certain operations connected with geometry

February 27, 2011

Liran Rotem "Low-M^* estimate for log concave functions

March 6, 2011

Alex Segal Characterizing duality via maps interchanging sections with projections

March 13, 2011

Hermann Koenig Solutions of the chain rule and Leibniz rule functional equations

March 27, 2011

Greg Kuperberg (UC Davis) Numerical cubature from geometry and coding theory

April 10, 2011

Ronen Eldan Extremal points of a high dimensional random walk and the convex hull of a high dimensional Brownian motion

May 15, 2011

Dima Faifman Convexity arising through the momentum mapping

May 22, 2011

Dan Florentin Santaló's inequality by complex interpolation

May 29, 2011

Dan Florentin The B-conjecture and versions of Santaló's inequality

June 5, 2011

Uri Grupel Thin spherical shell estimate for convex bodies with unconditional basis

July 10, 2011

Sasha Sodin Griffiths-Ginibre inequalities

July 17, 2011

Liran Rotem On the mean width of log-concave functions

July 24, 2011
Tomasz Tkocz (Warsaw) Volumes of projections of permutationally invariant convex bodies

July 28, 2011
Alexander Litvak (Alberta) Tail estimates for norms of submatrices of a random matrix and applications

August 7, 2011
Boaz Slomka   On shaking convex sets, Silvester's problem and related topics

September 4, 2011
Yanir Rubinstein   The Cauchy problem for the homogeneous Monge-Ampere equation

September 8, 2011
Alex Segal   Volume inequalities on convolution bodies

September 18, 2011
Uri Grupel   Central limit theorem in non convex bodies

November 6, 2011
Boaz Slomka   The Jacobian conjecture

November 13, 2011
Liran Rotem   Fixed points of the polarity transform for geometric convex functions

November 20, 2011
Hermann Koenig   Gaussian measure of sections of the n-cube

December 23, 2011
Christos Saroglou   The minimal surface area position is not always an M-position

December 4, 2011
Alex Segal   Projections of log-concave functions (joint work with Boaz Slomka)

Phenomena in High Dimensions
2nd Annual Conference, Paris, June 7–14, 2006

Wednesday, June 7, 2006

Mikhail Gromov   Linearized isoperimetry

Assaf Naor   Is there a local theory of metric spaces?

Semyon Alesker   Some results and conjectures on valuations invariant under a group

Dario Cordero-Erausquin   Entropy of marginals and Brascamp-Lieb inequalities

Boris Kashin   On some class of inequalities for orthonormal systems

Bo'az Klartag   A central limit theorem for convex sets

Radoslaw Adamczak   Concentration of measure for U-statistics with applications to the law of the iterated logarithm

Daniel Hug   Nakajima's problem: Convex bodies of constant width and constant brightness

Thursday, June 8, 2006

Noga Alon   Monotone graph properties

Carsten Schütt   On the minimum of several random variables

György Elekes   Incidences in Euclidean spaces

Laura Wisewell   Covering polygonal annuli by strips

Dimitris Gatzouras   Threshold for the volume spanned by random points with independent coordinates

Emmanuel Opshtein   Maximal symplectic packings

Ofer Zeitouni   Path enumeration and concentration for certain random matrices

Matthias Heveling   Does polynomial parallel volume imply convexity?

Van Vu   Random discrete structures

Friday, June 9, 2006

Apostolos Giannopoulos   Geometry of random polytopes

Vladimir Pestov   Lévy groups and the fixed point on compacta property

Shiri Artstein-Avidan   Randomness reduction results in asymptotic geometric analysis

Krzysztof Oleszkiewicz   On some deviation inequality for Rademacher sums

Bernard Maurey   Geometrical inequalities and Brownian motion

Matthias Reitzner   Elementary moves on triangulations

Witold Bednorz Regularity of processes with increments controlled by many Young functions

Andrea Colesanti Remarks about the spectral gap of log-concave measures on the real line

Stefan Valdimarsson Optimisers for the Brascamp-Lieb inequality

Stanislaw Szarek On the nontrivial projection problem

Monday, June 12, 2006

Marianna Csörnyei Structure of null sets, differentiability of Lipschitz functions, and other problems

Antonio Aviles-Lopez Homeomorphic classification of balls in the Hilbert space

Rafał Latała On the boundedness of Bernoulli processes

Guillaume Aubrun Approximating the inertia matrix of an unconditional convex body

Mark Rudelson Singular values of random matrices

Stefano Campi Volume inequalities for L_p-zonotopes

Sasha Sodin An isoperimetric inequality on l_p balls

Alexander Litvak On the vertex index of convex bodies

Franz Schuster Rotation intertwining additive maps—properties and applications

Aicke Hinrichs Normed spaces with extremal distance to the Euclidean space

Tuesday, June 13, 2006

Monika Ludwig Affine geometry of convex bodies

Emanuel Milman Gaussian marginals of uniformly convex bodies

Grigoris Paouris Concentration of mass on convex bodies

Jesus Bastero Asymptotic behaviour of typical k-marginals of strongly concentrated isotropic convex bodies

Nicole Tomczak-Jaegermann Banach-Mazur distances and projections on random subgaussian polytopes

Shahar Mendelson On weakly bounded empirical processes

Mariya Shcherbina On the volume of the intersection of a sphere with random half-spaces

Cyril Roberto Rigorous results for relaxation times in kinetically constrained spin models

Sergey Bobkov On isoperimetric constants for product probability measures

Wednesday, June 14, 2006

Gideon Schechtman Non-linear factorization of linear operators

Andrei Okounkov Algebraic geometry of random surfaces

Terence Tao The uniform uncertainty principle and applications

Phenomena in High Dimensions
3rd Annual Conference, Samos, June 25–29, 2007

Monday, 25 June, 2007

Endre Szemerédi   Some problems in extremal graph theory

Nicole Tomczak-Jaegermann   Embeddings under various notions of randomness

Zoltán Füredi   Almost similar configurations

Michael Krivelevich   Minors in expanding graphs

Greg Kuperberg   From the Mahler conjecture to Gauss linking integrals

Piotr Mankiewicz   How neighbourly an m-neighbourly symmetric polytope can be?

Rafał Latała   On the infimum convolution inequality

Emanuel Milman   Isoperimetric inequalities for uniformly log-concave measures and uniformly convex bodies

Olivier Guédon   Selections of arbitrary size of characters

Tuesday, 26 June, 2007

Roman Vershynin   Anti-concentration inequalities

Mark Rudelson   Invertibility of random matrices

Hermann König   Projecting l_∞ onto classical spaces

Tony Carbery   On equivalence of certain norms on sequence spaces

Mathieu Meyer   On some functional inequalities

Gabriele Bianchi   Determination of a set from its covariance: Complete confirmation of Matheron's conjecture

Martin Henk   Roots of Ehrhart polynomials

Stefano Campi   Estimating intrinsic volumes from finitely many projections

Wednesday, 27 June, 2007

Gady Kozma   Contracting Clusters of Critical Percolation

Alain Pajor   Marchenko-Pastur distribution for random vectors with log concave law

Elisabeth Werner   Geometry of sets of quantum states and super-operators

Stanisław Szarek   Sets of constant height and ppt states in quantum information theory

Guillaume Aubrun   Catalytic majorization in quantum information theory

Thursday, 28 June, 2007

Matthias Reitzner   Tail inequalities for random polytopes

Alexander Koldobsky   The complex Busemann-Petty problem

Wolfgang Weil   Projections and liftings on the sphere

Attila Pór   Density of ball packings and its application to the Hausdorff dimension of the residual set

Alexander Sodin   The non-backtracking random walk on a graph

Omer Friedland   Kahane-Khinchin type Averages

Franz Schuster   Valuations and Busemann-Petty type problems

Peter Pivovarov   Volume thresholds for Gaussian and spherical random polytopes and their duals

Christoph Haberl L_p intersection bodies

Joseph Lehec   A simple proof of the functional Santaló inequality

Vladyslav Yaskin   On strict inclusions in hierarchies of convex bodies

Boris Bukh   Measurable chromatic number and sets with excluded distances

Maryna Yaskina   Shadow boundaries and the Fourier transform

Stefan Valdimarsson   A multilinear generalisation of the Hilbert transform and fractional integration

Balázs Patkós   Equitable coloring of random graphs

Jesús Suárez   Twisting Schatten classes

Gergely Ambrus   On the maximal convex chains among random points in a triangle

Friday, 29 June, 2007

Michel Ledoux   Deviation inequalities on largest eigenvalues

Vitali Milman   On some recent achievements of Asymptotic Geometric Analysis

Alexander Litvak   Vertex index of convex bodies and asymmetry of convex polytopes

Mariya Shcherbina   Central limit theorem for linear eigenvalue statistics of orthogonally invariant matrix models

Krzysztof Oleszkiewicz   Gaussian concentration of vector valued random variables

Bo'az Klartag   Rate of convergence in the central limit theorem for convex bodies

Phenomena in High Dimensions
4th Annual Conference, Seville, 23–27 June 2008

Monday, 23 June, 2008

Hermann König   Measure of sections and slabs of the n-cube

Nathael Gozlan   Concentration of measure, Transport and Large Deviations

Stanislaw Szarek   Concentration for non-commutative polynomials in random matrices

Sergey G. Bobkov   Convex bodies and norms associated to convex measures

Elisabeth Werner   On L_p affine surface area

Yehoram Gordon   Generalizing the Johnson-Lindenstrauss lemma to k-dimensional affine subspaces

Alexander Litvak   Covering convex bodies by cylinders

Tuesday, 24 June, 2008

Ben Green   The Gowers norms

Mireille Capitaine   Asymptotic spectrum of large deformed Wigner matrices

Andrea Colesanti   The Minkowski problem for the torsional rigidity

Olivier Guedon   On the isotropic constant of random polytopes

Emanuel Milman   On the role of convexity in isoperimetry, spectral-gap and concentration

Matthieu Fradelizi   Some functional inverse Santaló inequalities

Andreas Winter   Some applications of measure concentration in quantum information theory

Rafał Latała   Bounds on moments of linear combinations of log-concave random vectors

Krzysztof Oleszkiewicz   Precise tail estimates for products and sums of independent random variables

Carsten Schütt   Uniform estimates for order statistics and Orlicz functions

Wednesday, 25 June, 2008

Daniel Hug   Random polytopes and mosaics: asymptotic results

Monika Ludwig   General affine surface areas

Wolfgang Weil   Generalized Averages of Section and Projection Functions

Franz Schuster   General L_p affine isoperimetric inequalities

M. Angeles Hernández Cifre   On the volume of inner parallel bodies

Nigel Kalton   The uniform structure of Banach spaces

Thursday, 26 June, 2008

Ofer Zeitouni   The single ring theorem

Alexander Soshnikov   On the largest eigenvalues of large random Wigner matrices with non-symmetrically distributed entries

Alexander Sodin   Universality on the edge in random matrix theory: Around Soshnikov's theorems

William B. Johnson   A characterization of subspaces and quotients of reflexive Banach spaces with unconditional bases

Ronen Eldan   Pointwise estimates for marginals of convex bodies

Gana Lytova   Central limit theorem for linear eigenvalue statistics of the Wigner and the sample covariance random matrices

Chiara Bianchini   Geometric inequalities for the Bernoulli constant

Jesús Suárez   On a problem of Lindenstrauss and Pelczynski

Pawel Wolff   Spectral gap for conservative spin systems: The case of gamma distribution

Joseph Lehec   Equipartitions and functional Santaló inequalities

Eugenia Saorín   From Brunn-Minkowski to Poincaré type inequalities

Marisa Zymonopoulou   Sections of complex convex bodies

David Alonso-Gutiérrez   An extension of the hyperplane conjecture and the Blaschke-Santaló inequality

Maryna Yaskina   Christoffel's problem and the Fourier transform

Peter Pivovarov   On the mean-width of isotropic convex bodies

Vladyslav Yaskin   Christoffel's problem and the Fourier transform. Part II.

Friday, 27 June, 2008

Gil Kalai   Fourier analysis of Boolean functions and applications

James R. Lee   Expander codes and pseudorandom subspaces

Semyon Alesker   A Fourier type transform on translation invariant valuations on convex sets

Nathan Linial   What is high-dimensional combinatorics?

François Bolley   Mean field limits of stochastic interacting particle systems

Subhash Khot   Inapproximability of NP-complete problems, discrete Fourier analysis, and geometry

Affine Convex Geometric Analysis
Banff, January 11–15, 2009
Organizers: *Monika Ludwig, Alina Stancu, Elisabeth Werner*

Monday, January 12, 2009

Rolf Schneider   The role of the volume product in stochastic geometry

Mathieu Meyer   Once again on Mahler's problem

Nicole Tomczak-Jaegermann   Random points uniformly distributed on an isotropic convex body

Alexander Litvak   On vectors uniformly distributed over a convex body

Joseph Lehec   A functional approach to the Blaschke-Santaló inequality

Matthieu Fradelizi   The volume product of convex bodies with many symmetries

Tuesday, January 13, 2009

Matthias Reitzner   A classification of $SL(n)$ invariant valuations

Fedor Petrov   Affine surface area and rational points on convex surfaces

Christoph Haberl   Blaschke valuations

Peter J. Olver   Moving frames, differential invariants and surface geometry

Franz Schuster   Valuations and affine Sobolev inequalities

Christian Steineder   The volume of the projection body and binary coding sequences

Carlo Nitsch   Affine isoperimetric inequalities and Monge-Amp'ere equations

Wednesday, January 14, 2009

Franck Barthe   Remarks on conservative spin systems and related questions in convexity

Andrea Colesanti   Recent contributions to the Christoffel problem coming from partial differential equations

Marina Yaskina   Non-symmetric convex bodies and the Fourier transform

Thursday, January 15, 2009

Mark Rudelson

Mark Meckes   On the measure-theoretic analogue of Dvoretzky's theorem

Eugenia Saorín Gómez   Approaching K by its form body and kernel

Vlad Yaskin   On strict inclusions in hierarchies of convex bodies

Marisa Zymonopoulou   New examples of non-intersection bodies

Olivier Guédon   Invertibility of matrices with iid columns

Efrén Morales Amaya   Characterization of ellipsoids by means of parallel translated sections

Daniel Hug   Random polytopes: Geometric and analytic aspects

State of Geometry and Functional Analysis
Tel Aviv University and Dead Sea, June 24–30, 2009
Conference in honour of the
70th Birthday of Professor Vitali Milman

Organizers: *S. Alesker, S. Artstein-Avidan, B. Klartag, S. Mendelson*
Scientific committee: *N. Alon, J. Bernstein, J. Bourgain, M. Gromov, M. Talagrand*

Wednesday, June 24, 2009

N. Alon Graph Eigenvalues and the interplay between discrete and continuous isoperimetric problems
Y. Eliashberg Leonard M. Blumenthal Lecture in Geometry: Symplectic topology of Stein manifolds
S.G. Bobkov Geometric and analytic problems in the theory of heavy-tailed convex measures
P. Enflo Extremal vectors and invariant subspaces
A. Wigderson Applications of TCS techniques to geometric problems
M. Gromov

Thursday, June 25, 2009

D. Kazhdan Satake isomorphism for symmetrizable Kac-Moody algebras
J. Lindenstrauss Frechet differentiability, porosity and asymptotic structure in Banach spaces
S. Argyros The solution of the "Scalar plus Compact" problem
I. Benjamini Random planar geometry
J. Bourgain Expansion in linear groups and applications

Mini symposium on Asymptotic Geometric Analysis

Organizers: *A. Litvak, V. Milman*
Speakers: *J. Bernues, M. Fradelizi, O. Maleva, T. Schlumprecht, S. Szarek, A. Tsolomitis*

Friday, June 26, 2009, Dead Sea

L. Pastur Universalities in random matrix theory
A. Giannopoulos Volume distribution on high-dimensional convex bodies
D. Preiss Lipschitz functions and negligible sets

Mini symposium in Convexity

Organizers: *S. Alesker, M. Ludwig*
Speakers: *I. Barany, A. Bernig, K. Boroczky, D. Hug, A. Koldobsky, D. Ryabogin, A. Stancu, W. Weil, E. Werner, A. Zvavitch*

Sunday, June 28, 2009

P. Gruber   A short survey of the geometry of numbers and modern applications of a classical idea

M. Ledoux   Measure concentration, functional inequalities, and curvature of metric measure spaces

N. Tomczak-Jaegermann   Random matrices with independent columns

Monday, June 29, 2009

O. Zeitouni   Limit laws for the eigenvalues of certain matrix ensembles

B. Bollobas   Convergent sequences of graphs

V. Bergelson   Ergodic theory along polynomials and combinatorial number theory

V. Pestov   Ramsey-Dvoretzky-Milman phenomenon: a survey

J.-M. Bismut   The hypoelliptic laplacian and its applications

Mini symposium on Probability in Convex Bodies

Organizers: *B. Klartag, R. Vershynin*

Speakers: *R. Latała, E. Milman, K. Oleszkiewicz, G. Paouris, M. Shcherbina, S. Sodin, A. Volcic*

Tuesday, June 30, 2009

G. Kalai   Noisy quantum computers above the "Threshold"

Y. Sinai   Non-standard Ergodic theorems and limit theorems

S. Novikov   New Discretization of Complex Analysis

M. Rudelson   Invertibility and spectrum of random matrices

R. Schneider   Classical convexity and stochastic geometry

Conference on Convex and Discrete Geometry
Technische Universität, Vienna, July 13–17, 2009
On the occasion of the retirement of Peter M. Gruber

Organizer: *Buchta, Haberl, Ludwig, Reitzner, Schuster*

Monday, July 13, 2009

Rolf Schneider (Freiburg) The many appearances of zonoids

Günter M. Ziegle (Berlin) Some centrally-symmetric polytopes

Paul Goodey (Norman) Some stereological results for non-symmetric convex bodies

Konrad Swanepoel (Chemnitz) Maximal pairwise touching families of translates of a convex body

Karoly Böröcz (Budapest) A characterization of balls

Ivan Netuka (Prague) Jensen measures and harmonic measures

Rajinder Hans-Gill (Chandigarh) On conjectures of Minkowski & Woods

Nikolai Dolbilin (Moscow)

Tuesday, July 14, 2009

Keith Ball (London) Superexpanders and Markov cotype in the work of Mendel and Naor

Apostolos Giannopoulos (Athens) Asymptotic shape of a random polytope in a convex body

Stefano Campi (Siena) Estimating intrinsic volumes of a convex body from finitely many tomographic data

Richard Gardner (Bellingham) Capacities, surface area, and radial sums

Carla Peri (Milan) On some new results in discrete tomography

Aljoša Volčič (Cosenza) Random symmetrizations of measurable sets

Luis Montejano (Mexico City) Topology and Transversal

Margarita Spirova (Chemnitz) Bodies of constant width in Minkowski geometry and related coverings problems

Wednesday, July 15, 2009

Peter McMullen (London) Translation tilings by polytopes

Semyon Alesker (Tel Aviv) Product on valuations

Andreas Bernig (Fribourg) Integral geometry of transitive group actions

Bo'az Klartag (Tel Aviv) On nearly radial marginals of high-dimensional probability measures

Thursday, July 16, 2009

Imre Bárány (Budapest) Simultaneous partitions by k-fans

Iskander Aliev (Cardiff) On feasibility of integer knapsacks

Martin Henk (Magdeburg) Successive minima type inequalities

Chuanming Zong (Beijing) Geometry of numbers in Vienna

Nikolai Dolbilin & Arkadii Maltsev (Moscow) Sergey Ryshkov

Gábor Fejes Tót (Budapest) László Fejes Tóth

David Larman (London) Victor Klee and Ambrose Rogers

Jörg Will (Siegen) Edmund Hlawka

Friday, July 17, 2009

Vitali Milman (Tel Aviv) Duality and rigidity for families of convex functions

Wolfgang Weil (Karlsruhe) Integral geometry of translation invariant functionals

Daniel Hug (Karlsruhe) Typical cells and faces in Poisson hyperplane mosaics

Idzhad Kh. Sabitov (Moscow) On the notion of combinatorial p-parametricity of polyhedra

Serguei Novikov (Moscow) New discretization of complex analysis (DCA)

Thematic Program on Asymptotic Geometric Analysis
Fields Institute, July–December 2010
Concentration Period on Convexity
September 9–10, 2010
Organizer: *Monika Ludwig*
Workshop on Asymptotic Geometric Analysis and Convexity
September 13–17, 2010
Organizers: *Monika Ludwig, Vitali Milman and Nicole Tomczak-Jaegermann*
Concentration Period on Asymptotic Geometric Analysis
September 20, 2010

Thursday, September 9, 2010

Dan Klain (University of Massachusetts Lowell) If you can hide behind it, can you hide inside it?

C. Hugo Jimenez (University of Seville) On the extremal distance between two convex bodies

Maria A. Hernandez Cifre (Universidad de Murcia) On the location of roots of Steiner polynomials

Vitali Milman (Tel Aviv University) Convexity in windows

Wolfgang Weil (Karlsruhe Institute of Technology) Valuations and local functionals

Friday, September 10, 2010

Alina Stancu (Concordia University, Montreal) Affine Differential Invariants for Convex Bodies

Deping Ye (The Fields Institute) On the homothety conjecture

Mathieu Meyer (Université de Paris Est Marne-la-Vallée) Functional inequalities related to Mahler's conjecture

Monday, September 13, 2010

Hermann Koenig (University of Kiel, Germany) The chain rule as a functional equation

Maryna Yaskina (University of Alberta) Shadow boundaries and the Fourier transform

Peter Pivovarov (Fields Institute) On the volume of random convex sets

Deane Yang (Polytechnic Institute of NYU) Towards an Orlicz Brunn-Minkowski theory

Alexander Litvak (University of Alberta) On the Euclidean metric entropy

Emanuel Milman (University of Toronto) Properties of isoperimetric, spectral-gap and log-Sobolev inequalities via concentration

Elizabeth Meckes (Case Western Reserve University) Another observation about operator compressions

Gautier Berck   Invariant distributions in integral geometry

Tuesday, September 14, 2010

Radoslaw Adamczak (University of Warsaw/Fields Institute) Random matrices with independent log-concave rows

Kei Funano (Kumamoto University) Concentration of measure phenomenon and eigenvalues of Laplacian

Andrea Colesanti (Universitá di Firenze) Integral functionals verifying a Brunn-Minkowski type inequality

Franz Schuster (Vienna University of Technology) Translation invariant valuations

Artem Zvavitch (Kent State University) Some geometric properties of intersection body operator

Distinguished Lecture Series
September 14–16, 2010

Avi Wigderson (Institute for Advanced Study) Randomness, pseudorandomness and derandomization

Wednesday, September 15, 2010

Daniel Hug (Karlsruhe Institute of Technology) Volume and mixed volume inequalities in stochastic geometry

Christoph Haberl (Vienna University of Technology) Minkowski valuations intertwining the special linear group

Matthias Reitzner (Univ. Osnabrueck) Poisson-Voronoi approximation

Olivier Guedon (Université Paris-Est Marne-La-Vallée) Embedding from l_p^n into l_r^N for $0 < r < p < 2$

Roman Vershynin (University of Michigan) Estimation of covariance matrices

Jie Xiao (Memorial University) Volume integral means of holomorphic mappings

Martin Henk (University of Magdeburg) The average Frobenius number

Thursday, September 16, 2010

Krzysztof Oleszkiewicz (Institute of Mathematics: University of Warsaw and Polish Academy of Sciences) Feige's inequality

David Alonso (Universidad de Zaragoza, Fields Institute) Volume of L_p-zonotopes and best best constants in Brascamp-Lieb inequalities

Dmitry Faifman (Tel Aviv University) The Poisson summation formula uniquely characterizes the Fourier transform

Grigoris Paouris (Texas A&M University) On the existence of subgaussian directions for log-concave measures

Alexander Barvinok (University of Michigan) Gaussian and almost Gaussian formulas for volumes and the number of integer points in polytopes

Carla Peri (Università Cattolica – Piacenza) On the reconstruction of inscribable sets in discrete tomography

Mikhail Ostrovskii (St. Johns University, Queens, NY) Properties of metric spaces which are not coarsely embeddable into a Hilbert space

Friday, September 17, 2010

Rafał Latała (University of Warsaw) Moments of unconditional logarithmically concave vectors

Vladyslav Yaskin (University of Alberta) The geometry of p-convex intersection bodies

Nikolaos Dafnis (University of Athens) Small ball probability estimates, ψ_2-behavior and the hyperplane conjecture

Fedor Nazarov (University of Wisconsin at Madison) Hormander's proof of the Bourgain-Milman theorem

Mark Rudelson (University of Missouri) Spectral properties of random conjunction matrices

Coxeter Lecture Series
September 17,20,21, 2010

Shiri Artstein-Avidan [Sept. 17] (Tel Aviv University) Abstract duality, the Legendre transform and a new duality transform
[Sept. 20]   Order isomorphisms and the fundamental theorem of affine geometry
[Sept. 21]   Multiplicative transforms and characterization of Fourier transform

Monday, September 20, 2010

Gideon Schechtman (Weizmann Institute of Science, Rehovot) Tight embedding of subspaces of L_p in l_p^n for even p

Julio Bernues (University of Zaragoza) Factoring Sobolev inequalities through classes of functions

Stanislaw J. Szarek (Université Paris 6 and Case Western Reserve University, Cleveland) Peculiarities of Dvoretzky's Theorem for Schatten classes

Yehoram Gordon (Technion, Haifa) Large distortion dimension reduction using random variables

Workshop on the Concentration Phenomenon, Transformation Groups and Ramsey Theory

October 12–15, 2010

Organizers: *Eli Glasner, V. Pestov and S. Todorcevic*

Tuesday, October 12, 2010

Slawomir Solecki (University of Illinois) Finite Ramsey theorems

Claude Laflamme (University of Calgary) The hypergraph of copies of countable homogeneous structures

Ilijas Farah (York University) Classification of nuclear C^*-algebras and set theory

Kei Funano (Kumamoto University) Concentration of maps and group actions

Matthias Neufang (The Fields Institute and Carleton University) Topological centres for group algebras, actions, and quantum groups

Vladimir Pestov (University of Ottawa) Ergodicity at identity of measure-preserving actions of Polish groups

Wednesday, October 13, 2010

Lewis Bowen (Texas A&M University) Entropy of group actions on probability spaces

Arkady Leiderman (Ben-Gurion University) On topological properties of the space of subgroups of a discrete group

Hanfeng Li (SUNY at Buffalo) Entropy for actions of sofic groups

Kostyantyn Slutskyy (University of Illinois at Urbana-Champaign) Classes of topological similarity in Polish groups

Sergey Bezuglyi (Institute for Low Temperature Physics) Homeomorphic measures on a Cantor set

Eli Glasner (Tel Aviv University) Topological groups with Rohlin properties

Thursday, October 14, 2010

Norbert Sauer (University of Calgary) On the oscillation stability of universal metric spaces

Aleksandra Kwiatkowska (University of Illinois at Urbana-Champaign) Point realizations of near-actions of groups of isometries

Lionel Nguyen Van Thé (Université Aix-Marseille 3, Paul Cézanne) Universal flows for closed subgroups of the permutation group of the integers

Jan Pachl (Fields Institute) Uniform measures and ambitable groups

Todor Tsankov (Paris 7) Unitary representations of oligomorphic groups

Yonatan Gutman (Université Paris-Est Marne-la-Vallée) Minimal hyperspace actions of Homeo$(bw\backslash w)$

Friday, October 15, 2010

Jose Iovino (The University of Texas at San Antonio) From discrete to continuous arguments in logic. What is needed and why

Miodrag Sokic (California Institute of Technology) Posets with linear orderings

Julien Melleray (Univ. Lyon 1) Applications of continuous logic to the theory of Polish groups

Jakub Jasinski (University of Toronto) Ramsey degrees of boron tree structures

Pandelis Dodos (University of Athens) Density Ramsey theory of trees

Vadim Kaimanovich (University of Ottawa) Continuity of the asymptotic entropy

Workshop on Geometric Probability and Optimal Transportation

November 1–5, 2010

Concentration Periods

October 25–19 and November 8–10

Organizers: *B. Klartag and R. McCann*

Monday, October 25, 2010

Frank Morgan (Williams College) Densities from geometry to Poincaré

Israel Michael Sigal (University of Toronto) Singularity formation under the mean-curvature flow

Tuesday, October 26, 2010

Aldo Pratelli (Pavia, Italy) About the approximation of orientation-preserving homeomorphisms via piecewise affine or smooth ones

Mariya Shcherbina (Institute for Low Temperature Physics Ukr. Ac. Sci.) Orthogonal and symplectic matrix models: universality and other properties

Scott Armstrong (University of Chicago) An introduction to infinity harmonic functions (in the framework of a Graduate Minicourse, part 1)

Charles Smart (New York University) An introduction to infinity harmonic functions (in the framework of a Graduate Minicourse, part 2)

Wednesday, October 27, 2010

Tony Gomis (NBI) A domain decomposition method, Cherruault transformations, homotopy perturbation method, and nonlinear dynamics: theories and comparative applications to frontier problems

Jerome Bertrand (University Paul Sabatier Toulouse) Wasserstein space over Hadamard space

Amir Moradifam (University of Toronto) Isolated singularities of polyharmonic inequalities

Magda Czubak (University of Toronto) Topological defects in the abelian Higgs model

Thursday, October 28, 2010

Scott Armstrong (University of Chicago) An introduction to infinity harmonic functions (in the framework of a Graduate Minicourse, part 3)

Charles Smart (New York University) An introduction to infinity harmonic functions (in the framework of a Graduate Minicourse, part 4)

Jiakun Liu (Princeton University) Global regularity of the reflector problem

Friday, October 29, 2010

Philippe Delanoe (CNRS at University of Nice Sophia Antipolis) Remarks on the regularity of optimal transport

Young-Heon Kim (University of British Columbia) Regularity of optimal transport maps on multiple products of spheres

Monday, November 1, 2010

Alexander Plakhov (University of Aveiro) Optimal mass transportation and billiard scattering by rough bodies

Brendan Pass (University of Toronto) The multi-marginal optimal transportation problem

Alexander Kolesnikov (MSUPA and HSE (Moscow)) Sobolev regularity of optimal transportation

Gabriel Maresch (TU Vienna) Optimal and better transport plans

Emil Saucan (Department of Mathematics, Technion, Haifa, Israel) Triangulation and discretizations of metric measure spaces

Jun Kitagawa (Princeton University) Regularity for the optimal transport problem with Euclidean distance squared cost on the embedded sphere

Cédric Villani (Institut Henri Poincaré, Université Claude Bernard Lyon 1) What is the fate of the solar system? (in the framework of the Distinguished Lecture Series)

Tuesday, November 2, 2010

Nassif Ghoussoub (University of British Columbia) Homogenization, inverse problems and optimal control via selfdual variational calculus

Yi Wang (Princeton University) The Aleksandrov-Fenchel inequalities of k+1-convex domains

Vitali Milman (Tel Aviv University) Overview of asymptotic geometric analysis; an introduction (in the framework of a Graduate Course)

Alexander Koldobsky (University of Missouri-Columbia) Positive definite functions and stable random vectors

Nicola Gigli (University of Nice) The Heat Flow as Gradient Flow

Wednesday, November 3, 2010

Irene M. Gamba (The University of Texas at Austin) Convolution inequalities for Boltzmann collision operators and applications

Micah Warren (Princeton University) Parabolic optimal transport equations on compact manifolds

Dmitry Jakobson (McGill University, Department of Mathematics and Statistics) Curvature of random metrics

Alfredo Hubard (NYU) Can you cut a convex body into five convex bodies, with equal areas and equal perimeters?

Cédric Villani (l'Institut Henri Poincaré, Université Claude Bernard Lyon 1) Particle systems and Landau damping (in the framework of the Distinguished Lecture Series)

Thursday, November 4, 2010

Alessio Figalli (University of Texas at Austin) The geometry of the Ma-Trudinger-Wang condition (in the framework of a Graduate Course)

Cédric Villani (l'Institut Henri Poincaré, Université Claude Bernard Lyon 1) From echo analysis to nonlinear Landau damping (in the framework of the Distinguished Lecture Series)

Michael Loss (School of Mathematics, Georgia Tech) Hardy-Littlewood-Sobolev Inequalities via Fast Diffusion Flows

Mark Meckes (Case Western Reserve University) The magnitude of a metric space

Shaghayegh Kordnoori (Islamic Azad University North Tehran Branch) Central limit theorem and geometric probability

Marjolaine Puel (IMT Toulouse, France) Jordan-Kinderlehrer-Otto scheme for a relativistic cost

Joseph Lehec (Universit Paris-Dauphine) A stochastic formula for the entropy and applications

Friday, November 5, 2010

Luigi Ambrosio (Scuola Normale Superiore) Elements of geometric measure theory in Wiener spaces

Paul Woon Yin Lee (UC Berkeley) Generalized Ricci curvature bounds for three dimensional contact subriemannian manifolds

Elton Hsu (Department of Mathematics, Northwestern University) Mass transportation and optimal coupling of Brownian motions

Emanuel Milman (University of Toronto) A generalization of Caffarelli's contraction theorem via heat-flow

Wilfrid Gangbo (Georgia Institute of Technology) Sticky particle dynamics with interactions

Monday, November 8, 2010

Konstantin Khanin (University of Toronto) Lagrangian dynamics on shock manifolds

Walter Craig (Department of Mathematics and Statistics, McMaster University) On the size of the Navier-Stokes singular set

Marina Chugunova (University of Toronto) Finite speed propagation of the interface and blow-up solutions for long-wave unstable thin-film equations

Gideon Simpson (University of Toronto) Coherent structures in the nonlinear Maxwell equations

Stephen W. Morris (University of Toronto) Icicles, washboard road and meandering syrup

Wednesday, November 10, 2010

Boris Khesin (University of Toronto) Optimal transport and geodesics for H^1 metrics on diffeomorphism groups

Seminar

Wednesday, September 22, 2010

Rafał Latała   Gaussian approximation of moments of sums of independent subexponential random variables

Monday, September 27, 2010

Krzysztof Oleszkiewicz A convolution inequality which implies Khinchine inequality with optimal constants (joint with Piotr Nayar)

Wednesday, October 6, 2010

Karsten Schütt (Kiel) A note on Mahler's conjecture

Thursday, October 7, 2010

Eli Glasner (Tel Aviv) Representations of dynamical systems on Banach spaces

Friday, October 8, 2010

Norbert Sauer (Calgary) Introduction to homogeneous structures and their partitions

Wednesday, October 20, 2010

Leonid Pastur (Academy of Sciences of Ukraine) "YES" and "NO" for the validity of Central Limit Theorem for spectral statistics of random matrices

Thursday, October 21, 2010

Sergey Bezuglyi (Institute for Low Temperature, Kharkov) Full groups in Cantor, Borel and measurable dynamics

Wednesday, November 17, 2010

Vladimir Pestov (Universite d'Ottawa) Non-locally compact Polish groups: some examples, techniques, results, and open problems, I

Thursday, November 18, 2010

Vladimir Pestov (Université d'Ottawa) Non-locally compact Polish groups: some examples, techniques, results, and open problems, II

Wednesday, November 24, 2010

Nicole Tomczak-Jaegermann   Singular numbers of random matrices in asymptotic non-limit regime

Thursday, November 25, 2010

Radek Adamczak   Bernstein type inequalities for geometrically ergodic Markov chains

Wednesday, December 1, 2010

Rafał Latała

Young Researchers Seminar

Steven Taschuk (University of Alberta) Milman's other proof of the Bourgain–Milman theorem

Tuesday, September 28, 2010

Susanna Spektor (University of Alberta) Bernstein-type inequality for independent random matrices

Tuesday, October 5, 2010

Kei Funano (Kumamoto University) Can we capture the asymptotic behavior of λ_k?

Monday, October 18, 2010

Yonatan Gutman (Université Paris-Est Marne-la-Vallée) Universal minimal spaces

Tuesday, October 19, 2010

Anastasios Zouzias (University of Toronto) Low rank matrix-valued Chernoff bounds and approximate matrix multiplication

Tuesday, October 26, 2010

Kostya Slutsky (University of Illinois at Urbana-Champaign) Ramsey action of Polish groups

Tuesday, November 9, 2010

Alexander Segal (Tel-Aviv University) Stability of the Brunn-Minkowski inequality

Tuesday, November 16, 2010

Dominic Dotterrer (University of Toronto) Triangles in a box

Tuesday, November 30, 2010

David Alonso-Guttierez On the projections of polytopes and their isotropy constant

Thursday, December 2, 2010

Peter Pivovarov Isoperimetric problems for random convex sets

Tuesday, December 7, 2010

Deping Ye   Additivity conjecture via Dvoretzky's theorem

Conference on Geometry and the Distribution of Volume on Convex bodies
Kibbutz Hagoshrim Guest House and Tel Aviv University Israel, March 31–April 5, 2011

Organizers: *Shiri Artstein-Avidan, Bo'az Klartag, Shahar Mendelson, Emanuel Milman, Vitali Milman, Rolf Schneider*

Thursday, March 31, 2011

Rolf Schneider (Freiburg) Diametrically complete sets in normed spaces

Dan Florentin (Tel Aviv) Convexity on windows,

Friday, April 1, 2011

Boaz Klartag (Tel Aviv) The Logarithmic Laplace Transform in Convex Geometry

Nikolaos Dafnis (Athens) Estimates for the affine and dual affine quermassintegrals of convex bodies

Alex Segal (Tel Aviv) Duality through exchange of section/projection property

Daniel Hug (Essen) Random tessellations in high dimensions

Boaz Slomka (Tel Aviv) Order isomorphisms for cones and ellipsoids

Dmitrii Faifman (Tel Aviv) Some invariants of Banach spaces through associated Finsler manifolds

Sunday, April 3, 2011

Wolfgang Weil (Karlsruhe) Curvature measures and generalizations

Liran Rotem (Tel Aviv) A la finite volume ratio and low M^* estimate for log-concave functions

Monday, April 4, 2011

Emanuel Milman (Haifa) Volumetric properties of log-concave measures and Paouris' q^* parameter

Petros Valettas (Athens) Distribution of the ψ_2-norm of linear functionals on isotropic convex bodies

Antonis Tsolomitis (Samos) Random polytopes in convex bodies

Tuesday, April 5, 2011

Apostolos Giannopoulos (Athens) Problems about positions of convex bodies

Ronen Eldan (Tel Aviv) On extremal points of a high dimensional random walk and the convex hull of a high dimensional Brownian motion

Grigoris Paouris (College Station) Estimates for distances and volumes of k-intersection bodies

Wednesday, April 6, 2011
Dimitris Gatzouras (Heraclion) Random 0–1 polytopes
Semyon Alesker (Tel Aviv) Valuations and integral geometry
Omer Friedland (Tel Aviv) Sparsity and other random embeddings

Conference on Convex and Integral Geometry
Goethe-University Frankfurt, September 26–30, 2011
Organizers: *Andreas Bernig, Monika Ludwig, Franz Schuster*

Monday, September 9, 2011

Richard Gardner (Western Washington University) Fundamental properties of operations between sets

Dmitry Ryabogin (Kent State University) A problem of Klee on inner section functions of a convex body

Wolfgang Weil (Karlsruhe Institute of Technology) Projection formulas for valuations

Joseph Fu (University of Georgia) Results and prospects in integral geometry

Gil Solanes (Universitat Autónoma de Barcelona) Integral geometry of complex space forms

Thomas Wannerer (Vienna University of Technology) Centroids of Hermitian area measures

Tom Leinster (University of Glasgow) Integral geometry for the 1-norm

Tuesday, September 27, 2011

Shiri Artstein-Avidan (Tel Aviv University) Dirential analysis of polarity

Andrea Colesanti (University of Florence) Derivatives of integrals of log-concave functions

Daniel Hug (Karlsruhe Institute of Technology) Mixed volumes, projection functions and flag measures of convex bodies

Alina Stancu (Concordia University) On some equi-affine invariants for convex bodies

Gaoyong Zhang (Polytechnic Institute of NYU) The Minkowski problem

Manuel Weberndorfer (Vienna University of Technology) Simplex inequalities for asymmetric Wulff shapes

María A. Hernández Cifre (University of Murcia) Coverings and compressed lattices

Wednesday, September 28, 2011

Christoph Haberl (University of Salzburg) Body valued valuations

Judit Abardia (Goethe-University Frankfurt) A characterization of complex projection bodies

Lukas Parapatits (University of Salzburg) Linearly intertwining L_p-Minkowski valuations

Semyon Alesker (Tel Aviv University) Theory of valuations and integral geometry

Thursday, September 29, 2011

Eva B. Vedel Jensen (University of Aarhus) Rotational integral geometry

Jan Rataj (Charles University Prague) Critical values and level sets of distance functions

Georges Comte (University of Savoie) Real singularities and local Lipschitz-Killing curvatures

Martina Zähle (Friedrich-Schiller-University Jena) Local and global curvatures for classes of fractals

Matthias Reitzner (University of Osnabrück) Real-valued valuations

Eugenia Saorín Gómez (University of Magdeburg) Inner quermassintegrals inequalities

Eberhard Teufel (University of Stuttgart) Linear combinations of hypersurfaces in hyperbolic space

Ivan Izmestiev (TU Berlin) Infinitesimal rigidity of convex surfaces through variations of the Hilbert-Einstein functional

Friday, September 30, 2011

Rolf Schneider (Albert-Ludwigs-University Freiburg) Random tessellations and convex geometry – a synopsis

Christoph Thäle (University of Osnabrück) Integral geometry and random tessellation theory

Gautier Berck (Polytechnic Institute of NYU) Regularized integral geometry

Vitali Milman (Tel Aviv University) The reasons behind some classical constructions in analysis

Colloquium on the Occasion of the 70th Birthday of Peter M. Gruber
Vienna, October 20–21, 2011
Organizers: *Monika Ludwig, Gabriel Maresch, Franz Schuster, Christian Steineder*

Thursday, October 20, 2011
Martin Henk (Magdeburg) Frobenius numbers and the geometry of numbers

Rajinder J. Hans-Gil (Chandigarh) On Conjectures of Minkowski and Woods

Jörg Wills (Siegen) Successive minima and the zeros of the Ehrhart polynomial

Friday, October 21, 2011
Peter Gruber (Vienna) Some flowers from my mathematical garden

Matthias Reitzner (Osnabrück) Approximation of convex bodies by polytopes

Vitali Milman (Tel Aviv) Operator and functional equations characterizing some constructions in analysis

Chuanming Zong (Beijing) Classification of convex lattice polytopes

Rolf Schneider (Freiburg) Constant width and diametrical completeness

LECTURE NOTES IN MATHEMATICS  Springer

Edited by J.-M. Morel, B. Teissier; P.K. Maini

Editorial Policy (for Multi-Author Publications: Summer Schools / Intensive Courses)

1. Lecture Notes aim to report new developments in all areas of mathematics and their applications - quickly, informally and at a high level. Mathematical texts analysing new developments in modelling and numerical simulation are welcome. Manuscripts should be reasonably selfcontained and rounded off. Thus they may, and often will, present not only results of the author but also related work by other people. They should provide sufficient motivation, examples and applications. There should also be an introduction making the text comprehensible to a wider audience. This clearly distinguishes Lecture Notes from journal articles or technical reports which normally are very concise. Articles intended for a journal but too long to be accepted by most journals, usually do not have this "lecture notes" character.

2. In general SUMMER SCHOOLS and other similar INTENSIVE COURSES are held to present mathematical topics that are close to the frontiers of recent research to an audience at the beginning or intermediate graduate level, who may want to continue with this area of work, for a thesis or later. This makes demands on the didactic aspects of the presentation. Because the subjects of such schools are advanced, there often exists no textbook, and so ideally, the publication resulting from such a school could be a first approximation to such a textbook. Usually several authors are involved in the writing, so it is not always simple to obtain a unified approach to the presentation.

   For prospective publication in LNM, the resulting manuscript should not be just a collection of course notes, each of which has been developed by an individual author with little or no coordination with the others, and with little or no common concept. The subject matter should dictate the structure of the book, and the authorship of each part or chapter should take secondary importance. Of course the choice of authors is crucial to the quality of the material at the school and in the book, and the intention here is not to belittle their impact, but simply to say that the book should be planned to be written by these authors jointly, and not just assembled as a result of what these authors happen to submit.

   This represents considerable preparatory work (as it is imperative to ensure that the authors know these criteria before they invest work on a manuscript), and also considerable editing work afterwards, to get the book into final shape. Still it is the form that holds the most promise of a successful book that will be used by its intended audience, rather than yet another volume of proceedings for the library shelf.

3. Manuscripts should be submitted either online at www.editorialmanager.com/lnm/ to Springer's mathematics editorial, or to one of the series editors. Volume editors are expected to arrange for the refereeing, to the usual scientific standards, of the individual contributions. If the resulting reports can be forwarded to us (series editors or Springer) this is very helpful. If no reports are forwarded or if other questions remain unclear in respect of homogeneity etc, the series editors may wish to consult external referees for an overall evaluation of the volume. A final decision to publish can be made only on the basis of the complete manuscript; however a preliminary decision can be based on a pre-final or incomplete manuscript. The strict minimum amount of material that will be considered should include a detailed outline describing the planned contents of each chapter.

   Volume editors and authors should be aware that incomplete or insufficiently close to final manuscripts almost always result in longer evaluation times. They should also be aware that parallel submission of their manuscript to another publisher while under consideration for LNM will in general lead to immediate rejection.

4. Manuscripts should in general be submitted in English. Final manuscripts should contain at least 100 pages of mathematical text and should always include

 – a general table of contents;
 – an informative introduction, with adequate motivation and perhaps some historical remarks: it should be accessible to a reader not intimately familiar with the topic treated;
 – a global subject index: as a rule this is genuinely helpful for the reader.

Lecture Notes volumes are, as a rule, printed digitally from the authors' files. We strongly recommend that all contributions in a volume be written in the same LaTeX version, preferably LaTeX2e. To ensure best results, authors are asked to use the LaTeX2e style files available from Springer's web-server at
ftp://ftp.springer.de/pub/tex/latex/svmonot1/ (for monographs) and
ftp://ftp.springer.de/pub/tex/latex/svmultt1/ (for summer schools/tutorials).
Additional technical instructions, if necessary, are available on request from:
lnm@springer.com.

5. Careful preparation of the manuscripts will help keep production time short besides ensuring satisfactory appearance of the finished book in print and online. After acceptance of the manuscript authors will be asked to prepare the final LaTeX source files and also the corresponding dvi-, pdf- or zipped ps-file. The LaTeX source files are essential for producing the full-text online version of the book. For the existing online volumes of LNM see:
http://www.springerlink.com/openurl.asp?genre=journal&issn=0075-8434.
The actual production of a Lecture Notes volume takes approximately 12 weeks.

6. Volume editors receive a total of 50 free copies of their volume to be shared with the authors, but no royalties. They and the authors are entitled to a discount of 33.3 % on the price of Springer books purchased for their personal use, if ordering directly from Springer.

7. Commitment to publish is made by letter of intent rather than by signing a formal contract. Springer-Verlag secures the copyright for each volume. Authors are free to reuse material contained in their LNM volumes in later publications: a brief written (or e-mail) request for formal permission is sufficient.

Addresses:
Professor J.-M. Morel, CMLA,
École Normale Supérieure de Cachan,
61 Avenue du Président Wilson, 94235 Cachan Cedex, France
E-mail: morel@cmla.ens-cachan.fr

Professor B. Teissier, Institut Mathématique de Jussieu,
UMR 7586 du CNRS, Équipe "Géométrie et Dynamique",
175 rue du Chevaleret,
75013 Paris, France
E-mail: teissier@math.jussieu.fr

For the "Mathematical Biosciences Subseries" of LNM:

Professor P. K. Maini, Center for Mathematical Biology,
Mathematical Institute, 24-29 St Giles,
Oxford OX1 3LP, UK
E-mail : maini@maths.ox.ac.uk

Springer, Mathematics Editorial I,
Tiergartenstr. 17,
69121 Heidelberg, Germany,
Tel.: +49 (6221) 4876-8259
Fax: +49 (6221) 4876-8259
E-mail: lnm@springer.com